Linux Kernel Debugging

Leverage proven tools and advanced techniques to effectively debug Linux kernels and kernel modules

Kaiwan N Billimoria

BIRMINGHAM—MUMBAI

Linux Kernel Debugging

Group Product Manager: Vijin Boricha
Publishing Product Manager: Riyan Khan
Content Development Editor: Romy Dias
Technical Editor: Arjun Varma
Copy Editor: Safis Editing
Project Coordinator: Ashwin Dinesh Kharwa
Proofreader: Safis Editing
Indexer: Rekha Nair
Production Designer: Roshan Kawale
Marketing Coordinator: Sourodeep Sinha
Senior Marketing Coordinator: Hemangi Lotlikar

First published: June 2022
Production reference: 1010722

Published by Packt Publishing Ltd.
Livery Place
35 Livery Street
Birmingham
B3 2PB, UK.

ISBN 978-1-80107-503-9

www.packt.com

To the brave Covid warriors who led from the front; to the open source and Linux communities at large; to the wonderful Runner's High community that I am lucky to be a part of, with Santhosh leading from the front.

Contributors

About the author

Kaiwan N Billimoria taught himself programming on his dad's PC in 1983. By the early 90s, he had discovered the joys of programming on Unix, and by 1997, on Linux!

Kaiwan has worked on many aspects of the Linux system programming stack, including Bash, system programming in C, kernel internals, device drivers, and embedded Linux. He has actively worked on commercial/FOSS projects. His contributions include drivers for the mainline Linux OS and many smaller projects hosted on GitHub. His Linux passion feeds well into his passion for teaching these topics to engineers, which he has done for close to three decades now. He's the author of *Hands-On System Programming with Linux* and *Linux Kernel Programming*. He is a recreational ultrarunner too.

> *First, to my wonderful family – my parents, Nads and Diana; my wife, Dilshad; my children, Sheroy and Danesh; my bro, Darius; and the rest of the family. Thanks for being there. The Packt team has shepherded me through this work with patience and excellence, as usual. A special call out to Romy Dias – thanks for all the timely and super editing!*

About the reviewer

Chi-Thanh Hoang is currently working as a principal software technologist at Dell Technologies, currently developing an O-RAN 5G radio. He has over 29 years of software development experience, specializing mostly in embedded networking systems (switches, routers, Wi-Fi, and mobile networks) from chipsets up to communication protocols and of course kernel/RTOS. His first experience with the Linux kernel was in 1993. He still does hands-on debugging inside the kernel. He has a bachelor's degree in electrical engineering from Sherbrooke University, Canada. He is also an avid tennis player and invariably tinkers with electronics and software.

Table of Contents

2

Approaches to Kernel Debugging

Part 2: Kernel and Driver Debugging Tools and Techniques

3

Debug via Instrumentation – printk and Friends

4

Debug via Instrumentation – Kprobes

5

Debugging Kernel Memory Issues – Part 1

6

Debugging Kernel Memory Issues – Part 2

7

Oops! Interpreting the Kernel Bug Diagnostic

8

Lock Debugging

Part 3: Additional Kernel Debugging Tools and Techniques

9

Tracing the Kernel Flow

10

Kernel Panic, Lockups, and Hangs

11

Using Kernel GDB (KGDB)

12

A Few More Kernel Debugging Approaches

Preface

Linux Kernel Debugging is a modern, up-to-date take on the key topic of kernel and kernel module debugging. It covers in detail various powerful open source tools, as well as many advanced techniques (far beyond printk!), to debug the kernel, kernel modules, and device drivers. This is a key skill that the professional developer must learn and possess.

Who this book is for

This book is for Linux kernel developers, module, device driver authors, and testers interested in debugging and enhancing their Linux systems at the level of the kernel. System administrators who want to understand and debug the internal infrastructure of their Linux kernels will also find this book useful. A good hold on C programming and the Linux command line is necessary. Some experience with kernel (module) development will certainly benefit you.

What this book covers

Chapter 1, *A General Introduction to Debugging Software*, begins this journey by covering what debugging software actually entails, how it's really a mix of science and art. A few select software "horror stories" will serve to underline the importance of careful design, good (and secure) coding, and the ability to debug issues. On the more practical side, you will then set up the required workspace on your Linux system (or VM) so that you can – very importantly – work upon examples and assignments that will follow later.

Chapter 2, *Approaches to Kernel Debugging*, covers various approaches that can be taken to perform debugging at the level of kernel code. This will give you the insight to select the best, or the most viable, approach(es) depending on your particular situation and system constraints.

Chapter 3, *Debug via Instrumentation – printk and Friends*, refreshes the basics of using the common kernel printk() API. Next, we go into specifics of how to leverage it for the express purpose of kernel/driver debug via the instrumentation approach. The heart of this chapter – the kernel's powerful dynamic debug framework and how you can leverage it even in production – is then covered in detail.

Chapter 4, Debug via Instrumentation – Kprobes, explains the kernel's powerfull Kprobes framework, a means to – among other things – instrument the kernel and modules, by hooking into pretty much any kernel or module function, even in production. This can prove to be a practically useful way to debug systems during production.

Chapter 5, Debugging Kernel Memory Issues – Part 1, looks at memory bugs and corruption – a very common issue when working with a language such as C. First, you'll learn why this is, and, importantly, about the typical types of memory issues that tend to arise in such systems. Next, you will learn how to tackle these memory issues head-on, using the powerful compiler-based KASAN technology, as well as the kernel's compiler-based UBSAN technology.

Chapter 6, Debugging Kernel Memory Issues – Part 2, continues the coverage of debugging kernel memory issues. We delve in depth into the details of catching common memory issues on slab (SLUB) memory and then detecting difficult kernel memory leakage bugs with kmemleak. A detailed comparison between various memory corruption issues and the appropriate tooling to detect them rounds off these two chapters.

Chapter 7, Oops! Interpreting the Kernel Bug Diagnostic, covers a key topic – what a kernel "Oops" diagnostic message really is and, very importantly, how to interpret it in depth. Along this interesting journey, you will generate a simple kernel Oops and understand exactly how to interpret it. Further, several tools and techniques to help with this task will be shown. Getting to the bottom of an Oops often helps pinpoint the root cause of the kernel bug! A few actual Oops messages will also be pointed out.

Chapter 8, Lock Debugging, looks at an integral part of writing robust kernel or driver code: locking. Unfortunately, it's really quite easy to land up with errors – deadlocks and such – that are difficult to debug after the fact. This chapter skims over the basics of lock debugging, instead spending the bulk of it on a really powerful modern tool that helps uncover deep locking issues (data races) – the **Kernel Concurrent Sanitizer (KCSAN)**. Here, you'll learn how to configure the (debug) kernel for KCSAN, and how to use it in detail. We round it off by delving into several actual instances of kernel bugs whose root cause is locking issues.

Chapter 9, Tracing the Kernel Flow, introduces powerful technologies that allow you to trace the flow of kernel code in detail, at the granularity of every function call made! Usage of the primary kernel tracing infrastructure – ftrace – is covered first. You will then learn how to use powerful frontends to ftrace: trace-cmd, the KernelShark GUI, and the perf-tools collection. We wrap up this topic with an introduction to using LTTng (and visualization with the TraceCompass GUI!) to perform kernel-level tracing and analysis.

Chapter 10, Kernel Panic, Lockups, and Hangs, explains what kernel panic means precisely, and about the code paths executed within the kernel when it panics. More importantly, you'll learn how to write a custom kernel panic handler routine so that your code (also) runs if and when the kernel does panic. Associated topics – detecting lockups and CPU / work queue stalls, and hangs within the kernel – are covered as well.

Chapter 11, Using Kernel GDB (KGDB), introduces the powerful KGDB kernel source-level debug framework. You will learn how to configure and set up KGDB, after which, you'll see how to make use of it practically to debug kernel/module code at the level of the source, setting breakpoints, hardware watchpoints, leveraging GDB Python scripts, and more.

Chapter 12, A Few More Kernel Debugging Approaches, rounds off this vast topic of kernel debugging by introducing other approaches you can – and at times should – use. This includes understanding what the powerful (though resource-intensive) Kdump/crash tooling is, which can at times be a lifesaver. Then, we introduce you to why static analysis is key, and the available tools for analyzing Linux kernel/module/driver code. An introduction to code coverage and kernel testing frameworks follows. We round off the discussion with an introduction to logging (via journalctl), kernel assertions, and warning macros.

To get the most out of this book

We assume you're familiar with programming in C and are comfortable working on the Linux command line (the shell). Even minimal experience writing kernel code or modules (drivers) will certainly help as well, though it isn't mandatory.

In terms of the software workspace setup, we cover this in detail in *Chapter 1, A General Introduction to Debugging Software*, in the *Setting up the workspace section*.

If you are using the digital version of this book, we advise you to type the code yourself or access the code from the book's GitHub repository (a link is available in the next section). Doing so will help you avoid any potential errors related to the copying and pasting of code.

Being software, and a pretty hands-on activity like debugging it, we highly recommend you read the book in a hands-on manner, trying out the demos and examples covered as you go. Be sure to work on the few exercises mentioned within the chapters as well. When you finish this book, your real journey will just begin! With the key knowledge we're hoping, indeed betting, you gain herein, it should be that much easier and sweeter an experience.

Download the example code files

You can download the example code files for this book from GitHub at `https://github.com/PacktPublishing/Linux-Kernel-Debugging`. If there's an update to the code, it will be updated in the GitHub repository.

We also have other code bundles from our rich catalog of books and videos available at `https://github.com/PacktPublishing/`. Check them out!

Download the color images

We also provide a PDF file that has color images of the screenshots and diagrams used in this book. You can download it here: `https://packt.link/2zUIX`.

Conventions used

There are a number of text conventions used throughout this book.

`Code in text`: Indicates code words in text, database table names, folder names, filenames, file extensions, pathnames, dummy URLs, user input, and Twitter handles. Here is an example: "The end result is the kernel configuration file saved as `.config` in the root of the kernel source tree."

A block of code is set as follows:

```
#include <linux/init.h>
#include <linux/module.h>
#include <linux/kernel.h>
[...]
static int __init printk_loglevels_init(void)
```

When we wish to draw your attention to a particular part of a code block, the relevant lines or items are set in bold:

```
    if (lotype & (1<<5)) {
        pr_emerg("CPU#%d: Possible thermal failure (CPU on fire
?).\n", smp_processor_id());
    }
```

Any command-line input or output is written as follows:

```
sudo apt update
sudo apt upgrade
sudo apt install build-essential dkms linux-headers-$(uname -r)
ssh -y
```

Bold: Indicates a new term, an important word, or words that you see onscreen. For instance, words in menus or dialog boxes appear in **bold**. Here is an example: "Select **System info** from the **Administration** panel."

> **Tips or Important Notes**
> Appear like this.

Get in touch

Feedback from our readers is always welcome.

General feedback: If you have questions about any aspect of this book, email us at customercare@packtpub.com and mention the book title in the subject of your message.

Errata: Although we have taken every care to ensure the accuracy of our content, mistakes do happen. If you have found a mistake in this book, we would be grateful if you would report this to us. Please visit www.packtpub.com/support/errata and fill in the form.

Piracy: If you come across any illegal copies of our works in any form on the internet, we would be grateful if you would provide us with the location address or website name. Please contact us at copyright@packt.com with a link to the material.

If you are interested in becoming an author: If there is a topic that you have expertise in and you are interested in either writing or contributing to a book, please visit authors.packtpub.com.

Share Your Thoughts

Once you've read *Linux Kernel Debugging*, we'd love to hear your thoughts! Scan the QR code below to go straight to the Amazon review page for this book and share your feedback.

https://packt.link/r/1801075034

Your review is important to us and the tech community and will help us make sure we're delivering excellent quality content.

Part 1: A General Introduction and Approaches to Kernel Debugging

In this section, you'll get an introduction to what debugging software really is about, along with several real-world examples of software going wrong. This is followed by the setup of the Linux kernel debugging workspace (including building debug and production kernels). Then, we delve into various approaches to kernel debugging.

The following chapters will be covered in this section:

- *Chapter 1, A General Introduction to Debugging Software*
- *Chapter 2, Approaches to Kernel Debugging*

1

A General Introduction to Debugging Software

Hello there! Welcome on this journey of learning how to go about debugging a really sophisticated, large, and complex piece of software that's proven absolutely critical to big businesses as well as tiny embedded systems and everything in between – the **Linux kernel**.

Let's begin this very first chapter, and our journey of kernel debugging, by first understanding a little more about what a **bug** really is, and the origins and myths of the term **debugging**. Next, a glimpse at some actual *real-world* software bugs will (hopefully) provide the required inspiration and motivation (to firstly avoid bugs and then to find and fix bugs, of course). You will be guided on how to set up an appropriate workspace to work on a custom kernel and debug issues, including setting up a full-fledged *debug* kernel. We'll wrap up with some useful tips on debugging.

In this chapter, we're going to cover the following main topics:

- Software debugging – what it is, origins, and myths
- Software bugs – a few actual cases
- Setting up the workspace
- Debugging – a few quick tips

Technical requirements

You will require a modern and powerful desktop or laptop. We tend to use **Ubuntu 20.04 LTS** running on an x86_64 Oracle VirtualBox **Virtual Machine (VM)** as the primary platform for this book. Ubuntu Desktop specifies the *recommended minimum system requirements* (`https://help.ubuntu.com/community/Installation/SystemRequirements`) for the installation and usage of the distribution; do refer to those specifications to verify that your system (even a guest) is up to it. The *Running Linux as a guest OS* section covers the details.

Cloning this book's code repository

The complete source code for this book is freely available on GitHub at `https://github.com/PacktPublishing/Linux-Kernel-Debugging`. You can work on it by cloning the Git tree using the following command:

```
git clone https://github.com/PacktPublishing/Linux-Kernel-Debugging
```

The source code is organized chapter-wise. Each chapter is represented as a directory in the repository – for example, `ch1/` has the source code for this chapter. A detailed description of installing a viable system is covered in the *Setting up the workspace* section.

Software debugging – what it is, origins, and myths

In the context of a software practitioner, a bug is a defect or an error within code. A key, and often large, part of our job as software developers is to hunt them down and fix them, so that, as far as is humanely possible, the software is defect-free and runs precisely as designed.

Of course, to fix a bug, you first have to find it. Indeed, with non-trivial bugs, it's often the case that you aren't even aware there is a bug (or several) until some event occurs to expose it! Shouldn't we have a disciplined approach to finding bugs before shipping a product or project? Of course we should (and do) – it's the **Quality Assurance (QA)** process, more commonly known as **testing**. Though glossed over at times, testing remains one of the – if not *the* – most important facets of the software life cycle. (Would you voluntarily fly in a new aircraft that's never been tested? Well, unless you're the lucky test pilot...)

Okay, back to bugs; once identified (and filed), your job as a software developer is to then identify what exactly is causing them – what the actual underlying *root cause* is. A large portion of this book is devoted to tools, techniques, and just thinking about how to do this exactly. Once the root cause is identified, and you have clearly understood the underlying issue, you will, in all probability, be able to fix it. Yay!

This process of identifying a bug – using tools, techniques, and some hard thinking to figure out its root cause – and then fixing it is subsumed into the word debugging. Without bothering to go into details, there's a popular story regarding the origin of the word debugging: on a Tuesday at Harvard University (on September 9, 1947), Admiral Grace Hopper's staff discovered a moth caught in a relay panel of a Mark II computer. As the system malfunctioned because of it, they removed the moth, thus *de-bugging* the system! Well, as it turns out: one, Admiral Hopper has herself stated that she didn't coin the term, debugging; two, its origins seem to be rooted in aeronautics. Nevertheless, the term debugging has stuck.

The following figure shows the picture at the heart of this story – the unfortunate but posthumously famous moth that inadvertently caught itself in the system that had to be debugged!

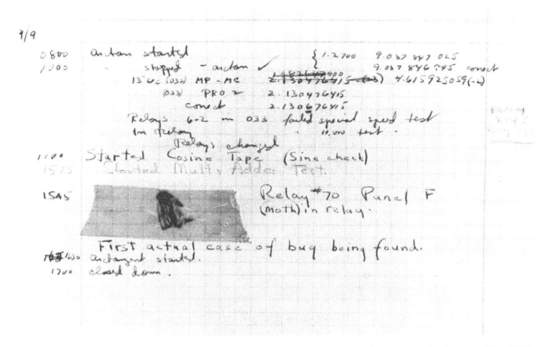

Figure 1.1 – The famous moth (by courtesy of the Naval Surface Warfare Center, Dahlgren, VA., 1988. - U.S. Naval Historical Center Online Library Photograph NH 96566-KN. Public Domain, https://commons.wikimedia.org/w/index.php?curid=165211)

Having understood what a bug and debugging basically are, let's move on to something both interesting and important – we'll briefly examine a few real-world cases where a software bug (or bugs) has been the cause of some unfortunate and tragic accidents.

Software bugs – a few actual cases

Using software to control electro-mechanical systems is not only common, it's pretty much all-pervasive in today's world. The unfortunate reality though, is that software engineering is a relatively young field and that we humans are naturally prone to making mistakes. These factors can combine to create unfortunate accidents when software doesn't execute conforming to its design (which, of course, leads to it being called **buggy**).

Several *real-world* examples of this occurring exist; we highlight a few of them in the following sub-sections. The brief synopsis given here is really just that – (too) brief: *to truly understand the complex issues behind failures like this, you do need to take the trouble to study the technical crash (or failure) investigation reports in detail* (do see the links in the *Further reading* section of this chapter). Here, I briefly mention and summarize these cases to, one, underline the fact that software failures, even in large, heavily tested systems, can and do occur, and two, to motivate all of us involved in any part of the software life cycle to pay closer attention, to stop making assumptions, and to do a better job of designing, implementing, and testing the software we work on.

Patriot missile failure

During the Gulf War, the US deployed a Patriot missile battery in Dharan, Saudi Arabia. Its job was to track, intercept, and destroy incoming Iraqi Scud missiles. But, on February 25, 1991, a Patriot system failed to do so, causing the death of 28 soldiers and injury to about 100 others. An investigation revealed that the problem's root was at the heart of the software tracking system. Briefly, the system uptime was tracked as a monotonically increasing integer value. It was converted to a *real* – floating-point – value by multiplying the integer by 1/10 (which is a recurring binary expression evaluating to 0.000110011001 1001100110011001100110011001100... ; a quick online calculator's available here: http://www.easysurf.cc/fracton2.htm). The trouble is, the computer used a 24-bit (integer) register for this conversion, resulting in the computation being truncated at 24 bits. This caused a loss of precision, which only became significant when the time quantity was sufficiently large.

This was exactly the case that day. The Patriot system had been up for about 100 hours, thus, the loss of precision during the conversion translated to an error of approximately 0.34 seconds. Doesn't sound like much, except that a Scud missile's velocity is about 1,676 meters per second, thus resulting in a tracking error of about 570 meters. This was large enough for the Scud to be outside the Patriot tracking system's *range gate* and it was thus not detected. Again, a case of loss of precision during conversion from an integer value to a real (floating-point) number value.

The ESA's unmanned Ariane 5 rocket

On the morning of June 4, 1996, the **European Space Agency's** (**ESA's**) Ariane 5 unmanned rocket launcher took off from the Guiana Space Centre, off the South American coast of French Guiana. A mere 40 seconds into its flight, the rocket lost control and exploded. The final investigation report revealed that the primary cause ultimately came down to a software overflow error.

It's more complex than that; a brief summary of the chain of events leading to the loss of the rocket follows. (In most cases like this, it's not one single event that causes an accident; rather, it's a chain of several events.) The overflow error occurred during the execution of code converting a 64-bit floating-point value to a 16-bit signed integer; an unprotected conversion gave rise to an exception (**Operand Error**; the programming language was **Ada**). This, in turn, occurred due to a much higher than expected internal variable value (BH – Horizontal Bias). The exception caused the shutdown of the **Inertial Reference Systems** (**SRIs**). This caused the primary **onboard computer** (**OBC**) to send erroneous commands to the nozzle deflectors resulting in full nozzle deflection of the boosters and the main Vulcain engine, which caused the rocket to veer dramatically off its flight path.

The irony is that the SRIs were, by default, not even supposed to function after launch; but due to a delay in the launch window, the design specified that they remain active for 50 seconds after launch! An interesting analysis of why this software exception wasn't caught during development and testing (https://archive.eiffel.com/doc/manuals/technology/contract/ariane/) boils down to concluding that the fault lay in a **reuse error**:

"The SRI horizontal bias module was reused from a 10-year-old software, the software from Ariane 4."

Mars Pathfinder reset issue

On July 4, 1997, NASA's Pathfinder lander touched down on the surface of Mars and proceeded to deploy its smaller robot cousin – the Sojourner rover, the very first wheeled device to embark upon another planet! The lander suffered from periodic reboots; the problem was ultimately diagnosed as being a classic case of **priority inversion** – a situation where a high-priority task is made to wait for lower-priority tasks. As such, this by itself may not cause an issue; the trouble is that the high-priority task was left off the CPU long enough for the *watchdog timer* to expire, causing the system to reboot.

An irony here was that there exists a well-known solution – enabling the **priority inheritance** feature of the semaphore object (allowing the task *taking* the semaphore lock to have its priority raised to the highest on the system for the duration for which it holds the lock, thus enabling it to complete its critical section and release the lock quickly, preventing starvation of higher-priority tasks). The VxWorks RTOS employed here defaulted to having the priority inheritance attribute turned off and the **Jet Propulsion Laboratory (JPL)** team left it that way. As they (very deliberately) allowed the robot to continuously stream telemetry debug data to Earth, they were able to correctly determine the root cause and thus fix it – by enabling the semaphore priority inheritance feature. An important lesson here is this one, as the team lead Glenn Reeves put it:

"We test what we fly and we fly what we test."

I'd venture that these articles (see the *Further reading* section) are a must-read for any system software developer!

The Boeing 737 MAX aircraft – the MCAS and lack of training of the flight crew

Two unfortunate accidents, taking 346 lives in all, put the Boeing 737 MAX under the spotlight: the crash of Lion Air Flight 610 from Jakarta into the Java Sea (October 29, 2018) and the crash of Ethiopian Airlines Flight 302 from Nairobi into the desert (March 10, 2019). These incidents occurred just 13 and 6 minutes after take-off, respectively.

Of course, the situation is complex. At one level, this is what likely caused these accidents: once Boeing determined that the aerodynamic characteristics of the 737 MAX left something to be desired, they worked on fixing it via a *hardware* approach. When that did not suffice, engineers came up with (what seemed) an elegant and relatively simple software fix, christened the **maneuvering characteristics augmentation system (MCAS)**. Two sensors on the aircraft's nose continually measure the aircraft's **angle of attack (AoA)**. When the AoA is determined to be too high, this typically entails a pending stall (dangerous!). The MCAS kicks in, (aggressively) moving control surfaces on the tail elevator, causing the nose to go down, and stabilizing the aircraft. But, for whatever reasons, the MCAS was designed to use only one of the sensors! If the sensor failed, the MCAS could automatically activate, causing the nose to go down and the aircraft to rapidly lose altitude; this is what seems to have actually occurred in both crashes.

Further, many pilot crews weren't explicitly trained in managing the MCAS (some claimed they weren't even aware of it!). The luckless flights' pilots apparently did not manage to override the MCAS, even when no actual stall occurred.

Other cases

A few other examples of such cases are as follows:

- June 2002, Fort Drum: a US Army report maintained that a software issue contributed to the death of two soldiers. This incident occurred when they were training to fire artillery shells. Apparently, unless the target altitude is explicitly entered into the system, the software assumes a default of zero. Fort Drum is apparently 679 feet above sea level.

- In November 2001, a British engineer, John Locker, noticed that he could easily intercept American military satellite feeds – live imagery of US spy planes over the Balkans. The almost unbelievable reason was the stream was being transmitted unencrypted, enabling pretty much anyone in Europe with a regular satellite TV receiver to see it! In today's context, many IoT devices have similar issues...

- Jack Ganssle, a veteran and widely known embedded systems developer and author, brings out the excellent TEM – The Embedded Muse – newsletter bi-monthly. Every issue has a section entitled *Failure of the Week*, typically highlighting a hardware and/or software failure. Do check it out!

- Read the web page on *Software Horror Stories* here (`http://www.cs.tau.ac.il/~nachumd/horror.html`); though old, it provides many examples of software gone wrong with, at times, tragic consequences.

- A quick Google search on Linux kernel bug stories yields interesting results: `https://www.google.com/search?q=linux+kernel+bug+story`.

Again, if interested in digging deeper, I urge you to read the detailed official reports on these accidents and faults; the *Further reading* section has several relevant links.

By now, you should be itching to begin debugging on Linux! Let's do just that – begin – by first setting up the workspace.

Setting up the workspace

Firstly, you'll have to decide whether to run your test Linux system as a native system (*on the bare metal*) or as a guest OS. We'll cover the factors that will help you decide. Next, we (briefly) cover the installation of some software (the guest additions) for a case where you use a guest Linux OS, followed by the required software packages to install.

Running Linux as a native or guest OS

Ideally, you should run a recent Linux distribution (Ubuntu, Fedora, and so on) on native hardware. We tend to use Ubuntu 20.04 LTS in this book as the primary system to experiment upon. The more powerful your system – in terms of RAM, processing power, and disk space – the better! Of course, as we shall be debugging at the level of the kernel, crashes and even data loss (the chances of the latter are small, but nevertheless...) can occur; hence, the system should be a *test* one with no valuable data on it.

If running Linux on native hardware – on the bare metal, as it were – isn't feasible for you, then a practical and convenient alternative is to install and use the Linux distribution as a guest OS, a VM. It's important to install a recent Linux distribution.

Running a Linux guest as a VM is certainly feasible but (there's always a *but* isn't there?!), it will almost certainly feel a lot slower than running Linux natively. Still, if you must run a Linux guest, it certainly works. It goes without saying that the more powerful your host system, the better the experience. There's also an arguable advantage to running your test system as a guest OS: even if it does crash (please do expect that to happen, especially with the deliberate (de)bugging we'll do in this book!), you don't even need to reboot the hardware; merely reset the hypervisor software running the guest (typically Oracle VirtualBox).

Alternate Hardware – Using Raspberry Pi (and Other) ARM-Based Systems

Though we specified that you can run a recent Linux distro either as a native system or as a guest VM, the assumption was that it's an x86_64 system. While that suffices, to get more out of the experience of this book (and simply to have more fun), I highly recommend you also try out the sample code and run the (buggy) test cases on alternate architectures. With many, if not most, modern embedded Linux systems being ARM-based (on both 32-bit ARM and 64-bit AArch64 processors), the Raspberry Pi hardware is extremely popular, relatively cheap, and has tremendous community support, making it an ideal test bed. I do use it every now and then within this book, in the chapters that follow; I'd recommend you do the same!

All the details – installation, setup, and so on – are amply covered in the well-documented Raspberry Pi documentation pages here: `https://www.raspberrypi.org/documentation/`.

Ditto for another popular embedded prototyping board - TI's BeagleBone Black (affectionately, the *BBB*). This site is a good place to get started with the BBB: `https://beagleboard.org/black`.

Running Linux as a guest OS

If you do decide to run Linux as an x86_64 guest, I'd recommend using Oracle VirtualBox 6.x (or the latest stable version) as a comprehensive and powerful all-in-one GUI hypervisor application appropriate for a desktop PC or laptop. Other virtualization software, such as VMware Workstation or QEMU, should also be fine. All of these are freely available and open source. It's just that the code for this book has been tested on Oracle VirtualBox 6.1. Oracle VirtualBox is considered **Open Source Software** (**OSS**) and is licensed under the GPL v2 (the same as the Linux kernel). You can download it from `https://www.virtualbox.org/wiki/Downloads`. Its documentation can be found here: `https://www.virtualbox.org/wiki/End-user_documentation`.

The host system should be either MS Windows 10 or later (of course, even Windows 7 will work), a recent Linux distribution (for example, Ubuntu or Fedora), or macOS.

The guest (or native) Linux distribution can be any sufficiently recent one. For the purpose of following along with the material and examples presented in this book, I'd recommend installing **Ubuntu 20.04 LTS**. This is what I primarily use for the book.

How can you quickly check which Linux distribution is currently installed and running?

On Debian/Ubuntu, the `lsb_release -a` command should do the trick; for example, on my guest Linux:

```
$ lsb_release -a 2> /dev/null
Distributor ID: Ubuntu
Description:    Ubuntu 20.04.2 LTS
Release:        20.04
Codename:       focal
$
```

How can you check if the Linux currently running is on native hardware or is a guest VM (or a container)? There are many ways to do so. The script `virt-what` is one (we will be installing it). Other commands include `hostnamectl(1)`, `dmidecode(8)` (on x86), `systemd-detect-virt(1)` (if systemd is the initialization framework), `lshw(1)` (x86, IA-64, PPC), *raw* ways via `dmesg(1)` (grepping for `Hypervisor detected`), and via `/proc/cpuinfo`.

In this book, I shall prefer to focus on setting up what is key from a kernel debug perspective; hence, we won't discuss the in-depth details of installing a guest VM (typically on a Windows host running Oracle VirtualBox) here. If you require some help, please refer to the many links to tutorials on precisely this within the *Further reading* section of this chapter. (FYI, these installation details, and a lot more, are amply covered in my previous book *Linux Kernel Programming, Chapter 1, Kernel Workspace Setup*).

Tip – Using Prebuilt VirtualBox Images

The **OSBoxes** project allows you to freely download and use prebuilt VirtualBox (as well as VMware) images for popular Linux distributions. See their site here: `https://www.osboxes.org/virtualbox-images/`.

In our case, you can download a prebuilt x86_64 Ubuntu 20.04.3 (as well as others) Linux image here: `https://www.osboxes.org/ubuntu/`. It comes with the guest additions preinstalled! The default username/password is `osboxes/osboxes.org`.

(Of course, for more advanced readers, you'll realize it's really up to you. Running an as-light-as-possible custom Linux system on a Qemu (emulated) standard PC is a choice as well.)

Note that if your Linux system is installed natively on the hardware platform, you're using an OSBoxes Linux distro with the VirtualBox guest additions preinstalled, or you're using a Qemu-emulated PC, simply skip the next section.

Installing the Oracle VirtualBox guest additions

The *guest additions* are essentially software (para-virtualization accelerators) that quite dramatically enhance the performance, as well as the *look and feel*, of the experience of running a guest OS on the host system; hence, it's important to have it installed. (Besides acceleration, the guest additions provide conveniences such as the ability to nicely scale the GUI window and share facilities such as folders, the clipboard, and to drag and drop between the host and the guest.)

Before doing this though, please ensure you have already installed the guest VM (as mentioned previously). Also, the first time you log in, the system will likely prompt you to update and possibly restart; please do so. Then, follow along:

1. Log in to your Linux guest VM (I'm using the login name letsdebug; guess why!) and first run the following commands within a Terminal window (on a shell):

   ```
   sudo apt update
   sudo apt upgrade
   sudo apt install build-essential dkms linux-headers-
   $(uname -r) ssh -y
   ```

   (Ensure you run each of the preceding commands on one line.)

2. Install the Oracle VirtualBox Guest Additions now. Refer to *How to Install VirtualBox Guest Additions in Ubuntu*: https://www.tecmint.com/install-virtualbox-guest-additions-in-ubuntu/.

3. On Oracle VirtualBox, to ensure that you have access to any shared folders you might have set up, you need to set the guest account to belong to the vboxsf group; you can do so like this (once done, you'll need to log in again, or sometimes even reboot, to have this take effect):

   ```
   sudo usermod -G vboxsf -a ${USER}
   ```

The commands (*step 1*), after updating, have us install the build-essential package along with a couple of others. This ensures that the compiler (gcc), make, and other essential build utility programs are installed so that the Oracle VirtualBox Guest Additions can be properly installed straight after (in step 2).

Installing required software packages

To install the required software packages, perform the following steps (do note that, here, we assume the Linux distribution is our preferred one, Ubuntu 20.04 LTS):

1. Within your Linux system (be it a native one or a guest OS), first do the following:

   ```
   sudo apt update
   ```

   Now, to install the remaining required packages for the kernel build, run the following command in a single line:

   ```
   sudo apt install bison flex libncurses5-dev ncurses-dev
   xz-utils libssl-dev libelf-dev util-linux tar -y
   ```

(The -y option switch has apt assume a *yes* answer to all prompts; careful though, this could be dangerous in other circumstances.)

2. To install the packages required for work we'll do in other parts of this book, run the following command in a single line:

```
sudo apt install bc bpfcc-tools bsdmainutils clang cmake
cppcheck cscope curl \
dwarves exuberant-ctags fakeroot flawfinder git gnome-
system-monitor gnuplot \
hwloc indent kernelshark libnuma-dev libjson-c-dev linux-
tools-$(uname -r) \
net-tools numactl openjdk-16-jre openssh-server perf-
tools-unstable psmisc \
python3-distutils rt-tests smem sparse stress sysfsutils
tldr-py trace-cmd \
tree tuna virt-what -y
```

A point to mention: all the packages mentioned above aren't strictly required in order for you to work on this book; some are those we encounter just once or twice.

> **Tip – A Script to Auto-Install Required Packages**
>
> To make the (immediately above-mentioned) package install task simpler, you can make use of a simple bash script that's part of the GitHub repository for this book: `pkg_install4ubuntu_lkd.sh`. It's been tested on an x86_64 OSBoxes Ubuntu 20.04.3 LTS VM (running on Oracle VirtualBox 6.1).
>
> FYI, to check which packages are taking up the most space, install the `wajig` package and run this: `sudo wajig large`.

Great; now that the packages are installed, let's proceed with understanding the next portion of our workspace setup – the need for two kernels!

A tale of two kernels

When working on a project or product, there obviously will be a Linux kernel that will be deployed as part of the overall system.

> **Requirements for a Working Linux System**
>
> A quick aside: a working Linux system minimally requires a bootloader, a kernel, and root filesystem images. Additionally, typical arm/arm64/ppc systems require a **Device Tree Blob (DTB)** as well.

This system that's deployed to the outside world is, in general, termed the **production system** and the kernel the **production kernel**. Here, we'll limit our discussion to the kernel only. The configuration, build, test, debug, and deployment of the production kernel is, no doubt, a key part of the overall project.

Do note though, that in many systems (especially the enterprise-class ones), the production kernel is often simply the default kernel that's supplied by the vendor (Red Hat, SUSE, Canonical, or others). On most embedded Linux projects and products, this is likely not the case: the platform (or **Board Support Package** (**BSP**)) team or a vendor will select a base mainline kernel (typically from kernel.org) and customize it. This can include enhancements, careful configuration, and deployment of the custom-built production kernel.

For the purpose of our discussion, let's assume that we need to configure and build a custom kernel.

A production and a debug kernel

However (and especially after having read the earlier *Software bugs – a few actual cases section*), you will realize that there's always the off-chance that even the kernel – more likely the code you and your team added to it (the kernel modules, drivers, interfacing components) – has hidden faults (bugs). With a view to catching them before the system hits the field, thorough testing/QA is of prime importance!

Now, the issue is this: unless certain deeper checks are enabled within the kernel itself, it's entirely possible that they can escape your test cases. So, why not simply enable them? Well, one, these deeper checks are typically switched off by default in the production kernel's configuration. Two, when turned on, they do result in performance degradation, at times quite significantly.

So, where does that leave us? Simple, really: you should plan on working with at least two, and possibly three, kernels:

- One, a carefully tuned production kernel, geared toward efficiency, security, and performance.

- Two, a carefully configured **debug kernel**, geared toward catching pretty much all kinds of (kernel) bugs! Performance is not a concern here, catching bugs is.

- Three (optional, case by case): a production kernel with one or more very specific debug options enabled and the rest turned off.

The second one, the so-called **debug kernel**, is configured in such a way that all required or recommended **debug options** are turned on, enabling you to (hopefully!) catch those hidden bugs. Of course, performance might suffer as a result, but that's okay; catching – and subsequently fixing – kernel-level bugs is well worth it. Indeed, in general, during development and (unit) testing, performance isn't paramount; catching and fixing deeply hidden bugs is! We do understand that, at times, bugs need to be reproduced and identified on the production kernel itself. The third option mentioned above can be a life-saver here.

The debug kernel is only used during development, testing, and very possibly later when bugs do actually surface. How exactly it's used later is something we shall certainly cover in the course of this book.

Also, this point is key: it usually is the case that the mainline (or *vanilla*) kernel that your custom kernel is based upon is working fine; the bugs are generally introduced via custom enhancements and **kernel modules**. As you will know, we typically leverage the kernel's **Loadable Kernel Module** (**LKM**) framework to build custom kernel code – the most common being device drivers. It can also be anything else: custom network filters/firewalls, a new filesystem, or I/O scheduler. These are *out-of-tree kernel components* (typically some .ko files) that become part of the root filesystem (they're usually installed in /lib/modules/$(uname -r)). The debug kernel will certainly help catch bugs in your kernel modules as their test cases are executed, as they run.

The third kernel option – an in-between of the first two – is optional of course. From a practical real-world point of view, it may be exactly what's required on a given setup. With certain kernel debug systems turned on, to catch specific types of bugs that you're hunting (or anticipate) and the rest turned off, it can be a pragmatic way to debug even a production system, keeping performance high enough.

For practical reasons, in this book, we'll configure, build, and make use of the first two kernels – a custom production one and a custom debug one, only; the third option is yours to configure as you gain experience with the kernel debug features and tools as well as your particular product or project.

Which kernel release should you use?

The Linux kernel project is often touted as the most successful open source project ever, with literally hundreds of releases and a release cadence that's truly phenomenal for such an enormous project (it averages a new release every 6 to 8 weeks!). Among all of them, which one should we use (as a starting point, at least)?

It's really important to use the *latest stable kernel version*, as it will include all the latest performance and security fixes. Not just that, the kernel community has different release *types*, which determine how long a given kernel release will be maintained (with bug and security fixes being applied, as they become known). For typical projects or products, selecting the latest **Long Term Stable** (LTS) kernel release thus makes the best sense. Of course, as already mentioned, on many projects – typically, the server-/enterprise-class ones – the vendor (Red Hat, SUSE, and others) might well supply the production kernel to be used; here, for the purpose of our learning, we'll start from scratch and configure and build a custom Linux kernel ourselves (as is often the case on embedded projects).

As of when I wrote this chapter, the latest LTS Linux kernel is 5.10 (particularly, version 5.10.60); I shall use this kernel throughout this book. (You will realize that by the time you're reading this, it's entirely possible – in fact pretty much guaranteed – that the latest LTS kernel has evolved to a newer version).

> **Important – Security**
>
> It's already happened of course. Now, it's March 2022, I'm writing the tenth chapter, the latest 5.10 LTS kernel is 5.10.104, and guess what? A serious and critical vulnerability (*vuln*) has emerged in recent Linux kernels – including 5.10.60! – christened *Dirty Pipe*. Details: *New Linux bug gives root on all major distros, exploit released*, Mar 2022: `https://www.bleepingcomputer.com/news/security/new-linux-bug-gives-root-on-all-major-distros-exploit-released/`. Here's an explanation by the person who found and reported the vuln (a must read!): *The Dirty Pipe Vulnerability*, Max Kellerman: `https://dirtypipe.cm4all.com/`.
>
> (It's also very interesting – the fix comes down to 2 lines – initializing a local variable to 0!: `https://lore.kernel.org/lkml/20220221100313.1504449-1-max.kellermann@ionos.com/`.)
>
> The upshot of all this: I recommend you use a fixed kernel (as of now, kernel versions 5.16.11, 5.15.25, and 5.10.102 are fixed). Since this book is based on the 5.10 LTS kernel, I thus highly recommend **you use a version 5.10 LTS kernel, specifically, version 5.10.102 or later**. (The material, of course, continues to be based on 5.10.60; besides the security implication, which of course really matters on actual production systems, the technical details remain identical.)

Besides, a key point in our favor – *the 5.10 LTS kernel will be supported by the community until December 2026, thus keeping it relevant and valid for a pretty long time!*

So, great – let's get to configuring and building both our custom production and debug 5.10 kernels! We'll begin with the production one.

Setting up our custom production kernel

Here, I shall have to assume you are familiar with the general procedure involved in building a Linux kernel from source: obtaining the kernel source tree, configuring it, and building it. In case you'd like to brush up on this, the *Linux Kernel Programming* book covers this in a lot of detail. As well, do refer to the tutorials and links in the *Further reading* section of this chapter.

Though this is meant to be our production kernel, we'll begin with a rather simplistic default that's based on the existing system (this approach is sometimes called the **tuned kernel config via the localmodconfig** one. FYI, this, and a lot more, is covered in depth in the *Linux Kernel Programming* book). Once we've got a reasonable starting point, we'll further tune the kernel for security. Let's begin by performing some base configuration:

1. Create a new directory in which you'll work upon the upcoming production kernel:

    ```
    mkdir -p ~/lkd_kernels/productionk
    ```

    Bring in the kernel source tree of your choice. Here, as mentioned in *Which kernel release should you use?*, we shall use the latest (at the time of writing this) **LTS Linux kernel, version 5.10.60**:

    ```
    cd ~/lkd_kernels
    wget https://mirrors.edge.kernel.org/pub/linux/kernel/
    v5.x/linux-5.10.60.tar.xz
    ```

    Notice that here we have simply used the wget utility to bring in the compressed kernel source tree; there are several alternate ways (including using git).

    > **Note**
    >
    > As you'll know, the number in parentheses following the command name – for example, wget (1) – is the section within the manual or man pages where documentation on this command can be found.

2. Extract the kernel source tree:

    ```
    tar xf linux-5.10.60.tar.xz --directory=productionk/
    ```

3. Switch to the directory it's just been extracted into (using cd linux-5.10.60)
    and briefly verify the kernel version information as shown in the following
    screenshot:

```
$ pwd
/home/letsdebug/lkd_kernels/productionk/linux-5.10.60
$ ls
COPYING         Kbuild      MAINTAINERS     README   certs/     fs/        ipc/      mm/        scripts/    tools/
CREDITS         Kconfig     Makefile        arch/    crypto/    include/   kernel/   net/       security/   usr/
Documentation/  LICENSES/   Module.symvers  block/   drivers/   init/      lib/      samples/   sound/      virt/
$ head Makefile
# SPDX-License-Identifier: GPL-2.0
VERSION = 5
PATCHLEVEL = 10
SUBLEVEL = 60
EXTRAVERSION =
NAME = Dare mighty things

# *DOCUMENTATION*
# To see a list of typical targets execute "make help"
# More info can be located in ./README
$
```

Figure 1.2 – Screenshot of the LTS kernel source tree

Every kernel version is christened with a (rather exotic) name; our 5.10.60 LTS
kernel has an appropriately nice name – *Dare mighty things* – don't you think?

4. Configure for appropriate defaults. This is what you can do to obtain a decent, tuned
    starting point for the kernel config based on the current config:

    ```
    lsmod > /tmp/lsmod.now

    make LSMOD=/tmp/lsmod.now localmodconfig
    ```

> **Note**
>
> The preceding command might interactively ask you to specify some choices;
> just selecting the defaults (by pressing the *Enter* key) is fine for now. The end
> result is the kernel configuration file saved as .config in the root of the
> kernel source tree (the current directory).

We back up the config file as follows:

```
cp -af .config ~/lkd_kernels/kconfig_prod01
```

> **Tip**
>
> You can always do make help to see the various options (including config)
> available; experienced readers can use alternate config options that better suit
> their project.

Before jumping into the building of our production kernel, it's really important to consider the security aspect. Let's first configure our kernel to be more secure, *hardened*.

Securing your production kernel

With security being a major concern, the modern Linux kernel has many security and kernel hardening features. The thing is, there always tends to be a trade-off between security and convenience/performance. Thus, many of these hardening features are *off* by default; several are designed as an opt-in system: if you want it, turn it on by selecting it from the kernel config menu (via the familiar `make menuconfig` UI). It makes sense to do this, especially on a production kernel.

The question is: *how will I know exactly which config features regarding security to turn on or off?* There's literature on this and, better, some utility scripts that examine your existing kernel config and can make recommendations based on existing state-of-the-art security best practices! One such tool is Alexander Popov's `kconfig-hardened-check` Python script (`https://github.com/a13xp0p0v/kconfig-hardened-check`). Here's a screenshot of installing and running it plus a portion of its output, when I ran it against my custom kernel configuration file:

```
$ git clone https://github.com/a13xp0p0v/kconfig-hardened-check
Cloning into 'kconfig-hardened-check'...
remote: Enumerating objects: 1339, done.
remote: Counting objects: 100% (123/123), done.
remote: Compressing objects: 100% (87/87), done.
remote: Total 1339 (delta 62), reused 90 (delta 35), pack-reused 1216
Receiving objects: 100% (1339/1339), 1.57 MiB | 880.00 KiB/s, done.
Resolving deltas: 100% (806/806), done.
$
$ ls kconfig-hardened-check/
LICENSE.txt  MANIFEST.in  README.md  bin/  contrib/  default.nix  kconfig_hardened_check/  setup.cfg  setup.py*
$ ls kconfig-hardened-check/bin/
kconfig-hardened-check*
$
$ cd kconfig-hardened-check
$ bin/kconfig-hardened-check -p X86_64 -c ~/lkd_kernels/kconfig_prod01
[+] Config file to check: /home/letsdebug/lkd_kernels/kconfig_prod01
[+] Detected architecture: X86_64
[+] Detected kernel version: 5.10
```

option name	desired val	decision	reason	check result
CONFIG_BUG	y	defconfig	self_protection	OK
CONFIG_SLUB_DEBUG	y	defconfig	self_protection	OK
CONFIG_GCC_PLUGINS	y	defconfig	self_protection	FAIL: not found
CONFIG_STACKPROTECTOR_STRONG	y	defconfig	self_protection	OK
CONFIG_STRICT_KERNEL_RWX	y	defconfig	self_protection	OK
CONFIG_STRICT_MODULE_RWX	y	defconfig	self_protection	OK
CONFIG_REFCOUNT_FULL	y	defconfig	self_protection	OK: version >= 5.5
CONFIG_IOMMU_SUPPORT	y	defconfig	self_protection	OK
CONFIG_RANDOMIZE_BASE	y	defconfig	self_protection	OK
CONFIG_THREAD_INFO_IN_TASK	y	defconfig	self_protection	OK
CONFIG_VMAP_STACK	y	defconfig	self_protection	OK

Figure 1.3 – Partial screenshot – truncated output from the kconfig-hardened-check script

(We won't be attempting to go into details regarding the useful `kconfig-hardened-check` script here, as it's beyond this book's scope. Do look up the GitHub link provided to see details.) Having followed most of the recommendations from this script, I generated a kernel config file:

```
$ ls -l .config
-rw-r--r-- 1 letsdebug letsdebug 156781 Aug 19 13:02
.config
$
```

> **Note**
> My kernel config file for the production kernel can be found in the
> book's GitHub code repository here: `https://github.com/`
> `PacktPublishing/Linux-Kernel-Debugging/blob/main/`
> `ch1/kconfig_prod01`. (FYI, the custom debug kernel config file that
> we'll be generating in the following section can be found within the same folder
> as well.)

Now that we have appropriately configured our custom production kernel, let's build it. The following commands should do the trick (with `nproc` helping us determine the number of CPU cores onboard):

```
$ nproc
4
$ make -j8
[ ... ]
BUILD    arch/x86/boot/bzImage
Kernel: arch/x86/boot/bzImage is ready  (#1)
$
```

> **Cross Compiling the kernel**
> If you're working on a typical embedded Linux project, you will need to
> install an appropriate toolchain and cross-compile the kernel. As well, you'd
> set the environment variable ARCH to the machine type (for example,
> ARCH=arm64) and the environment variable CROSS_COMPILE to the
> cross-compiler prefix (for example, CROSS_COMPILE=aarch64-none-
> linux-gnu-). Your typical embedded Linux builder systems – Yocto and
> Buildroot being very common – pretty much automatically take care of this.

As you can see, as a rule of thumb, we set the number of jobs to execute as twice the number of CPU cores available via the make option switch -j. The build should complete in a few minutes. Once done, let's check that the compressed and uncompressed kernel image files have been generated:

```
$ ls -lh arch/x86/boot/bzImage vmlinux
-rw-r--r-- 1 letsdebug letsdebug 9.1M Aug 19 17:21
arch/x86/boot/bzImage
-rwxr-xr-x 1 letsdebug letsdebug  65M Aug 19 17:21 vmlinux
$
```

Note that it's always only the first file, bzImage – the compressed kernel image – that we shall boot from. Then what's the second image, vmlinux, for? *Very relevant here*: it's what we shall (later) often require when we need to perform kernel debugging! It's the one that holds all the symbolic information, after all.

Our production kernel config will typically cause several kernel modules to be generated within the kernel source tree. They have to be installed in a well-known location (/lib/ modules/$(uname -r)); this is achieved by doing the following, as root:

```
$ sudo make modules_install
[sudo] password for letsdebug: xxxxxxxxxxxxxxxx
  INSTALL arch/x86/crypto/aesni-intel.ko
  INSTALL arch/x86/crypto/crc32-pclmul.ko
[ ... ]
   DEPMOD  5.10.60-prod01
$ ls /lib/modules/
5.10.60-prod01/  5.11.0-27-generic/  5.8.0-43-generic/
$ ls /lib/modules/5.10.60-prod01/
build@  modules.alias.bin modules.builtin.bin modules.dep.
bin  modules.softdep source@ kernel/       modules.builtin
modules.builtin.modinfo  modules.devname   modules.symbols
modules.alias   modules.builtin.alias.bin   modules.dep modules.
order modules.symbols.bin
$
```

For the final step, we make use of an internal script to generate the `initramfs` image and set up the bootloader (in this case, on our x86_64, it's GRUB) by simply running the following:

```
sudo make install
```

For details and a conceptual understanding of the initial RAM disk, as well as some basic GRUB tuning, do see the *Linux Kernel Programming* book. We also provide useful references within the *Further reading* section of this chapter.

Now all that's left to do is reboot your guest (or native) system, interrupt the bootloader (typically by holding the *Shift* key down during early boot; this can vary if you're booting via UEFI though), and select the newly built production kernel:

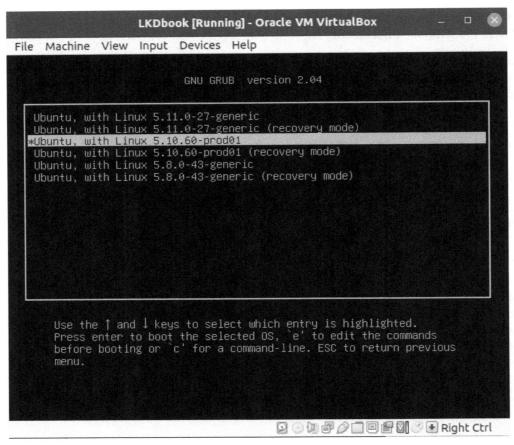

Figure 1.4 – Screenshot showing the GRUB bootloader screen and the new production kernel to boot from

As you can see from the preceding screenshot, I'm running the system as a guest OS via Oracle VirtualBox. I selected the new production kernel and pressed Enter to boot into it.

Voila, we're now running our (guest) system with our brand new production kernel:

```
$ uname -a
Linux dbg-LKD 5.10.60-prod01 #1 SMP PREEMPT Thu Aug 19 17:10:00
IST 2021 x86_64 x86_64 x86_64 GNU/Linux
$
```

> **Working on the guest over SSH**
>
> The new Linux kernel should run just fine with the existing root filesystem
> – the libraries and applications are loosely coupled with the OS, allowing
> different versions of the kernel (one at a time, of course) to simply mount the
> root filesystem and use them. Also, you may not get all the bells and whistles;
> for example, on my guest OS with our new production kernel, the screen
> resizing, shared folders, and such, features may not work. How come? They
> depend on the guest additions whose kernel modules haven't been built for this
> custom kernel. In this case, I find it *a lot easier to work on the guest using the
> console over SSH*. To do so, I installed the **dropbear** lightweight SSH server on
> the guest and then logged in over SSH from my host system. Windows users
> might like to try an SSH client such as **putty**. (In addition, you might need to
> set up another bridged mode network adapter on the Linux guest.)

You can (re)check the current kernel's configuration by looking up /boot/config-$(uname -r). In this case, it should be that of our production kernel, tuned towards security and performance.

> **Tip**
>
> To have the GRUB bootloader prompt always show up at boot: make a copy
> of /etc/default/grub (to be safe), then edit it as root, adding the line
> GRUB_HIDDEN_TIMEOUT_QUIET=false and (possibly) commenting
> out the line GRUB_TIMEOUT_STYLE=hidden.
>
> Change the GRUB_TIMEOUT value from 0 to 3 (seconds). Run sudo
> update-grub to have the changes take effect, and reboot to test.

So, good, you now have your guest (or native) Linux OS running a new production kernel. During the course of this book, you will encounter various kernel-level bugs while running this kernel. Identifying the bug(s) will often – though not always – involve you booting via the debug kernel instead. So, let's now move on to creating a custom debug kernel for the system. Read on!

Setting up our custom debug kernel

As you have already set up a production kernel (as described in detail in the previous section), I won't repeat every step in detail here, just the ones that differ:

1. Firstly, ensure you have booted into the production kernel that you built in the previous section. This is to ensure that our debug kernel config uses it as a starting point:

    ```
    $ uname -r
    5.10.60-prod01
    ```

2. Create a new working directory and extract the same kernel version again. It's important to build the debug kernel in a separate workspace from that of the production one. True, it takes a lot more disk space but it keeps them clean and from stepping on each other's toes as you modify their configs:

    ```
    mkdir -p ~/lkd_kernels/debugk
    ```

3. We already have the kernel source tree (we earlier used wget to bring in the 5.10.60 compressed source). Let's reuse it, this time extracting it into the debug kernel work folder:

    ```
    cd ~/lkd_kernels
    tar xf linux-5.10.60.tar.xz --directory=debugk/
    ```

4. Switch to the debug kernel directory and set up a starting point for kernel config – via the localmodconfig approach – just as we did for the production kernel. This time though, the config will be based on that of our custom production kernel, as that's what is running right now:

    ```
    cd ~/lkd_kernels/debugk/linux-5.10.60
    lsmod > /tmp/lsmod.now
    make LSMOD=/tmp/lsmod.now localmodconfig
    ```

5. As this is a debug kernel, we now *configure it with the express purpose of turning on the kernel's debug infrastructure* as much as is useful. (Though we do not care that much for performance and/or security, the fact is that as we're inheriting the config from the production kernel, the security features are enabled by default.)

The interface we use to configure our debug kernel is the usual one:

```
make menuconfig
```

Much (if not most) of the kernel debug infrastructure can be found in the last main menu item here – the one named `Kernel hacking`:

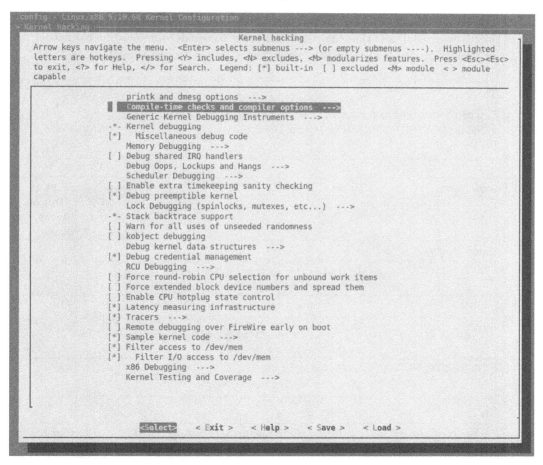

Figure 1.5 – Screenshot: make menuconfig / Kernel hacking – the majority of kernel debug options live here

There are just too many kernel configs relating to debugging to discuss individually here and now; several of them are important kernel debug features that we will explain and make use of in the chapters that follow. The following table (*Table 1.1*) summarizes some of the kernel config variables that we set or clear, depending on whether the config is for the debug or the production kernel. It is by no means exhaustive.

Not all of the config changes we make are within the `Kernel hacking` menu; others are changed as well (see the merged column rows in the table – for example, the first one is *General setup: init/Kconfig*, which specifies from which menu they originate as well as the `Kconfig` file(s) that they originate from).

Further, the `<D>` in the `Typical, value, ...` columns indicates that the *decision* is left to you (or the platform/BSP team) as the particular value to use does depend on the actual product or project, its **High Availability** (**HA**) characteristics, security posture, and so on.

> **Tip**
>
> You can search within the `make menuconfig` UI for a given config variable (`CONFIG_XXX`) by typing the key / (just as in `vi`!) and then typing the string to search for.

Kernel Config item	Meaning in brief	Typical value on production kernel	Typical value on debug kernel
General setup: `init/Kconfig`			
CONFIG_LOCALVERSION	Append a string to kernel version; useful (for example, `-kdbg01`)	`<D>`	`<D>`
CONFIG_IKCONFIG	Allow complete kernel config to be stored in-kernel; can extract via `scripts/extract-ikconfig` (or, see the next one); very useful	On (as m: module)	On
CONFIG_IKCONFIG_PROC	Access the in-kernel config file via `/proc/config.gz` (for example, extract with `gunzip -c /proc/config.gz`)	On	On
CONFIG_KALLSYMS_ALL	Loads all symbols into the kernel image	`<D>`	On

Kernel Config item	Meaning in brief	Typical value on production kernel	Typical value on debug kernel
Processor type and features: `arch/<arch>/Kconfig`			
CONFIG_CRASH_DUMP	Enable a crash-dump capable kernel which gets triggered if the kernel ever crashes	<D>	On
CONFIG_RANDOMIZE_BASE	The **Kernel Address Space Layout Randomization** (**KASLR**) feature support; randomizes the physical address at which the kernel image is decompressed, a security feature that can help deter exploit attempts	On	<D>
General architecture-dependent options: `arch/Kconfig`			
CONFIG_KPROBES	General architecture-dependent options / Kprobes: allow you to hook into almost any kernel function/address; useful for non-intrusive instrumentation and debugging	<D>	On
CONFIG_STACKPROTECTOR_STRONG	Intelligently add stack-protection canary logic via compiler; useful to detect **Buffer overFlow** (**BoF**) attacks	<D>	On
CONFIG_ARCH_MMAP_RND_BITS	Number of bits to use to determine the random offset to the base address of process memory regions; a higher number is good for security	32	28
CONFIG_VMAP_STACK	Enable vmapped (`vmalloc()` allocated) kernel stacks with guard pages	On	On
Executable file formats / CONFIG_COREDUMP	Enable core dumping	<D>	On
Enable loadable module support: `init/Kconfig`			
CONFIG_MODULE_SIG_FORCE	Only loads modules with a valid signature; used with a kernel lockdown LSM	<D>	Off
CONFIG_MODULE_SIG_ALL	Auto sign all kernel modules during the `make modules_install` step	On	On
CONFIG_UNUSED_SYMBOLS	Enable unused but exported kernel symbols; a bridge that should soon get removed	Off	On

Kernel Config item	Meaning in brief	Typical value on production kernel	Typical value on debug kernel
Device Drivers / Network device support: `drivers/net/Kconfig`			
`CONFIG_NETCONSOLE`	Network console (netconsole) logging support; log kernel printk's over the network	On	On
`CONFIG_NETCONSOLE_DYNAMIC`	Ability to dynamically reconfigure logging targets	<D>	On
Kernel Hacking: `lib/Kconfig.debug`, `lib/Kconfig.*`			
printk and dmesg options: `lib/Kconfig.debug`			
`CONFIG_DYNAMIC_DEBUG`	Dynamic debug feature for debug printk + logging; (this auto-selects the core `CONFIG_DYNAMIC_DEBUG_CORE` feature as well)	On	On
Compile-time checks and compiler options: `lib/Kconfig.debug`			
`CONFIG_DEBUG_INFO`	Compile the kernel and modules with debug info (`gcc -g`)	Off	On
`CONFIG_DEBUG_BUGVERBOSE`	`BUG()` macro prints filename, line number, instruction pointer register and Oops trace	<D>	On
`CONFIG_DEBUG_INFO_BTF`	Generates dedup-ed BPF Type Format (BTF) info; useful for running eBPF in future (requires `pahole` v1.16 or later installed; see note [1] below)	<D>	On
Generic Kernel Debugging Instruments: `lib/Kconfig[.debug|.kgdb|.ubsan]`			
`CONFIG_MAGIC_SYSRQ_DEFAULT_ENABLE`	Enable all magic SysRq functionality by setting this bitmask (`0x0`=off, `0x1`=all on, default is `0x01b6`)	<D>	0x1
`CONFIG_DEBUG_FS_DISALLOW_MOUNT`	Debugfs additional protection layer on production; with this, the APIs work but the debugfs filesystem isn't visible (not mounted)	On	Off
`CONFIG_KGDB`	Remote debug kernel via GDB	Off	On
`CONFIG_UBSAN`	Enables the Undefined Behavior sanity checker	<D>	On
`CONFIG_KCSAN`	Enables the Kernel Concurrency Sanitizer, a powerful way of detecting data races in the kernel at runtime; conditional (refer to *Chapter 8, Lock Debugging* for details)	<D>	On

Kernel Config item	Meaning in brief	Typical value on production kernel	Typical value on debug kernel
Memory Debugging: `lib/Kconfig[.debug|.kasan|.kgdb]`			
CONFIG_DEBUG_PAGEALLOC	Page memory allocations are tracked; can'help in detecting some types of memory corruption	Off	On
CONFIG_DEBUG_WX	Warn of any `W+X` memory mappings seen at boot (writeable memory should, in general, not be executable, that is, `W^X` should apply)	On	On
CONFIG_DEBUG_KMEMLEAK	Kernel memory leak detector	Off	On
CONFIG_SCHED_STACK_END_CHECK	Stack overrun check when `schedule()` is called; minimal runtime overhead	On	On
CONFIG_KASAN	Enable the Kernel Address SANitizer – adds compile-time instrumentation; extremely useful in catching many memory bugs (OOB, UAF, and so on)	Off	On
Debug Oops, Lockups, and Hangs: `lib/Kconfig.debug`			
CONFIG_PANIC_ON_OOPS	Enables kernel panic on any Oops (kernel bug)	On	Off
CONFIG_PANIC_TIMEOUT	Timeout (n, in seconds) after which system reboots; requires arch-level reboot support. Value `n == 0` implies wait forever, n positive implies reboot after n seconds	<D>	0
CONFIG_BOOTPARAM_SOFTLOCKUP_PANIC and CONFIG_BOOTPARAM_HARDLOCKUP_PANIC	Enable kernel panic on soft/hard lockups (implies that soft/hard lockup detection is enabled)	<D>	Off
CONFIG_BOOTPARAM_HUNG_TASK_PANIC	Enable kernel panic on hung task detection (implies that hung task detection is enabled)	<D>	Off

Kernel Config item	Meaning in brief	Typical value on production kernel	Typical value on debug kernel
Lock debugging (spinlocks, mutexes, and so on): `lib/Kconfig.debug`			
CONFIG_PROVE_LOCKING	Prove locking correctness via the very sophisticated lockdep lock validator; also detects possibility of deadlock	Off	On
CONFIG_LOCK_STAT	Track lock contention code regions	<D>	On
CONFIG_DEBUG_ATOMIC_SLEEP	Noisy warnings on any sleep performed within an atomic section of code	<D>	On
<Various>: `lib/Kconfig.debug`			
CONFIG_BUG_ON_DATA_CORRUPTION	Debug kernel data structures: If data corruption is found in a kernel structure, call BUG()	On	On
CONFIG_DEBUG_CREDENTIALS	Debug credential management: Debug checks to struct cred; useful for security as well	On	On
CONFIG_LATENCYTOP	Latency measuring infrastructure: Enable to use the latencyTOP tool	<D>	On
CONFIG_STRICT_DEVMEM	Filter access to /dev/mem: If off, userspace apps can memory map any memory region – user and kernel	On	On
Tracers: `kernel/trace/Kconfig`			
CONFIG_FUNCTION_TRACER	Enable ftrace – tracing of every kernel function; disabled at runtime by default	<D>	On
CONFIG_FUNCTION_GRAPH_TRACER	Enables tracing of the call graph as well; useful to profile/debug	<D>	On
CONFIG_DYNAMIC_FTRACE	Dynamic function tracing; no performance overhead when function tracing is disabled (the default)	<D>	On
<Early printk> : arch-dependent; CONFIG_EARLY_PRINTK on x86	x86 debugging: Useful to see early kernel printk's before console device is initialized	Off	On

Kernel Config item	Meaning in brief	Typical value on production kernel	Typical value on debug kernel
Kernel Testing and Coverage: `lib/Kconfig*`, `lib/kunit/Kconfig`			
`CONFIG_FAULT_INJECTION`	Enable the kernel's fault-injection framework	Off	On
Security: `security/Kconfig`, `security/*/Kconfig*`			
`CONFIG_SECURITY_DMESG_RESTRICT`	If on, only root can read kernel printk's via `dmesg`(1)	On	Off
`LOCK_DOWN_KERNEL_FORCE_CONFIDENTIALITY`	Kernel in lockdown (via an LSM), mode set to confidentiality; if on, modules can't be (un)loaded	\<D\>	Off
\<several\>	Several kernel configs regarding security as pointed out by (for example) the `kconfig-hardened-check` script	\<D\>	Off

Table 1.1 – Summary of a few kernel config variables, their meaning, and value

Besides the \<D\> value, the other values shown in the preceding table are merely my recommendations: *they may or may not be suitable for your particular use case.*

[1] Installing `pahole` v1.16 or later: `pahole` is part of the `dwarves` package. However, on Ubuntu 20.04 (or older), it's version 1.15, which causes the kernel build – when enabled with `CONFIG_DEBUG_INFO_BTF` – to fail. This is because `pahole` version 1.16 or later is required. To address this on Ubuntu 20.04, we've provided the v1.17 Debian package in the root of the GitHub source tree. Install it manually as follows:

```
sudo dpkg -i dwarves_1.17-1_amd64.deb
```

Viewing the current kernel config

Being able to view (query) the currently running kernel's configuration can prove to be a very useful thing, especially on production systems. This can be done by looking up (grepping) `/proc/config.gz` (a simple `zcat /proc/config.gz | grep CONFIG_<FOO>` is typical). The pseudofile `/proc/config.gz` contains the entire kernel config (it's practically equivalent to the `.config` file within the kernel source tree). Now, this pseudofile is only generated by setting `CONFIG_IKCONFIG=y`. As a safety measure on production systems, we set this config to the value m in production, implying that it's available as a kernel module (called `configs`). Only once you load this module does the `/proc/config.gz` file become visible; and of course, to load it up you require root access...

Here's an example of loading the `configs` kernel module and then querying the kernel config (for this very feature!):

```
$ ls -l /proc/config.gz
ls: cannot access '/proc/config.gz': No such file or directory
```

OK, to begin with (in production), it doesn't show up. So do this:

```
$ sudo modprobe configs
$ ls -l /proc/config.gz
-r--r--r-- 1 root root 34720 Oct  5 19:35 /proc/config.gz
$ zcat /proc/config.gz |grep IKCONFIG
CONFIG_IKCONFIG=m
CONFIG_IKCONFIG_PROC=y
```

Ah, it now works just fine!

Food for Thought

Did you notice? In *Table 1.1*, I set the production kernel's value for CONFIG_ KALLSYMS_ALL as <D>, implying it's up to the system architects to decide whether to keep it on or off. Why? Shouldn't the ability to view *all kernel symbols* be disabled (off) in a production system? Well, yes, that's the common decision. Recall, though, our brief on the Mars Pathfinder mission, where it initially failed due to a priority inversion issue. The tech lead of the software team at JPL, Glenn Reeves, made a very interesting statement in his now-famous response to Mike Jones (https://www. cs.unc.edu/~anderson/teach/comp790/papers/mars_ pathfinder_long_version.html): *The software that flies on Mars Pathfinder has several debug features within it that are used in the lab but are not used on the flight spacecraft (not used because some of them produce more information than we can send back to Earth). These features were not "fortuitously" left enabled but remain in the software by design. We strongly believe in the "test what you fly and fly what you test" philosophy.*

Sometimes, keeping debug features (and of course, logging) turned on in the production version of the system, can be immensely helpful!

For now, don't stress too much about exactly what each of these kernel debug options means and how you're to use them; we shall cover most of these kernel debug options in the coming chapters. The entries in *Table 1.1* are meant to kickstart the configuration of your production and debug kernels and give you a brief idea of their effect.

Once you're done generating the new debug kernel config, let's back it up as follows:

```
cp -af .config ~/lkd_kernels/kconfig_dbg01
```

Build it, as before: make -j8 all (adjust the parameter to -j based on the number of CPU cores on your box). When done, check out the compressed and uncompressed kernel image files:

```
$ ls -lh arch/x86/boot/bzImage vmlinux
-rw-r--r-- 1 letsdebug letsdebug  18M Aug 20 12:35 arch/x86/
boot/bzImage
-rwxr-xr-x 1 letsdebug letsdebug 1.1G Aug 20 12:35 vmlinux
$
```

Did you notice? The size of the vmlinux uncompressed kernel binary image file is huge. How come? All the debug features plus all the kernel symbols account for this large size...

Finish off by installing the kernel modules, initramfs, and bootloader update, as earlier:

```
sudo make modules_install && sudo make install
```

Great; now that you're done configuring both the production and debug kernels, let's briefly examine the difference between the configurations.

Seeing the difference – production and debug kernel config

It's enlightening – and really, it's the key thing within this particular topic – to see the differences between our original production and the just-built debug kernel configuration. This is made easy via the convenience script scripts/diffconfig; from within the debug kernel source tree, simply do this to generate the difference:

```
scripts/diffconfig ~/lkd_kernels/kconfig_prod01 ~/lkd_kernels/
kconfig_dbg01 >  ../../kconfig_diff_prod_to_debug.txt
```

View the output file in an editor, seeing for yourself the changes we wrought in configuration. There are indeed many deltas – on my system, the diff file exceeds 200 lines. Here's a partial look at the same on my system (I use the ellipse [...] to denote skipping some output):

```
$ cat kconfig_diff_prod_to_debug.txt
-BPF_LSM y
-DEFAULT_SECURITY_APPARMOR y
-DEFAULT_SECURITY_SELINUX n
-DEFAULT_SECURITY_SMACK n
[ ... ]
```

The - (minus sign) prefixing each of the preceding lines indicates that we removed this kernel config feature from the debug kernel. The output continues as follows:

```
DEBUG_ATOMIC_SLEEP n -> y
DEBUG_BOOT_PARAMS n -> y
DEBUG_INFO n -> y
DEBUG_KMEMLEAK n -> y
DEBUG_LOCK_ALLOC n -> y
DEBUG_MUTEXES n -> y
DEBUG_PLIST n -> y
DEBUG_RT_MUTEXES n -> y
DEBUG_RWSEMS n -> y
DEBUG_SPINLOCK n -> y
[ ... ]
```

In the preceding code snippet, you can clearly see the change made from the production to the debug kernel; for example, the first line tells us that the kernel config named DEBUG_ATOMIC_SLEEP was disabled in the production kernel and we've not enabled it (n->y) in the debug kernel! (Note that it will be prefixed with CONFIG_, that is, it will show up as CONFIG_DEBUG_ATOMIC_SLEEP in the kernel config file itself.)

Here, we can see how the suffix to the name of the kernel – the config directive named CONFIG_LOCALVERSION – has been changed between the two kernels, besides other things:

```
LKDTM n -> m
LOCALVERSION "-prod01" -> "-dbg01"
LOCK_STAT n -> y
MMIOTRACE n -> y
MODULE_SIG y -> n
[ ... ]
```

The + prefix to each line indicates the feature that has been added to the debug kernel:

```
+ARCH_HAS_EARLY_DEBUG y
+BITFIELD_KUNIT n
[ ... ]
+IKCONFIG m
+KASAN_GENERIC y
[ ... ]
```

In closing, it's important to realize these things:

- The particulars of the kernel configuration we're performing here – for both our production and debug kernels – *is merely representative*; your project or product requirements might dictate a different config.

- Many, if not most, modern embedded Linux projects typically employ a sophisticated builder tool or environment; Yocto and Buildroot are two common de facto examples. In such cases, you will have to adapt the instructions given here to cater to using these build environments (in the case of Yocto, this can become a good deal of work in specifying an alternate kernel configuration via a BB-append-style recipe).

By now, I am furtively hoping you've absorbed this material and, indeed, built yourself two custom kernels – a production and a debug one. If not, I request that you please do so before proceeding further.

So, great – well done! By now, you have both a custom 5.10 LTS production and debug kernel ready to rip. We'll certainly make use of them in the coming chapters. Let's finish this chapter with a few debug "tips" that I hope you'll find useful.

Debugging – a few quick tips

I'll start off by saying this: debugging is both a science and an art, refined by experience – the mundane hands-on slogging through to reproduce and identify a bug and its root cause, and (possibly) fix it. I'm of the opinion that the following few debug tips are really nothing new; that said, we do tend to get caught up in the moment and often miss the obvious. The hope is that you'll find these tips useful and return to them time and again!

- **Assumptions – just say NO!**

  Churchill famously said, *"Never, never, never, give up"*. We say *"Never, never, never, make assumptions"*.

  Assumptions are, very often, the root cause behind many, many bugs and defects. Think back, re-read the *Software bugs – a few actual cases section*!

  In fact (hey, I am partially joking here), just look at the word *assume: it just begs saying, "Don't make an ASS out of U and ME"*!

  Using assertions in your code is a great way to catch assumptions. The userspace way is to use the `assert()` macro. It's well documented in the man page. (We cover more on using macros within the kernel in *Chapter 12, A Few More Kernel Debugging Approaches*, in the *Assertions, warnings and BUG() macros section*).

- **Don't lose the forest for the trees!**

  At times, we do get lost in the twisted mazes of complex code paths. In these circumstances, it's really easy to lose sight of the big idea, the objective of the code. Try and *zoom out* and think of the bigger picture. It often helps spot the faulty assumption(s) that led to the error(s). Well-written documentation can be a lifesaver.

- **Think small**

  When faced with a difficult bug, try this: build/configure/get the *smallest possible version* of your problem (statement) to execute causing the issue or bug you're currently facing to surface. This often helps you track down the root cause of the problem. In fact, very often (in my own experience), the mere act of doing this – or even just the detailed jotting down of the problem you face – triggers you seeing the actual issue and its solution in your mind!

- **"It requires twice the brainpower to debug a piece of code than to write it"**

This paraphrased quote is by Brian Kernighan in the book *The Elements of Programming Style*. So, should we not use our full brainpower while writing code? Ha, of course you should... But, debugging *is* typically harder than writing code. The real point is this: take the trouble to first carefully do your groundwork: write a brief very high-level design document and write what you expect the code to do, at a high level of abstraction. Then move on to the specifics (with a so-called low-level design doc). Good documentation will save you one day (and blessings shall be showered upon you!).

That reminds me of another quote: *An ounce of design is worth a pound of refactoring* – Karl Wiegers.

- **Employ "Zen Mind, Beginner's Mind"**

Sometimes, the code can become too complex (spaghetti-like; it just *smells*). In many cases, just giving up and starting from scratch again, if viable, is perhaps the best thing to do.

This *Zen-Beginner's Mind* state also implies that we at least temporarily stop our (perhaps over-egotistical) thought patterns (*I wrote this so well, how can it be wrong!?*) and look at the situation from the point of view of somebody completely new to it. It is, in fact, one key reason why a colleague reviewing your code can spot bugs you'd never see! Plus, a good night's rest can do wonders.

- **Variable naming, comments**

I recall a Q&A on Quora revealing that *the hardest thing a programmer does* is name variables well! This is truer than it might appear at first glance. Variable names stick; choose yours carefully. As with commenting, don't go overboard either: a local variable for a loop index? `int i` is just fine (`int theloopindex` is just painful). The same goes for comments: they're there to explain the rationale, the design behind the code, what it's designed and implemented to achieve, *not* how the code works. Any competent programmer can figure that out.

- **Ignore logs at your peril!**

It's self-evident perhaps, but we can often miss the obvious when under pressure... Carefully checking kernel (and even app) logs often reveals the source of the issue you might be facing. Logs are usually able to be displayed in reverse-chronological order and give you a view of what actually occurred; Linux's systemd `journalctl(1)` utility is powerful; learn how to leverage it!

- **Testing can reveal the presence of errors but not their absence**

  A truism, unfortunately. Still, testing and QA is simply one of the most critical parts of the software process; ignore it at your peril! The time and the trouble taken to write exhaustive test cases – both positive and negative – pays off in large dividends in the long run, helping make the product or project a grand success. Negative test cases and *fuzzing* are critical for exposing (and subsequently fixing) security vulnerabilities in the code base. Then again, runtime testing only tests the portions of code actually executed. Take the trouble to perform *code coverage analysis*; 100% code coverage – and runtime testing it is the objective! (Again, we cover more on these key points in *Chapter 12, A Few More Kernel Debugging Approaches*, in the *An introduction to kernel code coverage tools and testing frameworks section*).

- **Incurring technical debt**

  Every now and then, you realize deep down that though what you've coded works, it's not been done well enough (perhaps there still exist corner cases that will trigger bugs or undefined behavior); that nagging feeling that perhaps this design and implementation simply isn't the best. The temptation to quickly check it in and hope for the best can be high, especially as deadlines loom! Please don't; there is really a thing called **technical debt**. It will come and get you.

- **Silly mistakes**

  If I had a penny for each time I've made really silly mistakes when developing code, I'd be a rich man! For instance, I once spent nearly half a day racking my head about why my C program would just refuse to work correctly until I realized I was editing the correct code but compiling an old version of it – performing the build in the wrong directory! (I am certain you've faced your share of such pesky frustrations.) Often, a break, a good night's sleep, can do wonders.

- **Empirical model**

  The word *empirical* means to validate something (anything) by actual and direct observation or experience rather than relying on theory.

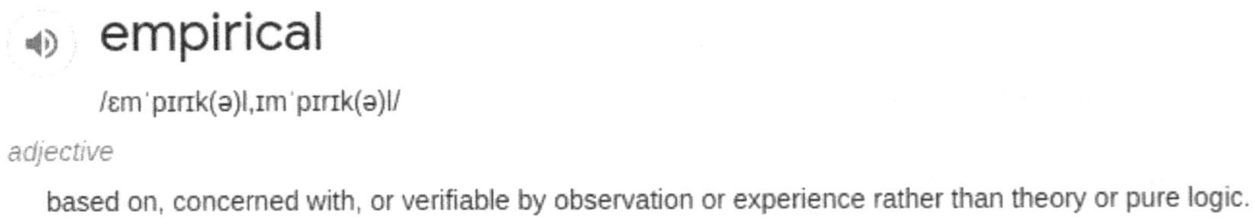

  🔊 empirical

  /ɛmˈpɪrɪk(ə)l,ɪmˈpɪrɪk(ə)l/

  adjective

  based on, concerned with, or verifiable by observation or experience rather than theory or pure logic.

  Figure 1.6 – Be empirical!

So, don't believe the book (this one is an exception of course!), don't believe the tutorial, the article, blog, tutor, or author: *be empirical – try it out and see for yourself!*

Years (decades, actually) back, on my very first day of work at a company I joined, a colleague emailed me a document that I still hold dear: *The Ten Commandments for C Programmers*, by Henry Spencer (`https://www.electronicsweekly.com/ open-source-engineering/linux/the-ten-commandments-for-c- programmers-2009-04/`). Do check it out. In a similar, albeit clumsier, manner, I present a quick checklist for you.

A programmer's checklist – seven rules

Very important! Did you remember to do the following?:

- Check all APIs for their failure case.

- Compile with warnings on (definitely with `-Wall` and possibly `-Wextra` or even `-Werror`; yes, treating warnings as errors is going to make its way into the kernel!); eliminate all warnings as far as is possible.

- Never trust (user) input; validate it.

- Eliminate unused (or dead) code from the code base immediately.

- Test thoroughly; 100% code coverage is the objective. Take the time and trouble to learn how to use powerful tools: memory checkers, static and dynamic analyzers, security checkers (`checksec`, `lynis`, and several others), fuzzers, code coverage tools, fault injection frameworks, and so on. Don't ignore security!

- With regard to kernels and especially drivers, after eliminating software issues, be aware that (peripheral) hardware issues could be the root cause of the bug. Don't discount it out of hand! (You'll learn this the hard way.)

- Do *not* assume anything (*assume: make an ASS out of U and ME*); using assertions helps catch assumptions, and thus bugs.

We shall elaborate on several of these points in the coming material.

Summary

Firstly, congratulations on completing this, our first chapter. Getting started is half the battle! You began by learning a bit about how the word **debug** came to be – equal parts myth, legend, and truth...

A key section was the brief description of some complex real-world cases of software gone wrong (several of them very unfortunate tragedies), where a software bug (or bugs) proved to be a key factor behind the disaster.

You understood that we're using the latest (at the time of this writing) 5.10 LTS kernel and how to set up the workspace (on x86_64, using either a native Linux system or Linux running as a guest OS). We covered the configuring and building of two custom kernels – a **production** and a **debug** one, with the production kernel geared toward high performance and security whereas the debug one was configured with several (most) kernel debug features turned on, in order to help catch bugs. I will assume you've done this for yourself, as future chapters will depend on it.

Finally, and I think very importantly, a few debugging tips and a small checklist wrapped up this chapter. I urge you to read through the tips and checklist often.

In the next chapter, you will learn that there can be, and are indeed, various approaches to debugging the Linux kernel (and its modules); you'll learn about them and which to use.

Further reading

- Real-world stories of software going wrong – software horror stories:

  - *SOFTWARE HORROR STORIES*: An old page, but still (mostly) valid and very interesting! Many, many incidents have been covered here; do take a gander: `http://www.cs.tau.ac.il/~nachumd/horror.html`

  - Patriot missile battery failure: `https://www-users.cse.umn.edu/~arnold/disasters/patriot.html`

  - Ariane 5 launcher crash:

    - The official report – *ARIANE 5 – Flight 501 Failure*, by the Inquiry Board: `http://sunnyday.mit.edu/nasa-class/Ariane5-report.html`

    - An excellent article on the same thing: *Design by Contract: The Lessons of Ariane*, Jean-Marc Jézéquel, Bertrand Meyer (the creator of the Eiffel programming language): `https://archive.eiffel.com/doc/manuals/technology/contract/ariane/`

  - Mars Pathfinder reset issues:

    - *Priority inversion*: `https://en.wikipedia.org/wiki/Priority_inversion`

- *What really happened on Mars?*, Glenn Reeves' detailed reply to Mike Jones' summary of the issue: `https://www.cs.unc.edu/~anderson/teach/comp790/papers/mars_pathfinder_long_version.html`

- *What the Media Couldn't Tell You About Mars Pathfinder*, Tom Durkin, 1998; PDF: `https://people.cis.ksu.edu//~hatcliff/842/Docs/Course-Overview/pathfinder-robotmag.pdf`

- *Now showing on satellite TV: secret American spy photos*, The Guardian, June 13, 2002: `https://www.theguardian.com/media/2002/jun/13/terrorismandthemedia.broadcasting`

- *Software problem kills soldiers in training incident*, June 13, 2002: `http://catless.ncl.ac.uk/Risks/22.13.html#subj2.1`

- Boeing 737 MAX and the MCAS:

  - *The inside story of MCAS: How Boeing's 737 MAX system gained power and lost safeguards*, The Seattle Times, June 22, 2019: `https://www.seattletimes.com/seattle-news/times-watchdog/the-inside-story-of-mcas-how-boeings-737-max-system-gained-power-and-lost-safeguards/`

  - *Boeing 737 Max: why was it grounded, what has been fixed and is it enough?*, The Conversation, Nov 28, 2020: `https://theconversation.com/boeing-737-max-why-was-it-grounded-what-has-been-fixed-and-is-it-enough-150688`

  - As an aside, do watch Nat Geo's 'Air Crash Investigation' series: `https://www.natgeotv.com/in/air-crash-investigation/about`

  - Recent: *DOWNFALL: The Case Against Boeing | Official Trailer | Netflix*, Feb 2022: `https://www.youtube.com/watch?v=vt-IJkUbAxY`

- Jack Ganssle's *TEM (The Embedded Muse)* newsletter – back issues: `http://www.ganssle.com/tem-back.htm`; excellent newsletters, do check it out

- Kernel and system workspace setup:

  - Various good online articles and tutorials on installing Linux as a guest VM on Oracle VirtualBox can be found at `https://github.com/PacktPublishing/Linux-Kernel-Programming/blob/master/Further_Reading.md#chapter-1-kernel-development-workspace-setup---further-reading`

- *Easy way to determine the virtualization technology of a Linux machine?*, StackExchange: `https://unix.stackexchange.com/questions/89714/easy-way-to-determine-the-virtualization-technology-of-a-linux-machine`

- Ubuntu Linux – the System Requirements page: `https://help.ubuntu.com/community/Installation/SystemRequirements`

- Kernel documentation: *Configuring the kernel* (`https://www.kernel.org/doc/html/latest/admin-guide/README.html#configuring-the-kernel`)

- Article: *How to compile a Linux kernel in the 21st century*, S Kenlon, Aug 2019: `https://opensource.com/article/19/8/linux-kernel-21st-century`

- Information on initrd / initramfs and the GRUB bootloader: from the *Further reading* notes from the *Linux Kernel Programming* book's GitHub repository: `https://github.com/PacktPublishing/Linux-Kernel-Programming/blob/master/Further_Reading.md#chapter-3-building-the-linux-kernel-from-source---further-reading`

- Customizing the GRUB bootloader: *How do I add a kernel boot parameter?* `https://askubuntu.com/questions/19486/how-do-i-add-a-kernel-boot-parameter`. Do realize, this tends to be x86_64- and Ubuntu-specific...

- *The Ten Commandments for C Programmers*, Henry Spencer: `https://www.electronicsweekly.com/open-source-engineering/linux/the-ten-commandments-for-c-programmers-2009-04/`

- Interesting:

  - A MUST-READ book: *The Mythical Man-Month*, Fred Brooks, 1975

  - *What is a coder's worst nightmare?*, Quora; answer by Mick Stute: `https://www.quora.com/What-is-a-coders-worst-nightmare`

  - *Reflections on Trusting Trust*, Ken Thompson: `https://www.cs.cmu.edu/~rdriley/487/papers/Thompson_1984_ReflectionsonTrustingTrust.pdf`

2
Approaches to Kernel Debugging

Even a casual perusal of topics related to kernel debugging will quickly have you realize that there are many approaches to it, and correspondingly, many tools and techniques that can and are brought to bear on the problem. In this relatively short chapter, we'll first check out some ways of classifying bugs by type. Classifying defects or bugs by type will help you gain a high-level understanding of them and where they fall, and at times overlap. We shall classify bugs by various types or views: the classic view – by memory issues, the security-related view, and finally, by typical issues caused within the Linux kernel.

Next, we consider why there are various approaches to kernel debugging, and then summarize exactly what these approaches are and when it's generally appropriate to use which approach. These topics will help lay the foundation for the remainder of the book, where we'll delve into learning how to employ each of these kernel debug approaches or techniques.

In this chapter, we're going to cover the following main topics:

- Classifying bug types
- Kernel debugging – why there are different approaches to it
- Summarizing the different approaches to kernel debugging

Technical requirements

The technical requirements and workspace remain identical to what's described in *Chapter 1, A General Introduction to Debugging Software*, under the *Technical requirements* and *Setting up the workspace* sections.

Classifying bug types

As you will know, defects (bugs) can be quite readily classified into different types. Here, I attempt to do so (it's nothing new, really), with an added twist: we shall look at common bug classes through different *lenses* or viewpoints – first, in the classic (typical, academic) manner, then focused on memory-type bugs, and then the security-related view of bugs. Having seen this, we'll further refine this classification to what you'll typically see when working with the Linux kernel. Do note that there can be, and often are, overlaps within these classifications. Let's begin with the first one – classifying defects/bugs in a classic manner.

Types of bugs – the classic view

The classic way of viewing the types of defects or bugs that can occur in a software program is as follows:

- **Logic or implementation errors**:

  - Includes off-by-one errors, infinite loops/recursion.

  - **Arithmetic errors**: Includes loss of precision errors (recall the Patriot missile and the Ariane 5 incidents!), arithmetic underflow or overflow, division by zero.

  - **Syntax defects**: This is quite obvious; defects such as (in C) using the equals = operator instead of the == operator (modern compilers and static analysis tools should certainly catch these).

- **Resource leakage and generic defects on resources**:

  - Includes the classic NULL pointer dereference bug and memory bugs: **Uninitialized Memory Reads** (**UMR**), leakage, double-free, **Use After Free** (**UAF**), **Out Of Bounds** (**OOB**) buffer overflow errors – read/write underflow/overflow, stack memory overflow, access violations, and so on.

- **Hardware**: Don't forget about the hardware! Faulty RAM, DMA issues, hardware freezes, microcode bugs, hardware interrupt misses/spurious interrupts, key bouncing, data endian errors, data packing/padding issues, instruction faults, and so on (see the *Further reading* section for an interesting post).

- **Races**: Data races, locking issues causing deadlock, livelocks (as in too many hardware interrupts in too short a time period; network driver layers often use the so-called **New API** (**NAPI**) to mitigate precisely this).

- **Performance defects**:

  - Includes data (cache line) alignment issues, data races, deadlocks, and livelocks.

  - Poorly chosen APIs (for example, blind usage of the kernel's page/slab allocator APIs – such as `__get_free_pages()` / `kmalloc()` – can lead to highly suboptimal memory usage due to wild amounts of internal fragmentation issues (*wastage*, really). Another: using moderate to highly contended locks with long critical sections is just begging for performance issues. (The usage of lock-free algorithms and APIs will help! Perhaps via the Linux kernel's **percpu** and **Read Copy Update** (**RCU**) lock-free primitives.)

  - Data races (mentioned in the earlier bullet point; as noted, overlaps in the bug classification can and do occur).

  - **Input/Output** (**I/O**): Suboptimal and heavy reads and writes cause major performance bottlenecks; this applies to both the filesystem and network layers; it's important to realize that often the actual performance bottlenecks are to do with suboptimal I/O usage and typically aren't CPU-related.

- There are more ways to classify bugs; we shan't go into any detail here, we shall merely mention them: According to interface-based and even teamwork-based side effects.

An interesting paper presented at a conference on software engineering in Melbourne (ICSE, 1992) proposed that defects (bugs) are introduced and removed at various rates at different points in the SDLC; interestingly, relatively high bug insertion rates occur in both the design and coding phases; it helps highlight the need for better design/architecture of the system (see the *Further reading* section for the link to the paper, and more).

Let's move along to another way, or viewport, of viewing bugs, the *memory defects* one.

Types of bugs – the memory view

As defects due to memory-related bugs are simply so common with procedural (and non-managed) languages such as C, we'll now view defects from the viewpoint of **memory corruption**:

- Incorrect memory access:
  - Using variables uninitialized; aka UMR bugs
  - Out-of-bounds memory accesses (read/write underflow/overflow bugs)
  - Use-after-free/use-after-return (out-of-scope) bugs
  - Double-free
- Memory leakage
- Data races
- Fragmentation (internal implementation) issues:
  - Internal
  - External

All these common memory issues (except fragmentation) are generally classified as **Undefined Behavior** (**UB**). Though fragmentation is a memory issue, it's not a bug in the sense that we're concerned with, so we won't delve into it further.

You'll have noticed that many bug classes are repeated from the previous classification. The reason I re-classify defects via memory corruption is to highlight it – it's definitely among the more common root causes of software issues!

Next, let's view bugs through the viewport of security-related ones.

Types of bugs – the CVE/CWE security-related view

There's an open database of publicly disclosed security vulnerabilities (*vulns*) and issues; it's used by security researchers, academicians, and by the industry to track security-related defects/bugs and helps folks to study and discuss them, build mitigations (fixes, patches) and thus respond to them in a consistent manner. Each security bug (and at times a whole bunch of them, forming a class) is assigned a number called a **Common Vulnerabilities and Exposures** (**CVE**) or **Common Weaknesses and Enumeration** (**CWE**) number.

There are several websites that categorize CWEs and CVEs; among them the US-based **National Institute of Standards and Technology (NIST)** with the **National Vulnerability Database (NVD)** (`https://nvd.nist.gov/vuln/full-listing`). It (among other things) provides a comprehensive categorization of software defects; I urge you to look up the site, and especially the page showing a subset of the CWE structure (`https://nvd.nist.gov/vuln/categories/cwe-layout`).

It's not just the NIST NVD; several other sites categorize CVEs too. Among them are the **CVE Details** site (`https://www.cvedetails.com/`); it provides excellent explanations alongside the CVE number. MITRE provides this service as well; this is its FAQ page: `https://www.cve.org/ResourcesSupport/FAQs`.

As a good example, many security-related bugs boil down to nothing but an implementation weakness or vulnerability, of the well-known (very often stack-based) **Buffer Overflow (BoF)**. The CWE MITRE site carries a detailed explanation of that here: *CWE-120: Buffer Copy without Checking Size of Input ('Classic Buffer Overflow')* (`https://cwe.mitre.org/data/definitions/120.html`), along with example code that demonstrates the vulnerabilities! Do check it out...

It's really important to realize that, at heart, many *security issues are mostly software defects – bugs!*

Types of bugs – the Linux kernel

It's also useful and relevant to look at bug types from the viewpoint of the Linux kernel itself. Paraphrasing from Sergio Prado's presentation *Linux Kernel Debugging: Going Beyond Printk Messages* (`https://www.youtube.com/watch?v=NDXYpR_m1CU`), he classifies Linux kernel bugs as follows:

- Defects (bugs) that cause the system to lock up or hang
- Defects that cause the system to crash and/or panic
- Logic or implementation defects
- Resource leakage defects
- Performance issues

All right, we've done the (perhaps rather dry) task of classifying bugs. I know what you're perhaps thinking: this is all rather academic, and perhaps a bit pointless? Well, the idea is that now that you understand how bugs can be classified, we shall get to the important point: based on classification (and other methods), which tools/techniques can you employ to debug them.

But first, we need to also understand that not all debugging techniques or approaches may be suitable for the task; the following section briefs you on this.

Kernel debugging – why there are different approaches to it

When the kernel has an error, a bug, no matter how trivial or non-trivial, the entire system is considered to be in a bad, unrecoverable state and a **kernel panic** typically ensues – a fatal condition wherein the system generates a brief diagnostic and then simply halts (or, it can be configured to reboot after a timeout). Debugging these scenarios is inherently hard, as, at least on the surface, it appears as though there is no diagnostic information to work with, and even if there were, the system is unresponsive, essentially dead. So how do you retrieve diagnostic information in order to analyze it?

What you will soon realize is that even though there are several techniques, tools, and approaches to kernel debugging, not all of them are suitable for any and all scenarios – the tools or techniques you use are often dictated by the particular scenario you find yourself in.

So what are these scenarios? Broadly speaking, they include the following:

- **The development phase of the project**: You are in the process of developing the code and active development is ongoing. This involves the usage of both the custom debug and production kernels.

- **Unit or individual developer testing and QA (integration/systems/acceptance) test phases**: You have developed a module or component and need to test it. This involves the usage of both the custom debug and production kernels.

- **Post-mortem analysis**: The kernel has crashed; you need to try and figure out the root cause and fix it. This involves the usage of both the custom debug and production kernels.

- **In-field or production**: The system is suffering from bugs and/or performance issues. You need to use appropriate tools to understand the underlying causes. This involves usage of the custom production kernel (and debug kernel – where symbols are required – for some of the tools).

Finally, let's get to the nitty-gritty: the following section gives you a summary of the different approaches, actual tools, techniques, and APIs (if appropriate) to debug the Linux kernel.

Summarizing the different approaches to kernel debugging

There are many approaches to kernel debugging. The one (or ones) to use depends upon the scenario. Here are the aforementioned scenarios and some approaches to kernel debugging in them.

The development phase

Are you currently in the **development phase** of the project? If yes, the following approaches and techniques can help:

- Code-based debugging techniques can immediately help (although they're even useful later). These include the following:

    - Code-level instrumentation with `printk()` and friends

    - Dynamic debug printk

    - Generating a kernel-mode stack dump and interpreting it

    - Using assertions within the code

    - Setting up and leveraging debug hooks within the code base – there are two typical ways to do this:

        - Via the **debugfs** pseudo filesystem

        - Via a special **ioctl(2)** hook function meant for debug purposes

- Single-stepping through the kernel (or module's) C code, setting breakpoints, watchpoints, examining the content of data, and so on: via the well-known **Kernel GDB (KGDB)** framework.

Unit testing and/or QA phases

In the **unit testing and/or QA phases** (both unit and integration/systems/acceptance tests), you, in your capacity as an individual developer on the project, typically run unit tests against the code you've developed. Besides that, your team and/or a dedicated QA team might run a complete (perhaps automated) test suite against the project (interim) release and discover and report bugs back to the development team. The following tools and techniques should be used to try and catch possible bugs in these phases:

- **Dynamic analysis**: You run tools that run on the live system, which perform checks on code paths as they're executed. These include the following:

  - Memory checkers: detecting memory issues or memory corruption (often the root cause of bugs) is critical.

  - **Undefined Behavior** (**UB**) checkers: UB includes things such as arithmetic underflows/overflows (including the well-known **Integer overFlow (IoF)** defect), invalid bit shifts, misaligned accesses, and so on.

  - Lock debugging tools and instrumentation.

- **Static analysis**: Involves employing tools that work upon the source code of the project (similar to the compiler). They can provide a great deal of insight into overlooked and possibly buggy, as well as security-wise risky, code.

- **Code coverage analysis**: This isn't really a debug technique; it's *to ensure that every line of code is actually exercised while testing is being done. This is critical – only then can we have high confidence in the product.* Here, you typically employ code coverage tools (such as **gcov**) to check which lines of code are actually executed during a given test run. (These techniques are typically more applicable to individual developer unit testing than to system-level testing, though they can certainly be applied there as well.)

- **Monitoring and tracing tools**: These can be employed in the development and testing/QA phases, and possibly even in production (in the field):

  - Kernel tracing infrastructure – this is a big area and includes the following:

    - **Ftrace and trace-cmd**

    - **Event Tracing**

    - The **Linux Tracing Toolkit: next generation** (**LTTng**), the Trace Compass and KernelShark GUIs

- **Perf**
- **Enhanced Berkeley Packet Filter (eBPF)**
- **SystemTap**
- User-mode tracing infrastructure (often employing the powerful **strace** and **ltrace** utilities)

- **Kernel probes (Kprobes)** – both static and dynamic
- Watchdogs
- Custom kernel panic handler
- Detection of soft and hard lockups
- Detection of hung tasks
- Magic SysRq handlers

- **Post-mortem analysis**: One of the common cases for most developers is the *after-a-crash* case: (the capture and) analysis of a kernel diagnostic – called the kernel **Oops**:

  - Oops (kernel log files) analysis
  - Using **kdump** to collect a kernel dump image (loosely equivalent to the *core dump* produced by a process when it crashes), and the powerful **crash** application to interpret it

- **In production in-field runtime (mentioned for completeness)**:

  - Any (or all) of the *monitoring and tracing tools* (mentioned in the previous bullet point)
  - Debug hooks within the code (via debugfs, ioctl)
  - Regular and dynamic debug printks
  - Logging (via the **systemd** journal and app-based logging)
  - A custom panic handler

Which kernel debug technology you should use is not only dependent on the phases of the software life cycle; some kernel debug techniques or technologies demand a significant amount of hardware and/or software resource availability:

- **Hardware constraints**: Some kernel debug techniques require significant amounts of hardware resource availability, which you may or may not be able to afford! For example, using the **kdump** technology requires significant amounts of RAM, network bandwidth, and/or disk space. Some tightly constrained embedded Linux systems just cannot afford this, whereas your typical server system can easily do so (the same goes for the well-known userspace Valgrind suite of tools; **Address Sanitizer (ASAN)** uses fewer resources...).

- **Software constraints**: Just as with hardware, some systems have a self-imposed design limitation on what can be enabled in the kernel config, which might preclude some debugging techniques. Again, kdump, and tracing infrastructure, are good examples of this.

A key point: dynamic analysis tools can only catch bugs within the code that they actually *see* and run. This leads us to understand that having test cases that cover all the code is extremely critical, as mentioned before (in *Chapter 1, A General Introduction to Debugging Software*), *100% code coverage is the objective!*

Please do note, that though I've definitively categorized the tools and techniques, cases will certainly arise where you can (and perhaps should) use a technique in a different scenario than has been shown above. *Keep it flexible and use what's appropriate to the situation at hand.*

Categorizing into different scenarios

The tables that follow are an attempt at a catch-all of kernel debug approaches, tools, and techniques, categorized by different scenarios to use them in.

Do note the following:

- For now, just look at the available tools/techniques/technologies/APIs for different scenarios and use cases; don't worry about how exactly to use them. That, of course, is really at the heart of the book and the coming chapters. The intent is to cover most of the ones mentioned here.

- As already mentioned, the scenarios in which to use a given tool or technique are typical, not absolute. You might come across a use case that's different. I suggest you adapt and make use of whichever techniques seem appropriate.

We'll begin with a summary table (*Table 2.1*) for the scenario where **you're developing the kernel (or kernel/driver module) code** – the coding phase:

Debugging Approach	Tool(s)/Technique(s)	Specifics/APIs/tool names/frontends
Code-based debug techniques: used within the kernel or module code	Instrumentation via the `printk()` and friends APIs	`printk()`, `printk_ratelimited()`, `trace_printk()`, and so on...
	Dynamic debug printk (`CONFIG_DYNAMIC_DEBUG`)	`pr_debug()`/`dev_dbg()` : can be dynamically enabled/disabled per call site
	Generating a kernel-mode stack dump and interpreting it	`[trace_]dump_stack()`; kernel mode stack call trace interpretation
	Using custom assertions	`WARN[_ON[_ONCE](), WARN_TAINT[_ONCE]()`, or a custom `assert()` macro
	Debug hooks	Via debugfs APIs, via a custom `ioctl()` method
Interactive debug	Interactive debug using GDB, KGDB, KDB	Single-step through the kernel (or kernel module) C code, set breakpoints, set hardware watchpoints, examine data structures, and so on via the kernel's KGDB (and/or KDB) framework. Use `CONFIG_GDB_SCRIPTS` for additional helpful Python scripts (`lx*` command verbs) within GDB

Table 2.1 – Summary of kernel debug techniques for the development/coding phase

Now let's check out a summary table of the kernel debugging tools and techniques that can be employed during the **testing and QA phases**:

Debugging Approach	Tool(s)/Technique(s)	Specifics/APIs/tool names/frontends
Dynamic analysis	Kernel memory checker tools	**Kernel Address SANitizer (KASAN), Undefined Behavior SANitizer (UBSAN), SLUB** debug techniques, **kmemleak**
	Undefined Behavior (UB) checkers (arithmetic over/underflows, and so on)	**UBSAN**
	Lock debugging tools	**Lockdep** (kernel lock validator), various other kernel configs for lock debugging, locking statistics
Static analysis	Perform static analysis on kernel (or kernel module) source code	**checkpatch.pl, sparse, smatch, Coccinelle, cppcheck, flawfinder, gcc, and so on**
Code coverage	Perform code coverage analysis	**kcov** and **gcov**

Table 2.2 – Summary of kernel debug techniques for the unit testing/QA phases

Now that we've viewed several kernel debug tools and techniques by scenario, let's view them in another few categories – **tracing, monitoring, and profiling tools**:

Tool(s)/Technique(s)	Specifics/APIs/tool names/frontends
Kernel tracing infrastructure	**Ftrace, LTTng, perf-events**, and **Perf, eBPF, SystemTap**
User space tracing tools	`strace` and `ltrace`, `uprobe*`
Profiling tools	`perf`, `perf-tools`, and `*bpfcc` (eBPF)
In-production instrumentation	Static and dynamic probes (**Kprobes** and **kretprobes**); **kprobe-perf** utility script; kernel event tracing
System-wide monitoring, panic handlers	Kernel watchdog, userspace watchdog daemon process (**watchdogd**); custom kernel panic handler code
	Magic SysRq handlers

Table 2.3 – Summary of kernel debug techniques to do with system monitoring and tracing

This leaves only the **kernel image capture, Oops, and post-mortem crash analysis tools and techniques**:

Tool(s)/Technique(s)	Specifics/APIs/tool names/frontends
Kernel image dump generation and capture	**kdump**
Kernel image dump analysis (or live kernel analysis)	**crash** and **GDB** (limited)
Analysis of kernel Oops within the kernel logs	Kernel **Oops** analysis
Logging	Kernel and user mode log analysis – captured via **systemd** journal; **journalctl** frontend
	Netconsole: sends kernel log messages over the network
Kernel/driver live patching	**KGraft**; for example, can use live patching to add instrumentation

Table 2.4 – Summary of kernel debug techniques to do with kernel image capture, Oops, post-mortem crash analysis, and logging

I've included logging (and log analysis) as well in *Table 2.4* (instead of allocating an unnecessary separate table for it); *logs are a really important means* of ascertaining what happened on the system after a crash (for user as well as kernel-level debugging).

Finally, the following table shows which kernel debug tools or techniques (or APIs) are effective for which types of kernel defects:

Type of kernel defect versus debugging techniques and tools to employ	Lockup/ Hang	Crash/ Panic	Logic/ Implementation	Resource Leakage	Performance Issues
Code-based/interactive debugging ((dynamic) printk, netconsole, assertions, debugfs hooks, GDB, KGDB)	Y	Y	Y	?	N
Dynamic analysis (memory checkers: KASAN, UBSAN, SLUB debug, kmemleak; locking: lockdep, lock stats, and so on)	Y	N	N	Y	Y
Static analysis tooling (Coccinelle, `checkpatch.pl`, `sparse`, `smatch`, `cppcheck`)	N	N	Y	?	N

Type of kernel defect versus debugging techniques and tools to employ	Lockup/ Hang	Crash/ Panic	Logic/ Implementation	Resource Leakage	Performance Issues
Monitoring and tracing tools (Ftrace, event tracing, LTTng, perf, eBPF, Kprobes; watchdog, panic handler, magic SysRq)	Y	?	?	?	Y
Post-mortem analysis (logging: kernel log analysis, systemd logs; Oops interpretation, kdump, crash, GDB)	?	Y	?	N	N

Table 2.5 – Summary of kernel debug tools/techniques versus types of kernel defects

Legend:

- **Y**: Yes, can/should be used

- **N**: No, avoid using it

- **?**: It depends... **Your Mileage May Vary (YMMV)**

Again, these guidelines are definitely not written in stone; you should use your judgement and try different techniques as required.

Which kernel debug tools and techniques you enable on your production kernel is typically something you (or the platform/BSP team) have to decide, based on system constraints (both hardware and software), performance considerations, and so on. A good starting point for this is what we already covered back in *Chapter 1, A General Introduction to Debugging Software*, under the *A tale of two kernels* section, which described in some detail how to go about configuring both a custom production and a custom debug kernel.

Summary

In this chapter, you learned that there are many approaches to debugging the kernel. We even categorized them in a suitable manner to help you quickly decide which to use in what situation. This was one of the key points – *not every tool or technique will be useful in every scenario or situation*. For example, employing a powerful memory checker such as KASAN to help find memory bugs is really useful during the development and unit testing phases but typically impossible during systems testing and production (as the production kernel will not be configured with KASAN enabled, but the debug kernel will).

You will also realize that both hardware and software constraints play a role in determining which kernel debug features can be enabled.

Further, we showed the various approaches, tools, and techniques (even at times the API or tool names) for kernel debugging categorized in several tables. This can aid you in narrowing down your armory: which of them to use in which situation.

It's important to not be too rigid when deciding which kernel debug tools/techniques to use for your project based solely on the tables we've shown here; keep it flexible and try different approaches for your situation until you find what clicks.

Good job on completing this chapter! In the next one, we'll get down to brass tacks and learn how to debug via the instrumentation approach!

Further reading

- *Estimating software fault content before coding*, Eick, Loader, et al, Proceedings of the International Conference on Software Engineering, Melbourne, June 1992: `https://dl.acm.org/doi/10.1145/143062.143090`

- A very in-depth and interesting academic article on UB: *Undefined Behavior in 2017*, Regehr, Cuoq, July 2017: `https://blog.regehr.org/archives/1520`

- *NASA Study on Flight Software Complexity*, March 2009: `https://www.nasa.gov/pdf/418878main_FSWC_Final_Report.pdf`. A deep and interesting read.

- Security-related defect tracking via CWE/CVE; very useful to track security-related defects and gain an understanding of them:

  - NIST NVD database – full listing: `https://nvd.nist.gov/vuln/full-listing`

  - CVE details: `https://www.cvedetails.com/`

  - CVE MITRE: `https://cve.mitre.org/`

  - *2021 CWE Top 25 Most Dangerous Software Weaknesses*: `https://cwe.mitre.org/top25/archive/2021/2021_cwe_top25.html`

- Hardware bug! *How a broken memory module hid in plain sight — and how I blamed the Linux Kernel and two innocent hard drives*, C Hollinger, Feb 2020: `https://towardsdatascience.com/how-a-broken-memory-module-hid-in-plain-sight-and-how-i-blamed-the-linux-kernel-and-two-innocent-ef8ce7560ecc`

- *Linux Kernel Debugging: Going Beyond Printk Messages - Sergio Prado, Embedded Labworks*, OSS/ELC Europe, May 2020, YouTube: `https://www.youtube.com/watch?v=NDXYpR_m1CU`. Excellent classification (and more) on kernel-level bugs

- *Debugging kernel and modules via gdb*, Linux kernel documentation: `https://www.kernel.org/doc/html/latest/dev-tools/gdb-kernel-debugging.html#debugging-kernel-and-modules-via-gdb`

- *The kernel debugging techniques for a device driver developer on arm64*, Christina Jacob, Oct 2019, Medium: `https://medium.com/@christina.jacob.koikara/the-kernel-debugging-techniques-for-a-device-driver-developer-on-arm64-fa984e4d2a09`

Part 2: Kernel and Driver Debugging Tools and Techniques

In this portion of the book, you will learn – in a hands-on fashion – several powerful kernel- and driver-level debugging tools and techniques. They'll span from leveraging the humble printk to using Kprobes, debugging kernel memory corruption, generating and interpreting an Oops, and finish up with powerful lock debugging techniques.

The following chapters will be covered in this section:

3
Debug via Instrumentation – printk and Friends

Quick, think: how often have you interspersed `printf()` instances (or the equivalent) in your program in order to follow its progress as it executes code, and indeed, to see at approximately which point it (perhaps) crashes? Often, I'm guessing! Don't feel bad at all, this is a really good debugging technique! It has a fancy name to boot: **instrumentation**.

What you've been doing is *instrumenting* your code, allowing you to see the flow (depending on the granularity of your print statements); this allows you to understand where it's been. Often enough, this is all that's required to debug many situations. Do recollect, though, what we discussed in the previous chapter – a technique like instrumentation is typically useful in certain circumstances, not all. For example, a resource leak (such as a memory leak) defect is difficult, if not impossible, to debug with instrumentation. For most other situations though, it's a really useful technique!

In this chapter, we're going to understand how to instrument kernel (or driver) code, primarily using the powerful `printk()` – and friends – APIs. Further, we shall continue along this path in the following chapter as well, focusing our efforts on another kernel technology than can be used for instrumentation on production systems – **kprobes**.

In this chapter, we will focus on covering the following main topics:

- The ubiquitous kernel printk
- Leveraging the kernel printk for debug purposes
- Using the kernel's powerful dynamic debug feature

These very practical topics are important: knowing how to efficiently debug via instrumentation can result in a quick cure for annoying bugs!

Technical requirements

The technical requirements and workspace remain identical to what's described in *Chapter 1, A General Introduction to Debugging Software*. The code examples can be found within the book's GitHub repository here: `https://github.com/PacktPublishing/Linux-Kernel-Debugging`.

The ubiquitous kernel printk

There's a good reason the famous and familiar **Kernighan and Ritchie (K&R)** *Hello, world* C program employs the `printf()` API: it's the preferred API via which any output is written to the screen (well, technically, to the **standard output** channel **stdout** of the calling process). After all, it's how we can actually *see* that our program is really doing something, right?

You will surely recall using this API when writing your very first C program. Did you write the code that incorporates the `printf()` function? No, of course not; then where is it? You know: it's part of the (typically rather large) standard C library – **GNU libc (glibc)** on Linux. Pretty much every binary executable program on a Linux box automatically and dynamically links into this library; thus `printf()` is pretty much always available! (On x86, doing `ldd $(which ps)` will have the useful `ldd` script show you the libraries that the `ps` app links into; one of them will be the standard C library `libc.so.*` – try it.)

Except `printf()` isn't available within the kernel! Why? This itself is a key point: the Linux kernel does not use libraries – dynamically or statically – in the way userspace applications do. There are what could perhaps pass as the equivalent: the `lib/` branch of the kernel source tree (peek at it here if you wish: `https://github.com/torvalds/linux/tree/master/lib`) contains many useful APIs that get built into the kernel image itself. Also, the kernel's framework for writing modules – the **Loadable Kernel Module (LKM)** – has facilities that kind of mimic the user mode library: the **module stacking** approach and the ability to link together several source files into a single kernel module object (`.ko`) file.

> **Important Note**
>
> These facilities – writing kernel modules with the LKM framework, the module stacking approach, the usage of the `printk()` API, and so on, are covered in detail in my earlier book *Linux Kernel Programming*.

So, how is the kernel or driver developer expected to emit a message that can be seen and, even better, logged? Via the ubiquitous `printk()` API, that's how! We say this because the `printk()` (and friends) APIs can be used *anywhere* – within interrupt handlers (all sorts – hardirq/softirq/tasklets), process context, while holding a lock; they're SMP-safe.

For you, the reader of this book, I do assume that you understand the basic usage of the useful `printk()` API, so I'll mostly skip over the very basics and instead explain a summary of typical basic usage, along with a few examples from the kernel code base.

The `printk()` API's signature is as follows:

```
// include/linux/printk.h
int printk(const char *fmt, ...);
```

If you're curious, the actual implementation is here within the kernel source: `kernel/printk/printk.c:printk()`.

> **Tip – Browsing Source Trees**
>
> Efficiently browsing large code bases is an important skill; the modern Linux kernel source tree's **Source Lines Of Code (SLOCs)** are in excess of 20 million lines! Though you could go with the typical `find <ksrc>/ -name "*.[ch]" |xargs grep –Hn "<pattern>"` approach, it quickly gets tiresome.
>
> Instead, please do yourself a big favor and learn to use powerful and efficient purpose-built code-browsing tools like (exuberant!) **ctags** and **cscope** (you installed them when following directions in *Chapter 1, A General Introduction to Debugging Software*). In fact, for the Linux kernel, they're built-in targets to the top-level `Makefile`; here's how you can build their index files for the kernel:
>
> `cd <kernel-src-tree>`
>
> `make –j8 tags`
>
> `make –j8 cscope`
>
> To build the indices for a particular architecture, set the environment variable `ARCH` to the architecture name; for example, to build `cscope` indices for AArch64 (ARM 64-bit):
>
> `make ARCH=arm64 cscope`
>
> You'll find links to tutorials on using `ctags` and `cscope` in the *Further reading* section of this chapter.

Great – let's actually make use of the famous `printk()`; to do so, we'll begin by checking out the logging levels at which messages can be emitted.

Using the printk API's logging levels

Syntax-wise, the printk API usage is almost identical to that of the familiar `printf(3)`; the immediately visible difference is the usage of a *logging level* prefixed to the format specifier, the KERN_`<foo>` as the first token. Here's a sample printk with the logging level set to KERN_INFO:

```
printk(KERN_INFO "Hello, kernel debug world\n");
```

First off, notice that KERN_INFO is not a separate parameter; it's part of the format string being passed as the argument. Next, it's not a priority level; it's merely a marker to specify that this printk is being logged as an *informational* one. Utilities to view logs – such as `dmesg(1)`, `journalctl(1)`, and even GUI tools such as `gnome-logs(1)` – can subsequently be used *to filter log messages* by logging level.

The printk has eight available log levels (from 0 to 7); you're expected to use the one appropriate to the situation at hand. We'll show them to you direct from the source. The comment to the right of each log level specifies the typical circumstances under which you're expected to use it:

```
// include/linux/kern_levels.h
[...]
#define KERN_EMERG    KERN_SOH "0" /* system is unusable */
#define KERN_ALERT    KERN_SOH "1" /* action must be taken
immediately */
#define KERN_CRIT     KERN_SOH "2" /* critical conditions */
#define KERN_ERR      KERN_SOH "3" /* error conditions */
#define KERN_WARNING  KERN_SOH "4" /* warning conditions */
#define KERN_NOTICE   KERN_SOH "5" /* normal but significant
condition */
#define KERN_INFO     KERN_SOH "6" /* informational */
#define KERN_DEBUG    KERN_SOH "7" /* debug-level messages */
#define KERN_DEFAULT  ""           /* the default kernel log
level */
[...]
```

You can see that the KERN_<FOO> log levels are merely strings ("0", "1", ..., "7") that get prefixed to the kernel message being emitted by printk, nothing more. KERN_SOH is simply the kernel **Start Of Header (SOH)**, which is the value \001. The man page on the ASCII code, ascii(1), shows that the numeric 1 (or \001) is the SOH character, a convention that is followed here.

What's the printk Default Log Level?

Within printk(), if the log level is not explicitly specified, what log level is the print emitted at? It's 4 by default, that is, KERN_WARNING. Note, though, that you are expected to always specify a suitable log level when using printk or, even better, use the convenience wrapper macros of the form pr_<foo>() where <foo> specifies the log level (it's coming right up).

Further, the kern_levels.h header contains integer equivalents of the string loglevel we've just seen (KERN_<FOO>) as the macro's LOGLEVEL_<FOO> (fear not, we shall make use of it in the first example code that comes up soon!).

A quick introduction to the pr_*() convenience macros will get us closer to the code. Let's go!

Leveraging the pr_<foo> convenience macros

For convenience, the kernel provides simple wrapper macros over the printk of the form pr_<foo> (or pr_*()) where <foo> specifies the log level; for example, in place of writing the code as follows:

```
printk(KERN_INFO "Hello, kernel debug world\n");
```

You can – and indeed should! – instead use the following:

```
pr_info("Hello, kernel debug world\n");
```

The kernel header include/linux/printk.h defines the following pr_<foo> convenience macros; you're encouraged to use them in place of the traditional printk():

- pr_emerg(): printk() at log level KERN_EMERG
- pr_alert(: printk() at log level KERN_ALERT
- pr_crit: printk() at log level KERN_CRIT
- pr_err(): printk() at log level KERN_ERR
- pr_warn(): printk() at log level KERN_WARNING
- pr_notice(): printk() at log level KERN_NOTICE
- pr_info(): printk() at log level KERN_INFO
- pr_debug() or pr_devel(): printk() at log level KERN_DEBUG

Here's an example of using the emergency printk:

```
// arch/x86/kernel/cpu/mce/p5.c
[...]
/* Machine check handler for Pentium class Intel CPUs: */
static noinstr void pentium_machine_check(struct pt_regs *regs)
{
    [...]
    if (lotype & (1<<5)) {
        pr_emerg("CPU#%d: Possible thermal failure (CPU on fire
?).\n", smp_processor_id());
    }
[...]
```

Is the processor on fire!? Whoops! The point: the above message is logged at level
KERN_EMERG.

While on the subject of using the pr_*() macros, there's one called pr_cont().
Its job is to act as a continuation string, continuing the previous printk! This can be
useful... here's an example of its usage:

```
// kernel/module.c
    if (last_unloaded_module[0])
        pr_cont(" [last unloaded: %s]",
                last_unloaded_module);
    pr_cont("\n");
```

We typically ensure that only the final pr_cont() contains the newline character. Right,
let's now learn how to automatically prefix every printk we emit!

Fixing the prefix

In addition, there's a rather special macro, pr_fmt(). It's used to generate a uniform
format string for the pr_*() macros (and indeed for any printk()). So, by overriding
its definition, by (re)defining it as the *very first (non-comment) line* of a source file, you
can guarantee prefixing a given format to all subsequent pr_*() macro and printk()
API invocations. This can be very useful, especially in a debug context, allowing us
to automatically prefix, say, the kernel module name, the function name, and the line
number to every single printk!

Let's check out an example: our very simple printk_loglevels kernel module
demonstrates a couple of things:

- Using the pr_fmt() macro to prefix a custom string to every single printk
- Using the pr_<foo>() macros to emit printks at different logging levels

> **Don't Forget**
>
> The code for this, and all kernel/driver modules and demos presented in this book, is available in its GitHub repo. For this particular demo, you can find the code here: `https://github.com/PacktPublishing/Linux-Kernel-Debugging/tree/main/ch3/printk_loglevels`
>
> Next, when trying out the kernel modules here, please ensure that you *have booted into the custom debug kernel* (or even the default distro kernel is okay for now). Attempting to use our custom production kernel may not work – why not? This is possibly because its security configuration is tight: it may not even allow you to try out a kernel module that isn't signed or if the signature can't be verified (more on this in the *Trying our kernel module on the custom production kernel section*).

Let's quickly check out the relevant code from the `ch3/printk_loglevels/printk_loglevels.c` file:

```
#define pr_fmt(fmt) "%s:%s():%d: " fmt, KBUILD_MODNAME, __
func__, __LINE__
#include <linux/init.h>
#include <linux/module.h>
#include <linux/kernel.h>
[...]
static int __init printk_loglevels_init(void)
{
        pr_emerg("Hello, debug world @ log-level KERN_EMERG
[%d]\n", LOGLEVEL_EMERG);
        pr_alert("Hello, debug world @ log-level KERN_ALERT
[%d]\n", LOGLEVEL_ALERT);
        pr_crit("Hello, debug world @ log-level KERN_CRIT
[%d]\n", LOGLEVEL_CRIT);
        pr_err("Hello, debug world @ log-level KERN_ERR
[%d]\n", LOGLEVEL_ERR);
        pr_warn("Hello, debug world @ log-level KERN_WARNING
[%d]\n", LOGLEVEL_WARNING);
        pr_notice("Hello, debug world @ log-level KERN_NOTICE
[%d]\n", LOGLEVEL_NOTICE);
        pr_info("Hello, debug world @ log-level KERN_INFO
[%d]\n", LOGLEVEL_INFO);
        pr_debug("Hello, debug world @ log-level KERN_DEBUG
[%d]\n", LOGLEVEL_DEBUG);
```

```
        pr_devel("Hello, debug world via the pr_devel() macro
(eff @KERN_DEBUG) [%d]\n", LOGLEVEL_DEBUG);
        return 0;                        /* success */
}
static void __exit printk_loglevels_exit(void)
{
        pr_info("Goodbye, debug world @ log-level KERN_INFO
[%d]\n", LOGLEVEL_DEBUG);
}
```

A (partial) screenshot of trying out this code is shown as follows – do study the output:

```
------------------------------------
sudo insmod ./printk_loglevels.ko && lsmod|grep printk_loglevels
------------------------------------

Message from syslogd@dbg-LKD at Sep  8 16:23:49 ...
 kernel:[53143.115411] printk_loglevels:printk_loglevels_init():34: Hello, debug world @ log-level KERN_EMERG   [0]
printk_loglevels       20480  0
------------------------------------
sudo dmesg
------------------------------------
[53143.115411] printk_loglevels:printk_loglevels_init():34: Hello, debug world @ log-level KERN_EMERG    [0]
[53143.115629] printk_loglevels:printk_loglevels_init():35: Hello, debug world @ log-level KERN_ALERT    [1]
[53143.115802] printk_loglevels:printk_loglevels_init():36: Hello, debug world @ log-level KERN_CRIT     [2]
[53143.115975] printk_loglevels:printk_loglevels_init():37: Hello, debug world @ log-level KERN_ERR      [3]
[53143.116148] printk_loglevels:printk_loglevels_init():38: Hello, debug world @ log-level KERN_WARNING [4]
[53143.116154] printk_loglevels:printk_loglevels_init():39: Hello, debug world @ log-level KERN_NOTICE   [5]
[53143.116160] printk_loglevels:printk_loglevels_init():40: Hello, debug world @ log-level KERN_INFO     [6]
[53143.116167] printk_loglevels:printk_loglevels_init():41: Hello, debug world @ log-level KERN_DEBUG    [7]
[53143.116173] printk_loglevels:printk_loglevels_init():42: Hello, debug world via the pr_devel() macro (eff @KERN_DEBUG) [7]
$ sudo rmmod printk_loglevels ; sudo dmesg |tail -n1
[53160.019525] printk_loglevels:printk_loglevels_exit():49: Goodbye, debug world @ log-level KERN_INFO    [6]
$
```

Figure 3.1 – Screenshot showing output from our printk_loglevels kernel module

(By the way, I often use a simple wrapper bash script named lkm – in the root of our source tree – to automate the build, load (insmod(8)), lsmod(8), and dmesg(1)) of the kernel module. The invocation of the script isn't seen in the preceding screenshot though.)

In the preceding code and screenshot, do notice the following:

- Due to our pr_fmt() macro (in the first line of code), every printk is prefixed with the module name, function name, and line number.

- The pr_<foo>() macros have emitted a printk at the relevant log level. Even the log level integer equivalent is printed within parentheses on the extreme right.

- Any printk at log level *emergency* (KERN_EMERG) is immediately displayed on all console devices. You can see the output in the preceding screenshot (see the line in the upper portion Message from syslog@dbg-LKD at ...).

- The `dmesg` utility has the ability to conveniently color-code the log output, helping our eyes to catch the more important kernel messages (so too does the powerful `journalctl` utility).

- To prevent serious **information leakage** security issues, many recent distros configure CONFIG_SECURITY_DMESG_RESTRICT to be on by default, thus requiring us to either use `sudo(8)` or have the appropriate capability bits set to view kernel logs via `dmesg`.

All right, now that we understand how to use the `printk()` API as well as the `pr_*()` macros, let's move on to figuring out a key point: once emitted, where exactly is the `printk() / pr_*() / dev_*()` output visible?

Understanding where the printk output goes

Without going into too many of the details (they're covered in my earlier *Linux Kernel Programming* book), let's quickly summarize this key point: we have issued several printks, but where does the output actually go? The following table precisely shows this.

The first important thing to understand is unlike the printf userspace family of APIs, the printk output does *not* go to `stdout`:

printk() (and friends) outputs to	When	Additional info
Log buffer in memory (RAM)	Always	`static char __log_buf[__ LOG_BUF_LEN]`; in RAM, volatile; designed as a ring buffer; gets overwritten when overflowed. Configurable via `CONFIG_LOG_BUF_ SHIFT` (init/Kconfig); the default value of 17 yields a log buffer size of 128 KB (also affected by `CONFIG_LOG_CPU_ MAX_BUF_SHIFT`)
Log file(s): modern	Always, by default on most systems (requires configuration)	The modern **systemd** framework's logging facility; `journalctl(1)` frontend; non-volatile
Log file(s): traditional	Always, by default on most systems (requires configuration)	The traditional **system logger daemon (syslogd)** along with the **kernel log daemon (klogd)** does logging; `dmesg(1)` frontend; non-volatile

printk() (and friends) outputs to	When	Additional info
Console device (we'll cover more on this in the *What exactly is the console device?* section)	On by default for log levels < 4 (that is, for emerg/alert/crit/err) on most systems (requires configuration)	Controlled via kernel tunable /proc/sys/kernel/printk

Table 3.1 – Summary of printk output locations

With modern Linux distros (including our x86_64 Ubuntu 20.04 LTS), **system daemon (systemd)** is the initialization framework used. Systemd is a pretty powerful (and intrusive!) framework, taking over many tasks on the OS. These include bringing up system services, logging, core dump manipulation, the kernel/userspace udev feature, and more. The logging framework includes sophisticated features such as log rotation, archival, and so on.

As well, on many modern distros, the traditional style logging does work along with the modern one. Here, the files logged into for kernel printks depend on the broad type of distro:

- Debian/Ubuntu type distros: /var/log/syslog
- Red Hat/Fedora/CentOS type distros: /var/log/messages

I'll also mention that the output of the kernel printk to the console device depends upon the log level that it's emitted at. The first number output by /proc/sys/kernel/printk specifies that all messages less than this value will appear on the console device (or devices). Recall that the lower the numeric value of the log level, the higher its relative importance. For example, this is on our x86_64 Ubuntu 20.04 LTS:

```
$ cat /proc/sys/kernel/printk
4    4    1    7
```

The first number is 4, representing the log level below which messages will appear on the console (as well as getting logged into the kernel log buffer and log files). In this case, we can conclude that all printks at a logging level less than 4 – KERN_WARNING – will appear on the console. In other words, all printks emitted at log levels KERN_EMERG, KERN_ALERT, KERN_CRIT, and KERN_ERR. This is useful as it displays only the more important log messages. Of course, as root, you can change this as you please.

Practically using the printk format specifiers – a few quick tips

Here are a few top-of-mind common printk *format specifiers* to keep in mind when writing portable code:

- For the size_t and ssize_t typedefs (which represent signed and unsigned integers respectively), use the %zu and %zd format specifiers respectively.

- When printing an address in kernel-space (a pointer):

  - *Very Important*: use %pK for security (it will emit only hashed values and helps prevent info leaks, a serious security issue).

  - Use %px for actual pointers, to see the actual address (*don't do this in production!*).

  - Use %pa for printing a physical address (must pass it by reference).

- To print a raw buffer as a string of hex characters, use %*ph (where * is replaced by the number of characters; use this for buffers with fewer than 65 characters and use the print_hex_dump_bytes() routine for buffers with more). Variations are available (see the kernel doc link that follows).

- To print IPv4 addresses, use %pI4, to print IPv6 addresses use %pI6 (a few variations exist).

An exhaustive list of printk format specifiers, which to use when (with examples!) is part of the official kernel documentation here: https://www.kernel.org/doc/Documentation/printk-formats.txt. I urge you to browse through it!

Right, now that you understand the basics of using `printk()` (and the related `pr_*()` / `dev_*()` macros), let's move on to specifics on using the printk for the purpose of debugging.

Leveraging the printk for debug purposes

You might imagine that all you have to do to emit a debug message to the kernel log is simply to issue a printk at the log level KERN_DEBUG. Though there's (a lot) more to it, the `pr_debug()` (and `dev_dbg()`) macros are actually designed to be more than mere printers when the kernel's **dynamic debug** option is enabled. We will learn about this powerful aspect in the coming *Using the kernel's powerful dynamic debug feature* section.

In this section, let's first learn more about issuing a debug print, followed by slightly more advanced ways that help in the issuing of debug messages to the kernel log.

Writing debug messages to the kernel log

In the simple kernel module we covered in the previous section (`printk_loglevel`), let's relook at the couple of lines of code that emitted a kernel printk at the debug log levels:

```
pr_debug("Hello, debug world @ log-level KERN_DEBUG    [%d]\n",
LOGLEVEL_DEBUG);
pr_devel("Hello, debug world via the pr_devel() macro (eff @
KERN_DEBUG) [%d]\n", LOGLEVEL_DEBUG);
```

Both macros `pr_debug()` and `pr_devel()` issue a print to the kernel log at log level KERN_DEBUG *but only when the symbol (macro) DEBUG is defined*! If it isn't defined, they remain silent – no debug output appears. This is precisely what's required!

Module authors should avoid using the `pr_devel()` macro. It's meant to be used for kernel-internal debug printk instances whose output should never be visible in production systems.

Figure 5.1 revealed that the messages from the pr_debug() and pr_devel() macros did indeed make it to the kernel log, but, recall, in order for this to work, the DEBUG symbol needs to be defined. Where was this done? Especially as it isn't defined in the code. The answer: we defined it within the module's Makefile. Check it out (I've highlighted the key portion here); the following Makefile snippet is kept simple, it unconditionally sets ccflags-y; in the code, we use a variable MYDEBUG to conditionally set ccflags-y):

```
$ cd ch3/printk_loglevels ; cat Makefile
[ ... ]
# Set FNAME_C to the kernel module name source filename
(without .c)
FNAME_C := printk_loglevels
PWD            := $(shell pwd)
obj-m          += ${FNAME_C}.o
# EXTRA_CFLAGS deprecated; use ccflags-y
  ccflags-y    += -DDEBUG -g -ggdb -gdwarf-4 -Og -Wall -fno-
omit-frame-pointer -fvar-tracking-assignments
   # man gcc: "...-Og may result in a better debugging
experience"
[ ... ]
```

By appending the value -DDEBUG to the ccflags-y variable, it gets defined in effect. The -D implies **define this symbol** – useful. Likewise, -U implies *undefine this symbol*. We typically employ these in the Makefile targets for the debug and production versions of the app, respectively, or, as in this case, the kernel module. So, here, to generate the production version, simply change the value of the Makefile variable MYDEBUG from n to y to enable debug mode.

Important – Building a Kernel Module for Debug or Production

The way your kernel module gets built is heavily influenced by the value that the DEBUG_CFLAGS variable gets set to. This variable is primarily set within the kernel's top-level Makefile. Here, the value it obtains depends upon the kernel config CONFIG_DEBUG_INFO. When it's on (implying a debug kernel), various debug flags make their way into DEBUG_CFLAGS, and thus your kernel module gets built with them. In effect, what I'm trying to emphasize here, is that the presence or absence of the -DDEBUG string within your kernel module's Makefile (as we do here) does not much influence the way that your kernel module is built.

In effect, when you boot via your debug kernel and build your kernel modules, they're automatically built with symbolic info and various kernel debug options turned on. On the other hand, when booted via the production kernel, and (re) built therein, your kernel modules end up without debug information/symbols.

As an example, when I built this kernel module (ch3/printk_loglevels) when on the debug kernel, the printk_loglevels.ko file size was 221 KB, but when built on the production kernel, the size dropped to under 8 KB! (The lack of debug symbols and info, KASAN instrumentation, and so on, account for this major difference.)

Quick tips: Doing make V=1 to actually see all options passed to the compiler can be very enlightening!

Further, and *very useful*, you can leverage readelf(1) to determine the DWARF format debug information embedded within the binary **Executable and Linker Format (ELF)** file. This can be particularly useful to figure out *exactly which compiler flags* your binary executable or kernel module has been built with. You can do so like this:

```
readelf --debug-dump <module_dbg.ko> | grep
producer
```

Note that this technique typically works only when debug info is enabled; further, when working with a different target architecture (for example, ARM), you'll need to run that toolchain's version: ${CROSS_COMPILE}readelf. Do see the *Further reading* section for links to a series of articles on the **GNU Debugger (GDB)**, which describe this (and more) in detail (the second part in the series mentioned is the relevant one here).

Let's see an example of actual usage of dev_dbg() within the kernel (drivers). An interesting, easy, and very cool way of emitting output on typical embedded projects is via an **Organic Light-Emitting Diode (OLED)** device. They typically work over an **Inter-Integrated Circuit (I2C)** bus, pretty much always available on embedded devices (such as the popular Raspberry Pi or the BeagleBone). We'll take the SSD1307 OLED framebuffer driver as an example from this driver source file within the kernel source tree:

```
// drivers/video/fbdev/ssd1307fb.c
static int ssd1307fb_init(struct ssd1307fb_par *par)
{
   [...]
        /* Enable the PWM */
        pwm_enable(par->pwm);
        dev_dbg(&par->client->dev, "Using PWM%d with a %lluns
period.\n",
            par->pwm->pwm, pwm_get_period(par->pwm));
        }
```

As you can see, the first parameter to the dev_dbg() macro is a pointer to a device structure. Here, it happens to be embedded within an i2c_client structure (as this device is being driven over the popular I2C protocol), which itself is embedded within the driver's *context* structure (named ssd1307fb_par). This sort of thing is quite typical in drivers.

To make it more interesting, here's a photo of an SSD1306 OLED display panel in action (which the `ssd1307fb` driver can drive as well):

Figure 3.2 – An SSD1306 OLED display panel

As hinted at, there's much more we can do to leverage the kernel's dynamic debug framework... Before that though, and now that you know the basics of using the printk for debug, let's round this off with a few more practical tips on debugging with the printk and friends.

Debug printing – quick and useful tips

When working on a project or product, you'll perhaps need to generate some debug printk. The `pr_debug()` macro will get the job done (as long as the symbol DEBUG is defined of course). But think about this: to look up the debug prints, you will need to run `dmesg` over and over. Several tips on what you can do in this situation follow:

1. Clear the kernel log buffer (in RAM) with `sudo dmesg -C`. Alternatively, `sudo dmesg -c` will first print the content and then clear the ring buffer. This way, stale messages don't clog the system and you see only the latest ones when you run `dmesg`.

2. Use `journalctl -f` to keep a *watch* on the kernel log (in a fashion similar to how `tail -f` on a file is used). Try it out!

3. **Make the printk behave like the printf and see its output on the console!** We can do this by setting the console log level to the value 8, thus ensuring that *all* printks (log levels 0 to 7) will be displayed on the console device:

    ```
    sudo sh -c "echo \"8 4 1 7\" > /proc/sys/kernel/printk "
    ```

    I often do this within a startup script when debugging kernel stuff. For example, on my Raspberry Pi, I keep a startup script that contains the following line:

    ```
    [ $(id -u) -eq 0 ] && echo "8 4 1 7" > /proc/sys/kernel/
    printk
    ```

Thus, when it runs as root, this takes effect and all printk instances directly appear on the `minicom(1)` (or whichever) console, just as printf output would.

Useful, yes!? But what about a very common case – when you're working on a device driver? The next section delves into the recommended way – using the dev_dbg() macro.

Device drivers – using the dev_dbg() macro

A key point for driver authors: when writing a device driver, you are expected to make use of the `dev_dbg()` macro to emit a debug message (and not the usual `pr_debug()`).

Why? The first parameter to this macro is `struct device *dev`, a pointer to `struct device`. This device structure is always present when writing a driver and serves to describe the device in detail. It's often embedded in a wrapper structure particular to the kind of driver being written. Printing via the `dev_dbg()` macro not only gets the debug printk across and into the kernel log (and possibly the console), but it also typically has useful information prefixed to the message (such as the name and (sometimes) the class of the device, the major:minor numbers if appropriate, and so on).

An example from the kernel's **Network Block Device (nbd)** driver will serve to show how it's used. I searched, via `cscope`, for kernel code that calls `dev_dbg()` on the 5.10.60 kernel and got over 22,000 hits! An important reason is that it's used for *dynamic debug* as we'll shortly learn:

```
// drivers/block/nbd.c
dev_dbg(nbd_to_dev(nbd), "request %p: got reply\n", req);
```

Here, the `nbd_to_dev()` inline function retrieves the device structure pointer from the `nbd_device` structure, where it's embedded.

Remember, *when writing a driver, in place of the* pr_*() *macros, please use the equivalent* dev_*() *macros!* The header include/linux/dev_printk.h contains their definitions – the dev_emerg(), dev_crit(), dev_alert(), dev_err(), dev_warn(), dev_notice(), dev_info(), and of course, as already covered, dev_dbg(). Everything remains as with the pr_*() macros except that the first parameter is a pointer to the device structure.

Trying our kernel module on the custom production kernel

As an experiment, boot into the custom production kernel we built back in *Chapter 1, A General Introduction to Debugging Software.* While running on this production kernel, let's build and then attempt to load the kernel module (notice we're running as root):

```
# make
[...]
# dmesg -C; insmod ./printk_loglevels.ko ; dmesg
insmod: ERROR: could not insert module ./printk_loglevels.ko:
Operation not permitted
[ 1933.232266] Lockdown: insmod: unsigned module loading is
restricted; see man kernel_lockdown.7
#
```

It fails due to the fact that, in our custom production kernel's configuration, we enabled the kernel **lockdown** mode (a recent kernel feature, from the 5.4 kernel, enabled via CONFIG_SECURITY_LOCKDOWN_LSM=y). This (and related) config options *disallow the loading of any kernel module that isn't signed or if the signature cannot be validated by the kernel.*

This implies that we can't even test our kernel module on the production kernel. You can, in one of two ways:

- Actually sign the kernel module (official kernel documentation – Kernel module signing facility: https://www.kernel.org/doc/html/v5.0/admin-guide/module-signing.html#kernel-module-signing-facility).

  (Also, FYI, with CONFIG_MODULE_SIG_ALL=y, all kernel modules are auto-signed upon installation, during the make modules_install step of the kernel build.)

- Or, you can always disable these kernel configs, rebuild the kernel, reboot with it, and then test. We do precisely this in the *Disabling the kernel lockdown section*, which follows.

FYI, here's a link to the man page on the kernel lockdown feature: `https://man7.org/linux/man-pages/man7/kernel_lockdown.7.html`.

All good with the debug prints, except what are you to do when there are multiple voluminous printks, especially in a high-volume code path? The following section has you covered.

Rate limiting the printk

Let's take a plausible scenario: you're writing a device driver for some chipset or peripheral device... Often, especially during development, and sometimes in order to debug in production, you of course intersperse your driver code with the now-familiar `dev_dbg()` (or similar) macro. This works well until your code paths containing the debug prints turn out to run (very) often. What will happen? It's quite straightforward:

- The kernel ring (circular) buffer isn't very large (typically between 64 KB and 256 KB, configurable at kernel build time). Once full, it wraps around. This causes you to lose perhaps precious debug prints.

- Debug (or other) prints in a very high-volume code path (within interrupt handler routines and timers, for instance), can dramatically slow things down (especially on an embedded system with prints traveling across a serial line), even leading to a *livelock* situation (a situation where the system becomes unresponsive as the processor(s) are tied up working on logging stuff – console output, framebuffer scrolling, log file appends, and so on).

- The very same debug (or other) printk message being repeated over and over again umpteen times (for example, a warning or debug message within a loop) doesn't really help anyone.

- Also, do realize that it's not just the printk (and similar) APIs that can lead to logging issues and failures; the usage of kprobes or indeed any kind of event tracing on high-volume code paths can cause this same issue to crop up (we cover kprobes in the following chapter and tracing in later ones).

In such situations, you'll notice this message or similar (typically from the `systemd-journald` process):

```
/dev/kmsg buffer overrun, some messages lost.
```

(By the way, if you're wondering what the /dev/kmsg character device node is all about, please do refer to the kernel documentation here: https://www.kernel.org/doc/Documentation/ABI/testing/dev-kmsg.)

To mitigate exactly these situations, the community came up with the *rate-limited printk – a means to throttle down and not emit prints (the same or different) when certain (tunable) thresholds have been exceeded*!

We'll discuss these thresholds in just a moment... The kernel provides the following macros to help you rate-limit your prints/logging (#include <linux/kernel.h>):

- printk_ratelimited(): *Warning!* Do *not* use it – the kernel warns against this.
- pr_*_ratelimited(): Where the wildcard * is replaced by the usual (emerg, alert, crit, err, warn, notice, info, debug).
- dev_*_ratelimited(): Where the wildcard * is replaced by the usual (emerg, alert, crit, err, warn, notice, info, dbg).

Ensure you use the pr_*_ratelimited() macros in preference to printk_ratelimited(); driver authors should use the dev_*_ratelimited() macros.

But how exactly are the prints rate limited? The kernel provides two tunable thresholds via the usual control file interfaces within procfs (under the /proc/sys/kernel folder), named printk_ratelimit and printk_ratelimit_burst for this purpose. Here, we directly reproduce the sysctl documentation (from https://www.kernel.org/doc/Documentation/sysctl/kernel.txt) that explains the precise meaning of these two (pseudo) files:

```
printk_ratelimit:
Some warning messages are rate limited. printk_ratelimit
specifies the minimum length of time between these messages (in
jiffies), by default we allow one every 5 seconds.
A value of 0 will disable rate limiting.

================================================================
printk_ratelimit_burst:
While long term we enforce one message per printk_ratelimit
seconds, we do allow a burst of messages to pass through.
printk_ratelimit_burst specifies the number of messages we can
send before ratelimiting kicks in.
```

On my x86_64 Ubuntu 20.04 LTS guest system, we find that their (default) values are as follows:

```
$ cat /proc/sys/kernel/printk_ratelimit
5
$ cat /proc/sys/kernel/printk_ratelimit_burst
10
```

This implies that, by default, *a burst of up to 10 printk messages occurring within a 5-second time interval can make it through* before rate-limiting kicks in and further messages are suppressed (until the next time interval).

The printk rate-limiter code, when it does suppress kernel printk instances, emits a helpful message mentioning exactly how many earlier printk callbacks were suppressed.

We write a simple kernel module to test printk rate limiting (again, only the relevant snippets are shown here):

```
// ch3/ratelimit_test/ratelimit_test.c
#define pr_fmt(fmt) "%s:%s():%d: " fmt, KBUILD_MODNAME, __
func__, __LINE__
[...]
#include <linux/kernel.h>
#include <linux/delay.h>
[...]
static int num_burst_prints = 7;
module_param(num_burst_prints, int, 0644);
MODULE_PARM_DESC(num_burst_prints, "Number of printks to
generate in a burst (defaults to 7).");
static int __init ratelimit_test_init(void)
{
    int i;
    pr_info("num_burst_prints=%d. Attempting to emit %d printks
in a burst:\n", num_burst_prints, num_burst_prints);
    for (i=0; i<num_burst_prints; i++) {
        pr_info_ratelimited("[%d] ratelimited printk @ KERN_
INFO [%d]\n", i, LOGLEVEL_INFO);
        mdelay(100); /* the delay helps magnify the rate-
limiting effect, triggering the kernel's "'n' callbacks
suppressed" message... */
```

```
    }
    return 0;    /* success */
}
```

If you build and run this module with defaults, not modifying the `num_burst_prints` module parameter (it defaults to the value 7), you can see that we emit seven rate-limited printks in a short time interval. This, in spite of the 100-millisecond delay (the delay is deliberate – you will soon see its effect).

Let's push it a bit: we test by passing the module parameter `num_burst_prints`, setting its value to some number greater than the maximum allowed burst (the value of `/proc/sys/kernel/printk_ratelimit_burst` – 10 by default). We set it to 60. The screenshot shows what happens at runtime:

```
# make; rmmod ratelimit_test; dmesg -C; insmod ./ratelimit_test.ko num_burst_prints=60 ; dmesg ; echo -n "# of printk's actually se
en: " ; dmesg |grep "ratelimited printk @"|wc -l

--- Building : KDIR=/lib/modules/5.10.60-prod01/build ARCH= CROSS_COMPILE= EXTRA_CFLAGS=-DDEBUG -g -ggdb -gdwarf-4 -Wall -fno-omit-
frame-pointer -DDYNAMIC_DEBUG_MODULE ---

make -C /lib/modules/5.10.60-prod01/build M=/home/letsdebug/Linux-Kernel-Debugging/ch5/ratelimit_test modules
make[1]: Entering directory '/home/letsdebug/lkd_kernels/productionk/linux-5.10.60'
make[1]: Leaving directory '/home/letsdebug/lkd_kernels/productionk/linux-5.10.60'
[14855.679081] ratelimit_test:ratelimit_test_init():40: num_burst_prints=60. Attempting to emit 60 printk's in a burst:
[14855.681387] ratelimit_test:ratelimit_test_init():44: [0] ratelimited printk @ KERN_INFO [6]
[14855.782887] ratelimit_test:ratelimit_test_init():44: [1] ratelimited printk @ KERN_INFO [6]
[14855.883286] ratelimit_test:ratelimit_test_init():44: [2] ratelimited printk @ KERN_INFO [6]
[14855.983924] ratelimit_test:ratelimit_test_init():44: [3] ratelimited printk @ KERN_INFO [6]
[14856.084340] ratelimit_test:ratelimit_test_init():44: [4] ratelimited printk @ KERN_INFO [6]
[14856.184749] ratelimit_test:ratelimit_test_init():44: [5] ratelimited printk @ KERN_INFO [6]
[14856.285232] ratelimit_test:ratelimit_test_init():44: [6] ratelimited printk @ KERN_INFO [6]
[14856.385645] ratelimit_test:ratelimit_test_init():44: [7] ratelimited printk @ KERN_INFO [6]
[14856.486079] ratelimit_test:ratelimit_test_init():44: [8] ratelimited printk @ KERN_INFO [6]
[14856.586458] ratelimit_test:ratelimit_test_init():44: [9] ratelimited printk @ KERN_INFO [6]
[14860.688772] ratelimit_test_init: 40 callbacks suppressed
[14860.688773] ratelimit_test:ratelimit_test_init():44: [50] ratelimited printk @ KERN_INFO [6]
[14860.789403] ratelimit_test:ratelimit_test_init():44: [51] ratelimited printk @ KERN_INFO [6]
[14860.889742] ratelimit_test:ratelimit_test_init():44: [52] ratelimited printk @ KERN_INFO [6]
[14860.990279] ratelimit_test:ratelimit_test_init():44: [53] ratelimited printk @ KERN_INFO [6]
[14861.090667] ratelimit_test:ratelimit_test_init():44: [54] ratelimited printk @ KERN_INFO [6]
[14861.191045] ratelimit_test:ratelimit_test_init():44: [55] ratelimited printk @ KERN_INFO [6]
[14861.291560] ratelimit_test:ratelimit_test_init():44: [56] ratelimited printk @ KERN_INFO [6]
[14861.391897] ratelimit_test:ratelimit_test_init():44: [57] ratelimited printk @ KERN_INFO [6]
[14861.492243] ratelimit_test:ratelimit_test_init():44: [58] ratelimited printk @ KERN_INFO [6]
[14861.592568] ratelimit_test:ratelimit_test_init():44: [59] ratelimited printk @ KERN_INFO [6]
# of printk's actually seen: 20
```

Figure 3.3 – Screenshot showing our ratelimit_test LKM in action

The preceding screenshot should make it clear: we attempt to emit 60 printks in a burst – but of course, it's the rate-limited version of the printk (via the `pr_info_ratelimited()` macro). The kernel's limit gets hit after just 10 printk (the default value of `/proc/sys/kernel/printk_ratelimit_burst`), thus, the kernel now prevents or suppresses further prints. This is clearly seen: you can see prints `[0]` to `[9]` – 10 of them being issued and then the message:

40 callbacks suppressed

After that, sufficient time elapsed (5 seconds here, as the `/proc/sys/kernel/printk_ratelimit` value is 5 by default) that the prints resumed! Our using `mdelay(100)` helped create a sufficient delay so that prints could resume... So, out of the 60 attempted prints, only 20 actually made it to the log (or console). This is a good thing and clearly demonstrates the point. As root, you can modify the rate-limit sysctl parameters to suit your requirements.

> **The ftrace trace_printk() API**
>
> The kernel's powerful ftrace subsystem (which we shall cover in detail in *Chapter 9, Tracing the Kernel Flow*) provides another way to mitigate high-volume logging issues: the `trace_printk()` API. The syntax is identical to the regular `printf()` (not `printk()`!) API. It has two major advantages over the typical printk: one, it's very fast (as it only writes to a RAM buffer); two, the size of the trace buffer is large by default and tunable by root.

So, in conclusion, if you have a code path with a high volume of printks, you can mitigate the potential ill effects by either employing the rate-limiting printk (and/or macros) or by using `trace_printk()` (more on the latter in *Chapter 9, Tracing the Kernel Flow*, in the *Using the trace_printk() for debugging* section).

So, by now you have the skills and knowledge to emit a debug printk (typically via the `pr_*[_ratelimited]()` or `dev_*[_ratelimited]()` macros)! It seems this is sufficient but only until you learn about and start using the kernel's pretty awesome dynamic debug framework. This is precisely what follows – read on and learn!

Using the kernel's powerful dynamic debug feature

The **instrumentation** approach to debugging – interspersing your kernel (and module) code with many printk is indeed a good technique. It helps you narrow things down and debug them! But as you've no doubt realized, there can be a (pretty high) cost to this:

- It eats into your disk (or flash) space as logs get filled in. This can be especially problematic on constrained embedded systems. Also, writing to disk is *much* slower than writing to RAM.

- It's fast in RAM, but the ring buffer is not that large and would thus quickly get overwhelmed; older prints will soon be lost.

- Even more important, on many production systems, a high volume of printks would have an adverse performance impact, creating bottlenecks and even possible livelocks! Rate limiting helps with this, to some extent...

A solution would be to use the `pr_debug()` and/or the `dev_dbg()` APIs! They're especially useful during development and testing as it's really easy to turn these debug printk on or off: the presence of the `DEBUG` symbol implies the debug printk will run (and be logged); its absence implies it won't.

That's great; however, think about this: when running in production (using the production kernel), the `DEBUG` symbol will almost certainly be undefined by default. Now let's say you have a situation while running in production where you want your debug prints for a given kernel module to appear, and thus get logged. Changing the code (or `Makefile`) to define the `DEBUG` symbol, then recompiling and re-installing it is very unlikely to be allowed during production.

So, what do you do (besides giving up)? There are two broad approaches to *dynamically* toggling debug prints: one, via module parameters, and two, via the kernel's powerful built-in dynamic debug facility – the latter being the superior one and very much the focus of this section. First, though, let's briefly check out the first option.

Dynamic debugging via module parameters

One approach is to use a **module parameter** to hold a `debug` predicate; keep it off by default (the value `0`). You can define it like this:

```
static int debug;
module_param(debug, int, 0644);
```

This has the kernel set up the module's parameter named `debug` under the sysfs pseudo filesystem (at `/sys/module/<module_name>/parameters/debug`, with the owner and group as root and the octal permissions as specified in the third parameter to the `module_param` macro).

Interestingly, the i8042 keyboard and mouse controller driver (very often found in x86-based laptops) does precisely this; it defines this module parameter:

```
// drivers/input/serio/i8042.c
static bool i8042_debug;
module_param_named(debug, i8042_debug, bool, 0600);
MODULE_PARM_DESC(debug, "Turn i8042 debugging mode on and
off");
```

This has the OS set up a module parameter named `debug` (notice the usage of the `module_param_named()` macro to achieve this), which is a Boolean and off (false) by default. A given module's parameters can be easily seen by leveraging the `modinfo(8)` utility; for example, let's look up the parameters you can supply to the kernel's `hid` driver:

```
$ modinfo -p /lib/modules/5.10.60-prod01/kernel/drivers/hid/
hid.ko
debug:toggle HID debugging messages (int)
ignore_special_drivers:Ignore any special drivers and handle
all devices by generic driver (int)
```

Okay, back to the i8042 driver; once loaded up, you can spot it's `debug` parameter under sysfs as follows:

```
$ ls -l /sys/module/i8042/parameters/debug
-rw------- 1 root root 4096 Oct  3 07:42 /sys/module/i8042/
parameters/debug
```

Of course, this sysfs-based pseudo file will only be seen after the module has been loaded into memory, for its lifetime.

Notice the permissions. In this case, only root can read or write to the debug pseudo file:

```
$ sudo cat /sys/module/i8042/parameters/debug
[sudo] password for letsdebug: xxxxxxxxxx
N
```

The root user can always turn it on dynamically, by writing the value `Y` (or `1`) into the sysfs pseudo-file representing it! This way, you can dynamically turn on or off debugging. So, here, to turn debugging on at runtime, do the following, as root of course:

```
# echo "Y" > /sys/module/i8042/parameters/debug
```

And turn it off again with the following:

```
# echo "N" > /sys/module/i8042/parameters/debug
```

Simple. In fact, think about this – you can easily extend this idea: one way to do so is to use an integer debug parameter, which, depending on its value, will have the module emit debug messages at various levels of verbosity. (For example, 0 means all debug messages are off, 1 implies only a few key debug prints will be emitted, 2 implies more debug verbosity, and so on.)

This general approach does work, but with a significant drawback, especially when compared to the kernel's dynamic debug facility:

- Performance – you will require a conditional statement of some sort (an `if`, `switch`, and so on) to check whether a debug print should be emitted or not every time. With multiple levels of verbosity, more checking is required.

- With the kernel's dynamic debug framework (which is covered next), you get several advantages:

  - The formatting of debug messages with useful information prefixed is part of the feature set, with a gentle learning curve.

  - Performance remains high, with next to no overhead when debugging is off (typically the default in production). This is achieved by sophisticated dynamic code patching techniques that the kernel employs (as is the case for ftrace as well).

  - It's always part of the mainline kernel (from way back, since the 2.6.30 kernel), not requiring home-brewed solutions that may or may not be maintained, available or working.

So, for the remainder of this section, we shall focus on learning to use and leverage the kernel's powerful **dynamic debug** framework, available right since the 2.6.30 kernel. Read on!

When the kernel config option `CONFIG_DYNAMIC_DEBUG` is enabled, *it allows you to dynamically turn on or off debug prints that have been compiled in the kernel image as well as within kernel modules.* This is done by having the kernel always compile in all `pr_debug()` and `dev_dbg()` callsites. Now, the really powerful thing is that you can not only enable or disable these debug prints but do so at various levels of scope: at the scope of a given source file, a kernel module, function, or even a line number.

This does imply that the kernel image will be larger; it's not by too much, approximately a 2% increase in kernel text size. If this is a concern (on a tightly constrained embedded Linux, perhaps), you can always just set the kernel config `CONFIG_DYNAMIC_DEBUG_CORE`. This enables the core support for dynamic printks but it only takes effect on kernel modules that are compiled with the symbol `DYNAMIC_DEBUG_MODULE` defined. Thus, our module `Makefile` always defines it. You could always comment it out.... This is the relevant line within our module `Makefile`:

```
# We always keep the dynamic debug facility enabled; this
# allows us to turn dynamically turn on/off debug printks
```

```
# later... To disable it simply comment out the following
# line
ccflags-y   += -DDYNAMIC_DEBUG_MODULE
```

In fact, it's not just `pr_debug()`; *all the following APIs can be dynamically enabled/ disabled per callsite*: `pr_debug()`, `dev_dbg()`, `print_hex_dump_debug()`, and `print_hex_dump_bytes()`.

Specifying what and how to print debug messages

As with many facilities, control over the kernel's dynamic debug framework – deciding which debug printks are enabled and what extraneous information is prefixed to them – is determined via a **control file**. Where's this control file then? It depends. If the debugfs pseudo filesystem is enabled within the kernel config (typically it is, with `CONFIG_DEBUG_FS=y`) and the kernel configs `CONFIG_DEBUG_FS_ALLOW_ALL=y` and `CONFIG_DEBUG_FS_DISALLOW_MOUNT=n` – usually the case for a debug kernel – then the control file is here:

```
/sys/kernel/debug/dynamic_debug/control
```

On many production environments though, for security reasons, the debugfs filesystem is present (functional) but invisible (it can't be mounted) via `CONFIG_DEBUG_FS_ DISALLOW_MOUNT=y`.

In this case, the debugfs APIs work just fine but the filesystem isn't mounted (in effect, it's invisible). Alternately, debugfs might be disabled altogether by setting the kernel config `CONFIG_DEBUG_FS_ALLOW_NONE` to y. In either of these cases, an identical but alternate control file for dynamic debug under the pseudo proc filesystem (procfs) should be used:

```
/proc/dynamic_debug/control
```

As with other pseudo filesystems, this *control* file under debugfs or procfs is a pseudofile; it exists only in RAM. It gets populated and manipulated by kernel code. Reading its content will give you a comprehensive list of all debug printk (and/or `print_hex_ dump_*()`) callsites within the kernel. Thus, its output is typically pretty large (over here, we're on the custom debug kernel and can hence use the debugfs location for the control file). Let's begin to interrogate it:

```
# ls -l /sys/kernel/debug/dynamic_debug/control
-rw-r--r-- 1 root root 0 Sep 16 12:26 /sys/kernel/debug/
dynamic_debug/control
```

```
# wc -l /sys/kernel/debug/dynamic_debug/control
3217 /sys/kernel/debug/dynamic_debug/control
```

Notice it's only writable as root (and we're running as root). Let's look up the first few lines of output:

```
# head -n5 /sys/kernel/debug/dynamic_debug/control
# filename:lineno [module] function flags format
drivers/powercap/intel_rapl_msr.c:151 [intel_rapl_msr] rapl_msr_
probe =_ "failed to register powercap control_type.\012"
drivers/powercap/intel_rapl_msr.c:94 [intel_rapl_msr] rapl_msr_
read_raw =_ "failed to read msr 0x%x on cpu %d\012"
sound/pci/intel8x0.c:3160 [snd_intel8x0] check_default_spdif_
aclink =_ "Using integrated SPDIF DMA for %s\012"
sound/pci/intel8x0.c:3156 [snd_intel8x0] check_default_spdif_
aclink =_ "Using SPDIF over AC-Link for %s\012"
#
```

The format of each entry is shown first; it's reproduced here:

```
filename:lineno [module] function flags format
```

Besides the flags member, all are obvious. The last one, format, is the actual printf-style format string that the debug print uses. So, let's zoom into the first actual entry seen and examine it minutely, with a (hopefully) helpful diagram:

Figure 3.4 – The dynamic debug control file format specifier

Here's the detailed breakdown, as per the control file output format specifier:

- filename: drivers/powercap/intel_rapl_msr.c: This is the full pathname of the source file.

- lineno: 151: This is the line number within the source file, the place in code where the debug print lives (so complicated; yup, I can be sarcastic).

- [module]: [intel_rapl_msr]: The name of the kernel module where the debug print lives. It's optional: if the debug print callsite is in a kernel module, this – the module name – shows up in square brackets.

- function: rapl_msr_probe: The function containing the debug print.

- flags: =_: Ah, this is really the interesting, juicy bit. We explain it shortly (*Table 3.2*).

- format: "failed to register powercap control_type.\012": This is the actual printf-style format string that's to be printed/logged.

Just to fully verify this, here's the actual code snippet of this example from the kernel code base (version 5.10.60: I've highlighted the relevant line – # 151 – below):

```
// drivers/powercap/intel_rapl_msr.c
149      rapl_msr_priv.control_type = powercap_register_control_
type(NULL, "intel-rapl", NULL);
150      if (IS_ERR(rapl_msr_priv.control_type)) {
151          pr_debug("failed to register powercap control_
type.\n");
152          return PTR_ERR(rapl_msr_priv.control_type);
153      }
```

You can see how it perfectly matches the control file's understanding of it.

(Interestingly, you can use Bootlin's online kernel code browser to look it up as well: https://elixir.bootlin.com/linux/v5.10.60/source/drivers/powercap/intel_rapl_msr.c#L151 – useful!)

The real magic lies in the so-called flags specifier. Using flags, you can program the dynamic debug framework to emit the debug print (thus having it logged) along with various useful prefixes. The following table summarizes how to program and interpret the flags specifier:

Dynamic debug control file: flags specifier value =\<foo>	Meaning
–	The debug print is currently off; this is the typical default.
P	The debug print is currently on and will be printed and logged at runtime; the specifiers that now follow in the left column can be added along with this one:
m	Module name is prefixed (if it's within a kernel module).
f	Function name is prefixed.
l	Line number in source file is prefixed; line range can be specified in from-to format.
t	If in process context (not in any kind of interrupt), the PID of the thread that runs this code path is prefixed.

Table 3.2 – Dynamic debug framework flags specifier

In addition, quite intuitively, you can use the following symbols:

- +: Add the flag(s) specified.
- -: Remove the flag(s) specified.
- =: Set to the flag(s) specified.

A quick experiment: let's grep for the number of debug printk callsites currently enabled within the kernel (notice how I use sed to strip away the first line, as it's the format string explanatory line and not an actual entry):

```
# cat /sys/kernel/debug/dynamic_debug/control |sed '1d' |wc -l
3216
```

So, here and now, we have a total of 3,216 debug prints recognized by the kernel's dynamic debug framework. Now let's `grep` the flags, only matching the ones that are turned off:

```
# grep " =_ " /sys/kernel/debug/dynamic_debug/control |sed '1d'
|wc -l
3174
```

So, of the total 3,216 debug printks in the kernel right now, 3,174 of them are turned off, leaving only *3216 - 3174 = 42* turned on (by the kernel/drivers/whatever). Let's verify this, by negating the sense of the `grep`:

```
# grep -n -v " =_ " /sys/kernel/debug/dynamic_debug/control |wc
-l
42
```

It's verified. Of the ones that are turned on, here's the last three:

```
# grep -v " =_ " /sys/kernel/debug/dynamic_debug/control |tail
-n3
init/main.c:1340 [main]run_init_process =p "  with
arguments:\012"
init/main.c:1129 [main]initcall_blacklisted =p "initcall %s
blacklisted\012"
init/main.c:1090 [main]initcall_blacklist =p "blacklisting
initcall %s\012"
```

So, as their `flags` value is =p, (just) the debug print will be emitted and logged when the line of code is hit; nothing will be prefixed to it.

Next, how do you program the dynamic debug framework? Very simple: just write the command – often via a simple `echo` statement – into the control file! Needless to say, it will only go through with root access (or, with the better and modern capabilities model, having a capability bit such as CAP_SYS_ADMIN set). The command syntax essentially is the following:

```
echo -n <match-spec* flags> > <control-file>
```

`match-spec` is one of the following:

```
match-spec ::= 'func' string |
               'file' string |
               'module' string |
               'format' string |
```

```
                'line' line-range

line-range ::= lineno     | '-'lineno |
               lineno'-' | lineno'-'lineno
lineno ::= unsigned-int
```

The match-spec syntax shown is taken direct from the kernel documentation on dynamic debug here: https://www.kernel.org/doc/html/latest/admin-guide/dynamic-debug-howto.html#command-language-reference.

The flags specifier has already been covered – see *Table 3.2*. Here's a table summarizing how to use match-spec to form a command, with examples:

match-spec	Example of a command string [format: match-spec* flags]	Meaning
func string	func run_init_ process +p	Turn on all debug prints in the kernel function run_init_process()
file string	file init/main.c +pf	Turn on all debug prints in the kernel source file init/main.c; prefix the debug prints with the function name
module string	module usbhid =pmflt	Turn on all debug prints in the kernel module named usbhid; prefix the debug prints with the module name, function name, line number, and PID of the thread context (when run in process context)
format string	format "Parser recognised the format (ret %d)\012" +p	Turn on all debug prints where the printf-style format string is "Parser recognised the format (ret %d)\012" (\012 is the newline char, \n)
line string	file kexec_file.c line 90-446 +pf	Turn on all debug prints where the filename is kexec_file.c *and* the line number range is 90 to 446 (inclusive)

Table 3.3 – Dynamic debug framework match-spec specifiers with examples

Issue the command or program it like this:

```
# echo -n "<command string>" > <control-file>
```

Where the `<command string>` parameter to `echo` is the command formed in the `match-spec* flags` format and `<control-file>` is either `<debugfs-mount>/dynamic_debug/control` or `/proc/dynamic_debug/control`.

Adding to this, several match specifications can be given in a single command; you can think of them as being implicitly ANDed to form a match to a subset of debug prints. You can even batch several commands into a file and write the file to the control file. More examples are available on the kernel documentation page on dynamic debug here: `https://www.kernel.org/doc/html/latest/admin-guide/dynamic-debug-howto.html#examples`.

Exercising dynamic debugging on a kernel module on a production kernel

For most of us module authors, using this powerful dynamic debug framework on our kernel module when it's running in production will be a useful thing indeed. A demo will help you understand how to do so. To make the demo a bit more realistic, let's boot up via our custom production kernel to help mimic an actual production environment.

Disabling the kernel lockdown

However, what if – as recommended back in the first chapter – you've configured the custom production kernel for security, by enabling (among other things) the kernel lockdown mode by default (by setting `CONFIG_LOCK_DOWN_KERNEL_FORCE_CONFIDENTIALITY=y`). If this isn't the case, and you can load up your (and other third-party) kernel modules on the production kernel, then all's well for this experiment and you can skip this section.

This lockdown mode is good for security, preventing you from loading unsigned kernel modules (along with other safety measures). But now that we'd like to test our kernel module on the production kernel, we will have to tweak the production kernel's configuration, setting the following within the `make menuconfig` UI:

1. Under **Security options | Basic module for enforcing kernel lockdown**:

    A. **Enable lockdown LSM early in init**: Set it to n (off).

    B. **Kernel default lockdown mode**: Set it to None.

2. Next, save the config, rebuild, and reboot via the new production kernel.

3. On the (GRUB) bootloader screen, press a key and edit the kernel command-line parameters, appending `lockdown=none`. This disables kernel lockdown mode.

For more details, please refer to the man page on kernel lockdown: `https://man7.org/linux/man-pages/man7/kernel_lockdown.7.html`.

Now let's get that debug printk working dynamically!

Demonstrating dynamic debugging on a simple misc driver

For the purposes of this demo, we'll grab a simple `misc` class character device driver from my earlier book, *Linux Kernel Programming – Part 2* (the original code's here: `https://github.com/PacktPublishing/Linux-Kernel-Programming-Part-2/tree/main/ch1/miscdrv_rdwr`). Of course, we will keep a copy in this book's GitHub repo as well...

Looking at the code, you will notice several instances of the `dev_dbg()` macro being invoked. Obviously, these are the debug prints that will only get logged when DEBUG is defined or we use the kernel's dynamic debug facility – the latter being what this demo is all about!

Here's a sample of the debug prints in the driver (due to space constraints, I won't show all the code here of course, only a few relevant bits):

```c
// ch3/miscdrv_rdwr/miscdrv_rdwr.c
#define pr_fmt(fmt) "%s:%s(): " fmt, KBUILD_MODNAME, __func__
static int open_miscdrv_rdwr(struct inode *inode, struct file
*filp)
{
    struct device *dev = ctx->dev;
    char *buf = kzalloc(PATH_MAX, GFP_KERNEL);
    [...]
    dev_dbg(dev, " opening \"%s\" now; wrt open file:
            f_flags = 0x%x\n",
  file_path(filp, buf, PATH_MAX), filp->f_flags);
    kfree(buf);
    [...]
}
static ssize_t write_miscdrv_rdwr(struct file *filp, const char
__user *ubuf, size_t count, loff_t *off)
{
```

```
    int ret = count;
    void *kbuf = NULL;
    [...]
    dev_dbg(dev, "%s wants to write %zu bytes\n",
            get_task_comm(tasknm, current), count);
    [...]
    ret = count;
    dev_dbg(dev, " %zu bytes written, returning...
            (stats: tx=%d, rx=%d)\n",
            count, ctx->tx, ctx->rx);
    [...]
}
[...]
```

Note that the Makefile file for this module will conditionally set the DEBUG symbol to undefined (as we're building in production mode). Thus, the debug prints will *not* make it to the console or kernel logs.

A quick mount |grep -w debugfs shows no output, implying that the debugfs filesystem isn't visible. This, again, is intentional, a security feature we enabled for our custom production kernel by setting CONFIG_DEBUG_FS_DISALLOW_MOUNT=y. Don't panic (yet) – as mentioned, there's a solution. Simply make use of the control file available here: /proc/dynamic_debug/control.

Grepping it for our module before it's inserted into memory reveals no data, as expected:

```
# grep miscdrv_rdwr /proc/dynamic_debug/control
#
```

Okay, now we can get it running. The following screenshot shows the source files via `ls`, the build (via our convenience `lkm` script), the resulting `dmesg` output, and the device node this driver creates:

```
# ls
Makefile  miscdrv_rdwr.c  rdwr_test_secret.c
# ../../lkm miscdrv_rdwr
Version info:
Distro:        Ubuntu 20.04.3 LTS
Kernel: 5.10.60-prod01
-------------------------------
sudo rmmod miscdrv_rdwr 2> /dev/null
-------------------------------
-------------------------------
sudo dmesg -C
-------------------------------
-------------------------------
make || exit 1
-------------------------------

--- Building : KDIR=/lib/modules/5.10.60-prod01/build ARCH= CROSS_COMPILE= EXTRA_CFLAGS=-UDEBUG -DDYNAMIC_DEBUG_MODULE ---

make -C /lib/modules/5.10.60-prod01/build M=/home/letsdebug/Linux-Kernel-Debugging/ch5/miscdrv_rdwr modules
make[1]: Entering directory '/home/letsdebug/lkd_kernels/productionk/linux-5.10.60'
  CC [M]  /home/letsdebug/Linux-Kernel-Debugging/ch5/miscdrv_rdwr/miscdrv_rdwr.o
  MODPOST /home/letsdebug/Linux-Kernel-Debugging/ch5/miscdrv_rdwr/Module.symvers
  CC [M]  /home/letsdebug/Linux-Kernel-Debugging/ch5/miscdrv_rdwr/miscdrv_rdwr.mod.o
  LD [M]  /home/letsdebug/Linux-Kernel-Debugging/ch5/miscdrv_rdwr/miscdrv_rdwr.ko
make[1]: Leaving directory '/home/letsdebug/lkd_kernels/productionk/linux-5.10.60'
-------------------------------
sudo insmod ./miscdrv_rdwr.ko && lsmod|grep miscdrv_rdwr
-------------------------------
miscdrv_rdwr           20480  0
-------------------------------
sudo dmesg
-------------------------------
[ 9177.333822] miscdrv_rdwr:miscdrv_rdwr_init(): LLKD misc driver (major # 10) registered, minor# = 58, dev node is /dev/llkd_m
iscdrv_rdwr
# ls -l /dev/llkd_miscdrv_rdwr
crw-rw-rw- 1 root root 10, 58 Sep 16 10:17 /dev/llkd_miscdrv_rdwr
#
```

Figure 3.5 – Screenshot of our miscdrv_rdwr loading up on our custom production kernel

Also notice (in the preceding screenshot) the following:

- The kernel version is `5.10.60-prod01`, showing that we're running on our custom production kernel.

- The value of the `ccflags-y` (or the older `EXTRA_CFLAGS`) variable is `-UDEBUG -DDYNAMIC_DEBUG_MODULE`, as expected.

With the current settings, the debug prints do not get logged. Let's try this out and see (*remember: "be empirical"!*):

```
# echo "DEBUG undefined, no logging?" > /dev/llkd_miscdrv_rdwr
# dmesg
[ 9177.333822] miscdrv_rdwr:miscdrv_rdwr_init(): LLKD misc
driver (major # 10) registered, minor# = 58, dev node is /dev/
llkd_miscdrv_rdwr
#
```

As is expected, the kernel log (seen via `dmesg`) shows only the earlier printk (which, being `pr_info()` does show up); *none of the debug prints appear.* So, let's set up to make them appear!

Now that our kernel module is loaded up, let's `grep` the dynamic debug control file again:

```
# grep "miscdrv_rdwr" /proc/dynamic_debug/control
<...>/miscdrv_rdwr.c:303 [miscdrv_rdwr]miscdrv_rdwr_init =_ "A
sample print via the dev_dbg(): driver initialized\012"
<...>/miscdrv_rdwr.c:242 [miscdrv_rdwr]close_miscdrv_rdwr =_ "
filename: \042%s\042\012"
<...>/miscdrv_rdwr.c:239 [miscdrv_rdwr]close_miscdrv_rdwr =_
"%03d) %c%s%c:%d    |   %c%c%c%u    /* %s() */\012"
[...]
#
```

Clearly, the dynamic debug control is aware that our module has debug prints. It's currently off, the `flags` value of `=_` proving this (for readability, I've truncated the pathname and shown only the first few lines of output).

Now let's set it up such that any and all debug prints from our `miscdrv_rdwr` kernel module will get logged via the dynamic debug framework:

```
# echo -n "module.miscdrv_rdwr +p" > /proc/dynamic_debug/
control
```

You'll need to do this only once per session. The value is retained until the module is removed or a power cycle (or reboot) occurs. Now let's retry the `grep` command. The following screenshot shows our setting the debug prints on – by using the +p flags specifier syntax, the subsequent `grep` shows that this has been noticed and set up:

Figure 3.6 – Screenshot showing setting on of debug prints for our miscdrv_rdwr module

Let's reprint and study the first line of output:

```
# grep "miscdrv_rdwr" /proc/dynamic_debug/control
<...>/miscdrv_rdwr.c:303 [miscdrv_rdwr]miscdrv_rdwr_init =p "A
sample print via the dev_dbg(): driver initialized\012"
```

This shows us the following:

- Source line 303 is a debug print callsite. It also shows the source file pathname, the module and the function name, and then the actual print format string.

- More importantly, between the function name and the format string, you can see =p. This implies of course that this debug print's callsite is known and, when this line of code is hit at runtime, the print will be emitted and logged!

To verify that this works, let's exercise our driver a bit (lazy fellow):

```
# echo "DEBUG undefined, dynamic debug now ON for this module" > /dev/llkd_miscdrv_rdwr
# dmesg
[  608.317065] miscdrv_rdwr:miscdrv_rdwr_init(): LLKD misc driver (major # 10) registered, minor# = 58, dev node is /dev/llkd_miscdrv_rdwr
[ 1010.813690] miscdrv_rdwr:open_miscdrv_rdwr(): 001) bash :1080  |  ...0  /* open_miscdrv_rdwr() */
[ 1010.813705] misc llkd_miscdrv_rdwr:  opening "/dev/llkd_miscdrv_rdwr" now; wrt open file: f_flags = 0x8241
[ 1010.813744] miscdrv_rdwr:write_miscdrv_rdwr(): 001) bash :1080  |  ...0  /* write_miscdrv_rdwr() */
[ 1010.813750] misc llkd_miscdrv_rdwr: bash wants to write 54 bytes
[ 1010.813758] misc llkd_miscdrv_rdwr:  54 bytes written, returning... (stats: tx=0, rx=54)
[ 1010.813772] miscdrv_rdwr:close_miscdrv_rdwr(): 001) bash :1080  |  ...0  /* close_miscdrv_rdwr() */
[ 1010.813777] misc llkd_miscdrv_rdwr:  filename: "/dev/llkd_miscdrv_rdwr"
#
```

Figure 3.7 – Dynamic debug in action!

It works indeed! The preceding screenshot clearly shows the debug printks have actually run and been logged.

Now let's turn it off:

```
# echo -n "module miscdrv_rdwr -p" > /proc/dynamic_debug/
control
# grep "miscdrv_rdwr" /proc/dynamic_debug/control |head -n1
<...>/miscdrv_rdwr.c:303 [miscdrv_rdwr]miscdrv_rdwr_init =_  "A
sample print via the dev_dbg(): driver initialized\012"
```

And let's retry it:

```
# echo "DEBUG undefined, dynamic debug now OFF for this module"
> /dev/llkd_miscdrv_rdwr
# dmesg
[...]
[ 1010.813777] misc llkd_miscdrv_rdwr:  filename: "/dev/llkd_
miscdrv_rdwr"
```

As expected, no debug prints have appeared (the one in the log is the earlier one – see the timestamp).

One more experiment: we turn on the display of the module name (m) and thread context PID (t: shows the thread PID that runs this driver code in process context):

```
# echo -n "module miscdrv_rdwr +ptm" > /proc/dynamic_debug/
control
```

Write to the device node and check dmesg:

```
# echo "DEBUG undefined, dynamic debug now ON for this module"
> /dev/llkd_miscdrv_rdwr
# dmesg
[...]
[ 1010.813777] misc llkd_miscdrv_rdwr:  filename: "/dev/llkd_
miscdrv_rdwr"
[ 1457.376915] [1080] miscdrv_rdwr: miscdrv_rdwr:open_miscdrv_
rdwr(): 001)  bash :1080   |  ...0   /* open_miscdrv_rdwr() */
[ 1457.376931] [1080] miscdrv_rdwr: misc llkd_miscdrv_rdwr:
opening "/dev/llkd_miscdrv_rdwr" now; wrt open file: f_flags =
0x8241
[...]
#
```

Aha! This time you can see the PID of the thread that performed the write in square brackets ([1080] it's in fact the PID of our bash shell, as echo is a bash built-in!) followed by the name of the module.

Super – you now know how to activate and deactivate debug prints on a production system using the kernel's dynamic debug framework.

Activating debug prints at boot and module init

It's important to realize that any debug prints within the early kernel initialization (boot) code paths or the initialization code of a kernel module, will *not automatically be enabled*. To enable them, do the following:

- For core kernel code and any built-in kernel modules, that is, for activating debug prints during boot, pass the kernel command-line parameter dyndbg="QUERY" or module.dyndbg="QUERY", where QUERY is the dynamic debug syntax (explained earlier). For example, dyndng="module myfoo* +pmft" will activate all debug prints within the kernel modules named myfoo* with the display as set by the flags specifier pmft.

- To activate debug prints at kernel module initialization, that is, when modprobe myfoo is invoked (by systemd, perhaps), there are several ways, by passing along module parameters (with examples):

  - Via /etc/modprobe.d/*.conf (put this in the /etc/modprobe.d/ myfoo.conf file): options myfoo dyndbg=+pmft

  - Via the kernel command line: myfoo.dyndbg="file myfoobar.c +pmf; func goforit +mpt"

  - Via parameters to modprobe itself: modprobe myfoo dyndbg==pmft (this, the = and not the +, overrides any previous settings!)

Interesting: dyndbg is an always-available kernel module parameter, even though you don't see it (even in /sys/module/<modname>/parameters). You can see it by grepping the dynamic debug control file or /proc/cmdline.

(FYI, details on passing parameters to and auto-loading kernel modules have been fully covered in my earlier *Linux Kernel Programming* book.)

The official kernel documentation on dynamic debug is indeed very complete; be sure to have a look: https://www.kernel.org/doc/html/latest/admin-guide/ dynamic-debug-howto.html#dynamic-debug.

Kernel boot-time parameters

As an important aside, the kernel has an enormous (and useful!) number of kernel parameters that can be optionally passed to it at boot (via the bootloader). See the complete list here in the documentation: *The kernel's command-line parameters*: `https://www.kernel.org/doc/html/v5.10/admin-guide/kernel-parameters.html` (here, we've shown the link for the 5.10 kernel documentation).

While on the topic of the kernel command line, several other useful options with regard to printk-based debugging exist, enabling us to enlist the kernel's help for debugging issues concerned with kernel initialization. For example, the kernel provides the following parameters in this regard (taken directly from the link):

```
debug
  [KNL] Enable kernel debugging (events log level).
[...]
initcall_debug
  [KNL] Trace initcalls as they are executed. Useful for working
out where the kernel is dying during startup.
[...]
ignore_loglevel
  [KNL] Ignore loglevel setting - this will print /all/ kernel
messages to the console. Useful for debugging. We also add
it as printk module parameter, so users could change it
dynamically, usually by /sys/module/printk/parameters/ignore_
loglevel.
```

Useful indeed – do try them out! The sheer volume of information posted is surprising at first; try to carefully and patiently analyze it.

We're almost done here. Let's complete this chapter with some miscellaneous but useful printk-related logging functions and macros.

Remaining printk miscellany

By now, you're familiar with most of the typical and pragmatic means to leverage the kernel's powerful and ubiquitous printk and its related APIs, macros, and frameworks. Of course, innovation never stops (especially in the open source universe). The community has come up with more and more ways (and tooling) to use this simple and powerful tool. Without claiming to cover absolutely everything, here's what I think is the remaining and relevant tooling to do with the printk that we haven't had a chance to cover until now. Do check it out – it will probably turn out to be useful one day!

Printing before console init – the early printk

You understand that the printk output can be sent to the console device of course (we covered this in the *Understanding where the printk output goes* section(see *Table 3.1*). By default, on most systems, it's configured such that all printk messages of log level 3 and below (<4) are auto-routed to the console device as well (in effect, all kernel printks emitted at log levels `emerg/alert/crit/err` will find their way to the console device).

What exactly is the console device?

Before going any further, it's useful to understand what exactly the console device is... Traditionally, the console device is a pure kernel feature, the initial Terminal window that the superuser logs into (`/dev/console`) in a non-graphical environment. Interestingly, on Linux, we can define several consoles – a **teletype terminal** (**tty**) window (such as `/dev/console`), a text-mode VGA, a framebuffer, or even a serial port served over USB (this being common on embedded systems during development).

For example, when we connect a Raspberry Pi to an x86_64 laptop via a USB-to-RS232 TTL UART (USB-to-serial) cable (see the *Further reading* section of this chapter for a blog article on this very useful accessory and how to set it up on the Raspberry Pi!) and then use `minicom(1)` (or `screen(1)`) to get a serial console, this is what shows up as the `tty` device – it's the serial port:

```
rpi # tty
/dev/ttyS0
```

Now, what's the problem? Let's find out!

Early init – the issue and a solution

Via the printk, you can send messages to the console (and kernel log). Yes, but think about this: very early in the boot process when the kernel is initializing itself, the console device isn't ready, it's not initialized, and thus can't be used. Obviously, for any printk emitted at this early boot time, their output can't be seen on the *screen* – the console (even though it may be logged within the kernel log buffer, but we don't yet have a shell to look it up).

Pretty often (especially during things such as embedded board bring-up), hardware quirks or failures can cause the boot to hang, endlessly probe for some non-existent or faulty device, or even crash! The frustrating thing is that these issues become hard to debug (to say the least!) in the absence of console – printk – output, which, if visible, can instrument the kernel's boot process and pretty clearly show where the issue(s) is occurring (recall the kernel command-line parameters `debug` and `initcall_debug` can be really useful at times like this – look back at the *Kernel boot-time parameters section* if you need to).

Well, as we know, necessity is the mother of invention: the kernel community came up with a possible solution to this issue – the so-called **early printk**. With it configured, kernel printks are still able to be sent to the console device. How? Well, it's pretty arch and device-specific, but the broad and typical idea is that bare minimal console initialization is performed (this console device is called the `early_console`) and the string to be displayed on it is literally *bit-banged* out over a serial line one character at a time within a loop (with typical bitrates ranging between 9,600 and 115,200 bps).

To make use of the facility involves doing three things:

- Configure and build the kernel to support the early printk (set `CONFIG_EARLY_PRINTK=y`), one time only.
- Boot the target kernel with the appropriate kernel command-line parameter – `earlyprintk=<value>`.
- The API to use to emit the early printk is called `early_printk()`; the syntax is identical to that of `printf()`.

Let's check out each of the above points briefly, first, configuring the kernel for early printk.

The kernel config for this feature tends to be arch-dependent. On an x86, you'll have to configure the kernel with `CONFIG_EARLY_PRINTK=y` (it's under the `Kernel Hacking | x86 Debugging | Early printk` menu). Optionally, you can enable early printk via a USB debug port. The file that forms the UI – the menu system – for the kernel config (via the usual `make menuconfig`) for the kernel debug options is the file `arch/x86/Kconfig.debug`. We'll show a snippet of it here – the section where the early printk menu option is:

```
config EARLY_PRINTK
    bool "Early printk" if EXPERT
    default y
    help
        Write kernel log output directly into the VGA buffer or to a serial
        port.

        This is useful for kernel debugging when your machine crashes very
        early before the console code is initialized. For normal operation
        it is not recommended because it looks ugly and doesn't cooperate
        with klogd/syslogd or the X server. You should normally say N here,
        unless you want to debug such a crash.
```

Figure 3.8 – Screenshot showing the early printk portion of the Kconfig.debug file

Reading the `help` screen shown here is indeed helpful! As it says, this option isn't recommended by default as the output isn't well-formatted and can interfere with normal logging. You're typically only to use it to debug an early init issue. (If interested, you'll find the details on the kernel's *Kconfig* grammar and usage in my earlier *Linux Kernel Programming* book.)

On the other hand, on an ARM (AArch32) system, the kernel config option is under `Kernel Hacking|Kernel low-level debugging functions (read help!)` with the config option being called `CONFIG_DEBUG_LL`. As the kernel clearly insists, let's read the help screen:

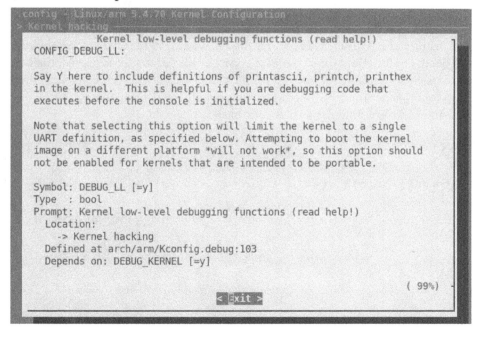

Figure 3.9 – Screenshot showing the make menuconfig UI menu for early printk on ARM-32

Do take note of what it says! Further, the sub-menu following it allows you to configure the low-level debug port (it's set to the EmbeddedICE DCC channel by default; you can change it to a serial UART if you have one available).

Okay, that's as far as kernel config goes, a one-time thing. Next, enable it by passing the appropriate kernel command-line parameter – `earlyprintk=<value>`. The official kernel documentation shows all possible ways to pass it (here: `https://www.kernel.org/doc/html/latest/admin-guide/kernel-parameters.html`):

```
earlyprintk=        [X86,SH,ARM,M68k,S390]
                    earlyprintk=vga
                    earlyprintk=sclp
                    earlyprintk=xen
                    earlyprintk=serial[,ttySn[,baudrate]]
                    earlyprintk=serial[,0x...[,baudrate]]
                    earlyprintk=ttySn[,baudrate]
                    earlyprintk=dbgp[debugController#]
        earlyprintk=pciserial[,force],bus:device.function[,baudrate
                    earlyprintk=xdbc[xhciController#]
```

(The remaining paragraphs of the kernel doc are useful to read as well.)

An optional `keep` parameter can be appended implying that printk messages sent via the early printk facility aren't disabled even after the VGA subsystem (or whatever the real console is) begins to function. Once the `earlyprintk=` parameter is passed, the kernel is primed to use it (essentially redirecting printk onto a serial, VGA, or whatever console you specified via this parameter). To emit a print, simply invoke the `early_printk()` API. Here's an example within the kernel code base:

```c
// kernel/events/core.c
    if (!irq_work_queue(&perf_duration_work)) {
        early_printk("perf: interrupt took too long (%lld >
%lld), lowering "
                "kernel.perf_event_max_sample_rate to %d\n",
                __report_avg, __report_allowed,
                sysctl_perf_event_sample_rate);
    }
```

What we've described above is mostly the arch-independent stuff. As an example, (only) on the x86, you can leverage a USB debug port (provided your system has one), as follows. Pass the kernel command-line parameter `earlyprintk=dbgp`. Note that it requires a USB debug port on the (x86) host system and a NetChip USB2 debug port key/cable (to connect to the client or target system). The kernel documentation details this facility here: `https://www.kernel.org/doc/html/latest/x86/earlyprintk.html#early-printk`.

Designating the printk to some known presets

The kernel provides macros that give you the ability to prefix – and thus designate – a printk as a firmware bug or warning, a hardware error, a message regarding a deprecated feature, and so on. This is specified via some kernel-defined macros. The value of the macro – a string (for example, `"[Firmware Bug]: "`) – is what will be prefixed to the message you're emitting:

```
// include/linux/printk.h
#define FW_BUG       "[Firmware Bug]: "
#define FW_WARN      "[Firmware Warn]: "
#define FW_INFO      "[Firmware Info]: "
[...]
/*
 * HW_ERR
 * Add this to a message for hardware errors, so that user
can report it to hardware vendor instead of LKML or software
vendor.
 */
#define HW_ERR       "[Hardware Error]: "
/*
 * DEPRECATED
 * Add this to a message whenever you want to warn userspace
about the use of a deprecated aspect of an API so they can stop
using it
 */
#define DEPRECATED   "[Deprecated]: "
```

Be sure to read the useful comments atop each of these.

An example or two of their usage follows:

```
// drivers/acpi/thermal.c
static int acpi_thermal_trips_update(struct acpi_thermal *tz,
int flag)
{
[...]
/*
 * Treat freezing temperatures as invalid as well; some
 * BIOSes return really low values and cause reboots at
startup. Below zero (Celsius) values clearly aren't right for
sure..
[...] */
} else if (tmp <= 2732) {
        pr_warn(FW_BUG "Invalid critical threshold (%llu)\n",
tmp);
```

Here's another example of a printk issuing a *deprecated* warning (notice the use of
rate-limiting as well!):

```
// net/batman-adv/debugfs.c
pr_warn_ratelimited(DEPRECATED "%s (pid %d) Use of debugfs file
\"%s\".\n%s", current->comm, task_pid_nr(current), name, alt);
```

Let's move along to the next point...

Printing exactly once

To emit a printk exactly once, use the macro `printk_once()`. It guarantees it will emit
the message exactly once, no matter how many times you call it (thus making it similar to
macros such as `WARN_[ON]_ONCE()`).

As with the usual `pr_*()` macros, their equivalents are defined for printing a message
exactly once: `pr_*_once()`. The wildcard * is replaced by the usual log levels (`emerg`,
`alert`, `crit`, `err`, `warn`, `notice`, `info`, and `debug`).

The IPv4 TCP code has an example of using pr_err_once():

```
// net/ipv4/tcp.c
[...]
if (unlikely(TCP_SKB_CB(skb)->tcp_flags & TCPHDR_SYN))
pr_err_once("%s: found a SYN, please report !\n", __func__);
[...]
```

Driver authors: you'll realize that the equivalent macros exist in include/linux/dev_printk.h. You're expected to use the dev_*_once() (in place of the pr_*_once()) macros. As one example, this i.MX53 **Real Time Clock (RTC)** chip driver uses it:

```
// drivers/rtc/rtc-mxc_v2.c
if (!--timeout) {
    dev_err_once(dev, "SRTC_LPSCLR stuck! Check your hw.\n");
    return;
```

Then – again, dear driver authors, do take note - there's the dev_WARN() and dev_WARN_ONCE() macros; the kernel comment explains it clearly:

```
// include/linux/dev_printk.h
/*
 * dev_WARN*() acts like dev_printk(), but with the key
difference of using WARN/WARN_ONCE to include file/line
information and a backtrace.
 */
```

Do think before using these [pr|dev]_*_once() macros; they're to be used when you want to emit a message exactly once.

> **Exercise**
>
> How many instances of using dev_WARN_ONCE() can you find within the 5.10.60 kernel code base? (Tip: use cscope!)

Emitting a printk from userspace

Testing is a critical part of the SDLC. While testing, it's often the case that you will run an automated batch (or suite) of test cases via a script. Now, say that your test (bash) script has initiated a test on your driver by invoking something like this:

```
echo "test data 123<...>" > /dev/mydevnode
```

That's fine, but you'd like to see the point at which the script initiated some action within our kernel module, by printing out a certain distinct (signature) message. As a concrete example, say we want the log to look something like this:

```
my_test_script:------------ start testcase 1
my_driver_module:
msg1, ..., msgn, msgn+1, ..., msgn+m
my_test_script:------------ end testcase 1
[...]
```

You can have your userspace test script write a message into the kernel log buffer, just like a kernel printk would, by writing the given message into the character device file /dev/kmsg:

```
sudo bash -c "echo \"my_test_script: start testcase 1\" > /dev/
kmsg"
```

(Note how we code it to run with root access.)

The message written to the kernel log via the special /dev/kmsg device file will be printed at the current default log level, typically, 4: KERN_WARNING. We can override this by actually prefixing the message with the required log level (as a number in string format, within angle brackets). For example, to write from the userspace into the kernel log at log level 6: KERN_INFO, use this:

```
sudo bash -c "echo \"<6>my_test_script: start testcase 1\" > /
dev/kmsg"
$ sudo dmesg --decode |tail -n1
user   :info  : [33561.862960] my_test_script: start testcase 1
```

Notice how I used the --decode option to dmesg to provide more human-readable output. Also, you can see that our latter message is emitted at log level 6 as specified within the echo statement.

There is really no way to distinguish between a user-generated kernel message and a kernel one generated by `printk()`; they look identical. So, of course, it could be as simple as prefixing some special signature byte or string within the message, such as @myapp@, in order to help you distinguish these user-generated prints from the kernel ones.

Easily dumping buffer content

Once, when working on a network driver, I wrote C code to quite painstakingly dump the content of the Ethernet (link) header, IP header, and so on, in order to analyze and understand exactly how things were working... My code did the typical thing: within a loop, dump each byte of the header structure, printing it in hexadecimal (and, if you're feeling adventurous, when printable, in ASCII as well on the right side). Sure, we can do these things, but don't waste time – the kernel provides!

The macro, `print_hex_dump_bytes()` is there for precisely this kind of work; it's a wrapper over a similar macro. The comments within its code clearly show you the meaning of each of its four parameters, and thus how to use it to efficiently dump memory buffer content:

```
// include/linux/printk.h
/**
 * print_hex_dump_bytes - shorthand form of print_hex_dump()
with default params
 * @prefix_str: string to prefix each line with; caller
supplies trailing spaces for alignment if desired
 * @prefix_type: controls whether prefix of an offset, address,
or none is printed (%DUMP_PREFIX_OFFSET, %DUMP_PREFIX_ADDRESS,
%DUMP_PREFIX_NONE)
 * @buf: data blob to dump
 * @len: number of bytes in the @buf
 * Calls print_hex_dump(), with log level of KERN_DEBUG,
rowsize of 16, groupsize of 1, and ASCII output included.
 */
#define print_hex_dump_bytes(prefix_str, prefix_type, buf, len)
\
    print_hex_dump_debug(prefix_str, prefix_type, 16, 1, buf,
len, true)
```

Great, but why does the macro invoke the debug version? Ah, it's tied into the kernel's dynamic debug circuitry! Thus (as we already mentioned in the *Using the kernel's powerful dynamic debug feature* section), every `print_hex_dump_debug()` and `print_hex_dump_bytes()` callsite is able to be dynamically toggled via the dynamic debug control file. Useful!

Here's an example of this macro in action (within a Qualcomm wireless network driver):

```
// drivers/net/wireless/ath/ath6kl/debug.c
void ath6kl_dbg_dump(enum ATH6K_DEBUG_MASK mask,
          const char *msg, const char *prefix,
          const void *buf, size_t len)
{
    if (debug_mask & mask) {
        if (msg)
            ath6kl_dbg(mask, "%s\n", msg);
        print_hex_dump_bytes(prefix,
            DUMP_PREFIX_OFFSET, buf, len);
    }
[...]
```

There – all done! Well, no, it's never actually all done, is it...?

Remaining points – bootloader log peeking, LED flashing, and more

A common problem when debugging kernel crashes is that once the kernel has crashed or panicked, the system is unusable (typically hung). Reading the kernel log will almost certainly help in debugging the (root) cause... But – I'm sure you see this – how can we see the kernel log if the system is hung?! Moreover, there's no iron-clad guarantee that the log data has been flushed from RAM into non-volatile log files before the system went into a tailspin...

For these reasons, more exotic debug techniques are required at times. One of them is this: after the system hangs (or panics), "warm" boot or reset back into the bootloader prompt (this is assuming that there is a way to do so – let's assume there is).

> **Warm Reset – How?**
>
> A warm reset or reboot is one where the board reboots but RAM content is preserved. I once worked on a prototyping project on a TI PandaBoard. It had a soft reset button; pressing it led to the board performing a warm reboot.
>
> The PC's *Ctrl + Alt + Delete* (the famous *"three-finger salute"* – the temptation to say "read between the lines" and put in a smiley here is great!) is the equivalent... But that's typically not configured on Linux; you can use the kernel's *Magic SysRq* facility (again, assuming it's so configured) to do so (fear not, we shall cover this in *Chapter 10, Kernel Panic, Lockups, and Hangs,* in the *What's this Magic SysRq thingy anyway?* section).

Once at the bootloader prompt, use its intelligence to dump the kernel log buffer memory region and you will see the kernel printks! (For example, many embedded systems use the powerful and elegant **Das U-Boot** as their bootloader; the command to dump a memory region is memory display (md).) Hang on though – a key point: even if you know the kernel log buffer address (that's easy: just do `sudo grep __log_buf /proc/kallsyms` to get it), it's not a physical address; it's a kernel virtual address. You will first have to figure out how to translate it to its physical counterpart – as that's all the bootloader sees. This is typically done by referring to the **Technical Reference Manual** (**TRM**) for the board or platform you're working on. Once you have the physical address, simply issue the md (or equivalent – GRUB has the dump command) command to dump the memory content. You will, in effect, see the kernel log!

I refer you to a few actual examples in this (older but excellent) information here: `https://elinux.org/Debugging_by_printing#Debugging_early_boot_problems`).

Flashing LEDs to debug

Sometimes, especially during the very early stages of board bring-up and likewise, all we really need to know is that some line of code got executed. You can do this by toggling an LED on/off or flashing it as lines of code are hit! Developers at times go to the extent of rigging up the system's GPIO pins (or equivalent) to do so and insert custom code in the kernel to trigger the LED. This is really nothing new – it's the *poor man's printf*.

(Interestingly, the Raspberry Pi does precisely this when it can't boot – it flashes an onboard LED a given number of times (short and long flashes)... Here's the documentation that explains how to interpret the LED flashes and thus understand what's causing the boot issue: `https://www.raspberrypi.com/documentation/computers/configuration.html#led-warning-flash-codes`).

Even better, you could perhaps even rig up a device such as the OLED display mentioned earlier to display debug messages. Of course, this will require that I2C initialization has been performed.

Another thing: you may have heard of the kernel's **netconsole** facility. Isn't that something to delve into? It certainly is – netconsole is a powerful thing, a means to send kernel printks across a network to a target system, which will store it for later perusal! (We cover it here: *Chapter 7, Oops! Interpreting the Kernel Bug Diagnostic,* in the *An Oops on an ARM Linux system and using netconsole* section; keep a keen eye out!).

Summary

Good going! You've just completed the first of many techniques for debugging the kernel. Instrumentation, though deceptively simple, almost always proves to be a useful and powerful debugging technique.

In this chapter, you began by learning the basics regarding the ubiquitous kernel `printk()`, `pr_*()`, and `dev_*()` routines and macros. We then went into more detail about the specific use of these routines to help in debug situations and tips and tricks that will prove useful in debugging your (driver) modules... This included leveraging the kernel's ability to rate-limit printks, often a necessity on high-volume code paths.

 The kernel's elegant and powerful dynamic debug framework was the highlight of this chapter. Here, you learned about it, and how to leverage it to be able to toggle your (and indeed the kernel's) debug prints even on production systems, with virtually no performance degradation when turned off.

We finished this chapter with a few remaining uses of printk macros that are sure to prove useful at some point in your kernel/driver journeys.

With these tools in hand, we'll move on to another powerful technology in the coming chapter: the kernel's kprobes framework where we'll of course focus on using it to aid us to debug things, primarily via the instrumentation approach. See you there!

Further reading

- Code browser tutorials:

    - Ctags Tutorial: `https://courses.cs.washington.edu/courses/cse451/10au/tutorials/tutorial_ctags.html`

    - The Vim/Cscope tutorial: `http://cscope.sourceforge.net/cscope_vim_tutorial.html`

- The printk and /dev/kmsg: `https://www.kernel.org/doc/Documentation/ABI/testing/dev-kmsg`

- *Debugging by printing*: `https://elinux.org/Debugging_by_printing`; covers useful info on debugging with the early printk facility, even debugging by dumping the kernel log from the bootloader as well!

- Signing kernel modules; official kernel documentation: *Kernel module signing facility*: `https://www.kernel.org/doc/html/v5.0/admin-guide/module-signing.html#kernel-module-signing-facility`

- Red Hat Developer series on GDB:

    - *The GDB developer's GNU Debugger tutorial, Part 1: Getting started with the debugger*, Seitz, RedHat Developer, Apr 2021: `https://developers.redhat.com/blog/2021/04/30/the-gdb-developers-gnu-debugger-tutorial-part-1-getting-started-with-the-debugger#`

    - *The GDB developer's GNU Debugger tutorial, Part 2: All about debuginfo*, Seitz, RedHat Developer, Jan 2022: `https://developers.redhat.com/articles/2022/01/10/gdb-developers-gnu-debugger-tutorial-part-2-all-about-debuginfo`

    - *Printf-style debugging using GDB, Part 3*, Buettner, RedHat Developer, Dec 2021: `https://developers.redhat.com/articles/2021/12/09/printf-style-debugging-using-gdb-part-3#`

- Dynamic debug:

    - Official kernel doc: *Dynamic debug*: `https://www.kernel.org/doc/html/latest/admin-guide/dynamic-debug-howto.html#dynamic-debug`

    - *The dynamic debugging interface*, Jon Corbet, LWN, March 2011: `https://lwn.net/Articles/434833/`

4
Debug via Instrumentation – Kprobes

A **kernel probe (kprobe)** is one of the powerful weapons in our debug/performance/observability armory! Here, you'll learn what exactly it can do for you and how to leverage it, with the emphasis being on debug scenarios. You will find that there's a so-called static and a dynamic probing approach to using them... We'll also cover using a way to figure out the return value of any function via a **kernel return probe (kretprobe)**!

Along the way, you'll learn what the **Application Binary Interface (ABI)** is and why it's important to know at least the basics of the processor ABI.

Don't miss delving into the section on dynamic kprobes or kprobe-based event tracing, as well as employing the `perf-tools` and (especially) the modern eBPF BCC frontends – it makes it all so much easier!

In this chapter, we're going to cover the following main topics:

- Understanding kprobes basics
- Using static kprobes – traditional approaches to probing
- Understanding the basics of the ABI
- Using static kprobes – demo 3 and demo 4
- Getting started with kretprobes
- Kprobes – limitations and downsides
- The easier way – dynamic kprobes or kprobe-based event tracing
- Trapping into the execve() API – via perf and eBPF tooling

Understanding kprobes basics

A kernel probe (**Kprobe, kprobe**, or simply **probe**) is a way to hook or trap into (almost) any function in the kernel proper or within a kernel module, including interrupt handlers. You can think of kprobes as a dynamic analysis/instrumentation toolset that can even be used on production systems to collect (and later analyze) debugging and/or performance-related telemetry.

To use it, kprobes have to be enabled in the kernel; the kernel config CONFIG_KPROBES must be set to y (you'll typically find it under the General architecture-dependent options menu). Selecting it automatically selects CONFIG_KALLSYMS=y as well. With kprobes, you can set up three – all optional – types of traps or hooks. To illustrate, let's say you want to trap into the kernel function do_sys_open() (which is the kernel function invoked when a userspace process or thread issues the open(2) system call; see the *System calls and where they land in the kernel* section for more details). Now, via the kernel kprobes infrastructure, you can set up the following:

- **A pre-handler routine**: Invoked just before the call to do_sys_open().
- **A post-handler routine**: Invoked just after the call to do_sys_open().
- **A fault-handler routine**: Invoked if, during the execution of the pre or post handler, a processor fault (exception) is generated (or if kprobes is single-stepping instructions); often, a page fault can occur triggering the fault handler.

They're optional – it's up to you to set up one or more of them. Further, there are two broad types of kprobes you can register (and subsequently unregister):

- **The regular kprobe**: via the [un]register_kprobe[s]() kernel APIs

- **A return probe or kretprobe**: via the [un]register_kretprobe[s]() kernel APIs, providing access to the probed function's return value

Let's first work with the regular kprobe and come to the kretprobe a bit later... To trap into a kernel or module function, issue the kernel API:

```
#include <linux/kprobes.h>
int register_kprobe(struct kprobe *p);
```

The parameter, a pointer to struct kprobe, contains the details; the key members we need to be concerned with are the following:

- const char *symbol_name: The name of the kernel or module function to trap into (internally, the framework employs the kallsyms_lookup() API – or a variation of it – to resolve the symbol name into a **kernel virtual address** (**KVA**), and store it in a member named addr). There are a few limitations on which functions you can and cannot trap into (we cover this in the *Kprobes – limitations and downsides section*).

- kprobe_pre_handler_t pre_handler: The pre-handler routine function pointer, called just before addr is executed.

- kprobe_post_handler_t post_handler: The post-handler routine function pointer, called just after addr is executed.

- kprobe_fault_handler_t fault_handler: The function pointer to a fault handling routine, which is invoked if executing addr causes a fault of any kind. You must return the value 0 to inform the kernel that it must actually handle the fault (typical) and return 1 if you handled it (uncommon).

Without going into the gory details, it's interesting to realize that you can even set up a probe to a specified *offset* within a function! This is achieved by setting the offset member of the kprobe structure to the desired value (watch out though: the offset should be used with care, especially on **Complex Instruction Set Computing** (**CISC**) machines).

Once done, you're expected to release the trap or probe (often on module exit), via the unregister routine:

```
void unregister_kprobe(struct kprobe *p);
```

Failing to do so will cause a kernel bug(s) and freeze when that kprobe is next hit; a resource leak failure of a sort.

> **How Do Kprobes Work under the Hood?**
>
> Unfortunately, this topic lies beyond the scope of this book. Interested readers can certainly refer to the excellent kernel documentation, which explains the fundamentals of how kprobes actually work: *Concepts: Kprobes and Return Probes*: https://www.kernel.org/doc/html/latest/trace/kprobes.html#concepts-kprobes-and-return-probes.

What we intend to do

This kind of methodology to set up a probe – where, if any change in the function to be probed or in the output format is required, it requires a recompile of the module code – is called a **static kprobe**. Is there any other way? Indeed there is: modern Linux kernels have the infrastructure – mostly via the deep ftrace and tracepoints framework, called **dynamic probing** or kprobe-based event tracing. There's no C code to write and deal with and no recompile necessary!

In the following sections, we'll show you different ways of setting things up, going from the traditional "manual" static kprobes interface approaches to the more recent and advanced dynamic kernel probes/tracepoints approach. To make this interesting, here's how we'll go about writing a few demos, most of which will trap into the kernel file open code path:

- **Demo 1 - Traditional and manual approach – simplest case**: attaching a static kprobe, hardcoding it to trap into the open system call; code: ch4/kprobes/1_kprobe

- **Demo 2 - Traditional and manual approach**: attaching a static kprobe, slightly better than our Demo 1, softcoding it via a module parameter (to the open system call); code: ch4/kprobes/2_kprobe

- **Demo 3 - Traditional and manual approach**: attaching a static kprobe via a module parameter (to the open system call), plus retrieving the pathname to the file being opened (useful!); code: ch4/kprobes/3_kprobe

- **Demo 4 - Traditional, semi-automated approach**: a helper script *generates a template* for both the kernel module C code and the `Makefile`, enabling attaching a static kprobe to any function specified via the module parameter; code: `ch4/kprobes/4_kprobe_helper`

- Next, we'll take a quick look at what a return probe is – the *kretprobe* – and how to use it (static).

- **Modern, easier, dynamic event tracing approach**: attaching a dynamic kprobe (as well as a kretprobe) to both the `open` and the `execve` system calls, retrieving the pathname to the file being opened/executed

- **Modern, easier, and powerful eBPF approach**: tracing the file `open` and the `execve` system calls.

Great – we'll begin with the traditional static kprobes approaches. Let's get going!

Using static kprobes – traditional approaches to probing

In this section, we'll cover writing kernel modules that can probe a kernel or module function in the traditional manner – statically. Any modifications will require a recompile of the source.

Demo 1 – static kprobe – trapping into the file open the traditional static kprobes way – simplest case

Right, let's see how we can trap into (or intercept) the `do_sys_open()` kernel routine by *planting* a kprobe. This code snippet will typically be within the *init* function of a kernel module. You'll find the code for this demo here: `ch4/kprobes/1_kprobe`:

```
// ch4/kprobes/1_kprobe/1_kprobe.c
#include "<...>/convenient.h"
#include <linux/kprobes.h>
[...]
static struct kprobe kpb;
[...]
/* Register the kprobe handler */
kpb.pre_handler = handler_pre;
```

```
kpb.post_handler = handler_post;
kpb.fault_handler = handler_fault;
kpb.symbol_name = "do_sys_open";
if (register_kprobe(&kpb)) {
    pr_alert("register_kprobe on do_sys_open() failed!\n");
    return -EINVAL;
}
pr_info("registering kernel probe @ 'do_sys_open()'\n");
```

An interesting use of kprobes is to figure out (approximately) how long a kernel/module function takes to execute. To figure this out... come on, you don't need me to tell you!:

1. Take a timestamp in the pre-handler routine (call it `tm_start`). We can use the `ktime_get_real_ns()` routine to do so.

2. As the first thing in the post-handler routine, take another timestamp (call it `tm_end`).

3. (`tm_end` - `tm_start`) is the time taken (do peek at our `convenient.h:SHOW_DELTA()` macro to see how to correctly perform the calculation).

The pre- and post-handler routines follow. Let's begin with the pre-handler routine:

```
static int handler_pre(struct kprobe *p, struct pt_regs *regs)
{
    PRINT_CTX(); // uses pr_debug()
    spin_lock(&lock);
    tm_start = ktime_get_real_ns();
    spin_unlock(&lock);
    return 0;
}
```

Here's our post-handler:

```
static void handler_post(struct kprobe *p, struct pt_regs
*regs, unsigned long flags)
{
    spin_lock(&lock);
    tm_end = ktime_get_real_ns();
    PRINT_CTX(); // uses pr_debug()
```

```
        SHOW_DELTA(tm_end, tm_start);
        spin_unlock(&lock);
}
```

It's pretty straightforward, right? We grab the timestamps and the SHOW_DELTA() macro calculates the difference. Where is it? In our *convenience* header file (named – surprise, surprise! – convenient.h). Similarly, the PRINT_CTX() macro defined there gives us a nice one-line summary of the state of the process/interrupt context in the kernel that executed the macro (details on interpreting this follows). The spinlock is used, of course, for concurrency control – as we're operating on shared writable data items.

As the comment next to the PRINT_CTX() macro says, it internally uses pr_debug() to emit output to the kernel log. Hence, it will only appear if either of the following apply:

- The symbol DEBUG is defined.

- More usefully, DEBUG is deliberately left undefined (as is typical in production) and you make use of the kernel's dynamic debug facility to turn on/off these prints (as discussed in detail in *Chapter 3, Debug via Instrumentation – printk and Friends* in the *Using the kernel's powerful dynamic debug feature section*).

A sample fault handler is defined too. We don't do anything much here, merely emit a printk specifying which fault occurred, leaving the actual fault handling – a complex task – to the core kernel (here, we simply copy the fault handler code from the kernel tree: samples/kprobes/kprobe_example.c):

```
static int handler_fault(struct kprobe *p, struct pt_regs
*regs, int trapnr)
{
    pr_info("fault_handler: p->addr = 0x%p, trap #%dn",
            p->addr, trapnr);
    /* Return 0 because we don't handle the fault. */
    return 0;
}
NOKPROBE_SYMBOL(handler_fault);
```

Notice a couple of things here:

- The third parameter to the fault handler callback, trapnr, is the numerical value of the trap that occurred; it's very arch-specific. For example, on x86, 14 implies it's a page fault (similarly, you can always look up the manual for other processor families to see their values and the meaning).

- The NOKPROBE_SYMBOL(foo) macro is used to specify that the function foo cannot be probed. Here, it's specified so that recursive or double faults are prevented.

Now that we've seen the code, let's give it a spin!

Trying it out

The test.sh and run bash scripts (within the same directory) are simple wrappers (run is a wrapper over the wrapper test.sh!) to ease testing these demo kernel modules. I'll leave it to you to check out how they work:

```
$ cd <lkd-src-tree>/ch4/kprobes/1_kprobe ; ls
1_kprobe.c  Makefile   run test.sh
$ cat run
KMOD=1_kprobe
echo "sudo dmesg -C && make && ./test.sh && sleep 5 && sudo
rmmod ${KMOD} 2>/dev/null ; sudo dmesg"
sudo dmesg -C && make && ./test.sh && sleep 5 && sudo rmmod
${KMOD} 2>/dev/null ; sudo dmesg
$
```

The run wrapper script invokes the test.sh wrapper script (which performs insmod and sets up the dynamic debug control file to enable our debug printks). We allow the probe to remain active for 5 seconds – plenty of file open system calls, resulting in the invocation of do_sys_open() and our resulting pre- and post- handlers running, can happen in that time span.

Let's give our first demo a spin on our x86_64 Ubuntu VM running our custom *production* kernel:

```
$ ./run
sudo dmesg -C && make && ./test.sh && sleep 5 && sudo rmmod 1_kprobe 2>/dev/null ; sudo dmesg

--- Building : KDIR=/lib/modules/5.10.60-prod01/build ARCH= CROSS_COMPILE= EXTRA_CFLAGS=-DDYNAMIC_DEBUG_MODULE ---

make -C /lib/modules/5.10.60-prod01/build M=/home/letsdebug/Linux-Kernel-Debugging/ch5/kprobes/1_kprobe modules
make[1]: Entering directory '/home/letsdebug/lkd_kernels/productionk/linux-5.10.60'
make[1]: Leaving directory '/home/letsdebug/lkd_kernels/productionk/linux-5.10.60'
Module 1_kprobe: function to probe: do_sys_open()

-- Module 1_kprobe now inserted, turn on any dynamic debug prints now --
Wrt module 1_kprobe, one or more dynamic debug prints are On
/home/letsdebug/Linux-Kernel-Debugging/ch5/kprobes/1_kprobe/1_kprobe.c:68 [1_kprobe]handler_post =p "\012"
/home/letsdebug/Linux-Kernel-Debugging/ch5/kprobes/1_kprobe/1_kprobe.c:65 [1_kprobe]handler_post =p "%03d) %c%s%c:%d    |   %c%c
%c%u   /* %s() */\012"
/home/letsdebug/Linux-Kernel-Debugging/ch5/kprobes/1_kprobe/1_kprobe.c:49 [1_kprobe]handler_pre =p "%03d) %c%s%c:%d    |   %c%c%
c%u   /* %s() */\012"
--  All set, look up kernel log with, f.e., journalctl -k -f  --
```

Figure 4.1 – Kprobes demo 1 – invoking the run script

You can see that the `run` script (invoking the `test.sh` script) sets things up... About 5 seconds later, here's a snippet of the output seen via `sudo dmesg`:

```
[81970.137707] 1_kprobe:handler_post(): 002)   rmmod :8183    |  ...1   /* handler_post() */
[81970.138152] 1_kprobe:handler_pre(): 003)  systemd-journal :395    |  ...1   /* handler_pre() */
[81970.138589] 1_kprobe:handler_post(): delta: 195 ns (~ 0 us ~ 0 ms)
[81970.139587] 1_kprobe:handler_post():
[81970.139588] 1_kprobe:handler_post(): 003)  systemd-journal :395    |  ...1   /* handler_post() */
[81970.139589] 1_kprobe:handler_post(): delta: 142 ns (~ 0 us ~ 0 ms)
[81970.141131] 1_kprobe:handler_post():
[81970.141752] 1_kprobe:handler_pre(): 003)  systemd-journal :395    |  ...1   /* handler_pre() */
[81970.142245] 1_kprobe:handler_post(): 003)  systemd-journal :395    |  ...1   /* handler_post() */
[81970.143010] 1_kprobe:handler_post(): delta: 100 ns (~ 0 us ~ 0 ms)
[81970.143545] 1_kprobe:handler_post():
[81970.175571] 1_kprobe:kprobe_lkm_exit(): bye, unregistering kernel probe @ 'do_sys_open()'
$ ▮
```

Figure 4.2 – Kprobes demo 1 – partial dmesg output

Great! Our static kprobe, being hit both before and after entering the `do_sys_call()` kernel function, executes the pre- and post-handlers in our module and produces the prints you see in the preceding screenshot. We need to interpret the `PINT_CTX()` macro's output.

Interpreting the PRINT_CTX() macro's output

In *Figure 4.2*, notice the useful output we obtain from our `PRINT_CTX()` macro (defined within our `convenient.h` header). I reproduce three of the relevant lines here, color-coding them to help you clearly understand them:

```
[81970.141752] 1_kprobe:handler_pre(): 003)  systemd-journal :395    |  ...1   /*
handler_pre() */
[81970.142245] 1_kprobe:handler_post(): 003)  systemd-journal :395
|  ...1   /* handler_post() */
[81970.143010] 1_kprobe:handler_post(): delta: 100 ns (~ 0 us ~ 0 ms)
```

Figure 4.3 – Kprobes demo 1 – pre- and post-handler sample kernel printk output

Let's get into what these three lines of output are:

- **First line**: Output from the pre-handler routine's `PRINT_CTX()` macro.

- **Second line**: Output from the post-handler routine's `PRINT_CTX()` macro.

- **Third line**: The delta – the (approximate, usually pretty accurate) time it took for `do_sys_open()` to run – is seen in the post-handler as well. It's fast, isn't it?!

- **Also notice**: The dmesg timestamp (the time from boot in seconds . microseconds – don't completely trust its absolute value though!) and – due to our enabling the debug printks and the fact that it employs the pr_fmt() overriding macro – the module name:function_name() is also prefixed. For example, here, the latter two lines are prefixed with the following:

  ```
  1_kprobe:handler_post():
  ```

Further, the following screenshot shows you how to fully interpret the output from our useful PRINT_CTX() macro:

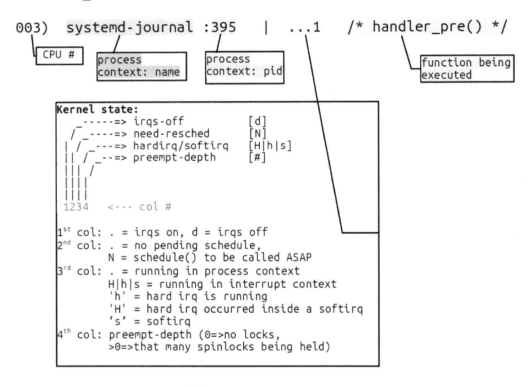

Figure 4.4 – Interpreting the PRINT_CTX() macro output

Do ensure you carefully study and understand this. It can be very useful in deep debug situations. (In fact, I've mostly mimicked this well-known format from the kernel's *ftrace* infrastructure's latency format display. It's explained in the ftrace documentation here as well: `https://www.kernel.org/doc/Documentation/trace/ftrace.txt`, under the heading *Output format*.)

Ftrace and the latency trace info fields

Worry not, we'll tackle ftrace in detail in *Chapter 9, Tracing the Kernel Flow*. Within it, the *Delving deeper into the latency trace info* section covers the interpretation of the so-called latency trace info fields in great detail. As a point of interest, ask yourself: what if the `PRINT_CTX()` macro runs in an interrupt context? What will be the values for the process context name and **Process Identifier (PID)** fields in that case? The short answer – it will be whichever process (or thread) happened to be *caught* in (preempted/ interrupted by) the interrupt!

Importantly, actually try this out yourself – and the following demo modules! It will go a long way in helping you experiment and learn how to use kprobes effectively.

The kprobes Blacklist – You Can't Trap This!

A few kernel functions cannot be trapped into via the kprobes interfaces (mainly because they're used internally within the kprobe implementation). You can quickly check which these are – they're available in the `blacklist` pseudofile here: `<debugfs-mount>/kprobes/blacklist` (the **debug filesystem (debugfs)** is typically mounted under `/sys/kernel/debug`, of course). The kernel documentation discusses this and other kprobe limitations here: `https://www.kernel.org/doc/html/latest/trace/kprobes.html#kprobes-features-and-limitations`. Do check it out.

You can even do cool (read dangerous) things such as modify parameters in the pre-handler! Careful though – it can result in freezes or even outright kernel panic if done incorrectly. This could be a useful thing for testing – a way to, for example, inject deliberate faults... While on this, FYI, the kernel has a sophisticated *fault injection* framework; more on this in a later chapter.

Demo 2 – static kprobe – specifying the function to probe via a module parameter

Our second kprobes demo is quite similar to the first, but it differs as follows.

One, we now add and make use of two module parameters – a string one for the name of the function to probe, and an integer determining verbosity:

```
// ch4/kprobes/2_kprobe/2_kprobe.c
#define MAX_FUNCNAME_LEN   64
static char kprobe_func[MAX_FUNCNAME_LEN];
module_param_string(kprobe_func, kprobe_func, sizeof(kprobe_
func), 0);
MODULE_PARM_DESC(kprobe_func, "function name to attach a kprobe
to");

static int verbose;
module_param(verbose, int, 0644);
MODULE_PARM_DESC(verbose, "Set to 1 to get verbose printks
(defaults to 0).");
```

The kprobe_func module parameter is useful! It allows us to pass any (valid) function as the probe target, avoiding hardcoding it. Of course, we now set the symbol_name member of the kprobe structure to the parameter:

```
kpb.symbol_name = kprobe_func;
```

In addition, the module's init code checks that the kprobe_func string is non-null.

The verbose parameter, if set, has the post-handler routine invoke the PRINT_CTX() macro. In this demo, we have our test.sh wrapper set the module parameters as follows:

```
// ch4/kprobes/2_kprobe/test.sh
FUNC_TO_KPROBE=do_sys_open
VERBOSE=1
[...]
sudo insmod ./${KMOD}.ko kprobe_func=${FUNC_TO_KPROBE}
verbose=${VERBOSE} || exit 1
```

One issue you'll quickly notice with kprobes instrumentation (indeed, it's quite common to many kinds of instrumentation and tracing) is the sheer volume of printks that can get generated! With a view to limiting it (thus trying to mitigate the overflow of the kernel ring buffer), we introduce a macro named SKIP_IF_NOT_VI. If defined, we only log information in the pre- and post-handlers when the process context is the vi process (of course, it's just a demo; feel free to change it to whatever's suitable or undefine it):

```
#ifdef SKIP_IF_NOT_VI
/* For the purpose of this demo, we only log information when
the process context is 'vi' */
    if (strncmp(current->comm, "vi", 2))
        return 0;
#endif
```

That's pretty much it. I leave it to you to try it out. (Don't forget to run vi so that some output's captured!).

> **Exercise**
>
> As a small exercise:
>
> a) Try passing other functions to probe via the module parameter kprobe_func.
>
> b) Convert the hardcoded SKIP_IF_NOT_VI macro into a module parameter.

Well, great, you now know how to write (simple) kernel modules that leverage the kernel's kprobes framework. To be able to go deeper within a typical debugging context, it is necessary to understand more in-depth things... Things such as how the processor's **General Purpose Registers (GPRs)** are typically used, how the processor interprets a stack frame and passes function parameters, and so on... This is the domain of the **Application Binary Interface**, the **ABI**! The following section helps you gain an introduction to it, information that can prove extremely valuable during a deep debug session. Read on!

Understanding the basics of the Application Binary Interface (ABI)

In order to gain access to the parameters of a function, you have to first understand at least the basics of how the compiler arranges for the parameters to be passed. At the level of the assembly (assuming the programming language is C), you'll realize that it's really the compiler that generates the required assembly that actually implements the function call, parameter passing, local variable instantiation, and return!

But how does the compiler manage to do this? Compiler authors need to understand how the machine works... Obviously, all of this is very arch-specific; the precise specification as to how exactly *function calling conventions, return value placement, stack and register usage*, and so on, is provided by the microprocessor documentation called the ABI document.

Briefly, the ABI conveys the underlying details at the level of the machine concerning the following:

- CPU register usage
- Function procedure calling and return conventions
- Precise stack frame layout in memory
- The details regarding data representation, linkage, object file formats, and so on

For example, x86-32 processors always use the stack to store the parameters to a function before issuing the CALL machine instruction. On the other hand, ARM-32 processors use both CPU **GPRs** as well as the stack (details follow).

Here, we shall only focus on one primary aspect of the ABI – the *function calling conventions (and related register usage)* on a few key architectures:

Arch (CPU) family	How parameters to a function are passed	Other useful info (typical usage)

IA-32 (x86-32)	All parameters passed via the stack.
	
	Access is via offsets to the Stack Pointer (SP) and Base Pointer (BP) registers.
		• Stack frame layout:
		
		   o Parameters <-- higher address
		
		   o RET address
		
		   o [SFP] : Optional, if Frame Pointer (FP) enabled
		
		   o Local variables <-- lower address (top of stack)
		
		• Return value: Typically in accumulator (EAX)

ARM-32 (Aarch32)	Follows the ARM Procedure Call Standard (APCS). The first four parameters to the function are passed in these 32-bit CPU registers: r0, r1, r2, r3. Any remaining parameters are passed via the stack (stack frame layout is as with x86-32).	• r4 to r9: Scratch registers (often used for local variables) • r7: If system call being issued, holds system call (syscall) # • r11: If enabled, the FP • r13: Stack Pointer (SP) • r14: Link Register (LR); holds the return address, used to return at function exit • r15: Program Counter (PC). • Return value: typically in r0
x86_64	The first six parameters to the function are passed in these 64-bit CPU registers: RDI, RSI, RDX, RCX, R8, R9. Any remaining parameters are passed via the stack (stack frame layout is as with the x86-32, except that 64-bit alignments are used and RBP acts as the base pointer register),	• RAX: Accumulator; holds syscall # when syscall being issued • RBP: Base pointer, start of stack • RSP: Stack pointer, current location on stack • Return value: Typically in accumulator (RAX) • PSR: Processor Status Register
Aarch64 (ARMv8)	APCS: the first eight parameters to the function are passed in these 64-bit CPU registers: X0 to X7 . Any remaining parameters are passed via the stack.	• X8: (indirect) Return value address • X9 to X15: Local variables, caller saved • X29 (FP): Frame pointer • X30 (LR): Link register, used to return at function exit • X31 (SP): Stack pointer or a zero register, depending on context • Return value: Typically in X0

Table 4.1 – Summary of function call and register usage ABI information for a few processor families

A couple of additional points:

- Pretty much all modern processors have a *downward-growing stack* – the stack *grows* from higher virtual addresses to lower virtual addresses. If interested (and I recommend you to be!), do look up more details in the blog article mentioned just after these points. Things are not always simple: in the presence of compiler optimization, the details seen in *Table 4.1* might not hold (for example, on the x86-32, gcc and Windows FASTCALL piggyback the first two function parameters into registers ECX and EDX). So, do check and recheck – **Your Mileage May Vary (YMMV)**...

- The ABI details mentioned here apply to how the C compiler (gcc/clang) typically works, thus for the C language, using integer or pointer parameters (not floating-point arguments or returns). Also, we won't go into more detail here (callee/caller-saved registers, the so-called red zone optimization, exception cases, and so on). Refer to the *Further reading* section for further links.

Links to the ABI documentation for various processor families and its basic details can be found in this (my) blog article: *APPLICATION BINARY INTERFACE (ABI) DOCS AND THEIR MEANING*: https://kaiwantech.wordpress.com/2018/05/07/application-binary-interface-abi-docs-and-their-meaning/.

Now that we have at least basic knowledge of the processor ABI and how the compiler (gcc/clang) uses it on Linux, let's put our newfound know-how to use. In the following section, we'll learn how to do something pretty useful – determine the pathname of the file being opened via our kprobe-based open system call trap. More generally, we'll learn in effect how to retrieve the parameters of the trapped (probed) function!

Using static kprobes – demo 3 and demo 4

Continuing to work via the traditional static kprobes approach (recall: the word static implies any change will require a code recompile), let's learn to do more with kprobes – useful and practical stuff that really helps when debugging. Retrieving the parameters of the probed function certainly qualifies as being a very useful skill!

The two demo programs that follow (demos 3 and 4), will show you how to do precisely this, with demo 4 using an interesting approach – we'll *generate* our kprobe C code (and Makefile file) via a bash script. Let's work on and understand these demos!

Demo 3 – static kprobe – probing the file open syscall and retrieving the filename parameter

You'll agree, I think, that the second demo is better than the first – it allows the passing of any function to be probed (as a module parameter). Now, continuing with our example of probing do_sys_open(), you've seen (from the first two demos) that we can indeed probe it. In a typical debugging/troubleshooting scenario, though, this isn't nearly enough: being able to **retrieve the parameters of the probed function** can be really important, and can prove to be the difference between figuring out the root cause of the issue or not.

> **Tip**
> The underlying cause of many bugs is the incorrect passing of parameters (often an invalid or corrupted pointer). Take care – check and recheck your assumptions!

In line with our demos, this is the signature of the to-be-probed routine, do_sys_open():

```
long do_sys_open(int dfd, const char __user *filename, int
flags, umode_t mode);
```

Gaining access to its parameters within the pre-handler can be tremendously helpful! The previous section on the basics of the ABI focused on how precisely to do this. We'll end up demonstrating that we can gain access to and print the pathname to the file being opened – the second parameter, filename.

> **Jumper probes (jprobes)**
> There's a set of kernel interfaces that enable direct access to any probed function's parameters – it's called a **jumper probe (jprobe)**. However, the jprobe interfaces were deprecated in the 4.15 kernel, the rationale being that you could gain access to a probed (or traced) function's parameters in other, simpler ways – essentially by leveraging the kernel's tracing infrastructure.
>
> We do cover the basics of using the kernel tracing infrastructure to do various useful things at different points in this book. Here, look out for capturing parameters the *manual* way in the material that follows, and a much simpler automated way in the section on kernel event tracing within this chapter: *The easier way – dynamic kprobes or kprobe-based event tracing*. It's worth mentioning that, if your project or product uses a kernel version below 4.15, leveraging the jprobes interfaces can be a useful thing indeed!. Here's the kernel doc on this: https://www.kernel.org/doc/html/latest/trace/kprobes.html?highlight=kretprobes#deprecated-features.

So, let's leverage our knowledge of the processor ABI, and now, in this, our third kprobes demo, gain access to the probed function's second parameter, the file being opened. Interesting stuff, right? Read on!

Retrieving the filename

Here's a few snippets from the code of our third demo (obviously, I won't show everything here due to space constraints – please install and look it up from the book's GitHub repo). Let's pick up the action at a key portion of the code, the kernel module's pre-handler code path:

```
// ch4/kprobes/3_kprobe/3_kprobe.c
static int handler_pre(struct kprobe *p, struct pt_regs *regs)
{
    char *param_fname_reg;
```

Notice the parameters to the pre-handler:

- First, a pointer to the kprobe structure
- Second, a pointer to a structure named pt_regs

Now, this struct pt_regs structure is of interest to us: it encapsulates – obviously in an arch-specific manner – the CPU registers. Its definition is thus within an arch-specific header file. Let's consider, as an example, that you're going to run this kernel module on an ARM-32 (AArch32) based system (for example, on a Raspberry Pi 0W or the BeagleBone Black). The pt_regs structure for ARM-32 is defined here: arch/arm/include/asm/ptrace.h (and/or in arch/arm/include/uapi/asm/ptrace.h). For ARM (AArch32), the processor's CPU registers are held in the array member named uregs. The ptrace.h header has a macro:

```
#define ARM_r1       uregs[1]
```

From the ABI for ARM-32 (refer to *Table 4.1*), we know that the first four parameters (arguments) to a function are passed in CPU GPRs r0 to r3. The second parameter is thus piggy-backed into the register r1; hence, our code to gain access to it is this:

```
#ifdef CONFIG_ARM
/* ARM-32 ABI:
* First four parameters to a function are in the foll GPRs: r0,
r1, r2, r3
```

```
* See the kernel's pt_regs structure - rendition of the CPU
registers here:
https://elixir.bootlin.com/linux/v5.10.60/source/arch/arm/
include/asm/ptrace.h#L135
*/
param_fname_reg = (char __user *)regs->ARM_r1;
#endif
```

In a completely analogous fashion, for the x86 and AArch64, we use conditional compilation based on the CPU architecture to retrieve the value for the *second parameter* into our local variable param_fname_reg as follows:

```
#ifdef CONFIG_X86
    param_fname_reg = (char __user *)regs->si;
#endif
[...]
#ifdef CONFIG_ARM64
/* AArch64 ABI:
* First eight parameters to a function (and return val) are in
the foll GPRs: x0 to x7 (64-bit GPRs)
* See the kernel's pt_regs structure - rendition of the CPU
registers here:
https://elixir.bootlin.com/linux/v5.10.60/source/arch/arm64/
include/asm/ptrace.h#L173
*/
    param_fname_reg = (char __user *)regs->regs[1];
#endif
```

Clearly (as *Table 4.1* reveals), on the x86_64, the second parameter is held in the [R]SI register, and in the register X1 on the ARM64 (AArch64); our code retrieves it per the ABI!

Now it's simply a matter of emitting a printk to reveal the name of the file being opened. But hang on... the intricacies of programming in the kernel imply that you cannot simply retrieve the memory at the pointer referred to by our local variable `param_fname_reg`. Why not? Careful, it's a pointer to *userspace* memory (and we're running in kernel space), hence, we employ the `strncpy_from_user()` kernel API to bring it (copy it) into kernel memory space in our already-allocated kernel buffer `fname` (which we allocate in the module's init code path via `kzalloc()`):

```
if (!strncpy_from_user(fname, param_fname_reg, PATH_MAX))
    return -EFAULT;
pr_info("FILE being opened: reg:0x%px   fname:%s\n",
(void *)param_fname_reg, fname);
```

As an interesting aside, only when we test this kernel module on our *debug kernel* does the `strncpy_from_user()` function throw a warning printk:

BUG: sleeping function called from invalid context at lib/ strncpy_from_user.c:117

The line of code at this point (`lib/strncpy_from_user.c:117` in the 5.10.60 kernel, as seen) is the **might_fault()** function. A bit simplistically, this function checks if the kernel config CONFIG_PROVE_LOCKING or CONFIG_DEBUG_ATOMIC_SLEEP is enabled, it calls the **might_sleep()** routine; the comments for this routine (`include/ linux/kernel.h`) clearly tell the story – *it's a debug aid, checking that a sleep does not occur in any kind of atomic context*:

```
/**
 * might_sleep - annotation for functions that can sleep
 * this macro will print a stack trace if it is executed in an atomic
 * context (spinlock, irq-handler, ...). Additional sections where blocking is
 * not allowed can be annotated with non_block_start() and non_ block_end()
 * pairs.
 * This is a useful debugging help to be able to catch problems early and not be bitten later when the calling function happens to sleep when it is not supposed to.
 */
```

I've highlighted the key part of the comment. We find that both CONFIG_PROVE_
LOCKING and CONFIG_DEBUG_ATOMIC_SLEEP are enabled in our debug kernel; that's
why this warning is emitted. Well, here and now, we can't do much about it; we simply
leave it at that – a warning to be acknowledged, a "To Do" on our list.

There, it's done; the remainder of the module code is mostly identical to that of our 2_
kprobe module, so we'll skip showing it here. Let's perform a sample run by executing
our wrapper run script. As before (in the 2_kprobe demo), to cut down on the volume,
we only emit printks when the process context is vi. The final sudo dmesg from our
wrapper script reveals the kernel log buffer content. The screenshot here (*Figure 6.5*)
shows the trailing portion of the output:

```
[138698.587054] 3_kprobe:handler_pre(): 003) vi :20612   | ...1  /* handler_pre() */
[138698.588181] 3_kprobe:handler_pre(): FILE being opened: reg:0x000061bfeaedda10   fname:/etc/vim/after/syntax/sh/
[138698.590315] 3_kprobe:handler_post(): delta: 190 ns (~ 0 us ~ 0 ms)
[138698.591400] 3_kprobe:handler_pre(): 003) vi :20612   | ...1  /* handler_pre() */
[138698.592480] 3_kprobe:handler_pre(): FILE being opened: reg:0x000061bfeaedda10   fname:/var/lib/vim/addons/after/syntax/sh/
[138698.594687] 3_kprobe:handler_post(): delta: 190 ns (~ 0 us ~ 0 ms)
[138698.595773] 3_kprobe:handler_pre(): 003) vi :20612   | ...1  /* handler_pre() */
[138698.596914] 3_kprobe:handler_pre(): FILE being opened: reg:0x000061bfeaefbc80   fname:/home/letsdebug/.vim/after/syntax/sh/
[138698.599127] 3_kprobe:handler_post(): delta: 176 ns (~ 0 us ~ 0 ms)
[138700.289318] 3_kprobe:handler_pre(): 003) vi :20612   | ...1  /* handler_pre() */
[138700.292977] 3_kprobe:handler_pre(): FILE being opened: reg:0x000061bfeaed7980   fname:/home/letsdebug/.viminfo
[138700.300213] 3_kprobe:handler_post(): delta: 855 ns (~ 0 us ~ 0 ms)
[138700.303410] 3_kprobe:handler_pre(): 003) vi :20612   | ...1  /* handler_pre() */
[138700.306711] 3_kprobe:handler_pre(): FILE being opened: reg:0x000061bfeaf06640   fname:/home/letsdebug/.viminfo.tmp
[138700.313252] 3_kprobe:handler_post(): delta: 552 ns (~ 0 us ~ 0 ms)
[138700.374248] 3_kprobe:kprobe_lkm_exit(): bye, unregistering kernel probe @ 'do_sys_open'
```

Figure 4.5 – Trailing portion of the dmesg kernel log buffer output from the 3_kprobe demo on an
x86_64 VM (filtered to show only vi process context)

Look at the preceding screenshot. The pathname of the file being opened – the second
parameter to the probed function do_sys_open() – is clearly displayed!

Trying it out on a Raspberry Pi 4 (AArch64)

For a bit of variety and fun, I also ran this kernel module on a Raspberry Pi 4 running
a 64-bit Ubuntu system (thus fully configured to exploit its AArch64 – arm64 –
architecture). We build the module and then insmod it:

```
rpi4 # sudo dmesg -C; insmod ./3_kprobe.ko kprobe_func=do_sys_
open ; sleep 1 ; dmesg|tail -n5
[ 3893.514219] 3_kprobe:kprobe_lkm_init(): FYI, skip_if_not_vi
is on, verbose=0
[ 3893.525200] 3_kprobe:kprobe_lkm_init(): registering kernel
probe @ 'do_sys_open'
```

The printks clearly show that the (new) module parameter skip_if_not_vi is on by default, implying that *only* the vi process context – when it opens files – will be captured by our module. Okay, let's do an experiment: let's change it by modifying the parameter on the fly, a useful thing. First, though, don't forget to dynamically turn on all our debug prints:

```
rpi4 # echo -n "module 3_kprobe +p" > /sys/kernel/debug/
dynamic_debug/control
rpi4 # grep 3_kprobe /sys/kernel/debug/dynamic_debug/control
<...>/3_kprobe.c:98 [3_kprobe]handler_pre =p "%03d) %c%s%c:%d
| %c%c%c%u   /* %s() */\012"
<...>/3_kprobe.c:158 [3_kprobe]handler_post =p "%03d) %c%s%c:%d
| %c%c%c%u   /* %s() */\012"
rpi4 #
```

Now we query and then modify the module parameter skip_if_not_vi to the value 0:

```
rpi4 # cat /sys/module/3_kprobe/parameters/skip_if_not_vi
1
rpi4 # echo -n 0 > /sys/module/3_kprobe/parameters/skip_if_not_
vi
```

Now, *all* file open system calls are trapped via our module. The following screenshot reveals this (you can clearly see both the dmesg and the systemd-journal processes opening various files):

```
[ 4410.773412] 3_kprobe:handler_pre(): 001)  dmesg :10746   |  d..1   /* handler_pre() */
[ 4410.779891] systemd-journald[890]: /dev/kmsg buffer overrun, some messages lost.
[ 4410.787758] 3_kprobe:handler_pre(): FILE being opened: reg:0x0000aaaac84c6be8   fname:/etc/terminal-color
s.d
[ 4410.787762] 3_kprobe:handler_post(): delta: 1888 ns (~ 1 us ~ 0 ms)
[ 4410.787859] 3_kprobe:handler_pre(): 001)  dmesg :10746   |  d..1   /* handler_pre() */
[ 4410.795365] 3_kprobe:handler_pre(): 003)  systemd-journal :890   |  d..1   /* handler_pre() */
[ 4410.805236] 3_kprobe:handler_pre(): FILE being opened: reg:0x0000aaaac84c5e60   fname:/dev/kmsg
[ 4410.811591] 3_kprobe:handler_pre(): FILE being opened: reg:0x0000aaab01cb3b10   fname:/run/log/journal/be
ef23d9925c4395a56932e79c3b6d4d/system.journal
[ 4410.819616] 3_kprobe:handler_post(): delta: 2407 ns (~ 2 us ~ 0 ms)
[ 4410.857187] 3_kprobe:handler_post(): delta: 2018 ns (~ 2 us ~ 0 ms)
[ 4410.863792] 3_kprobe:handler_pre(): 003)  systemd-journal :890   |  d..1   /* handler_pre() */
[ 4410.872539] 3_kprobe:handler_pre(): FILE being opened: reg:0x0000aaab01cb3b10   fname:/run/log/journal/be
ef23d9925c4395a56932e79c3b6d4d/system.journal
[ 4410.886218] 3_kprobe:handler_post(): delta: 5260 ns (~ 5 us ~ 0 ms)
[ 4410.892820] systemd-journald[890]: /dev/kmsg buffer overrun, some messages lost.
[ 4410.900428] 3_kprobe:handler_pre(): 003)  systemd-journal :890   |  d..1   /* handler_pre() */
rpi4 #
```

Figure 4.6 – Partial screenshot showing our 3_kprobe running on a Raspberry Pi 4 (AArch64), displaying all files being opened

Good, it runs flawlessly here as well – thanks to our taking the AArch64 architecture into account in our module code (recall the #ifdef CONFIG_ARM64 ... lines within the 3_kprobe.c module code)!

Voilà! We have the names of all files being opened. Make sure you try this out yourself (at least on your x86_64 Linux VM).

Demo 4 – semi-automated static kprobe via our helper script

This time, we'll make it more interesting! A shell (bash) script (kp_load.sh) takes parameters – including the name of the function we'd like to probe and, optionally, the kernel module that contains it (if the to-be-probed function lives within a kernel module). It then generates a template for both the kernel module C code and the Makefile, enabling attaching a kprobe to a given function via a module parameter.

Due to a scarcity of space, I won't attempt to show the code of the script and the kernel module (helper_kp.c) here; just its usage. Of course, I'd expect you to browse through the code (ch4/kprobes/4_kprobe_helper) and try it out.

The helper script will first perform a few sanity checks – it first verifies that kprobes is indeed supported on the current kernel. Running it (as root) without parameters has it display its usage or help screen:

```
$ cd ch4/kprobes/4_kprobe_helper
$ sudo ./kp_load.sh
[sudo] password for letsdebug: xxxxxxxxxxxx
[+] Performing basic sanity checks for kprobes support...  OK
kp_load.sh: minimally, a function to be kprobe'd has to be
specified (via the --probe=func option)
Usage: kp_load.sh [--verbose] [--help] [--mod=module-pathname]
--probe=function-to-probe
          --probe=probe-this-function  : if module-pathname
            is not passed, then we assume the function to be
            kprobed is in the kernel itself.
        [--mod=module-pathname]        : pathname of kernel
              module that has the function-to-probe
        [--verbose]                    : run in verbose mode;
                            shows PRINT_CTX() o/p, etc
```

```
        [--showstack]                          : display kernel-mode
                                        stack, see how we got here!
        [--help]                               : show this help
                                               screen
$
```

Let's do something interesting – *probe the system's network adapter's hardware interrupt handler.* The steps that follow perform this, using our kp_load.sh helper script to actually get things done (**Platform**: Ubuntu 20.04 LTS running our custom production kernel (5.10.60-prod01) on an x86_64 guest VM):

1. Identify the network driver on the device (on my system, its the enp0s8 interface). The ethtool utility can interrogate a lot of low-level details on the network adapter. Here, we use it to query the driver that's driving the **Network Interface Card** (**NIC**) or adapter for the enp0s8 interface:

   ```
   # ethtool -i enp0s8 |grep -w driver
   driver: e1000
   ```

The -i parameter to ethtool specifies the network interface. Further, lsmod verifies the e1000 device driver is indeed present in kernel memory (as it's been configured as a module):

   ```
   # lsmod |grep -w e1000
   e1000                    135168  0
   ```

2. Find the e1000 driver's code location within the kernel and identify the hardware interrupt handler function. Most (if not all) Ethernet **Original Equipment Manufacturers'** (**OEMs'**) NIC device driver code is within the drivers/net/ethernet folder. The e1000 network driver resides here as well: drivers/net/ethernet/intel/e1000/.

Okay, here's the code that sets up the network adapter's hardware interrupt:

```
// drivers/net/ethernet/intel/e1000/e1000_main.c
static int e1000_request_irq(struct e1000_adapter
*adapter)
{
    struct net_device *netdev = adapter->netdev;
    irq_handler_t handler = e1000_intr;
    [...]
    err = request_irq(adapter->pdev->irq, handler,
                irq_flags, netdev->name, netdev);
    [...]
```

(FYI, here's the convenient link to the code online: `https://elixir.bootlin.com/linux/v5.10.60/source/drivers/net/ethernet/intel/e1000/e1000_main.c#L253`. Bootlin's online kernel code browser tooling can be a life saver!)

We can see that the **hardware interrupt (hardirq)** handler routine is named `e1000_intr()`; this is its signature:

```
static irqreturn_t e1000_intr(int irq, void *data);
```

Its code is here: `https://elixir.bootlin.com/linux/v5.10.60/source/drivers/net/ethernet/intel/e1000/e1000_main.c#L3745`). Cool!

3. Let's probe it via our helper script:

```
# ./kp_load.sh --mod=/lib/modules/5.10.60-prod01/kernel/
drivers/net/ethernet/intel/e1000/e1000.ko --probe=e1000_
intr --verbose --showstack
```

Do carefully check and note the parameters we've passed to our `kp_load.sh` helper script. It runs... In the following screenshot, you can see how our helper script performs its sanity checks, validates the function to probe (and even shows its kernel virtual address via its `/proc/kallsyms` entry). It then creates a temporary folder (`tmp/`), copies in the C LKM template file (`helper_kp.c`), renaming it appropriately within there, generates the `Makefile` file (using a shell scripting technique called a HERE document), switches to the `tmp/` folder, builds the kernel module, and then loads it into kernel memory (via `insmod`). Whew!:

```
$ ls
Readme.txt  common.sh*  err_common.sh*  helper_kp.c  kp_load.sh*
$ sudo ./kp_load.sh --mod=/lib/modules/5.10.60-prod01/kernel/drivers/net/ethernet/intel/e1000/e1000.ko --probe=e1000_in
tr --verbose --showstack
[+] Performing basic sanity checks for kprobes support...  OK

FUNCTION=e1000_intr PROBE_KERNEL=0 TARGET_MODULE=/lib/modules/5.10.60-prod01/kernel/drivers/net/ethernet/intel/e1000/e1
000.ko ; VERBOSE=1 SHOWSTACK=1
Verbose mode is on
--------------------------------------------------------------------------------------
[ Validate the to-be-kprobed function e1000_intr ]
--------------------------------------------------------------------------------------
ffffffffc00a7b20 t e1000_intr   [e1000]
Target kernel Module: /lib/modules/5.10.60-prod01/kernel/drivers/net/ethernet/intel/e1000/e1000.ko
--------------------------------------------------------------------------------------
KPMOD=helper_kp-e1000_intr-11Oct21
--- Generating tmp/Makefile -------------------------------------------------------
--- make ------------------------------------------------------------------------
make -C /lib/modules/5.10.60-prod01/build  M=/home/letsdebug/Linux-Kernel-Debugging/ch6/kprobes/4_kprobe_helper/tmp mod
ules
make[1]: Entering directory '/home/letsdebug/lkd_kernels/productionk/linux-5.10.60'
--- Dynamic Makefile for helper_kprobes util ---
Building with KERNELRELEASE =
  CC [M]  /home/letsdebug/Linux-Kernel-Debugging/ch6/kprobes/4_kprobe_helper/tmp/helper_kp-e1000_intr-11Oct21.o
/home/letsdebug/Linux-Kernel-Debugging/ch6/kprobes/4_kprobe_helper/tmp/helper_kp-e1000_intr-11Oct21.c:61:12: warning: '
running_avg' defined but not used [-Wunused-variable]
   61 | static int running_avg=0;
      |
--- Dynamic Makefile for helper_kprobes util ---
Building with KERNELRELEASE =
  MODPOST /home/letsdebug/Linux-Kernel-Debugging/ch6/kprobes/4_kprobe_helper/tmp/Module.symvers
  CC [M]  /home/letsdebug/Linux-Kernel-Debugging/ch6/kprobes/4_kprobe_helper/tmp/helper_kp-e1000_intr-11Oct21.mod.o
  LD [M]  /home/letsdebug/Linux-Kernel-Debugging/ch6/kprobes/4_kprobe_helper/tmp/helper_kp-e1000_intr-11Oct21.ko
make[1]: Leaving directory '/home/letsdebug/lkd_kernels/productionk/linux-5.10.60'
-rw-r--r-- 1 root root 14640 Oct 11 10:32 helper_kp-e1000_intr-11Oct21.ko
--------------------------------------------------------------------------------------
 kernel module helper_kp-e1000_intr-11Oct21 is already inserted... proceeding...
/sbin/insmod ./helper_kp-e1000_intr-11Oct21.ko funcname=e1000_intr verbose=1 show_stack=1
$
$ journalctl -k > myklog
$ sudo rmmod helper_kp-e1000_intr-11Oct21
$ █
```

Figure 4.7 – Screenshot showing the kp_load.sh helper script executing and loading up the custom kprobe LKM

4. I save the kernel log to a file (`journalctl -k > myklog`), remove the LKM from kernel memory, and open the log file in the vi editor; the output is pretty large. Here's a partial screenshot (*Figure 4.8*), capturing our custom kprobe's pre-handler routine's printk, the output from our `PRINT_CTX()` macro, and mostly, the output from the `dump_stack()` routine! The last two lines of output are from the kprobe's post handler routine:

```
24872 Oct 11 06:52:03 dbg-LKD kernel: delta: 44593120 ns (~ 44593 us ~ 44 ms)
24873 Oct 11 06:52:03 dbg-LKD kernel: helper_kp_e1000_intr_11Oct21:handler_pre():Pre 'e1000 intr'.
24874 Oct 11 06:52:03 dbg-LKD kernel: 003) [kworker/3:3]:2086   | d.h1   /* handler_pre() */
24875 Oct 11 06:52:03 dbg-LKD kernel: CPU: 3 PID: 2086 Comm: kworker/3:3 Tainted: G          OE     5.10.60-prod01 #4
24876 Oct 11 06:52:03 dbg-LKD kernel: Hardware name: innotek GmbH VirtualBox/VirtualBox, BIOS VirtualBox 12/01/2006
24877 Oct 11 06:52:03 dbg-LKD kernel: Workqueue: events e1000_watchdog [e1000]
24878 Oct 11 06:52:03 dbg-LKD kernel: Call Trace:
24879 Oct 11 06:52:03 dbg-LKD kernel: <IRQ>
24880 Oct 11 06:52:03 dbg-LKD kernel: dump_stack+0x76/0x94
24881 Oct 11 06:52:03 dbg-LKD kernel: ? e1000_intr+0x1/0x110 [e1000]
24882 Oct 11 06:52:03 dbg-LKD kernel: handler_pre.cold+0x5/0xc4a [helper_kp_e1000_intr_11Oct21]
24883 Oct 11 06:52:03 dbg-LKD kernel: kprobe_ftrace_handler+0xf2/0x160
24884 Oct 11 06:52:03 dbg-LKD kernel: ? __handle_irq_event_percpu+0x45/0x1c0
24885 Oct 11 06:52:03 dbg-LKD kernel: ftrace_ops_assist_func+0x98/0x140
24886 Oct 11 06:52:03 dbg-LKD kernel: 0xffffffffc050e0e3
24887 Oct 11 06:52:03 dbg-LKD kernel: RIP: 0010:e1000_intr+0x1/0x110 [e1000]
24888 Oct 11 06:52:03 dbg-LKD kernel: Code: c2 77 bf 48 8b 87 80 03 00 00 f0 80 a0 90 00 00 00 fe 45 31 c0 83 87 48 0b 00 00 01 e
      b a4 66 66 2e 0f 1f 84 00 00 00 00 00 e8 <db> 64 46 00 48 8b 86 80 00 00 00 8b 80 c0 00 00 00 85 c0 0f 84 b2
24889 Oct 11 06:52:03 dbg-LKD kernel: RSP: 0018:ffffb63b80148f28 EFLAGS: 00000046 ORIG_RAX: 0000000000000000
24890 Oct 11 06:52:03 dbg-LKD kernel: RAX: ffffffffc00a7b20 RBX: ffff9b25e526e000 RCX: 0000000000000000
24891 Oct 11 06:52:03 dbg-LKD kernel: RDX: 0000000000010001 RSI: ffff9b25c5092000 RDI: 0000000000000010
24892 Oct 11 06:52:03 dbg-LKD kernel: RBP: ffffb63b80148f60 R08: ffff9b25f78da400 R09: 0000000000000000
24893 Oct 11 06:52:03 dbg-LKD kernel: R10: 0000000000000000 R11: 0000000000000000 R12: 0000000000000000
24894 Oct 11 06:52:03 dbg-LKD kernel: R13: ffffb63b80148f74 R14: 0000000000000010 R15: 0000000000000000
24895 Oct 11 06:52:03 dbg-LKD kernel: ? e1000_maybe_stop_tx+0x90/0x90 [e1000]
24896 Oct 11 06:52:03 dbg-LKD kernel: ? e1000_intr+0x5/0x110 [e1000]
24897 Oct 11 06:52:03 dbg-LKD kernel: ? __handle_irq_event_percpu+0x45/0x1c0
24898 Oct 11 06:52:03 dbg-LKD kernel: ? e1000_intr+0x5/0x110 [e1000]
24899 Oct 11 06:52:03 dbg-LKD kernel: ? __handle_irq_event_percpu+0x45/0x1c0
24900 Oct 11 06:52:03 dbg-LKD kernel: handle_irq_event_percpu+0x33/0x90
24901 Oct 11 06:52:03 dbg-LKD kernel: handle_irq_event+0x39/0x60
24902 Oct 11 06:52:03 dbg-LKD kernel: handle_fasteoi_irq+0xc5/0x1a0
24903 Oct 11 06:52:03 dbg-LKD kernel: ? handle_nested_irq+0x110/0x110
24904 Oct 11 06:52:03 dbg-LKD kernel: asm_call_irq_on_stack+0x12/0x20
24905 Oct 11 06:52:03 dbg-LKD kernel: </IRQ>
24906 Oct 11 06:52:03 dbg-LKD kernel: common_interrupt+0x136/0x1d0
24907 Oct 11 06:52:03 dbg-LKD kernel: asm_common_interrupt+0x1e/0x40
24908 Oct 11 06:52:03 dbg-LKD kernel: RIP: 0010:e1000_watchdog+0x19d/0x590 [e1000]
24909 Oct 11 06:52:03 dbg-LKD kernel: Code: 39 f0 0f 82 fa 01 00 00 41 83 bc 24 f8 fb ff ff 04 0f 87 51 01 00 00 49 8b 94 24 e0 f
      b ff ff b8 10 00 00 00 89 82 c8 00 00 00 <41> c6 84 24 f1 f9 ff ff 01 49 8b 44 24 c8 a8 04 0f 84 e6 01 00 00
24910 Oct 11 06:52:03 dbg-LKD kernel: RSP: 0018:ffffb63b81b67e20 EFLAGS: 00000297
24911 Oct 11 06:52:03 dbg-LKD kernel: RAX: 0000000000000010 RBX: ffff9b25c5092000 RCX: 0000000000000100
24912 Oct 11 06:52:03 dbg-LKD kernel: RDX: ffffb63b82160000 RSI: 0000000000000100 RDI: ffff9b25c5092d80
24913 Oct 11 06:52:03 dbg-LKD kernel: RBP: ffffb63b81b67e58 R08: ffff9b25c50931a8 R09: ffff9b263ddab9e0
24914 Oct 11 06:52:03 dbg-LKD kernel: R10: ffff9b25e45c366c R11: 0000000000000018 R12: ffff9b25c50931a0
24915 Oct 11 06:52:03 dbg-LKD kernel: R13: ffff9b25f662ad00 R14: ffff9b25c5092d80 R15: ffff9b25c5092900
24916 Oct 11 06:52:03 dbg-LKD kernel: process_one_work+0x1b8/0x3b0
24917 Oct 11 06:52:03 dbg-LKD kernel: worker_thread+0x50/0x3a0
24918 Oct 11 06:52:03 dbg-LKD kernel: ? process_one_work+0x3b0/0x3b0
24919 Oct 11 06:52:03 dbg-LKD kernel: kthread+0x154/0x180
24920 Oct 11 06:52:03 dbg-LKD kernel: ? kthread_unpark+0x80/0x80
24921 Oct 11 06:52:03 dbg-LKD kernel: ret_from_fork+0x22/0x30
24922 Oct 11 06:52:03 dbg-LKD kernel: helper_kp_e1000_intr_11Oct21:handler_post():kworker/3:3:2086. Post 'e1000_intr'.
24923 Oct 11 06:52:03 dbg-LKD kernel: delta: 25862601 ns (~ 25862 us ~ 25 ms)
                                                                        24922,1         20%
```

Figure 4.8 – (Partial) screenshot showing output from the kernel log as emitted by our helper script's custom kprobe within the pre-handler routine; the last two lines are from the post handler

Interesting! Our custom autogenerated kprobe has achieved this!

Don't fret regarding how exactly to interpret the kernel-mode stack right now; we shall cover all this in detail in the coming chapters. For now, I'll point out the following key things with regard to *Figure 4.8* (ignoring the line number and the first five columns on the left):

- Line `24873`: Output from our custom generated kprobe: as the `verbose` flag is set, a debug printk shows the call site – `helper_kp_e1000_intr_11Oct21:handler_pre():Pre 'e1000_intr'`.

- Line `24874`: Output from our `PRINT_CTX()` macro – `003) [kworker/3:3]:2086    |  d.h1  /* handler_pre() */`. The four-character `d.h1` sequence is interpreted per *Figure 4.4*: hardware interrupts are disabled (off), we're currently running in the hardirq context (of course we are, the probe is on the NIC's interrupt handler) and a (spin)lock is currently held.

- Lines `24875` to `24921`: Output from the `dump_stack()` routine; useful information indeed! For now, just read it bottom-up, ignoring all lines that begin with a `?`. Well, one key point: in this particular case, do you notice that there are actually *two* kernel-mode stacks on display here?

  - The upper portion is within the `<IRQ>` and `</IRQ>` tokens. This tells us it's the IRQ stack – a special stack region used to hold stack frames when a hardware interrupt is being processed (this is an arch-specific feature known as **interrupt** (or **IRQ**) stacks; most modern processors use it).

  - The lower portion of the stack, after the `</IRQ>`, is the regular kernel-mode stack. It's typically the (kernel) stack of the process context that happened to get rudely interrupted by the hardware interrupt (here, it happens to be a kernel thread named `kworker/3:3`).

Interpreting kthread names

By the way, how do you interpret kernel thread names (such as the `kworker/3:3` kthread seen here)? They're essentially cast in this format: `kworker/%u:%d[%s]  (kworker/cpu:id[priority])`.

Refer to this link for more details: `https://www.kernel.org/doc/Documentation/kernel-per-CPU-kthreads.txt`.

Nice, using the helper script does make things easier. There's a price to pay, of course – there's always a trade-off (as with life): our `helper_kp.c` LKM's C code template remains hardcoded for any and every probe we set up using it.

Now you know how to code static kprobes; more so, how you can leverage this technology to help you carefully instrument – and thus debug – kernel/module code, even on production systems! The other side of the coin is the kretprobe. Let's jump into learning how to use it.

Getting started with kretprobes

At the outset of this chapter, you learned how to use the basic kprobes APIs to set up a static kprobe (or two). Let's now cover an interesting counterpart to the kprobe – the **kretprobe**, *allowing us to gain access to any (well, most) kernel or module function's return value!* This – being able to dynamically look up a given function's return value – can be a gamechanger in a debug scenario.

> **Pro Dev Tip**
>
> **Don't assume**: If a function returns a value, always check for the failure case. One day it could fail – yes, even the `malloc()` or the `kmalloc()` APIs! Fail to catch the possible failure and you'll be flailing to figure out what happened!

The relevant kretprobe APIs are straightforward:

```
#include <linux/kprobes.h>
int register_kretprobe(struct kretprobe *rp);
void unregister_kretprobe(struct kretprobe *rp);
```

The `register_kretprobe()` function returns 0 on success and, in the usual kernel style (the 0/-E convention), a negative `errno` value on failure.

> **Tip – errno Values and Their Meaning**
>
> As you'll know, `errno` is an integer found in every process's uninitialized data segment (more recently, it's constructed to be *thread-safe* by employing the powerful **Thread Local Storage (TLS)** Pthreads feature, implemented via the compiler and the usage of the `__thread` keyword in the variable declaration). When a system call fails (typically returning −1), the programmer can query the error diagnostic by looking up `errno`. The kernel (or underlying driver) will return the appropriate negative `errno` integer. glibc glue code will set it to positive by multiplying it by −1. It serves as an index into a 2D array of English error messages, which can be conveniently looked up via the `[p]error(3)` or `strerror(3)` glibc APIs.
>
> I often find it useful to be able to quickly look up a given `errno` value. Use the userspace headers `/usr/include/asm-generic/errno-base.h` (covers `errno` values 1 to 34) and `/usr/include/asm-generic/errno.h` (covers `errno` values 35 to 133), as of this writing.
>
> For example, if you notice that a kernel/module function's return value is −101 in a log file, the corresponding positive `errno` value can be looked up; here, it's: `#define ENETUNREACH 101 /* Network is unreachable */`

The `kretprobe` structure internally contains the `kprobe` structure, allowing you to set up the probe point (the function to *return* the probe) via it. In effect, the probe point will be `rp->kp.addr` (where `rp` is the pointer to the `kretprobe` structure, kp the pointer to the `kprobe` structure, with the address being typically figured out via `rp->kp.symbol_name` – set to the name of the function to be probed). The `rp->handler` is the kretprobe handler function; its signature is this:

```
int kretprobe_handler(struct kretprobe_instance *ri, struct
pt_regs *regs);
```

Just as with kprobes, you will receive all CPU registers within the handler function via the second parameter, the `pt_regs` structure. The first parameter, the `kretprobe_instance` structure, holds (among other housekeeping fields), the following:

- `ri->ret_addr`: The return address
- `ri->task`: The pointer to the process context's task structure (which encapsulates all attributes of the executing task)
- `ri->data`: A means to gain access to a per-instance private data item

But what about the main feature, the return value from the probed function? Ah, recall our discussion on the processor ABI (in the *Understanding the basics of the Application Binary Interface (ABI) section*): the return value is again placed into a processor register, the particular register being of course very arch-specific. *Table 4.1* shows you the relevant details. But hang on, you don't have to manually look it up. There's an elegant, simpler way – a macro:

```
regs_return_value(regs);
```

This macro is a hardware-agnostic abstraction, separately defined for each processor family, that provides the return value from the appropriate register (the registers being passed via `struct pt_regs *regs` of course)! For example, here's the essential implementation of `regs_return_value()` on the following architectures (CPUs):

- ARM (AArch32) is: `return regs->ARM_r0;`
- A64 (AArch64) is: `regs->regs[0]`
- x86 is: `return regs->ax;`

It just works.

The kernel community has provided sample source code for some select kernel features; this includes the kprobe and kretprobe. Here are some relevant snippets from the sample code for the kretprobe (found here within the kernel code base: `samples/kprobes/kretprobe_example.c`) via the following bullet points:

- A module parameter `func` enables us to pass any function to probe (ultimately, for obtaining its return value, the whole point here):

  ```
  static char func_name[NAME_MAX] = "kernel_clone";
  module_param_string(func, func_name, NAME_MAX, S_IRUGO);
  MODULE_PARM_DESC(func, "Function to kretprobe; this
  module will report the function's execution time");
  ```

- The kretprobe structure definition:

  ```
  static struct kretprobe my_kretprobe = {
      .handler        = ret_handler,
      .entry_handler  = entry_handler,
      .data_size      = sizeof(struct my_data),
      /* Probe up to 20 instances concurrently. */
      .maxactive      = 20,
  };
  ```

Let's now delve into this `kretprobe` structure:

- `handler`: The member that specifies the function to run when the function we're probing completes, enabling us to fetch the return value; it's the *return handler*.

- `entry_handler`: The member that specifies the function to run when the function we're probing is entered; it gives you a chance to determine whether the return will be collected:

  - If you return 0 (implying success), the return function – `handler` – will be called upon the return of the probed function.

  - If you return non-zero, the k[ret]probe does not even happen; in effect, it gets disabled for this particular function instance (the official kernel doc here gives you the in-depth details on the entry handler and private data fields: `https://www.kernel.org/doc/html/latest/trace/kprobes.html?highlight=kretprobes#kretprobe-entry-handler`).

- `maxactive`: Used to specify how many instances of the probed function can be simultaneously probed; the default is the number of CPU cores on the box (`NR_CPUS`). If the `nmissed` field of the kretprobe structure is positive, it implies you missed that many instances (you can then increase `maxactive`). Again, the kernel doc here gives you the in-depth details: `https://www.kernel.org/doc/html/latest/trace/kprobes.html?highlight=kretprobes#how-does-a-return-probe-work`.

- In the module initialization code path, plant the return probe:

```
my_kretprobe.kp.symbol_name = func_name;
ret = register_kretprobe(&my_kretprobe);
```

- Here's the actual return handler code (skipping details):

```
static int ret_handler(struct kretprobe_instance *ri,
struct pt_regs *regs)
{
    unsigned long retval = regs_return_value(regs);
    struct my_data *data = (struct my_data *)ri->data;
    [...]
    delta = ktime_to_ns(ktime_sub(now, data->entry_
stamp));
    pr_info("%s returned %lu and took %lld ns to
execute\n", func_name, retval, (long long)delta);
```

```
        return 0;
    }
```

I've highlighted (in bold) the key lines – where the return address is obtained and printed.

- In the module cleanup code path, the kretprobe is unregistered (and the missed instances count is displayed):

```
unregister_kretprobe(&my_kretprobe);
pr_info("kretprobe at %p unregistered\n", my_kretprobe.
kp.addr);
/* nmissed > 0 suggests that maxactive was set too low.
*/
pr_info("Missed probing %d instances of %s\n", my_
kretprobe.nmissed, my_kretprobe.kp.symbol_name);
```

Do try it out...

Kprobes miscellany

A couple of remaining things to mention while on this topic of k[ret]probes:

- One, you can even set up multiple kprobes or kretprobes with a single API call, as follows:

```
#include <linux/kprobes.h>
int register_kprobes(struct kprobe **kps, int num);
int register_kretprobes(struct kretprobe **rps, int num);
```

As you might expect, these are convenience wrappers calling the underlying registration routine in a loop. The unregister_k[ret]probes() routine counterparts are used to unregister the probes. We won't delve further into those here.

- Two, a kprobe or kretprobe can be temporarily disabled via the following:

```
int disable_kprobe(struct kprobe *kp);
int disable_kretprobe(struct kretprobe *rp);
```

And later re-enabled via the corresponding and analogous enable_k[ret]probe() APIs. This can be useful: a way to throttle the amount of debug telemetry being logged.

> **Inner Workings**
>
> If you'd like to delve into the inner workings of how kprobes and kretprobes are internally implemented, the official kernel documentation covers it here: `https://www.kernel.org/doc/Documentation/kprobes.txt`. Check out the *Concepts: Kprobes and Return Probes* section.

Now that you know how to use both kprobes and kretprobes, it's also important to understand that they have some inherent limitations, even downsides. The following section covers just this.

Kprobes – limitations and downsides

We do realize that no single feature can do anything and everything – in the words of Frederick J Brooks (in his incomparable book *The Mythical Man Month*): *"there is no silver bullet"*.

As we've seen, certain kernel/module functions cannot be probed, including the following:

- Functions marked with the `__kprobes` or `nokprobe_inline` annotation.
- Functions marked via the `NOKPROBE_SYMBOL()` macro.
- The pseudofile `/sys/kernel/debug/kprobes/blacklist` holds the names of functions that can't be probed. (Incidentally, our `ch4/kprobes/4_kprobe_helper/kp_load.sh` script checks this against the function attempting to be probed). Also, some inline functions might not be able to be probed.

There's more to note on the point of using k[ret]probes on production systems due to the possibility of stability issues; the next section throws some light on this.

Interface stability

We know that kernel APIs can change at any point (this is a given with kernel development and maintenance in any case). So, you can imagine a situation where your kernel module sets up kprobes for some functions, say, `x()` and `y()`. In a later kernel release though, there's no telling what will happen – these functions might be deprecated, or their signatures (and thus parameters and return type) might change, leading to your k[ret]probe kernel module requiring constant maintenance. (Honestly, this is pretty much a given.)

A bit more on this last point. An important word of caution: it can be dangerous – from a stability and security viewpoint – to include third-party kernel modules on production systems, especially (and obviously), on mission-critical ones. Their presence can also void the warranty given by the OS vendors (such as Red Hat, **SUSE Linux Enterprise Server (SLES)**, Canonical, and so on). DevOps people are in general extremely wary of letting untested code into production systems, let alone kernel modules; they won't exactly be thrilled when you insert them.

Also, kprobes can cause kernel instability when attached to high-volume code paths (such as scheduling, interrupt/timer, or networking code). Avoid them, if possible. If not, at least mitigate the risk by reducing printk usage and using printk rate-limiting APIs (we covered rate-limiting the printk in the previous chapter).

How will I know if a certain kernel (or module) function runs very often? The **funccount** utility – via either the `perf-tools[-unstable]` package or the more recent *eBPF* tools packages – can profile and show you high-volume code paths within the kernel. (The utility script is typically named `funccount-perf` or `funccount-bpfcc`, depending on what you have installed.)

A modern, cleaner, and far more efficient approach to the static kprobe (or kretprobe) is to employ tracing mechanisms that are already built into the kernel fabric, and are thus tested and production-ready. These include using dynamic kprobes or kprobe-based event tracing (frontends such as `kprobe-perf` take advantage of these), kernel trace points (provided via ftrace), perf and eBPF frontends, and so on. They're also simply a lot easier to use; they don't require C coding and deep knowledge of kernel internals, and they're DevOps/sysad friendly as well! Let's get started exploring them!

The easier way – dynamic kprobes or kprobe-based event tracing

Similar, but much superior, to how I built a small script in demo 4, to make it easier for us to hook into any kernel function via kprobes, there is a package called `perf-tools` (or `perf-tools-unstable`). The creator and lead author is Brendan Gregg. Within the useful tools you'll find in this package, a bash script named `kprobe` (or `kprobe-perf`) is a fantastic wrapper, easily letting us set up kprobes (and kretprobes)!

Assuming you've installed the package (we specified it back in *Chapter 1, A General Introduction to Debugging Software*), let's go ahead and verify it's there and then run the script (by the way, on my x86_64 Ubuntu 20.04 LTS system, the package name is `perf-tools-unstable` and the script is called `kprobe-perf`):

```
# dpkg -l|grep perf-tools
ii   perf-tools-unstable    1.0.1~20200130+git49b8cdf-1ubuntu1
all       DTrace-like tools for Linux
# file $(which kprobe-perf)
/usr/sbin/kprobe-perf: Bourne-Again shell script, ASCII text
executable
#
```

Great; let's run it and see its help screen (you will eventually need to run it as root, so I just do so here):

```
# kprobe-perf
USAGE: kprobe [-FhHsv] [-d secs] [-p PID] [-L TID] kprobe_definition [filter]
                    -F              # force. trace despite warnings.
                    -d seconds      # trace duration, and use buffers
                    -p PID          # PID to match on events
                    -L TID          # thread id to match on events
                    -v              # view format file (don't trace)
                    -H              # include column headers
                    -s              # show kernel stack traces
                    -h              # this usage message

Note that these examples may need modification to match your kernel
version's function names and platform's register usage.
    eg,
        kprobe p:do_sys_open
                                # trace open() entry
        kprobe r:do_sys_open
                                # trace open() return
        kprobe 'r:do_sys_open $retval'
                                # trace open() return value
        kprobe 'r:myopen do_sys_open $retval'
                                # use a custom probe name
        kprobe 'p:myopen do_sys_open mode=%cx:u16'
                                # trace open() file mode
        kprobe 'p:myopen do_sys_open filename=+0(%si):string'
                                # trace open() with filename
        kprobe -s 'p:myprobe tcp_retransmit_skb'
                                # show kernel stacks
        kprobe 'p:do_sys_open file=+0(%si):string' 'file ~ "*stat"'
                                # opened files ending in "stat"

See the man page and example file for more info.
#
```

Figure 4.9 – Screenshot showing the help screen of the kprobe-perf script

The help screen does a great job of summing up how you can use this useful utility; do refer to its man page (as well as an online examples page here: `https://github.com/brendangregg/perf-tools/blob/master/examples/kprobe_example.txt`).

I recommend you first try out a few examples similar to those shown in *Figure 4.9*.

Next, and without further ado, let's leverage this powerful script to very easily do what we so painstakingly progressed to in the earlier four demos: set up a kprobe on `do_sys_open()` and print the pathname of the file being opened (as our earlier example in *Figure 4.5* shows!):

```
# kprobe-perf 'p:do_sys_open file=+0(%si):string'
Tracing kprobe do_sys_open. Ctrl-C to end.
      kprobe-perf-8171    [002] ...1  9159.540104: do_sys_open:
(do_sys_open+0x0/0xf0) file="/etc/ld.so.cache"
      kprobe-perf-8171    [002] ...1  9159.540259: do_sys_open:
(do_sys_open+0x0/0xf0) file="/lib/x86_64-linux-gnu/libc.so.6"
      kprobe-perf-8171    [002] ...1  9159.542030: do_sys_open:
(do_sys_open+0x0/0xf0) file=(fault)
      kprobe-perf-8171    [002] ...1  9159.542818: do_sys_open:
(do_sys_open+0x0/0xf0) file="trace_pipe"
       irqbalance-676     [000] ...1  9162.010699: do_sys_open:
(do_sys_open+0x0/0xf0) file="/proc/interrupts"
       irqbalance-676     [000] ...1  9162.011642: do_sys_open:
(do_sys_open+0x0/0xf0) file="/proc/stat"
[...]^C
```

Notice the syntax:

- The `'p:do_sys_open'` sets up a kprobe on the `do_sys_open()` kernel function.

- On the x86_64, the ABI tells us that the `[R]SI` register holds the second parameter to the function (recall *Table 4.1*) – in this case, it's the pathname of the file being opened. The script employs the syntax `+0(%si):string` to display its content as a string (prefixed with `file=`).

As easy as that! To minimally test it, I ran ps in another terminal window; immediately the kprobe-perf script dumped lines like this:

```
ps-8172     [000] ...1  9164.231685: do_sys_open: (do_sys_
open+0x0/0xf0) file="/lib/x86_64-linux-gnu/libdl.so.2"
ps-8172     [000] ...1  9164.232582: do_sys_open: (do_sys_
open+0x0/0xf0) file="/lib/x86_64-linux-gnu/libc.so.6"
ps-8172     [000] ...1  9164.233758: do_sys_open: (do_sys_
open+0x0/0xf0) file="/lib/x86_64-linux-gnu/libsystemd.so.0"
ps-8172     [000] ...1  9164.234776: do_sys_open: (do_sys_
open+0x0/0xf0) file="/lib/x86_64-linux-gnu/librt.so.1"
[...]
ps-8172     [000] ...1  9164.248680: do_sys_open: (do_sys_
open+0x0/0xf0) file="/proc/meminfo"
ps-8172     [000] ...1  9164.249511: do_sys_open: (do_sys_
open+0x0/0xf0) file="/proc"
ps-8172     [000] ...1  9164.260290: do_sys_open: (do_sys_
open+0x0/0xf0) file="/proc/1/stat"
ps-8172     [000] ...1  9164.260854: do_sys_open: (do_sys_
open+0x0/0xf0) file="/proc/1/status"
[...]
```

...and plenty more... You can literally gain insight into how ps works by doing this! (In fact, the wonderful strace utility – *it traces all system calls issued by a process* – can approach this level of detail as well! Don't ignore it.)

The point here of course is to simply show you how much easier it is to get the same valuable information – internally leveraging the kernel's kprobes framework and knowledge of the processor ABI – using this tool.

Further, the output format that the kprobe-perf script uses is as follows:

```
#                                  _-----=> irqs-off
#                                 / _----=> need-resched
#                                | / _---=> hardirq/softirq
#                                || / _--=> preempt-depth
#                                ||| /     delay
#           TASK-PID    CPU#    ||||    TIMESTAMP  FUNCTION
#             | |         |     ||||       |          |
           ps-8172     [000]  ...1  9164.260854: do_sys_open:
(do_sys_open+0x0/0xf0) file="/proc/1/status"
```

It's familiar, with good reason: it's again that of ftrace, and very similar to what we did with our PRINT_CTX() macro (recall *Figure 4.4*).

As you probably guessed, the kprobe-perf script, to get the job done, somehow sets up a kprobe; this is indeed easily verified by looking up the kprobes/list pseudo-file under your debugfs mount point. While the preceding command was running, I ran this in another terminal window:

```
# cat /sys/kernel/debug/kprobes/list
ffffffff965d1a60  k  do_sys_open+0x0     [FTRACE]
```

Clearly, a kprobe was set up on the do_sys_open() kernel function.

Kprobe-based event tracing – minimal internal details

So, how does the kprobe-perf script set up a kprobe? Ah, here's the really interesting thing: it does so by leveraging the kernel's *ftrace* infrastructure, which internally tracks key events within the kernel. This is known as the kernel's **event tracing** framework, and within it, the **kprobes events** framework. It can be considered to be a subset of the larger ftrace kernel system; that's why you see [FTRACE] to the right of the kprobes/list line! (We cover ftrace in depth in *Chapter 9, Tracing the Kernel Flow*.)

The kprobe events code was introduced into the kernel back in 2009 by Masami Hiramatsu. Essentially, via it, the kernel can toggle the tracing of select (with a few limitations) kernel functions.

Internally speaking, here's the bare minimum information on how a kprobe is set up: within the debugs tracing folder (typically here: /sys/kernel/debug/tracing/), there will exist a directory named events (this is assuming the kernel config CONFIG_KPROBE_EVENTS=y; it typically is, even on distro and many production kernels). Under it, there are folders representing various subsystems and/or well-known event classes that the kernel's event tracing infrastructure tracks.

Using the event tracing framework to trace built-in functions

The kernel's **event tracing** infrastructure also mirrors these tracepoints at this location: /sys/kernel/tracing. This can be particularly useful when, on a production system, debugfs is kept invisible (as a security measure).

Let's peek into it:

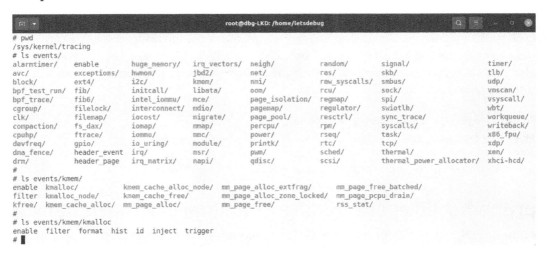

Figure 4.10 – Screenshot showing the kernel's event tracing - (pseudo) files and folders under /sys/kernel/tracing/events

All right, a huge number of event classes and subsystems can be easily and readily traced!

Take, for example, the common kernel memory allocator routine, the really popular kmalloc() slab API. In *Figure 4.10*, you can see the pseudo-files corresponding to tracing kmalloc() at the bottom (within the events/kmem/kmalloc directory).

Here, the format pseudofile really has the details on what gets reported (and how it's internally looked up); it essentially represents a structure that the kernel maintains and has the ability to look up. (Running kprobe-perf with the -v option switch will show you this format file and won't perform tracing.)

Writing 1 to the enable pseudo-file enables tracing and it runs under the hood. You can see the resulting output by reading the pseudo-file named /sys/kernel/[debug]/tracing/trace (or trace-pipe; reading from trace_pipe keeps a *watch* on the file, similar to doing a tail -f on a file – useful indeed).

Let's try this out. We'll give it a quick spin (here, on a Raspberry Pi 0W):

```
rpi # pwd
/sys/kernel/debug/tracing
rpi # cat events/kmem/kmalloc/enable
0
rpi # echo 1 > events/kmem/kmalloc/enable
rpi #
rpi # cat trace_pipe
          sshd-680     [000] ....   700.723280: kmalloc: call_site=__alloc_skb+0x70/0x164 ptr=236acdbb byte
s_req=576 bytes_alloc=1024 gfp_flags=GFP_KERNEL|__GFP_NOWARN|__GFP_NOMEMALLOC
          sshd-680     [000] ....   700.723391: kmalloc: call_site=pskb_expand_head+0x70/0x33c ptr=f5e025aa
 bytes_req=1024 bytes_alloc=1024 gfp_flags=GFP_ATOMIC|__GFP_NOWARN|__GFP_NOMEMALLOC
     kworker/u2:1-56   [000] ....   700.723674: kmalloc: call_site=__alloc_skb+0x70/0x164 ptr=a25030bf byte
s_req=352 bytes_alloc=512 gfp_flags=GFP_ATOMIC|__GFP_NOWARN|__GFP_NOMEMALLOC
     kworker/u2:1-56   [000] ....   700.725507: kmalloc: call_site=__alloc_skb+0x70/0x164 ptr=a25030bf byte
s_req=352 bytes_alloc=512 gfp_flags=GFP_ATOMIC|__GFP_NOWARN|__GFP_NOMEMALLOC
     kworker/u2:1-56   [000] ....   700.725607: kmalloc: call_site=__alloc_skb+0x70/0x164 ptr=a25030bf byte
s_req=384 bytes_alloc=512 gfp_flags=GFP_ATOMIC|__GFP_NOWARN|__GFP_NOMEMALLOC
```

Figure 4.11 – Truncated screenshot showing an example of easily tracing the kmalloc() routine

Voilà; and so easily achieved! Every single `kmalloc()` invocation – invoked by either the kernel or a module – is traced. The precise printk format specifier that details the `kmalloc` information (the content you see from `kmalloc: call site=...` onwards) is specified by the `events/kmem/kmalloc/format` pseudo-file.

Once done, turn the probe off with the following:

```
rpi # echo 0 > events/kmem/kmalloc/enable
```

And empty the kernel trace buffer with the following:

```
rpi # echo > trace
```

(By the way, event tracing via the `enable` pseudo-file is just one way to use the kernel's powerful event tracing framework; do refer to the official kernel documentation for more here: *Event Tracing*: https://www.kernel.org/doc/html/latest/trace/events.html#event-tracing.)

So, think about this: *Figure 4.10* shows us the automatically available tracepoints that the kernel makes available – in effect, the built-in kernel tracepoints. But what if you need to trace a function that isn't within there (in effect, that's not present under /sys/kernel/[debug]/tracing/events)? Well, there's always a way – the coverage of the next section!

Setting up a dynamic kprobe (via kprobe events) on any function

To set up a kprobe dynamically on *any* given kernel (or module) function (with a few exceptions, as mentioned in the *Kprobes – limitations and downsides section*), let's learn how to employ the kernel's dynamic event tracing framework and what's christened the *function-based kprobes* feature.

You should realize, though, that the kprobe can only be set up if the function to be probed is either of the following:

- Present in the kernel global symbol table (can be seen via `/proc/kallsyms`)
- Present within the ftrace framework's available function list (here: `<debugs_ mount>/tracing/available_filter_functions`)

What if the function to be probed is within a kernel module? That's no problem: once the module is loaded into kernel memory, the internal machinery will ensure that all symbols are by now part of the kernel's symbol table, and will thus be visible within `/proc/ kallsyms`; view it as root, of course. In fact, the section that follows shows precisely this.

To set up a *dynamic kprobe*, do as follows:

1. Initialize the dynamic probe point:

    ```
    # cd /sys/kernel/debug/tracing
    ```

    If, for whatever reason, this doesn't work – typically if debugfs is invisible on a production kernel (or ftrace is disabled), then change it to the following:

    ```
    # cd /sys/kernel/tracing
    ```

    Then set up the dynamic kprobe as follows:

    ```
    echo "p:<kprobe-name> <function-to-kprobe> [...]" >>
    kprobe_events
    ```

    The p: specifies you're setting up a (dynamic) kprobe. The name following the : character is any name you wish to give this probe (it will default to the function name if you don't pass anything). Then, put a space and the actual function to probe. Optional arguments can be used to specify more stuff – querying the probed function's parameter values being typical! We'll learn more as we go along...

> **Tip**
> On production systems that are configured with the kernel config option
> `CONFIG_DEBUG_FS_DISALLOW_MOUNT=y` – rendering the debugfs
> filesystem effectively invisible – the debugfs filesystem won't even have a mount
> point. In cases like this, make use of the `/sys/kernel/tracing` location
> (as shown earlier) and perform the dynamic kprobe work from therein.

Let's set this up with our usual example. Set up a simple kprobe (with no additional
info, such as open file pathname, generated) on the `do_sys_open()` function:

```
echo "p:my_sys_open do_sys_open" >> kprobe_events
```

Now that it's set up, under the `/sys/kernel/[debug]/tracing/events`
folder, you will now find a (pseudo) folder named `kprobes`; it – `/sys/kernel/`
`[debug]/tracing/events/kprobes/` – *will contain any and all dynamic*
kprobes that have been defined:

```
# ls -lR events/kprobes/
events/kprobes/:
total 0
drwxr-xr-x 2 root root 0 Oct  9 18:58 my_sys_open/
-rw-r--r-- 1 root root 0 Oct  9 18:58 enable
-rw-r--r-- 1 root root 0 Oct  9 18:58 filter
events/kprobes/my_sys_open:
total 0
-rw-r--r-- 1 root root 0 Oct  9 18:59 enable
-rw-r--r-- 1 root root 0 Oct  9 18:58 filter
-r--r--r-- 1 root root 0 Oct  9 18:58 format
[...]
```

2. The probe is disabled by default; let's enable it (as root):

    ```
    echo 1 > events/kprobes/my_sys_open/enable
    ```

    Now it's enabled and running, you can look up the trace data by simply doing the
    following:

    ```
    cat trace
    [...]
                cat-192796  [001] .... 392192.698410: my_
    sys_open: (do_sys_open+0x0/0x80) file="/usr/lib/locale/
    locale-archive"
    ```

```
          cat-192796   [001] .... 392192.698650: my_
sys_open: (do_sys_open+0x0/0x80) file="trace"
      gnome-shell-7441        [005] .... 392192.777608: my_
sys_open: (do_sys_open+0x0/0x80) file="/sys/class/net/
wlo1/statistics/rx_packets"
[...]
```

Doing a cat trace_pipe allows you to *watch* the file, feeding data as it becomes available – a very useful thing while using dynamic kprobe events interactively. Or, you can perhaps do something like this to save it to a file:

```
cp /sys/kernel/tracing/trace /tmp/mytrc.txt
```

3. To finish, first write 0 to the enable file to disable the kprobe and then do this to destroy it:

```
echo 0 > events/kprobes/my_sys_open/enable
echo "-: <kprobe-name>" >> kprobe_events
```

Alternatively, doing the following:

```
echo > /sys/kernel/tracing/kprobe_events
```

… clears *all* probe points.

So, here, let's disable and destroy our custom dynamic kprobe my_sys_open as follows:

```
echo 0 > events/kprobes/do_sys_open/enable
echo "-:my_sys_open" >> kprobe_events
```

Once all dynamic probe points (kprobes) are destroyed, the /sys/kernel/ [debug]/tracing/events/kprobe_events pseudo-file itself disappears.

Further, doing echo > trace empties the kernel trace buffer of all its trace data.

Even more in-depth details on how to use this powerful dynamic kprobes-based event tracing are beyond the scope of this book; I refer you to the excellent kernel documentation here: *Kprobe-based Event Tracing*: https://www.kernel.org/doc/html/latest/trace/kprobetrace.html#kprobe-based-event-tracing.

It's also very educative to read through the source of the kprobe-perf script itself: https://github.com/brendangregg/perf-tools/blob/master/kernel/kprobe.

Taking care not to overflow or overwhelm

Do keep this in mind though! Just as mentioned with regard to our manual usage of kprobes, the `kprobe-perf` script has a similar warning within it:

```
# WARNING: This uses dynamic tracing of kernel functions, and
could cause kernel panics or freezes, depending on the function
traced. Test in a lab environment, and know what you are doing,
before use.
```

Try and mitigate this by only tracing precisely what's required for as small a time window as is feasible. The `kprobe-perf` script's `-d` option – duration specifier – is useful in this regard. It has the kernel internally buffer the output into a per-CPU buffer; the size is fixed via `/sys/kernel/[debug]/tracing/buffer_size_kb`. If you still get overflows, try increasing its size.

Trying this on an ARM system

As with the x86, doing `echo "p:my_sys_open do_sys_open"` > `/sys/kernel/debug/tracing/kprobe_events` (as root) on an ARM system will work of course... But what if we'd like to display the filename parameter to the open system call as well? Well, we (think we) know how to, so let's try it out:

```
# echo "p:my_sys_open do_sys_open file=+0(%si):string" > /sys/
kernel/debug/tracing/kprobe_events
bash: echo: write error: Invalid argument
```

Whoops; why did it fail?

It should be obvious: the register holding the second argument – the pathname of the file being opened – is named `[R]SI` on the x86[_64], but not on the ARM processor! On ARM-32 the first four parameters to a function are piggy-backed on CPU registers `r0`, `r1`, `r2`, and `r3` (again, do refer to *Table 4.1*). So, take this arch-dependence into account:

```
echo "p:my_sys_open do_sys_open file=+0(%r1):string" > /sys/
kernel/debug/tracing/kprobe_events
```

Now it will work!

We can go further, printing out all arguments to the open call:

```
echo 'p:my_sys_open do_sys_open dfd=%r0 file=+0(%r1):string
flags=%r2 mode=%r3' > /sys/kernel/debug/tracing/kprobe_events
```

(Don't forget to enable the probe.)

It is even simpler to do with the wrapper kprobe[-perf] script (you need the perf-tools[-unstable] package installed):

```
rpi # kprobe-perf 'p:my_sys_open do_sys_open dfd=%r0
file=+0(%r1):string flags=%r2 mode=%r3'
Tracing kprobe my_sys_open. Ctrl-C to end.
        cat-1866     [000] d...  8803.206194: my_sys_open:
(do_sys_open+0x0/0xd8) dfd=0xffffff9c file="/etc/ld.so.preload"
flags=0xa0000 mode=0x0
        cat-1866     [000] d...  8803.206548: my_sys_open:
(do_sys_open+0x0/0xd8) dfd=0xffffff9c file="/usr/lib/arm-linux-
gnueabihf/libarmmem-v6l.so" flags=0xa0000 mode=0x0
        cat-1866     [000] d...  8803.207085: my_sys_open:
(do_sys_open+0x0/0xd8) dfd=0xffffff9c file="/etc/ld.so.cache"
flags=0xa0000 mode=0x0
        cat-1866     [000] d...  8803.207235: my_sys_open:
(do_sys_open+0x0/0xd8) dfd=0xffffff9c file="/lib/arm-linux-
gnueabihf/libc.so.6" flags=0xa0000 mode=0x0
        cat-1866     [000] d...  8803.209703: my_sys_open:
(do_sys_open+0x0/0xd8) dfd=0xffffff9c file="/usr/lib/locale/
locale-archive" flags=0xa0000 mode=0x0
        cat-1866     [000] d...  8803.210395: my_sys_
open: (do_sys_open+0x0/0xd8) dfd=0xffffff9c file="trace_pipe"
flags=0x20000 mode=0x0
^C
Ending tracing...
rpi #
```

Interesting, right? Do try it out yourself.

Exercise

Set up a kprobe to trigger whenever an interrupt handler's tasklet (bottom half) routine is scheduled to execute. Also display the kernel mode stack leading up to this point.

One solution

With traditional IRQ handling (top/bottom halves, rather than the modern thread-based IRQ handling), the top half runs with all interrupts disabled across all CPUs (guaranteeing it runs atomically) while the bottom half – the tasklet – runs with all interrupts enabled on all processors. In this context, this is what typically occurs. A driver author, within the hardware interrupt handler (the so-called top half) typically requests the kernel to schedule its tasklet by invoking the kernel API `schedule_tasklet()`. Let's look up its underlying kernel implementation:

```
# grep tasklet_schedule /sys/kernel/debug/tracing/available_
filter_functions
__tasklet_schedule_common
__tasklet_schedule
```

Okay, this tells us that we should set up a dynamic kprobe on the function named `__tasklet_schedule()`. Further, we pass the `-s` option switch to `kprobe-perf`, asking it to also provide a (kernel-mode) *stack trace* – in effect, telling us how exactly each instance of this function was invoked! This can be really useful when debugging:

```
# kprobe-perf -s 'p:mytasklets __tasklet_schedule'
Tracing kprobe mytasklets. Ctrl-C to end.
     kworker/0:0-1855    [000] d.h.  9909.886809: mytasklets:
(__tasklet_schedule+0x0/0x28)
     kworker/0:0-1855    [000] d.h.  9909.886829: <stack trace>
 => __tasklet_schedule
 => bcm2835_mmc_irq
 => __handle_irq_event_percpu
 => handle_irq_event_percpu
 => handle_irq_event
 => handle_level_irq
 => generic_handle_irq
 => __handle_domain_irq
 => bcm2835_handle_irq
 => __irq_svc
 => bcm2835_mmc_request
 => __mmc_start_request
 => mmc_start_request
 => mmc_wait_for_req
```

```
=> mmc_wait_for_cmd
=> mmc_io_rw_direct_host
=> mmc_io_rw_direct
=> process_sdio_pending_irqs
=> sdio_irq_work
=> process_one_work
=> worker_thread
=> kthread
=> ret_from_fork
[...]
```

Figure 4.4 helps us interpret the output of the kworker... lines: we can see from the d.h. four-character sequence that interrupts are currently disabled (off) and a hardirq (a hardware interrupt handler) is running.

The remaining output – the kernel mode stack content at the time (the upper portion being the *IRQ stack*) shows us how this particular interrupt came up and how it ended up running a tasklet (which itself internally becomes a softirq of type TASKLET_SOFTIRQ). Further, the stack trace (*always read it bottom up*) shows that this interrupt is likely generated by I/O being performed on the **Secure Digital MultiMedia Card (SD MMC)** card.

(Again folks, FYI, the in-depth details regarding interrupts and their handling are covered in my earlier, freely available e-book, *Linux Kernel Programming – Part 2*.)

Using dynamic kprobe event tracing on a kernel module

Note that we're trying this out on our custom production kernel (on our x86_64 Ubuntu guest) to mimic a production environment:

1. First, load up our test driver, the miscdrv_rdwr kernel module from *Chapter 3, Debug via Instrumentation – printk and Friends*:

    ```
    $ cd <lkd-src-tree>/ch3/miscdrv_rdwr
    $ ../../lkm
    Usage: lkm name-of-kernel-module-file (without the .c)
    $ ../../lkm miscdrv_rdwr
    Version info:
    Distro:    Ubuntu 20.04.3 LTS
    ```

```
Kernel: 5.10.60-prod01
[...]
sudo dmesg
-------------------------------
[ 1987.178246] miscdrv_rdwr:miscdrv_rdwr_init(): LLKD
misc driver (major # 10) registered, minor# = 58, dev
node is /dev/llkd_miscdrv_rdwr
$
```

2. A quick grep shows that its symbols are now part of the kernel global symbol table (as expected, even on our production kernel):

```
$ sudo grep miscdrv /proc/kallsyms
ffffffffc0562000 t write_miscdrv_rdwr [miscdrv_rdwr]
ffffffffc0562982 t write_miscdrv_rdwr.cold     [miscdrv_
rdwr]
ffffffffc0562290 t open_miscdrv_rdwr  [miscdrv_rdwr]
ffffffffc0562480 t close_miscdrv_rdwr [miscdrv_rdwr]
ffffffffc0562650 t read_miscdrv_rdwr  [miscdrv_rdwr]
ffffffffc05629b5 t read_miscdrv_rdwr.cold      [miscdrv_
rdwr]
[...]
```

> **The .cold Compiler Attribute**
>
> By the way, why are some functions suffixed with `.cold`? The short answer is that it's a compiler attribute specifying that the function is unlikely to be executed. These so-called cold functions are typically placed in a separate linker section to improve the code locality of the required-to-be-fast non-cold sections! *It's all about optimization.* Also notice that, above, there's both the normal version and the cold version of some functions (the read and write I/O routines of our driver).

3. In another terminal window, let's set up a dynamic kprobe on our `write_miscdrv_rdwr()` module function, as root:

```
cd /sys/kernel/tracing
echo "p:mymiscdrv_wr write_miscdrv_rdwr" >> kprobe_events
# cat kprobe_events
p:kprobes/mymiscdrv_wr write_miscdrv_rdwr
#
```

We give the probed function a name, `mymiscdrv_wr`. Now enable it:

echo 1 > events/kprobes/mymiscdrv_wr/enable

4. Run a test:

A. In one terminal window, within the tracing folder (`/sys/kernel/tracing`), run the following:

cat trace_pipe

B. In another terminal window, we run our userspace program to write to our misc class driver's device file. This will guarantee our probe point (the `write_miscdrv_rdwr()` module function) gets invoked:

$ ./rdwr_test_secret w /dev/llkd_miscdrv_rdwr "dyn kprobes event tracing is awesome"

The userspace process executes, writing into our device driver. The following screenshot shows both the execution of this userspace process as well as the dynamic kprobe being set up and traced:

Figure 4.12 – Screenshot showing testing a dynamic kprobe via the kprobe events framework on a device driver module function

Study the preceding screenshot carefully. The terminal window at the bottom is where we set up our dynamic probe (corresponding to steps 3 and 4A above). The terminal window on top is where we test by invoking the write functionality of our driver (twice; in effect, this corresponds to step 4B). You can see how, in the lower terminal, the dynamic kprobe is set up and enabled. Then it *watches* for trace data by doing a `cat` on the `trace_pipe` file. When data becomes available, we see it...

Disable the probe point and destroy it with the following:

```
# echo 0 > events/kprobes/mymiscdrv_wr/enable
# echo "-:mymiscdrv_wr" >> kprobe_events
# cat kprobe_events
# echo > trace
```

The last command empties out the kernel `trace` buffer.

In effect, you should by now realize that what we've done in this section is pretty much automated by the `kprobe-perf` bash script! It even has other interesting options to try. This makes it a powerful weapon in our debug/observability armory!

Before concluding this section, it's good to know that even userspace application processes can be traced via the kernel's dynamic event tracing framework – this feature is called **Uprobes** (as opposed to kprobes). I refer you to the official kernel documentation on it here: *Uprobe-tracer: Uprobe-based Event Tracing*: `https://www.kernel.org/doc/html/latest/trace/uprobetracer.html#uprobe-tracer-uprobe-based-event-tracing`.

Setting up a return probe (kretprobe) with kprobe-perf

With the `kprobe-perf` wrapper script, you can set up a return probe – a kretprobe – as well! Using it is simplicity itself. Here's our usual example, fetching the return value of the `do_sys_open()` kernel function:

```
rpi # kprobe-perf 'r:do_sys_open ret=$retval'
Tracing kprobe do_sys_open. Ctrl-C to end.
    kprobe-perf-2287    [000] d... 13013.021003: do_sys_open:
(sys_openat+0x1c/0x20 <- do_sys_open) ret=0x3
        <...>-2289    [000] d... 13013.027167: do_sys_open:
(sys_openat+0x1c/0x20 <- do_sys_open) ret=0x3
        <...>-2289    [000] d... 13013.027504: do_sys_open:
(sys_openat+0x1c/0x20 <- do_sys_open) ret=0x3
```

```
^C
Ending tracing...
rpi #
```

The key point here is the return value being fetched; it shows up as the following:

```
ret=0x3
```

This makes sense; the return to the open API is the file descriptor assigned within the process context's open file table. Here, it happens to be the value 3 (with 0, 1, and 2 typically being taken up by `stdin`, `stdout`, and `stderr`).

Next, let's examine the notation:

```
do_sys_open: (sys_openat+0x1c/0x20 <- do_sys_open)
```

This implies that our probed function `do_sys_open()` has been called by and is returning to the `sys_openat()` function. Further, the notation `<func>+0x1c/0x20` following the function name, or generically, the `<func>+off/len`, is interpreted as follows:

- `off`: The **offset** within the function `<func>` where the code returns
- `len`: What the kernel feels is the overall **length** of the function `<func>` (it's an approximation, usually correct)

The remainder of the output is in the usual ftrace format notation that you should be familiar with by now...

Before concluding this section, we'll mention that even more can be achieved by leveraging this powerful function-based dynamic kprobes framework within the kernel. Steven Rostedt's slides show how you can burrow ever deeper and extract pretty much any arguments to a function being probed and indeed, delve into relevant kernel structures (via offsets) to reveal their runtime values (`https://events19.linuxfoundation.org/wp-content/uploads/2017/12/oss-eu-2018-fun-with-dynamic-trace-events_steven-rostedt.pdf`). Do check it out.

Well, well, we're almost done! Let's complete this chapter with one more section where you'll learn something quite practical – briefly understanding and tracing the execution of processes on the system. This can help as an audit-like facility, allowing you to log whatever userspace processes executed.

Trapping into the execve() API – via perf and eBPF tooling

On Linux (and UNIX), user mode applications – processes – are launched or executed via a family of so-called exec C library (glibc) APIs: `execl()`, `execlp()`, `execv()`, `execvp()`, `execle()`, `execvpe()`, and `execve()`.

A quick couple of things to know about these seven APIs: the first six are merely glibc wrappers that transform their arguments and ultimately invoke the `execve()` API – *it is the actual system call*, the one that causes the process context to switch to kernel mode and run the kernel code corresponding to the system call. Also, FYI, `execvpe()` is a GNU extension (and thus practically only seen on Linux).

The point here is simply this: ultimately, pretty much all processes (and thus apps) are executed via the kernel code of `execve()`! Within the kernel, `execve()` becomes the `sys_execve()` function (in a bit of an indirect fashion, via the `SYSCALL_DEFINE3()` macro), which invokes the actual worker routine, `do_execve()`.

System calls and where they land in the kernel

This, in fact, is typical of many (but not all) system calls: the user-issued system call `foo()` becomes `sys_foo()`, which, if short enough performs the work itself, else invokes the actual worker routine, `do_foo()`.

For example, the `execve(2)` system call becomes `fs/exec.c:sys_execve()` in the kernel (technically via the `SYSCALL_DEFINE3()` macro, 3 being the number of parameters passed via the syscall), which in turn invokes the worker function `fs/exec.c:do_execve()`.

Caution though, this isn't always the case... For example, the `open(2)` system call's code path within the kernel's a bit different; the following screenshot sums this up:

```
Generically:
foo()  ------|------>        sys_foo() → do_foo()


getpgid() ---|------> sys_getpgid() → do_getpgid()

open() ------|------>        sys_open() → do_sys_open() → do_sys_openat2() → do_filp_open() → …

execve() ----|------> sys_execve() → do_execve()

User mode    |                           Kernel mode
```

Figure 4.13 – How user mode system calls map within the kernel

An aside, but a useful one: how does a non-privileged user mode task (a process or thread) actually manage to cross the boundary from user mode into privileged kernel mode (depicted by the vertical red line in *Figure 4.13*)? The short answer is that every processor supports one or more machine instructions that allow this to happen – these are often referred to as **call gates** or **traps** (we say that the process *traps* from user mode to kernel mode).

For example, the x86 traditionally used software interrupt `int 0x80` to perform the trap; modern versions use the `syscall` machine instruction; ARM-32 uses the `SWI` (software interrupt) machine instruction; and AArch64 (ARM64) uses the `SVC` (supervisor) instruction to do so. See the man page on `syscall(2)` for more details.

Again, FYI, there is an alternate almost equivalent system call to `execve()` – `execveat()`. The difference being that the first parameter to it is a directory relative to which the program – the second parameter – is executed.

Let's return to the main point: now that we know that processes are executed via `execve()`, won't it be cool to trap into it – perhaps via injecting a kprobe into the `sys_execve()` or `do_execve()` kernel APIs? Yes, but... (there's always a *but* isn't there?): on modern kernels, *it simply doesn't work* – trying it via the static kprobe approach, the `register_kprobe()` fails. Please try it yourself. Remember, always the empirical approach!

In fact, on my x86_64 Ubuntu 20.04 LTS VM, even the `execsnoop-perf(8)` wrapper tool, built for precisely this purpose, (which internally uses the kernel's ftrace `kprobe_events` pseudo-file), fails:

```
$ sudo execsnoop-perf
Tracing exec()s. Ctrl-C to end.
ERROR: adding a kprobe for execve. Exiting.
```

The more recent eBPF tooling solves this once and for all; install and employ (as root) `execsnoop-bpfcc(8)` and it just works! The following section has us peeking into the exec via an eBPF frontend.

Observability with eBPF tools – an introduction

An extension of the well-known **Berkeley Packet Filter** or BPF, **eBPF** is the **extended BPF**. Very briefly, BPF used to provide the supporting infrastructure within the kernel to effectively trace network packets. eBPF is a relatively recent kernel innovation – available only from the Linux 4.1 kernel (June 2015) onward. It extends the BPF notion, allowing you to trace much more than just the network stack. Also, it works for tracing both kernel space as well as userspace apps. In effect, *eBPF and its frontends are the modern approach to tracing and performance analysis on a Linux system.*

To use eBPF, you will need a system with the following:

- Linux kernel 4.1 or newer
- Kernel support for eBPF (do see `https://github.com/iovisor/bcc/blob/master/INSTALL.md#kernel-configuration`)

Using the eBPF kernel feature directly is considered to be very hard, so there are several easier *frontends* to use. Among them, the **BPF Compiler Collection** (**BCC**), **bpftrace**, and **libbpf+BPF CO-RE** (**Compile Once – Run Everywhere**) are regarded as being very useful. It's really simplest to install the `bcc` binary packages for these frontends. You'll find instructions regarding this here: `https://github.com/iovisor/bcc/blob/master/INSTALL.md#packages`.

Check out the following link to a picture that opens your eyes to just how many powerful BCC/BPF tools are available to help trace different Linux subsystems and hardware: `https://www.brendangregg.com/BPF/bcc_tracing_tools_early2019.png`.

Here, we don't intend to delve into details; instead, we'll give you a quick flavor of using a BCC frontend utility to track processes being executed. To try this, I assume you've installed the BCC frontend package (we did this back in *Chapter 1, A General Introduction to Debugging Software*). *Quick tip*: on Ubuntu, do `sudo apt install bpfcc-tools`, but do see the following callout.

> **eBPF BCC Installation**
>
> You can install the BCC tools package for your regular host Linux distro
> by reading the installation instructions here: `https://github.com/`
> `iovisor/bcc/blob/master/INSTALL.md`. Sometimes, though
> (especially on older distros, such as Ubuntu 18.04), this approach of installing
> the `bpfcc-tools` package will typically work only on a pre-built Linux
> distro (such as Ubuntu/Debian/RedHat, and so on) *but may not on a Linux that
> has a custom kernel*. The reason: the installation of the BCC toolset includes
> (and depends upon) the installation of the `linux-headers-$(uname`
> `-r)` package. This `linux-headers` package exists only for distro kernels
> (and not for our custom 5.10 kernel that we shall often be running on the
> guest). With Ubuntu 20.04 LTS, it *does* seem to work, even when running a
> custom kernel.

Once the `bpfcc-tools` package is installed, you can get a listing (and feel) for all the
frontend utilities by doing the following:

```
dpkg -L bpfcc-tools |grep "^/usr/sbin.*bpfcc$"
```

On my x86_64 Ubuntu 20.04 LTS guest VM (running our custom `5.10.60-prod01`
kernel), I find there are 112 `*-bpfcc` utilities installed (they're actually Python scripts).

In the section just prior to this one, we saw that the `execve()` (or `execveat()`) system
call is the one that actually executes processes. We attempted to trace its execution via the
`perf-tools` utilities (`execsnoop-perf`), but it just failed. Now, with the eBPF BCC
frontends installed, let's retry:

```
$ uname -r
5.10.60-prod01
$ sudo execsnoop-bpfcc 2>/dev/null
[...]
PCOMM            PID     PPID    RET ARGS
id               7147    7053      0 /usr/bin/id -u
id               7148    7053      0 /usr/bin/id -u
git              7149    7053      0 /usr/bin/git config --global
credential.helper cache --timeout 36000
cut              7151    7053      0 /usr/bin/cut -d= -f2
grep             7150    7053      0 /usr/bin/grep --color=auto
^PRETTY_NAME /etc/os-release
cat              7152    7053      0 /usr/bin/cat /proc/version
```

```
ip                 7157    7053     0 /usr/bin/ip a
sudo               7159    7053     0 /usr/bin/sudo route -n
route              7160    7159     0 /usr/sbin/route -n
[...]
```

It just works as processes get executed. The `execsnoop-bpfcc` script displays a line of output showing a few details regarding the process that just executed. Notice how all parameters to the command being executed are displayed as well! The help screen is definitely worth looking up (just run it with the -h option switch). The man pages should be installed as well. Both have one-liner example usage; do check it out.

As with the `perf-tools` utilities, all the *-bpfcc scripts need to be run as root. A fair amount of noise can be generated initially. We defeat it by redirecting `stderr` to the null device.

Our good old example – trapping into and tracing do_sys_open() – right from the beginning of this chapter, can, once again, be very easily achieved with BCC:

```
$ sudo opensnoop-bpfcc 2>/dev/null
PID     COMM                FD ERR PATH
1431    upowerd              9   0 /sys/devices/LNXSYSTM:00/
LNXSYBUS:00/PNP0A03:00/PNP0C0A:00/power_supply/BAT0/voltage_now
1431    upowerd              9   0 /sys/devices/LNXSYSTM:00/
LNXSYBUS:00/PNP0A03:00/PNP0C0A:00/power_supply/BAT0/capacity
1431    upowerd             -1   2 /sys/devices/LNXSYSTM:00/
LNXSYBUS:00/PNP0A03:00/PNP0C0A:00/power_supply/BAT0/temp
[...]
431     systemd-udevd       14   0 /sys/fs/cgroup/unified/system.
slice/systemd-udevd.service/cgroup.procs
431     systemd-udevd       14   0 /sys/fs/cgroup/unified/system.
slice/systemd-udevd.service/cgroup.threads
[...] ^C
```

Again, Brendan Gregg's page on eBPF tracing tools (https://www.brendangregg.com/ebpf.html) will help you see the depth of tools available and how to begin making use of them.

Summary

In this chapter, you learned what kprobes and kretprobes are and how to exploit them to add useful telemetry (instrumentation) to your project or product in a dynamic fashion. We saw that you can even use them on production systems (though you should be careful to not overload the system).

We first covered the traditional static approach to using k[ret]probes, one where any change will require a recompile of the code; we even provided a semi-automated script to generate a kprobe as required. We then covered the better, efficient, dynamic kprobe tracing facilities that are built into modern Linux kernels. Using these techniques is not only a lot easier but has other advantages – they're pretty much always built into the kernel, no new code is required at the last minute on production systems, and running them is more efficient under the hood. As a bonus, you learned how to leverage the kernel's ftrace-based event tracepoints – a large number of kernel subsystems and their APIs can be very easily traced.

We finished this large-ish chapter by delving a bit into a practical consideration – how to trace the execution of a process (as an example). You found that tracing or tracking process execution, the opening of files (and in a similar fashion, most other things), can be very easily done via the modern eBPF tooling (`bpfcc-tools` BCC frontends) and, to some extent, via the `perf-tools` frontend.

Just of June 2022, the very recent 5.18 kernel has a new feature: fprobes. The fprobe is similar in intent to the k[ret]probe, but faster and is based on ftrace (`https://www.kernel.org/doc/html/latest/trace/fprobe.html#fprobe-function-entry-exit-probe`).

The next chapter is bound to be very useful; we'll delve into kernel memory issues and how to find and debug them! I highly recommend you first take the time to practice (do the suggested exercises mentioned during the course of this chapter), get comfortable with the content of this and earlier chapters, and then, after a quick break, jump into the next one!

Further reading

- Official kernel documentation: *Kernel Probes (Kprobes)*: `https://www.kernel.org/doc/html/latest/trace/kprobes.html#kernel-probes-kprobes`

- *[Kernel] Kprobe*, Brian Pan, November 2020: `https://ppan-brian.medium.com/kernel-kprobe-5036d7a8455f`

- Kprobes via modern ftrace tracing, kprobe events:

  - *Taming Tracepoints in the Linux Kernel*, Keenan, Mar 2020: `https://blogs.oracle.com/linux/post/taming-tracepoints-in-the-linux-kernel`

  - *Fun with Dynamic Kernel Tracing Events, The things you just shouldn't be able to do!* Steven Rostedt, Oct 2018: `https://events19.linuxfoundation.org/wp-content/uploads/2017/12/oss-eu-2018-fun-with-dynamic-trace-events_steven-rostedt.pdf`

  - *Dynamic tracing in Linux user and kernel space, Pratyush Anand, July 2017*: `https://opensource.com/article/17/7/dynamic-tracing-linux-user-and-kernel-space` (includes coverage on userspace probing with uprobe as well)

- Brendan Gregg's perf-tools page: `https://github.com/brendangregg/perf-tools`

- Specific to kprobes: kprobes-perf examples: `https://github.com/brendangregg/perf-tools/blob/master/examples/kprobe_example.txt`

- Specific to kprobes: kprobes-perf and related tooling code: `https://github.com/brendangregg/perf-tools/tree/master/kernel`

- *Traps, Handlers* (x86 specific): `https://www.cse.iitd.ernet.in/~sbansal/os/lec/18.html`

- CPU ABI, function calling, and register usage conventions:

  - *APPLICATION BINARY INTERFACE (ABI) DOCS AND THEIR MEANING*: `https://kaiwantech.wordpress.com/2018/05/07/application-binary-interface-abi-docs-and-their-meaning/`

  - X86_64:

    - *x64 Cheat Sheet*: `https://cs.brown.edu/courses/cs033/docs/guides/x64_cheatsheet.pdf`

    - *X86 64 Register and Instruction Quick Start*: `https://wiki.cdot.senecacollege.ca/wiki/X86_64_Register_and_Instruction_Quick_Start`

    - ARM32 / Aarch32: *Overview of ARM32 ABI Conventions*, Microsoft, July 2018: `https://docs.microsoft.com/en-us/cpp/build/overview-of-arm-abi-conventions?view=msvc-160`

- ◆ ARM64 / Aarch64:

- ◆ ARMv8-A64-bit Android on ARM, Campus London, September 2015, Architecture Overview presentation: `https://armkeil.blob.core.windows.net/developer/Files/pdf/graphics-and-multimedia/ARMv8_Overview.pdf` (do check out the *ARMv8 terminology reference* on page 32)

- ◆ Overview of ARM64 ABI conventions, Microsoft, Mar 2019: `https://docs.microsoft.com/en-us/cpp/build/arm64-windows-abi-conventions?view=msvc-160`

- ◆ ARM Cortex-A Series Programmer's Guide for ARMv8-A / Fundamentals-of-ARMv8: `https://developer.arm.com/documentation/den0024/a/Fundamentals-of-ARMv8`

- ◆ ARMv8 Registers: `https://developer.arm.com/documentation/den0024/a/ARMv8-Registers`

- ◆ ARM64 Reversing and Exploitation Part 1 - ARM Instruction Set + Simple Heap Overflow, Sept 2020: `http://highaltitudehacks.com/2020/09/05/arm64-reversing-and-exploitation-part-1-arm-instruction-set-heap-overflow/`

- *How Linux kprobes works*, Dec 2016: `https://vjordan.info/log/fpga/how-linux-kprobes-works.html`

- eBPF:

  - ▪ Installing eBPF: `https://github.com/iovisor/bcc/blob/master/INSTALL.md`

  - ▪ BCC tutorial: `https://github.com/iovisor/bcc/blob/master/docs/tutorial.md`

  - ▪ *Linux Extended BPF (eBPF) Tracing Tools*, Brendan Gregg (see the pics as well!): `https://www.brendangregg.com/ebpf.html`

  - ▪ *How eBPF Turns Linux into a Programmable Kernel*, Jackson, October 2020: `https://thenewstack.io/how-ebpf-turnslinux-into-a-programmable-kernel/`

  - ▪ *A Gentle Introduction to eBPF, InfoQ, May 2021*: `https://www.infoq.com/articles/gentle-linux-ebpf-introduction/`

- (Kernel-level) *A thorough introduction to eBPF*, Matt Fleming, LWN, December 2017: `https://lwn.net/Articles/740157/`

- *How io_uring and eBPF Will Revolutionize Programming in Linux*, Glauber Costa, April 2020: `https://thenewstack.io/how-io_uring-and-ebpf-will-revolutionizeprogramming-in-linux/`

- Miscellaneous:

  - Old but interesting, mostly on using SystemTap: *Locating System Problems Using Dynamic Instrumentation*, Prasad, Cohen, et al, 2005: `https://sourceware.org/systemtap/systemtap-ols.pdf`

  - *Different Approaches to Linux Host Monitoring*, Kelly Shortridge, capsule8: `https://capsule8.com/blog/different-approaches-to-linux-monitoring/`

5
Debugging Kernel Memory Issues – Part 1

There's no doubt about it, C (and C++) is a really powerful programming language, one that allows the developer to straddle both high-level layered abstractions (after all, object-oriented languages such as Java and Python are written in C) as well as to work upon the bare metal, as it were. This is fantastic. Of course, there's a price to pay: the compiler will do only so much. You want to overflow a memory buffer? Go ahead, it doesn't care. Want to peek at or poke an unmapped memory region? No problem.

Well, no problem for the compiler, but big problems for us! This is nothing new really. We mentioned just this in *Chapter 2, Approaches to Kernel Debugging*. C being a procedural and non-managed programming language (in memory terms), it's ultimately the programmer's responsibility to ensure that runtime memory usage is correct and well behaved.

The Linux kernel is almost entirely written in C (over 98% of the code is in C, as of the time of this writing). You see the potential for problems, right? (In fact, there's a slowly growing effort to begin porting the kernel, or portions of it, to a more memory-safe language such as **Rust**. See the *Further reading* section for links on this). In a similar vein, compilers are getting smarter. The **Clang/Low Level Virtual Machine (LLVM)** compiler – with which you can certainly build the kernel and modules – seems superior to the well-known **GNU Compiler Collection** or **GCC** compiler in terms of intelligent code generation, avoiding **Out Of Bounds (OOB)** accesses, and more. We cover some introductory material on using Clang as well here, though the focus is on the most commonly used GCC compiler. Here, we'll attempt to tackle this all-too-common and stubborn bug source – memory issues! **The goal, after all, is to make your code memory safe.**

Due to the vast scope of material to be covered on kernel memory debugging, we've split the discussion into two chapters, this one and the next.

In this chapter, we shall focus on and cover the following main topics (look out for detailed coverage of the kernel's SLUB debug framework and catching memory leakage in the next one):

- What's the problem with memory anyway?
- Using KASAN and UBSAN to find memory bugs
- Building your kernel and modules with Clang
- Catching memory defects in the kernel – comparison and notes (Part 1)

Technical requirements

The technical requirements and workspace remain identical to what's described in *Chapter 1, A General Introduction to Debugging Software*. The code examples can be found within the book's GitHub repository here: `https://github.com/PacktPublishing/Linux-Kernel-Debugging`. The only thing new in terms of software installation is the usage of the powerful Clang compiler. We cover the details in the *Building your kernel and modules with Clang* section.

What's the problem with memory anyway?

The introductory paragraphs at the start of this chapter informed you of the annoying fact that though programming in C is like having a superpower (at least for your typical OS/driver/embedded domains), it's a double-edged sword: we humans inadvertently create defects and bugs. Memory bugs, especially, are simply all too common.

In fact, in *Chapter 2, Approaches to Kernel Debugging*, in the *Types of bugs – the memory view* section, we mentioned that among the different ways of classifying bug types is the *memory view*. For easy recollection – and to stress its importance here – I reproduce the short list of common memory corruption bug types:

- Incorrect memory accesses:

    - Using variables uninitialized, aka **Uninitalized Memory Read (UMR)** bugs

    - **Out-Of-Bounds (OOB)** memory accesses (read/write underflow/overflow bugs)

    - **Use-After-Free (UAF)** and **Use-After-Return (UAR)** (aka out-of-scope) bugs

    - Double-free bugs

- Memory leakage

- Data races

- (Internal) Fragmentation

These (except the last) are among the well-understood **Undefined Behavior (UB)** issues that a process, or even the OS, can blunder into. In this chapter, you'll learn about these issues – with the emphasis being within the kernel/driver code – and, more importantly, how to use various tools and approaches to catch them.

More precisely, within this chapter, we shall focus on the first two: **incorrect memory accesses** – which include all kinds of common memory bugs: UMR, OOB, UAF/UAR, and double-free. In the following chapter, we'll focus on catching memory defects in slab memory via the SLUB debug framework as well as detecting **memory leaks**. We'll cover data races and their complexities in *Chapter 8, Lock Debugging*, (as they are most commonly caused by incorrectly working with locks), and (internal) fragmentation, or wastage, will be mentioned in the next chapter, in the *Learning to use the slabinfo and related utilities* section.

> **It's Not Only about Bugs but Also about Security**
>
> Human error and C (and C++) create an unfortunate mix at times – **bugs**! But – and here's a key point – *security issues very often tend to be bugs or defects at heart*. This is why getting it right in the first place, and/or later hunting down and fixing bugs, is so critical to today's modern production systems and, indeed, the cloud (a huge portion of which is powered via the Linux kernel and its built-in hypervisor component – **Kernel Virtual Machine (KVM)**). Hackers currently have a pretty wide choice of OS-level exploits to choose from; this is especially true for older kernels. To see what I mean, take a peek here: `https://github.com/xairy/linux-kernel-exploitation`.
>
> If nothing else, remember: unless you're running the latest stable kernel (which will have the latest bugfix and security patches), and have configured it with security in mind as well, you're asking for trouble! Again, see (much) more on Linux kernel security via a link in the *Further reading* section.

The goal is to have your project or product achieve **memory safety**.

Tools to catch kernel memory issues – a quick summary

Let's get to the important thing: what tools and/or approaches are available to you when debugging kernel memory issues? Several exist; among them are the following:

- Directly with dynamic (runtime) analysis, specifically, memory checker tooling:

  - **Kernel Address Sanitizer (KASAN)**

  - **Undefined Behavior Sanitizer (UBSAN)**

  - SLUB debug techniques

  - **Kernel memory leak detector (kmemleak)**

- Indirectly with the following:

  - Static analysis tools: **checkpatch.pl**, **sparse**, **smatch**, **Coccinelle**, **cppcheck**, **flawfinder**, and **GCC**

  - Tracing techniques

  - K[ret]probes instrumentation

  - Post-mortem analysis tooling (logs, Oops analysis, kdump/crash, [and K]GDB)

The first bullet point above – the one using which you can more or less *directly catch kernel memory defects* – is of course what we shall primarily focus on here. Subsequent chapters in this book will cover the *indirect* techniques mentioned in the second bullet point. So, patience – you'll get there. Also, as implied by the *indirect* wording, these may or may not help you catch memory bugs.

Okay. I'll attempt to summarize this information with specifics on the tools you can use in the following table. More detailed tables will be presented later in this chapter.

Type of memory bug or defect	Tool(s)/techniques to detect it
Uninitialized Memory Reads (**UMR**)	Compiler (warnings) [1], static analysis
Out-of-bounds (**OOB**) memory accesses: read/write underflow/overflow defects on compile-time and dynamic memory (including the stack)	KASAN [2], SLUB debug
Use-After-Free (**UAF**) or dangling pointer defects (aka **Use-After-Scope** (**UAS**) defects)	KASAN, SLUB debug
Use-After-Return (**UAR**) aka UAS defects	Compiler (warnings), static analysis
Double-free	Vanilla kernel [3], SLUB debug, KASAN
Memory leakage	kmemleak

Table 5.1 – A summary of tools (and techniques) you can use to detect kernel memory issues

A few notes to match the numbers in square brackets in the second column:

- **[1]**: Modern GCC/Clang compilers definitely emit a warning for UMR, with recent ones even being able to auto-initialize local variables (if so configured).

- **[2]**: KASAN catches (almost!) all of them – wonderful. The SLUB debug approach can catch a couple of these, but not all. Vanilla kernels don't seem to catch any.

- **[3]**: By vanilla kernel, I mean that this defect was caught on a regular distro kernel (with no special config set for memory checking).

All right! Now you know – in theory – how to catch memory bugs in the kernel or your driver, but in practice? Well, that requires you to learn to use the tool(s) mentioned above and practice! As mentioned already, understanding, configuring, and learning to leverage KASAN and UBSAN (along with using Clang) is the focus of this chapter (SLUB debug and kmemleak will be that of the next one). So, let's get on with it.

Using KASAN and UBSAN to find memory bugs

The **Kernel Address Sanitizer** (**KASAN**) is a port of the **Address Sanitizer** (**ASAN**) tooling of the Linux kernel. The ASAN project proved to be so useful in detecting memory-related defects that having similar abilities within the kernel was a no-brainer. ASAN is one of the few tools that could detect the buffer overread defect that was at the root of the (in)famous so-called **Heartbleed** exploit! See the *Further reading* section for a very interesting XKCD comic link that superbly illustrates the bug at the heart of Heartbleed.

Understanding KASAN – the basics

A few points on KASAN will help you understand more:

- KASAN is a dynamic – runtime – analysis tool; it works while the code runs. This should have you realize that unless the code actually runs (executes), KASAN will not catch any bugs. *This underlines the importance of writing really good test cases (both positive and negative), and the use of fuzzing tools to catch rarely-run code paths!* More on this in later chapters, but it's such a key point that I am stressing it here as well.

- The technology behind KASAN is called **Compile-Time Instrumentation** (**CTI**) (aka **static instrumentation**). Here, we don't intend to go into the internals of how it works; please see the *Further reading* section for more on this. Very briefly, when the kernel is built with the GCC or Clang -fsanitize=kernel-address option switch, the compiler inserts assembly-level instructions to validate every memory access. Further, every byte of memory is *shadowed* (tracked) using 1 byte of shadow memory to track 8 bytes of actual memory.

- Overhead is relatively low (a factor of around *2x* to *4x*). This is low, especially when compared with dynamic instrumentation approaches such as Valgrind's, where the overhead can easily be *20x* to *50x*.

Well, in terms of overhead from KASAN, it's really the RAM (more than CPU) overheads that can hurt. It does all depend on where you're coming from. For an enterprise-class server system, using several megabytes of RAM as overhead for KASAN can be considered tolerable. This is likely not the case for a resource-constrained embedded system (your typical Android smartphone, TV, wearable devices, low-end routers, and similar products being good examples). For this key reason, the modern Linux kernel supports three types, or modes, of KASAN implementations:

- **Generic KASAN** (the one we're referring to and using here, unless mentioned otherwise): High overhead and debug-only.

- **Software tag-based KASAN**: Medium-to-low overhead on actual workloads. Currently ARM64 only.

- **Hardware tag-based KASAN**: Low overhead and production-capable. Currently, ARM64 only.

The first is the default and the one to use when actively debugging (or bug hunting). It has the largest relative overhead among the three, but is very effective at bug catching! The software tag-based approach has significantly lower overhead; it's appropriate for testing actual workloads. The third hardware tag-based version has the lowest overhead and is even suitable for production use!

> **Memory Checking on User-Mode Apps**
>
> The ASAN tooling was in fact first implemented (by Google engineers) as a GCC (and soon, Clang) patch for userspace applications. The suite includes ASAN, **Leak Sanitizer (LSAN)**, **Memory Sanitizer (MSAN)**, **Thread Sanitizer (TSAN)**, and **Undefined Behavior Sanitizer (UBSAN)**. They – especially ASAN – are really powerful and are simply a *must-use for userspace app memory checking*! My earlier book *Hands-On System Programming with Linux* does cover using ASAN (and Valgrind) in some detail.

In the discussion that follows, I assume that the Generic KASAN mode is being employed, primarily for the purpose of (memory) debugging. Actually, as you'll see in the following section, this is a bit of a moot point as the other tag-based modes are currently only supported on ARM64.

Requirements to use KASAN

Firstly, as KASAN (as well as UBSAN) are compiler-based technologies, which compiler should you use? Both GCC and Clang are supported. You will require a relatively recent version of the compiler to be able to leverage KASAN – as of this writing, you'll need the following:

- **GCC version**: 8.3.0 or later

- **Clang version**: Any. For detecting OOB accesses on global variables, Clang version 11 or later is required.

The following table neatly summarizes some key information about KASAN:

KASAN Mode	GCC	Clang	Internal working	Platforms supported	Suitable for
Generic KASAN	>= 8.3.0	Any (>= 11 for OOB on global variables)	CTI	x86_64, ARM, ARM64, Xtensa, S390, RISC-V	Development/debug only; global variables also instrumented; SLUB and SLAB implementation
Software tag-based KASAN	Not supported		CTI	Currently only on ARM64 (hardware tag-based: requires ARMv8.5 or later with **Memory Tagging Extension**)	Dev/debug and production; hardware tag-based requires SLUB implementation
Hardware tag-based KASAN	>= 10+	>= 11+	hardware tag-based		

Table 5.2 – Types of KASAN and compiler/hardware support requirements

> **The Kernel and Compilers**
>
> Traditionally, the Linux kernel has been very tightly coupled to the GCC compiler; that's slowly changing. Clang is now almost fully supported, and Rust is making an entry. In fact, FYI, Clang is typically used to compile **Android Open Source Project (AOSP)** kernels. We cover using Clang in the *Building your kernel and modules with Clang* section.

Next, hardware-wise, KASAN traditionally requires a 64-bit processor. Why? Recall that it uses a shadow memory region whose size is one-eighth of the kernel virtual address space. On an x86_64, the kernel VAS region is 128 TB (as is the user-mode **Virtual Address Space (VAS)** region). An eighth of this is significant – it's 16 terabytes. So, what platforms does KASAN actually work on? Quoting directly from the official kernel documentation: *Currently, Generic KASAN is supported for the x86_64, arm, arm64, xtensa, s390, and riscv architectures, and tag-based KASAN modes are supported only for arm64.*

Did you notice? Even **ARM** – the ARM 32-bit processor – is supported! This is a recent thing, as of the 5.11 kernel. Not only that, as of this writing at least, the lower overhead tag-based KASAN type is supported only for ARM64. Did you pause to wonder, why ARM64? Clearly, it's due to the incredible popularity of Android. Many, if not most, Android devices are powered via an ARM64 core(s) within a **System on Chip (SoC)**. Detecting memory defects on Android – both in userspace and within the kernel – is critical in today's information economy. Thus, tag-based KASAN modes work on this key platform!

In *Table 5.2*, I highlight **Generic KASAN** in bold as it's the one we're going to work with here.

Configuring the kernel for Generic KASAN mode

Of course, you need to configure your kernel to support Generic KASAN mode. It's straightforward: enable it by setting CONFIG_KASAN=y. When performing the kernel config (via the usual method, the make menuconfig UI), you'll find the menu option here:

```
Kernel hacking | Memory Debugging | KASAN: runtime memory
debugger
```

To make it a bit more interesting, let's configure the kernel for KASAN for ARM64:

```
make ARCH=arm64 menuconfig
```

The screenshot shows you how it looks (here, we've navigated to the KASAN sub-menu):

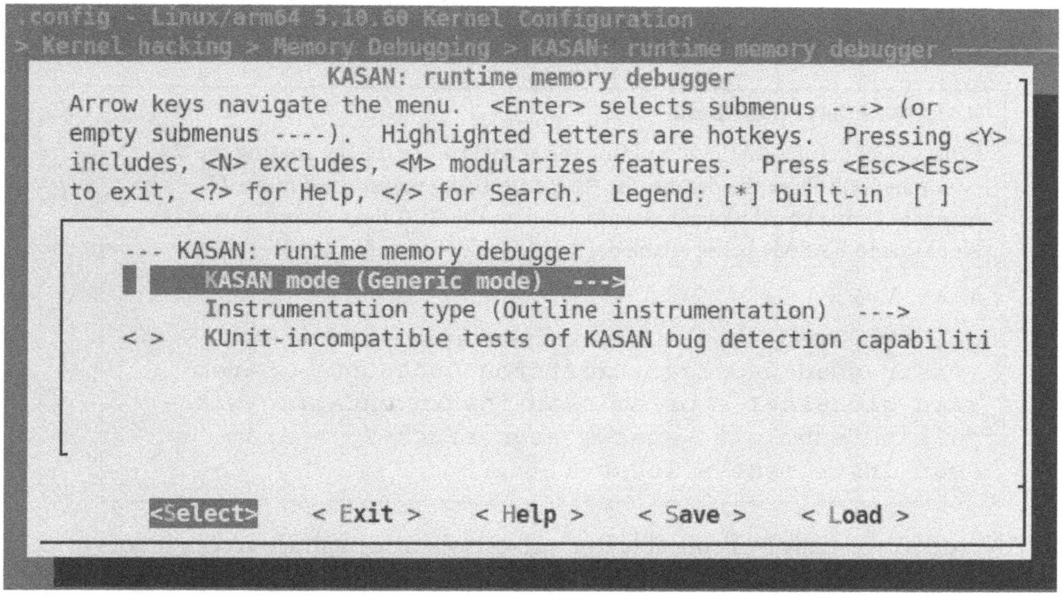

Figure 5.1 – Screenshot of kernel config enabling KASAN

Keep the mode as Generic mode. The < Help > button will show you that this corresponds to the kernel config CONFIG_KASAN_GENERIC=y. In fact, this Help display reveals some interesting information:

```
This mode consumes about 1/8th of available memory at kernel
start and introduces an overhead of ~x1.5 for the rest of the
allocations. The performance slowdown is ~x3.
```

Also, here, only because it's the ARM64 architecture, does the kernel config option
`CONFIG_HAVE_ARCH_KASAN_SW_TAGS` get initialized to y:

```
$ grep KASAN .config
CONFIG_KASAN_SHADOW_OFFSET=0xdfffffd000000000
CONFIG_HAVE_ARCH_KASAN=y
CONFIG_HAVE_ARCH_KASAN_SW_TAGS=y
CONFIG_CC_HAS_KASAN_GENERIC=y
CONFIG_KASAN=y
CONFIG_KASAN_GENERIC=y
[...]
```

In addition, you can see how the kernel configures the shadow memory region start offset
via the value allotted to the kernel config `CONFIG_KASAN_SHADOW_OFFSET` (it's a
kernel virtual address of course) and other configs.

> **KASAN – Effect on the Build**
>
> With `CONFIG_KASAN=y`, building the kernel source tree by passing the `V=1`
> parameter will show the details: the GCC flags being passed, and more. Here's a
> snippet of what you'd typically see, focused on the GCC flags passed during the
> build due to KASAN being enabled:
>
> ```
> make V=1
> gcc -Wp,-MMD,[...] -fsanitize=kernel-address
> -fasan-shadow-offset=0xdffffc0000000000 --param
> asan-globals=1 --param asan-instrumentation-with-
> call-threshold=0 --param asan-stack=1 --param
> asan-instrument-allocas=1 [...]
> ```

KASAN works essentially by being able to check every single memory access; it does
this by using a technique called **Compile Time Instrumentation (CTI)**. Put very
simplistically, the compiler inserts function calls (`__asan_load*()` and `__asan_store*()`) before every 1-, 2-, 4-, 8-, or 16-byte memory access. Thus, the runtime
can figure out whether the access is valid or not (by checking the corresponding
shadow memory bytes). Now, there are two broad ways the compiler can perform this
instrumentation: outline and inline. *Outline instrumentation* has the compiler inserting
actual function calls (as just mentioned); *inline instrumentation* achieves the same thing
but in a time-optimized manner by directly inserting the code (and not having the
overhead of a function call)!

You can set the kernel config option `Instrumentation type` to either `CONFIG_KASAN_OUTLINE` (the default) or `CONFIG_KASAN_INLINE`. It's the typical trade-off: the *outline* type, the default, will result in a smaller kernel image while the *inline* type will result in a larger image but is faster (by a factor of 1.1x to 2x).

Also, (especially for your debug kernel), it's worth enabling the kernel config `CONFIG_STACKTRACE`, so that you also obtain stack traces of the allocation and freeing of affected slab objects in the report when a bug is detected. Similarly, turning on `CONFIG_PAGE_OWNER` – here within the menu `Kernel hacking | Memory Debugging | Track page owner` – will get you the stack traces of the allocation and freeing of affected physical pages. It's off by default; you have to boot with the parameter `page_owner=on`.

Also, when configuring an x86_64 for KASAN, you'll find an additional kernel config regarding vmalloc memory corruption detection. The option shows up like this:

```
[*]    Back mappings in vmalloc space with real shadow memory
```

This helps detect vmalloc-related memory corruption issues (at the cost of higher memory usage during runtime).

So much for the theory and KASAN kernel config. Do configure and (re)build your (debug) kernel and we're good to give it a spin!

Bug hunting with KASAN

I'll assume that by now you've configured, built, and booted into your (debug) kernel that's enabled with KASAN (as the previous section has described in detail). On my setup – an x86_64 Ubuntu 20.04 LTS guest VM – this has been done.

To test whether KASAN works, we'll need to execute code that has memory bugs (I can almost hear some of you old-timers say *"Yeah? That shouldn't be too hard"*). We can always write our own test cases but why reinvent the wheel? This is a good opportunity to look at a part of the kernel's test infrastructure! The following section shows you how we'll leverage the kernel's **KUnit** unit testing framework to run KASAN test cases.

Using the kernel's KUnit test infrastructure to run KASAN test cases

Why take the trouble to write our own test cases to test KASAN when the community has already done the work for us? Ah, the beauty of open source.

The Linux kernel has by now evolved sufficiently to have many kinds of test infrastructure, including full-fledged test suites, built into it; testing various aspects of the kernel is now a matter of configuring the kernel appropriately and running the tests!

With regard to possible built-in test frameworks within the kernel, the two primary ones are the KUnit framework and the **kselftest** framework. FYI, the official kernel documentation, of course, has all the details. As a start, you can check this one: *Kernel Testing Guide*: `https://www.kernel.org/doc/html/latest/dev-tools/testing-overview.html#kernel-testing-guide` – it provides a rough overview of available testing frameworks and tooling (including dynamic analysis) within the kernel.

Again, FYI, there are several other related and useful frameworks: kernel fault injection, `notifier` error injection, the **Linux Kernel Dump Test Module** (**LKDTM**), and so on. You'll find them under the kernel config here: `Kernel hacking | Kernel Testing and Coverage`.

Again, we don't intend to delve into the details of how KUnit works here; the idea is to merely use KUnit to test KASAN as a practical example at this point. For details on using these test frameworks – it will probably prove useful! – do see the links within the *Further reading* section.

As a pragmatic thing to do, and to begin getting familiar with it, let's leverage the kernel's **KUnit** – **Unit Testing for the Linux kernel** – framework to execute KASAN test cases!

It's really very simple to do. First, ensure your debug kernel is configured to use KUnit: `CONFIG_KUNIT=y` (or `CONFIG_KUNIT=m`).

We intend to run KASAN test cases, thus, we must have the KASAN test module configured as well:

```
CONFIG_KASAN_KUNIT_TEST=m
```

The kernel's module code for the KASAN test cases we're going to run is here: `lib/test_kasan.c`. A quick peek will show you the various test cases (there are many of them – 38 as of this writing):

```
// lib/test_kasan.c
static struct kunit_suite kasan_kunit_test_suite = {
    .name = "kasan",
    .init = kasan_test_init,
    .test_cases = kasan_kunit_test_cases,
    .exit = kasan_test_exit,
};
kunit_test_suite(kasan_kunit_test_suite);
```

This sets up the suite of test cases to execute. The actual test cases are in the `kunit_suite` structure's member named `test_cases`. It's a pointer to an array of `kunit_case` structures:

```
static struct kunit_case kasan_kunit_test_cases[] = {
    KUNIT_CASE(kmalloc_oob_right),
    KUNIT_CASE(kmalloc_oob_left),
    [...]
    KUNIT_CASE(kmalloc_double_kzfree),
    KUNIT_CASE(vmalloc_oob),
    {}
};
```

The `KUNIT_CASE()` macro sets up the test case. To help understand how it works, here's the code for the first of the test cases:

```
// lib/test_kasan.c
static void kmalloc_oob_right(struct kunit *test)
{
    char *ptr;
    size_t size = 123;

    ptr = kmalloc(size, GFP_KERNEL);
    KUNIT_ASSERT_NOT_ERR_OR_NULL(test, ptr);
    KUNIT_EXPECT_KASAN_FAIL(test, ptr[size + OOB_TAG_OFF] =
'x');
    kfree(ptr);
}
```

Quite intuitively, the actual checking occurs within the `KUNIT_ASSERT|EXPECT_*()` macros seen above. The first macro asserts that the return from the `kmalloc()` API doesn't result in an error and isn't null. The second macro, `KUNIT_EXPECT_KASAN_FAIL()`, has the KUnit code expect failure – a negative test case. This is indeed what should be done here: we expect that writing *beyond* the *right* side of the buffer (a write overflow defect) should trigger KASAN to report a failure! I'll leave it to you to study the implementation of these macros if interested.

Furthermore, and quite interestingly, the name and exit members of the `kunit_suite` structure specify functions to execute before and after each test case is run, respectively. The module leverages this to ensure that the kernel sysctl `kasan_multi_shot` is temporarily enabled and to set `panic_on_warn` to 0 (else, only the first invalid memory access would trigger a report and a possible kernel panic!).

Finally, let's try it out:

```
$ uname -r
5.10.60-dbg01
$ sudo modprobe test_kasan
```

This will cause all test cases within the KASAN test module to execute! Looking up the kernel log (via `journalctl -k` or `dmesg`) will show you the detailed KASAN reports for each of the test cases. As they're voluminous, I show a sampling of the output. The very first test case – `KUNIT_CASE(kmalloc_oob_right)` – causes KASAN to generate this report (its output is truncated – see more of it below):

```
[  164.772135]    # Subtest: kasan
[  164.772149]    1..38
[  164.773166] ============================================================
[  164.776786] BUG: KASAN: slab-out-of-bounds in kmalloc_oob_right+0x159/0x260 [test_kasan]
[  164.780268] Write of size 1 at addr ffff8880316a45fb by task kunit_try_catch/1206

[  164.787155] CPU: 2 PID: 1206 Comm: kunit_try_catch Tainted: G        0        5.10.60-dbg01 #6
[  164.787166] Hardware name: innotek GmbH VirtualBox/VirtualBox, BIOS VirtualBox 12/01/2006
[  164.787170] Call Trace:
[  164.787204]  dump_stack+0xbd/0xfa
[  164.787232]  print_address_description.constprop.0.cold+0xd4/0x4db
[  164.787257]  ? trace_preempt_off+0x2a/0xf0
[  164.787303]  ? kmalloc_oob_right+0x159/0x260 [test_kasan]
[  164.787323]  kasan_report.cold+0x37/0x7c
[  164.787354]  ? kmalloc_oob_right+0x159/0x260 [test_kasan]
[  164.787384]  __asan_store1+0x6d/0x70
[  164.787402]  kmalloc_oob_right+0x159/0x260 [test_kasan]
[  164.787415]  ? kvm_sched_clock_read+0x9/0x20
[  164.787436]  ? kmalloc_oob_left+0x270/0x270 [test_kasan]
[  164.787449]  ? sched_clock_cpu+0x1b/0x1f0
[  164.787480]  ? kunit_binary_str_assert_format+0x100/0x100 [kunit]
[  164.787523]  ? lock_downgrade+0x3c0/0x3c0
[  164.787540]  ? mark_held_locks+0x29/0xa0
[  164.787558]  ? _raw_spin_unlock_irqrestore+0x55/0x70
[  164.787570]  ? __kthread_parkme+0x71/0x100
[  164.787585]  ? __this_cpu_preempt_check+0x13/0x20
[  164.787609]  ? trace_preempt_on+0x2a/0xf0
[  164.787614]  ? __kthread_parkme+0x71/0x100
[  164.787653]  kunit_try_run_case+0x8d/0x130 [kunit]
[  164.787672]  ? kunit_catch_run_case+0x120/0x120 [kunit]
[  164.787691]  ? kunit_try_catch_throw+0x40/0x40 [kunit]
[  164.787712]  kunit_generic_run_threadfn_adapter+0x2e/0x50 [kunit]
[  164.787733]  kthread+0x22a/0x260
[  164.787751]  ? kthread_cancel_delayed_work_sync+0x20/0x20
[  164.787777]  ret_from_fork+0x22/0x30

[  164.791168] Allocated by task 1206:
[  164.794501]  kasan_save_stack+0x23/0x50
[  164.794514]  __kasan_kmalloc.constprop.0+0xcf/0xe0
[  164.794526]  kasan_kmalloc+0x9/0x10
[  164.794537]  kmem_cache_alloc_trace+0x1a5/0x370
[  164.794553]  kmalloc_oob_right+0xa3/0x260 [test_kasan]
[  164.794568]  kunit_try_run_case+0x8d/0x130 [kunit]
[  164.794584]  kunit_generic_run_threadfn_adapter+0x2e/0x50 [kunit]
[  164.794597]  kthread+0x22a/0x260
[  164.794615]  ret_from_fork+0x22/0x30
```

Figure 5.2 – First part of the KUnit KASAN bug-catching example

Notice the following in the preceding screenshot:

- In the first two lines, KUnit shows the test title (as # Subtest: kasan) and that it will run test cases 1..38.

- KASAN successfully, as expected of it, detected the memory defect, the write overflow, and generated a report. The report begins with BUG: KASAN: [...] and the details follow.

- The following lines reveal the root cause: the format that KASAN displays the offending function in is func()+0xoff_from_func/0xsize_of_func, where, within the function named func(), the error occurred at an offset of 0xoff_from_func bytes from the start of the function, and the kernel estimates the function length to be 0xsize_of_func bytes. So here, the code in the kmalloc_oob_right() function, at an offset of 0x159 bytes from the start of it (followed by an educated guess of the function's length as 0x260 bytes), within the kernel module test_kasan (shown within square brackets on the extreme right), attempted to illegally write at the specified address. The defect, the bug, is an OOB write to a slab memory buffer, as seen by the slab-out-of-bounds token:

  ```
  BUG: KASAN: slab-out-of-bounds in kmalloc_oob_
  right+0x159/0x260 [test_kasan]
  Write of size 1 at addr ffff8880316a45fb by task kunit_
  try_catch/1206
  ```

- The following line reveals the process context within which this occurred (we'll cover the meaning of the tainted flags in the following chapter):

  ```
  CPU: 2 PID: 1206  Comm: kunit_try_catch Tainted: G
  O         5.10.60-dbg01 #6
  ```

- The next line shows the hardware detail (you can see it's a VM, VirtualBox).

- The majority of the output is the call stack (labeled Call Trace:). By reading it bottom up (and ignoring any lines prefixed with a ?), you can literally see how control came to this, the buggy code!

- The line Allocated by task 1206: and the following output reveals the call trace of the memory allocation code path. This can be very helpful, showing by whom and where the memory buffer was allocated to begin with.

The remainder of the output can be seen in the following screenshot:

```
[  164.797882]  The buggy address belongs to the object at ffff8880316a4580
                 which belongs to the cache kmalloc-128 of size 128
[  164.804507]  The buggy address is located 123 bytes inside of
                 128-byte region [ffff8880316a4580, ffff8880316a4600)
[  164.811106]  The buggy address belongs to the page:
[  164.814441]  page:000000001af581d3 refcount:1 mapcount:0 mapping:0000000000000000 index:0xffff8880316a6b00 pfn:0x316a4
[  164.814452]  head:000000001af581d3 order:2 compound_mapcount:0 compound_pincount:0
[  164.814464]  flags: 0xffffffc0010200(slab|head)
[  164.814478]  raw: 000fffffc0010200 ffffea0000cd0c08 ffff888001040ad0 ffff88800104f4c0
[  164.814491]  raw: ffff8880316a6b00 0000000000190018 00000001ffffffff 0000000000000000
[  164.814500]  page dumped because: kasan: bad access detected
```

Figure 5.3 – Second part of the KUnit KASAN bug-catching example

As `CONFIG_PAGE_OWNER=y` (as we suggested in the *Configuring the kernel for Generic KASAN mode* section), the following output turns up as well. It gives you insight into where the faulty-accessed page(s) is located and its ownership:

```
[  164.817779]  Memory state around the buggy address:
[  164.821195]   ffff8880316a4480: fc fc fc fc fc fc fc fc fc fc fc fc fc fc fc fc
[  164.824828]   ffff8880316a4500: fc fc fc fc fc fc fc fc fc fc fc fc fc fc fc fc
[  164.828377]  >ffff8880316a4580: 00 00 00 00 00 00 00 00 00 00 00 00 00 00 00 03
[  164.831826]                                                                  ^
[  164.835291]   ffff8880316a4600: fc fc fc fc fc fc fc fc fc fc fc fc fc fc fc fc
[  164.838802]   ffff8880316a4680: fc fc fc fc fc fc fc fc fc fc fc fc fc fc fc fc
[  164.842251]  ==================================================================
[  164.845747]  Disabling lock debugging due to kernel taint
[  164.846982]      ok 1 - kmalloc_oob_right
[  164.847514]  ==================================================================
[  164.850583]  BUG: KASAN: slab-out-of-bounds in kmalloc_oob_left+0x159/0x270 [test_kasan]
[  164.853608]  Read of size 1 at addr ffff88800df70a8f by task kunit_try_catch/1207
```

Figure 5.4 – Third (and final) part of the KUnit KASAN bug-catching example

In the preceding screenshot, you can see KASAN justifying itself. It shows the actual memory region where the defect occurred and even points out the precise byte where it did (via the > for the row and ^ for the column symbols)! As a side effect of this bug, the kernel now disables all lock debugging. Further, KUnit says that running this first test case went well: `ok 1    -    kmalloc_oob_right`.

Interpreting this information is important. It helps you drill down to what actually triggered the bug. We do just this in the section that follows!

Interpreting the KASAN shadow memory output

In *Figure 5.4*, you can see the KASAN shadow memory revealing the defect's cause. We print the key line – the one prefixed with a right arrow symbol >:

```
>ffff8880318ad980: 00 00 00 00 00 00 00 00 00 00 00 00 00 00 00
03
                 ^
```

These are the KASAN *shadow memory* bytes, each one represents 8 bytes of actual memory. The byte 03 is pointed at (by the symbol ^) telling us where the issue lies. What do the bytes 00, 03, and so on, mean? The details follow:

- Generic KASAN assigns one shadow byte to track 8 bytes of kernel memory (think of an 8-byte chunk as a *memory granule*).

- A granule (an 8-byte region) is encoded as being accessible, partially accessible, part of a red zone, or free.

- The encoding of a memory granule (8-byte region) by shadow byte tracking is done as follows:

  - **Shadow memory = 00**: All 8 bytes are accessible (no error).

  - **Shadow memory = N (where N can be a value between 1 and 7)**: The first N bytes are accessible (fine); the remaining (8-N) bytes aren't legally accessible.

  - **Shadow memory < 0**: A negative value implies the entire granule (8 bytes) is inaccessible. The particular (negative) values and their meaning (already freed-up memory, red zone region, and so on) are encoded in a header file (mm/kasan/kasan.h).

So, now you'll realize that the shadow byte 03 implies that the memory was partially accessible. The first 3 bytes (as here, *N = 3*) were legally accessible; the remaining 5 (*8 – 3 = 5*) bytes weren't. Let's take the trouble to verify this in detail. This is the line of code that triggers the bug, of course (it's here within the kernel code base):

```
// lib/test_kasan.c
static void kmalloc_oob_right(struct kunit *test)
    [...]
    size_t size = 123;
    ptr = kmalloc(size, GFP_KERNEL);
    [...]
    KUNIT_EXPECT_KASAN_FAIL(test, ptr[size + OOB_TAG_OFF] =
'x');
```

Now, the variable size is set to the value 123 and OOB_TAG_OFF is 0 when CONFIG_KASAN_GENERIC is enabled. So, in effect, the (buggy) code is this:

```
ptr[123] = 'x';
```

Now, Generic KASAN's memory granule size is 8 bytes. So, among the 123 bytes allocated, the fifteenth memory granule is the one being written to (as *8 * 15 = 120*). The diagram that follows clearly shows the memory buffer and how it's been overflowed:

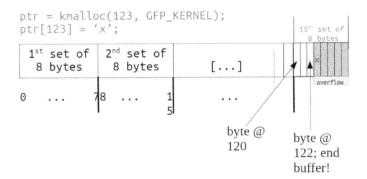

Figure 5.5 – The kmalloc'ed memory (slab) buffer that was overflowed

Check it out: Towards the right end, byte positions 120, 121, and 122 are valid and legal to read/write, but our KUnit KASAN test case deliberately wrote to byte position 123 – 1 byte beyond the end of the slab buffer, *a clear OOB write overflow violation*, and KASAN caught it! Not only that, as *Figure 5.4* and *Figure 5.5* clearly show, the kernel is intelligent enough to show the shadow value of 03 here, implying that the first 3 bytes are valid but the remaining 5 aren't – that's precisely the case!

Further, the surrounding bytes are set to the value 0xfc (see *Figure 5.4*). What does it mean? It's clear from the header – it's a red zone within the kernel SLUB object:

```
// mm/kasan/kasan.h
#ifdef CONFIG_KASAN_GENERIC
#define KASAN_FREE_PAGE          0xFF  /* page was freed */
#define KASAN_PAGE_REDZONE       0xFE  /* redzone for kmalloc_
large allocations */
#define KASAN_KMALLOC_REDZONE    0xFC  /* redzone inside slub
object */
#define KASAN_KMALLOC_FREE       0xFB  /* object was freed
(kmem_cache_free/kfree) */
#define KASAN_KMALLOC_FREETRACK 0xFA  /* object was freed and
has free track set */
```

Back to our interpretation of *Figure 5.4*: The next line (BUG: KASAN: [...]) is just to show you that this continues with the next test case... KASAN has now caught the second test case's bug (KUNIT_CASE(kmalloc_oob_left)). The kernel log contains the same info as for the first defect: the bug summary by KASAN, the output of dump_stack() – the stack(s) call frames, who performed the allocation, the page-ownership info, and the memory state around the buggy access. This continues all the way to the thirty-eighth test case – fantastic!

A quick check of the kernel log shows what we expect – how the kernel's KUnit KASAN test case module has caught all 38 test cases with memory defects:

```
$ journalctl -kb |grep -w "ok"
Oct 29 18:55:02 dbg-LKD kernel:    ok 1 - kmalloc_oob_right
Oct 29 18:55:02 dbg-LKD kernel:    ok 2 - kmalloc_oob_left
Oct 29 18:55:02 dbg-LKD kernel:    ok 3 - kmalloc_node_oob_right
Oct 29 18:55:02 dbg-LKD kernel:    ok 4 - kmalloc_pagealloc_oob_right
Oct 29 18:55:02 dbg-LKD kernel:    ok 5 - kmalloc_pagealloc_uaf
Oct 29 18:55:02 dbg-LKD kernel:    ok 6 - kmalloc_pagealloc_invalid_free
Oct 29 18:55:02 dbg-LKD kernel:    ok 7 - kmalloc_large_oob_right
Oct 29 18:55:02 dbg-LKD kernel:    ok 8 - kmalloc_oob_krealloc_more
Oct 29 18:55:02 dbg-LKD kernel:    ok 9 - kmalloc_oob_krealloc_less
Oct 29 18:55:02 dbg-LKD kernel:    ok 10 - kmalloc_oob_16
Oct 29 18:55:02 dbg-LKD kernel:    ok 11 - kmalloc_uaf_16
Oct 29 18:55:02 dbg-LKD kernel:    ok 12 - kmalloc_oob_in_memset
Oct 29 18:55:02 dbg-LKD kernel:    ok 13 - kmalloc_oob_memset_2
Oct 29 18:55:02 dbg-LKD kernel:    ok 14 - kmalloc_oob_memset_4
Oct 29 18:55:02 dbg-LKD kernel:    ok 15 - kmalloc_oob_memset_8
Oct 29 18:55:02 dbg-LKD kernel:    ok 16 - kmalloc_oob_memset_16
Oct 29 18:55:02 dbg-LKD kernel:    ok 17 - kmalloc_memmove_invalid_size
Oct 29 18:55:02 dbg-LKD kernel:    ok 18 - kmalloc_uaf
Oct 29 18:55:02 dbg-LKD kernel:    ok 19 - kmalloc_uaf_memset
Oct 29 18:55:02 dbg-LKD kernel:    ok 20 - kmalloc_uaf2
Oct 29 18:55:02 dbg-LKD kernel:    ok 21 - kfree_via_page
Oct 29 18:55:02 dbg-LKD kernel:    ok 22 - kfree_via_phys
Oct 29 18:55:03 dbg-LKD kernel:    ok 23 - kmem_cache_oob
Oct 29 18:55:03 dbg-LKD kernel:    ok 24 - memcg_accounted_kmem_cache
Oct 29 18:55:03 dbg-LKD kernel:    ok 25 - kasan_global_oob
Oct 29 18:55:03 dbg-LKD kernel:    ok 26 - kasan_stack_oob
Oct 29 18:55:03 dbg-LKD kernel:    ok 27 - kasan_alloca_oob_left
Oct 29 18:55:03 dbg-LKD kernel:    ok 28 - kasan_alloca_oob_right
Oct 29 18:55:03 dbg-LKD kernel:    ok 29 - ksize_unpoisons_memory
Oct 29 18:55:03 dbg-LKD kernel:    ok 30 - kmem_cache_double_free
Oct 29 18:55:03 dbg-LKD kernel:    ok 31 - kmem_cache_invalid_free
Oct 29 18:55:03 dbg-LKD kernel:    ok 32 - kasan_memchr
Oct 29 18:55:03 dbg-LKD kernel:    ok 33 - kasan_memcmp
Oct 29 18:55:03 dbg-LKD kernel:    ok 34 - kasan_strings
Oct 29 18:55:04 dbg-LKD kernel:    ok 35 - kasan_bitops_generic
Oct 29 18:55:04 dbg-LKD kernel:    ok 36 - kasan_bitops_tags
Oct 29 18:55:04 dbg-LKD kernel:    ok 37 - kmalloc_double_kzfree
Oct 29 18:55:04 dbg-LKD kernel:    ok 38 - vmalloc_oob
```

Figure 5.6 – Screenshot showing how the kernel's KUnit KASAN test case module has caught all 38 test cases with memory defects

As can be clearly seen from the preceding screenshot, all 38 test cases are reported as ok (passed).

> **Exercise**
> Do perform what we've just done – running the kernel's KUnit KASAN test cases – on your box. Note, from the kernel log, the various KASAN test cases and verify that all ran correctly.

By the way, notice this:

```
$ lsmod |egrep "kunit|kasan"
test_kasan           81920  0
kunit                49152  1 test_kasan
```

In my particular case, you can see from the lsmod output that KUnit has been configured as a kernel module.

You can learn how to write your own suite of KUnit test cases. Do see the *Further reading* section for more on using KUnit!

Remaining tests with our custom buggy kernel module

Did you notice, in spite of having run all the KASAN KUnit test cases, there appear to be a few remaining generic memory defects (as we identified both in *Chapter 4, Debug via Instrumentation – Using Kprobes*, as well as in the *What's the problem with memory anyway?* section of this chapter) that the KUnit test cases don't cover?

- The **uninitialized memory read** (**UMR**) bug
- The **use-after-return** (**UAR**) bug
- Simple memory leakage bugs (we'll discuss memory leakage in more detail later in this chapter)

So, I wrote a kernel module to exercise these test cases (when running the Generic KASAN-enabled debug kernel of course), along with some more interesting ones. To test against KASAN, remember to boot via your custom debug kernel, one that (obviously) has CONFIG_KASAN=y.

Due to space constraints, I won't show the entire code of our test module here (do refer to it on the book's GitHub repo and read the comments therein – you'll find it under the ch5/kmembugs_test folder). To get a flavor of it, let's take a peek at one of the test cases and how it's invoked. Here's the code of the UAR test case:

```
// ch5/kmembugs_test/kmembugs_test.c
/* The UAR - Use After Return - testcase */
static void *uar(void)
{
    volatile char name[NUM_ALLOC];
    volatile int i;
    pr_info("testcase 2: UAR:\n");
    for (i=0; i<NUM_ALLOC-1; i++)
        name[i] = 'x';
    name[i] = '\0';
    return name;
}
```

The module is designed to be loaded up via a bash script named load_testmod and the test cases are run interactively (via a bash wrapper script named run_tests). The run_tests script (which you must run as root) displays a menu of available tests and asks you to select any one by typing in its assigned number. You can see a screenshot of the menu – and thus all the test cases you can try out – in *Figure 5.8*, in the section that follows.

The script then writes this number to our debugfs pseudofile here: /sys/kernel/debug/test_kmembugs/lkd_dbgfs_run_testcase. The debugfs write hook function then receives this data from userspace, validates it, and invokes the appropriate test case routine (via a rather long if-else ladder). This design allows you to test interactively and execute any test case(s) as many times as you wish to.

Here's a code snippet showing how our debugfs module code invokes the preceding uar() test case:

```
// ch5/kmembugs_test/debugfs_kmembugs.c
static ssize_t dbgfs_run_testcase(struct file *filp, const char
__user *ubuf, size_t count, loff_t *fpos)
{
    char udata[MAXUPASS];
    volatile char *res1 = NULL, *res2 = NULL;
    [...]
```

```
    if (copy_from_user(udata, ubuf, count))
        return -EIO;
    udata[count-1]='\0';
    pr_debug("testcase to run: %s\n", udata);
    /* Now udata contains the data passed from userspace -
 the testcase # to run (as a string) */
    if (!strncmp(udata, "1", 2))
        umr();
    else if (!strncmp(udata, "2", 2)) {
        res1 = uar();
        pr_info("testcase 2: UAR: res1 = \"%s\"\n",
 res1 == NULL ? "<whoops, it's NULL; UAR!>" : (char *)res1);
    } else if (!strncmp(udata, "3.1", 4))
 . . .
```

Clearly, this – test case #2 – is a defect, a bug. You know that local variables are valid only for their lifetime – while the function's executing. This, of course, is because local (or automatic) variables are allocated on the (kernel mode) stack frame of the process context in execution. Thus, you must stop referencing a local variable once outside the scope of its containing function. We (deliberately) don't! We attempt to fetch it as a return. The trouble is, by that time, it's gone...

Right, before diving into running the test cases (though there's no reason you can't run them right now), we divert into an interesting dilemma: how a known bug (like our UAR one) can at times appear to work perfectly fine.

Stale frames – trouble in paradise

The amazing (or crazy) thing about bugs like this one – the UAR defect – is that the code will sometimes seem to work! How come? It's like this: the memory holding the content of the local (automatic) variable is on the stack. Now, though we colloquially say that the stack frames are allocated on function entry and destroyed on function return (the so-called function **prologue** and **epilogue**), the reality isn't quite so dramatic.

The reality is that memory is typically allocated at page-level granularity. This includes the memory for stack pages. Thus, once a page of memory for the stack is allocated, there's usually enough for several frames (this, of course, depends on the circumstances). Then, when more memory for the stack is needed, it's grown (by allocating more pages, downwards, as it's the stack). The system knows where the top of the stack is by having the **Stack Pointer (SP)** register track this memory location. Also, you'll realize that the so-called "top of the stack" is typically the lowest legal address. Thus, when frames are allocated and/or a function is invoked, the SP register value reduces. When a function returns, the stack shrinks by adding to the SP register (remember, it's a downward-growing stack!). The following diagram is a representation of a typical kernel-mode stack on a (32-bit) Linux system:

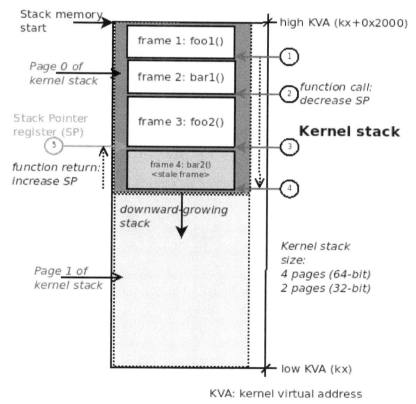

Figure 5.7 – A diagram of a typical kernel-mode stack on 32-bit Linux; function call chain: foo1() -> bar1() -> foo2() -> bar2()

So, it could well happen at some point that stale stack frames (and the corresponding data within them) exist underneath valid frames and could possibly be successfully referenced – without the system throwing a memory fault! – even later.

Carefully study *Figure 5.7*. As an example, we've shown the kernel-mode stack on a 32-bit Linux system, thus the size of the kernel stack will be 2 pages, typically 8 KB. Now, let's say the process context in execution (within the kernel) invoked these functions in this order (this is the call chain, shown as the circled steps 1 to 4 in the figure):

```
foo1() → bar1() --> foo2() --> bar2()
```

Imagine we're at the leaf node, function `bar2()` in this example. It returns (circled step 5 above). This causes the SP register to get incremented back to the address of the call frame representing the function `foo2()`. So, though it remains intact on the stack, the stack memory of the call frame for function `bar2()` is now actually invalid! *But an incorrect (read buggy) access to it might still succeed.*

This should ideally not happen, but hey, it's an imperfect world, right?! The moral here: we require tools – and clear thinking is the best one – to catch tricky bugs such as UAS ones!

Right, back to our test cases! To run the tests, follow these steps:

1. Run the following command:

   ```
   cd <book_src>/ch5/kmembugs_test
   ```

2. Load it up:

   ```
   ./load_testmod
   [...]
   ```

   This should have the kernel module built and loaded into memory with `dmesg` showing that the debugfs pseudofile here – `<debugfs_mountpt>/test_kmembugs/lkd_dbgfs_run_testcase` – has been created.

3. Run our bash script to test:

   ```
   sudo ./run_tests
   ```

Following is a screenshot showing that our `test_kmembugs` module is indeed loaded up (this was done via our `load_testmod` script), the menu shown via our `run_tests` script, and our running test case #2 – the UAR bug:

```
$ lsmod |grep test_kmembugs
test_kmembugs            61440  0
$ sudo ./run_tests
Debugfs file: /sys/kernel/debug/test_kmembugs/lkd_dbgfs_run_testcase

Generic KASAN: enabled
UBSAN: enabled
KMEMLEAK: enabled

Select testcase to run:
1  Uninitialized Memory Read - UMR
2  Use After Return - UAR

Memory leakage
3.1  simple memory leakage testcase1
3.2  simple memory leakage testcase2 - caller to free memory
3.3  simple memory leakage testcase3 - memleak in interrupt ctx

OOB accesses on static (compile-time) global memory + on stack local memory
4.1  Read  (right) overflow
4.2  Write (right) overflow
4.3  Read  (left) underflow
4.4  Write (left) underflow

OOB accesses on dynamic (kmalloc-ed) memory
5.1  Read  (right) overflow
5.2  Write (right) overflow
5.3  Read  (left) underflow
5.4  Write (left) underflow

6  Use After Free - UAF
7  Double-free

UBSAN arithmetic UB testcases
8.1  add overflow
8.2  sub overflow
8.3  mul overflow
8.4  negate overflow
8.5  shift OOB
8.6  OOB
8.7  load invalid value
8.8  misaligned access
8.9  object size mismatch

9  copy_[to|from]_user*() tests
10 UMR on slab (SLUB) memory

(Type in the testcase number to run):
2
Running testcase "2" via test module now...
[89638.348632] testcase to run: 2
[89638.350942] test_kmembugs:uar(): testcase 2: UAR:
[89638.352918] testcase 2: UAR: res1 = "<whoops, it's NULL; UAR!>"
$
```

Figure 5.8 – Partial screenshot showing both the build and output of our kmembugs_test LKM

Here's an example screenshot of our test case framework catching the left OOB write buggy access via KASAN:

```
206  */                                          |9  copy_[to|from]_user*() tests
207  int global_mem_oob_left(int mode, char *p)  |10 UMR on slab (SLUB) memory
208  {
209      volatile char w, x, y, z;               |(Type in the testcase number to run):
210      volatile char local_arr[20];            |4.4
211      char *volatile ptr = p - 3; // left OOB  Running testcase "4.4" via test module now...
212      */                                      [13372.544725] testcase to run: 4.4
213      if (mode == READ) {                      [13372.553282] =====================================================
214          /* Interesting: this OOB access isn't [13372.562448] BUG: KASAN: global-out-of-bounds in global_mem_oob_left+0x172/0x267 [test_kmembugs]
215          w = *(volatile char *)ptr;  // invali[13372.57100] Write of size 1 at addr ffffffffc09aaabd by task run_tests/21489
216
217          /* ... but these OOB accesses are cau[13372.581341] CPU: 0 PID: 21489 Comm: run_tests Tainted: G    B D    0     5.10.60-dbg02-gcc #17
218          * We conclude that *only* the index-8[13372.585154] Hardware name: innotek GmbH VirtualBox/VirtualBox, BIOS VirtualBox 12/01/2006
219          * And, KASAN compiled with clang 11 c[13372.588567] Call Trace:
220          */                                   [13372.591665]  dump_stack+0xbd/0xfa
221          x = p[-3];  // invalid, OOB left read [13372.594362]  print_address_description.constprop.0.cold+0x5/0x4db
222                                                [13372.597049]  ? trace_preempt_off+0x2a/0xf0
223          y = local_arr[-5];  // invalid, not w[13372.599451]  ? global_mem_oob_left+0x172/0x267 [test_kmembugs]
224          z = local_arr[5];   // valid, within [13372.601856]  kasan_report.cold+0x37/0x7c
225      } else if (mode == WRITE) {              [13372.603928]  ? global_mem_oob_left+0x172/0x267 [test_kmembugs]
226          /* Interesting: this OOB access isn't [13372.605993]  __asan_store1+0x6d/0x70
227          *(volatile char *)ptr = 'w';         [13372.607932]  global_mem_oob_left+0x172/0x267 [test_kmembugs]
```

Figure 5.9 – Partial screenshot showing KASAN catching a buggy left OOB on write to global memory

A few things to realize:

- Firstly, the compilers, both GCC and Clang, are clever enough to warn us regarding the (here, pretty obvious) bugs. Both the UAR and UMR defects are indeed caught by them (at the precise place in the code where they occur), albeit as *warnings*! Here's one of the warnings emitted by GCC, clearly with regard to our UAR bug:

```
<...>/ch5/kmembugs_test/kmembugs_test.c:115:9: warning:
function returns address of local variable [-Wreturn-
local-addr]
  115 |    return (void *)name;
      |           ^~~~~~~~~~~~
```

> **This Is Important**
>
> It is your job as the programmer to carefully heed all compiler warnings and – as far as is humanely possible! – fix them.

- The script interrogates the kernel config file to see whether your current kernel is configured for KASAN, UBSAN, and KMEMLEAK and displays what it finds. It also shows the path to the debugfs pseudofile where the test case number will be written (in order to invoke that test).

Here's a sample run of the UAR test case:

```
$ sudo ./run_tests
[...]
(Type in the testcase number to run):
2
Running testcase""2" via test module now...
[  144.313592] testcase to run: 2
[  144.313597] test_kmembugs:uar(): testcase 2: UAR:
[  144.313600] testcase 2: UAR: res1 = "<whoops,'it's
NULL; UAR!>"
$
```

- The output in the kernel log (seen via dmesg above) clearly tells the story: we've executed the UAR test case, and neither the kernel nor KASAN has caught it (if it had, we'd see plenty of complaints in the log!). Our own code checks for the variable res1 being NULL and concludes that a UAR bug occurred. We can do this as we specifically initialized it to NULL and check after it's supposedly set to the string returned by the function uar(); else, we'd have not caught it.

All right, we're now done with several test cases with KASAN enabled. What's KASAN's scorecard like? The following section shows you just this.

KASAN – tabulating the results

What memory corruption bugs (defects) does KASAN actually manage to, and not manage to, catch? From our test runs, we tabulated the results in the table that follows. Do study it carefully, along with the notes that go with it:

Testcase # [1]	Memory defect type (below) / Infrastructure used (right)	Distro kernel [2]	Compiler warning? [3]	With KASAN [4]	With UBSAN [5]
Defects not covered by the kernel's KUnit test_kasan.ko module					
1	Uninitialized Memory Read – UMR	N	Y [C1]	N	N
2	Use After Return – UAR	N	Y [C2]	N [SA]	N [SA]
3	Memory leakage [6]	N	N	N	N
Defects covered by the kernel's KUnit test_kasan.ko module					
4	OOB accesses on static global (compile-time) memory				
4.1	Read (right) overflow	N [V1]	N	Y [K1]	Y [U1,U2]
4.2	Write (right) overflow		N	Y [K1]	
4.3	Read (left) underflow		N	Y [K2]	
4.4	Write (left) underflow		N	Y [K2]	
4	OOB accesses on static global (compile-time) stack local memory				
4.1	Read (right) overflow	N [V1]	N	Y [K3]	Y [U1,U2]
4.2	Write (right) overflow		N	Y [K2]	
4.3	Read (left) underflow		N	Y [K2]	
4.4	Write (left) underflow				
5	OOB accesses on dynamic (kmalloc-ed slab) memory				
5.1	Read (right) overflow				
5.2	Write (right) overflow				
5.3	Read (left) underflow				
5.4	Write (left) underflow	N	N	Y [K4]	N
6	Use After Free - UAF	N	N	Y [K5]	N
7	Double-free	Y [V2]	N	Y [K6]	N

8	Arithmetic UB (*via the kernel's test_ubsan.ko module*)				
8.1	Add overflow				Y
8.2	Sub(tract) overflow				N
8.3	Mul(tiply) overflow				N
8.4	Negate overflow				N
	Div by zero	N	N	N	Y
8.5	Bit shift OOB	Y [U3]		Y [U3]	Y [U3]
Other than arithmetic UB defects (copied from the kernel's KUnit test_ubsan.ko module)					
8.6	OOB	Y [U3]		Y [U3]	Y [U3]
8.7	Load invalid value	Y [U3]	N	Y [U3]	Y [U3]
8.8	Misaligned access	N		N	N
8.9	Object size mismatch	Y [U3]		Y [U3]	Y [U3]
9	OOB on copy_ [to|from]_user*()	N	Y [C3]	Y [K4]	N

Table 5.3 – Summary of memory defect and arithmetic UB test cases caught (or not) by KASAN

You'll find the explanations for the footnote notations seen in the table (such as [C1], [U1], and so on) below.

Test environment

- [1] The test case number: do refer to the source of the test kernel module to see it – ch5/kmembugs_test/kmembugs_test.c, the debugfs entry creation and usage in debugfs_kmembugs.c, and the bash scripts load_testmod and run_tests, all within the same folder.

- [2] The compiler used here is GCC version 9.3.0 on x86_64 Ubuntu Linux. A later section - *Using Clang 13 on Ubuntu 21.10* - covers using the **Clang 13** compiler.

- [3] To test with KASAN, I had to boot via our custom debug kernel (5.10.60-dbg01) with CONFIG_KASAN=y and CONFIG_KASAN_GENERIC=y. We assume the Generic KASAN variant is being used.

- Test cases 4.1 through 4.4 work both upon static (compile-time allocated) global memory as well as stack local memory. That's why the test case numbers are 4.x in both.

Compiler warnings

- Version: this is for GCC version 9.3.0 on x86_64 Ubuntu:

  - [C1] The GCC compiler reports the UMR as a warning:

    ```
    warning: '<var>' is used uninitialized in this function
    [-Wuninitialized]
    ```

  - [C2] GCC reports the potential UAF defect as a warning:

    ```
    warning: function returns address of local variable
    [-Wreturn-local-addr]
    ```

  - [C3] GCC (quite cleverly!) catches the illegal copy_[to|from]_user() here. It figures out that the destination size is too small:

    ```
    * In function 'check_copy_size',
        inlined from 'copy_from_user' at ./include/linux/
    uaccess.h:191:6,
        inlined from 'copy_user_test' at <...>/ch5/kmembugs_
    test/kmembugs_test.c:482:14:
    ./include/linux/thread_info.h:160:4: error: call to
    '__bad_copy_to' declared with attribute error: copy
    destination size is too small
      160 |     __bad_copy_to();
          |     ^~~~~~~~~~~~~~~
    ```

- With the **Clang** 13 compiler (we cover using Clang to build the kernel and modules in the *Building your kernel and modules with Clang* section), the warnings are pretty much identical as with GCC. In addition, it emits variable 'xxx' set but not used [-Wunused-but-set-variable].

The following section delves into the details – don't miss out!

KASAN – detailed notes on the tabulated results

The footnote notations for KASAN ([K1], [K2], and so on) are explained in detail here. It's *really important* to read through all the notes, as we've mentioned certain caveats and corner cases as well:

- [K1] KASAN catches and reports the OOB access on global static memory as follows:

  ```
  global-out-of-bounds in <func>+0xstart/0xlen [modname]
  Read/Write of size <n> at addr <addr> by task <taskname/
  PID>
  ```

  The report will contain one of Read or Write depending upon whether a read or write buggy access occurred.

- [K2] Here, there are a number of caveats to note:

  - The **Out-Of-Bounds (OOB)** read/write left underflow on the global memory test case is caught *only when compiled with Clang version 11 or greater*. It isn't even caught by GCC 10 or 11, due to the way its red zoning works.

  - *KASAN only catches global memory OOB accesses when compiled with Clang 11 and later!* Thus, in my test runs with GCC 9.3 and Clang 10, I see it fails to catch the read/write underflow (left OOB) accesses on a global buffer (test cases 4.3 and 4.4)! Here, it does seem to catch the overflow defects on global memory, though you shouldn't take this for granted... (By the way, Clang is pronounced "clang" not "see-lang"). Also, though it's documented as supporting GCC from version 8.3.0 onward, this failed to catch (only) the read/write underflow bug test cases on global memory. Be sure to read the upcoming *Compiling your kernel and module with Clang* section!

  - However, even with GCC 9.3, the way the internal red zoning and padding seems to work, it appears that the first declared global (which variable exactly depends on how the linker sets it up) may not have a left red zone, causing left OOB buggy accesses to be missed... This is why – as a silly workaround for now, until GCC's fixed – we use three global arrays. We pass the middle one as the test buffer to work upon (any but the first) in the test cases. Hopefully, GCC will be fixed - properly red zoned – and any and all OOB accesses are caught. With our particular test runs, the buggy left OOB accesses are caught on global memory, even when compiled with GCC 9.3!

- These observations, caveats, and what-have-you are by their very nature at times a bit *iffy*. They can end up working in one way on one system and quite another on a differently configured system or architecture. Thus, *we heartily recommend you test your workload using an appropriately configured debug kernel with all tools at your disposal, including the usage of more recent compiler technology such as Clang, and the various tools and techniques covered in this book.* Yes, it's a lot of work and yes, it's worth it!

- [K3] KASAN catches and reports the OOB access on stack local memory as follows:

```
stack-out-of-bounds in <func>+0xstart/0xlen [modname]
Read/Write of size <n> at addr <addr> by task <taskname/
PID>
```

- [K4] KASAN catches and reports the OOB access on dynamic slab memory as follows:

```
BUG: KASAN: slab-out-of-bounds in <func>+0xstart/0xlen
[modname]
Read/Write of size <n> at addr <addr> by task <taskname/
PID>
```

- [K5] KASAN catches and reports the UAF defect as follows:

```
BUG: KASAN: use-after-free in <func>+0xstart/0xlen
[modname]
Read/Write of size <n> at addr <addr> by task <taskname/
PID>
```

- [K6] KASAN catches and reports the double-free as follows:

```
BUG: KASAN: double-free or invalid-free in
<func>+0xstart/0xlen [modname]
```

- In all the above cases, KASAN's report also shows the actual violation in detail along with the process context, (kernel-mode stack) call trace, and the shadow memory map, showing which variable the OOB memory access belongs to (if applicable) as well as the *memory state around the buggy address.*

> **Tip – the All-Results-in-One-Place Table**
>
> For your ready reference, in Part 2 of this key topic (the next chapter), in the *Catching memory defects in the kernel – comparisons and notes (Part 2)* section, *Table 6.4* tabulates our test case results for our test runs with all the tooling technologies – vanilla/distro kernel, compiler warnings, with KASAN, with UBSAN, and with SLUB debug – we employ in this chapter. In effect, it's a compilation of all the findings in one place, thus allowing you to make quick (and hopefully helpful) comparisons.

Did you notice regarding the kernel's built-in KUnit-based test cases on KASAN that the `test_kasan` kernel module does *not have* test cases for these three memory defects – the UMR, UAR, and memory leaks. Why? Simple: *KASAN does not catch these bugs!* Okay, so now what can we conclude? Well, the KUnit (and other) test suites are often run in an automated fashion where the expected end result is that all viable test cases are passed; in fact, they *must* pass. This wouldn't have happened had they contained these three defects, so they don't. Now, don't read it wrong – this is simply the way the test suites are designed. There certainly exist other means besides KASAN by which these defects will be caught. Relax – we'll get there and catch them.

Here and now, we're showing that KASAN itself doesn't catch these particular nasty bugs. Later in the book, we'll see which tools do.

FYI, KASAN is a key component to catching difficult-to-find bugs via the fuzzing approach. Syzkaller (aka syzbot) – the de facto powerful Linux kernel fuzzer – requires KASAN to be configured in the kernel! We cover fuzzing briefly in *Chapter 12, A few More Kernel Debug Approaches*, in the *What is fuzzing?* section. Be sure to check it out.

Good going – you now know how to leverage the power of KASAN to help catch those tricky memory bugs! Let's now move on to using UBSAN.

Using the UBSAN kernel checker to find Undefined Behavior

One of the serious issues with a language such as C is that the compiler produces code for the correct case, but when the source code does something unexpected or just plain wrong, the compiler often does not understand what to do – it simply and blithely ignores such cases. This actually helps in the generation of highly optimized code at the cost of (possible security) bugs! Examples of this are common: overflowing/underflowing an array, arithmetic defects (such as dividing by zero or overflowing/underflowing a signed integer), and so on. Even worse, at times the buggy code seems to work (as we saw with accessing stale stack memory in the *Stale frames – trouble in paradise* section). Similarly, bad code might work in the presence of optimization, or not. Thus, cases such as these cannot be predicted and are called **Undefined Behavior (UB)**.

The kernel's **Undefined Behavior Sanitizer (UBSAN)** catches several types of runtime UB. As with KASAN, it uses **Compile Time Instrumentation (CTI)** to do so. With UBSAN enabled fully, the kernel code is compiled with the `-fsanitize=undefined` option switch. The UB caught by UBSAN includes the following:

- Arithmetic-related UB:

  - Arithmetic overflow / underflow / divide by zero / and so on...

  - OOB accesses while bit shifting

- Memory-related UB:

  - OOB accesses on arrays

  - NULL pointer dereferences

  - Misaligned memory accesses

  - Object size mismatches

Some of these defects in fact overlap with what Generic KASAN catches as well. UBSAN instrumented code is certainly larger and slower (by a factor of 2 or 3 times). Still, it's very useful – especially during development and unit testing – to catch UB defects. In fact, enabling UBSAN on production systems is feasible if you can afford the larger kernel text size and processor overheads (on everything besides tiny embedded systems, you probably can).

Configuring the kernel for UBSAN

Within the make menuconfig UI, you'll find the menu system for UBSAN at Kernel hacking | Generic Kernel Debugging Instruments | Undefined behaviour sanity checker.

A screenshot of the relevant menu is seen here:

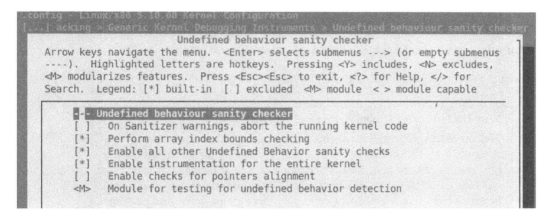

Figure 5.10 – Partial screenshot of the UBSAN menu for the Linux kernel config

To work with it, you should turn on the following kernel configs: CONFIG_UBSAN, CONFIG_UBSAN_BOUNDS (performs bound checking on array indices for static arrays – very useful!), CONFIG_UBSAN_MISC, and CONFIG_UBSAN_SANITIZE_ALL (you can look up the details for each here: lib/Kconfig.ubsan). Setting CONFIG_TEST_UBSAN=m has the lib/test_ubsan.c code built as a module.

> **UBSAN – Effect on the Build**
>
> With `CONFIG_UBSAN=y`, building the kernel source tree by passing the `V=1` parameter will show the details, the GCC flags being passed, and more. Here's a snippet of what you see focused on the GCC flags passed during the build due to UBSAN being enabled:
>
> ```
> make V=1
> gcc -Wp,-MMD,[...] -fsanitize=bounds
> -fsanitize=shift -fsanitize=integer-divide-by-zero
> -fsanitize=unreachable -fsanitize=signed-integer-
> overflow -fsanitize=object-size -fsanitize=bool
> -fsanitize=enum [...]
> ```

Hunting down UB with UBSAN

Detecting UB on OOB (static) array accesses (and the like) is where UBSAN shines. Take, for example, our test case #4.4. We define a few static global arrays like this:

```
static char global_arr1[10], global_arr2[10],  global_arr3[10];
```

> **Why Declare Three Global Arrays and Not Just One?**
>
> Well, as of this writing, there seems to be an issue with the way that the GCC compiler (at least as of version 9.3) sets up red zoning for global data. We observe that the red zone for the *first* global in a module may *not* have its left red zone correctly set up, causing the left OOB (underflow) buggy accesses to be missed as a side effect! So, by setting up three global arrays and passing the pointer to any but the first (we set up our test cases to pass the pointer to the second one), KASAN and UBSAN should be able to catch the buggy access! (Do note that the ordering of global variables within a module depends on the linker). This issue does not seem to occur with Clang 11+.
>
> Interestingly, our efforts on this will eventually pay off: due to my reporting the issue – left OOB failing with GCC – as well as pointing out that the kernel's `test_kasan` module doesn't test for it, *Marco Elver* (the current KCSAN maintainer) has investigated this and added a patch to include this test case – *add globals left-out-of-bounds test* – to the `test_kasan` module (17 Nov 2021 – see here: `https://lore.kernel.org/all/20211117110916.97944-1-elver@google.com/T/#u`). Further, this book's very able technical reviewer, *Chi-Thanh Hoang*, has figured out that this is essentially due to GCC's lack of a left red zone (as mentioned above) and added this information to the kernel Bugzilla (`https://bugzilla.kernel.org/show_bug.cgi?id=215051`). The hope is that GCC maintainers will pick this up and suggest or implement a fix.

Below, we show one of our buggy test cases – the right OOB accesses on global memory – accessing one of these global arrays, incorrectly of course, for both read and write (I only show a portion of its code here). Note that the parameter p is a pointer to a piece of global memory within this module, typically the second one, global_arr2[]:

Here's its invocation via our debugfs hook:

```
[...] else if (!strncmp(udata, "4.4", 4))
        global_mem_oob_left(WRITE, global_arr2);
```

Here's the (partial) code (note that the // style comments might spill over a line here; in the code they're fine):

```
int global_mem_oob_right(int mode, char *p)
{
    volatile char w, x, y, z;
    volatile char local_arr[20];
    char *volatile ptr = p + ARRSZ + 3; // OOB right
    [...]
    } else if (mode == WRITE) {
        *(volatile char *)ptr = 'x';   // invalid, OOB right
write
        p[ARRSZ - 3] = 'w'; // valid and within bounds
        p[ARRSZ + 3] = 'x'; // invalid, OOB right write
        local_arr[ARRAY_SIZE(local_arr) - 5] = 'y'; // valid
and within bounds
        local_arr[ARRAY_SIZE(local_arr) + 5] = 'z'; // invalid,
OOB right write
    } [...]
```

Once it detects a buggy access to memory (like the ones above), UBAN displays an error report like this to the kernel log:

```
array-index-out-of-bounds in <C-source-pathname.c>:<line#>
 index <index> is out of range for type '<var-type> [<size>]'
```

Here's a screenshot showing just this. The right window shows the kernel log. For this case, ignore the top portion of the log – it's part of the error report from KASAN. The remainder – what we're interested in – is from UBSAN:

```
167  * OOB on static (compile-time) mem: OOB read/write       [13676.756743] The buggy address belongs to the variable:
168  * Covers both read/write overflow on both static g        [13676.758424]  global_arr2+0xd/0xffffffffffffff6540 [test_kmembugs]
169  * The parameter p is a pointer to one of the globa
170  * this module.                                            [13676.761739] Memory state around the buggy address:
171  * Note: With gcc 10, 11 or clang < 11, KASAN isn't        [13676.763397]  ffffffffc09ba980: 00 00 00 00 00 00 00 00 00 00 00 00 00 00 00 00
172  * memory OOB on read/write underflow!                     [13676.765267]  ffffffffc09baa00: 00 00 00 00 00 00 00 00 00 00 00 00 00 00 00 00
173  */                                                        [13676.767093] >ffffffffc09baa80: 00 02 f9 f9 f9 f9 f9 00 02 f9 f9 f9 f9 f9 f9 f9
174  int global_mem_oob_right(int mode, char *p)               [13676.768736]                     ^
175  {                                                         [13676.770492]  ffffffffc09bab00: 00 02 f9 f9 f9 f9 f9 01 f9 f9 f9 f9 f9 f9 f9 f9
176      volatile char w, x, y, z;                             [13676.772293]  ffffffffc09bab80: 00 f9 f9 f9 f9 f9 f9 00 00 00 00 00 00 00 00 00
177      volatile char local_arr[20];                          [13676.774952] ===============================================================
178      char *volatile ptr = p + ARRSZ + 3; // OOB righ        [13676.776211]
179                                                            [13676.778116] UBSAN: array-index-out-of-bounds in /home/letsdebug/Linux-Kernel-Debu
180      if (mode == READ) {                                   t/kmembugs_test.c:194:12
181          w = *(volatile char *)ptr;  // invalid, OOB       [13676.781758] index 25 is out of range for type 'char [20]'
182          ptr = p + 3;                                      [13676.783505] CPU: 5 PID: 21522 Comm: run_tests Tainted: G    B D    0      5.10.60
183          x = *(volatile char *)ptr;  // valid              [13676.785334] Hardware name: innotek GmbH VirtualBox/VirtualBox, BIOS VirtualBox 12
184                                                            [13676.787197] Call Trace:
185          y = local_arr[ARRAY_SIZE(local_arr) - 5];         [13676.789020]  dump_stack+0xbd/0xfa
186          z = local_arr[ARRAY_SIZE(local_arr) + 5];         [13676.790882]  ubsan_epilogue+0x9/0x45
187      } else if (mode == WRITE) {                           [13676.792723]  __ubsan_handle_out_of_bounds+0x70/0x80
188          *(volatile char *)ptr = 'x';   // invalid,        [13676.794687]  global_mem_oob_right+0x1de/0x266 [test_kmembugs]
189                                                            [13676.796543]  ? leak_simple2+0x19b/0x19b [test_kmembugs]
190          p[ARRSZ - 3] = 'w'; // valid and within bou       [13676.796693]  ? __might_sleep+0x22d/0x2f0
191          p[ARRSZ + 3] = 'x'; // invalid, OOB right w        [13676.800734]  ? __kasan_check_write+0x14/0x20
192                                                            [13676.802041]  dbgfs_run_testcase+0x257/0x51a [test_kmembugs]
193          local_arr[ARRAY_SIZE(local_arr) - 5] = 'y';       [13676.804503]  ? _sub_I_65535_1+0x17/0x17 [test_kmembugs]
194          local_arr[ARRAY_SIZE(local_arr) + 5] = 'z';       [13676.806376]  ? rcu_read_lock_held_common+0x1e/0x60
195      }                                                     [13676.808157]  ? rcu_read_lock_any_held+0x60/0x110
196      return 0;
```

Figure 5.11 – Partial screenshot 1 of 3 showing UBSAN catching the right OOB write
to a stack local variable

Here you can see how UBSAN has precisely caught the UB on line 194 – the attempt to write after the end legal index of the local (stack-based) array! Of course, it's entirely possible the line number you see here might change over time due to modifications to the code.

After this, test case # 4.3 intentionally, adventurously – and disastrously – now attempts a read underflow on a local stack memory variable. This too is cleanly caught by UBSAN! The following partial screenshot shows you the juicy bit:

```
211  char *volatile ptr = p - 3; // left OOB                   [13959.698401] afffffffc09baa80: 00 02 f9 f9 f9 f9 f9 00 02 f9 f9 f9 f9 f9 f9
212                                                            [13959.700017]                   ^
213  if (mode == READ) {                                       [13959.701726]  ffffffffc09baa80: 00 02 f9 f9 f9 f9 f9 01 f9 f9 f9 f9 f9 f9 f9
214      /* Interesting: this OOB access isn't caught          [13959.703366]  ffffffffc09baa80: 00 f9 f9 f9 f9 f9 f9 00 00 00 00 00 00 00 00 00
215      w = *(volatile char *)ptr;  // invalid, OOB           [13959.705303] ===============================================================
216                                                            [13959.797343]
217      /* ... but these OOB accesses are caught by           [13959.709187] UBSAN: array-index-out-of-bounds in /home/letsdebug/Linux-Kernel-Debugging/ch7/kmembugs_tes
218       * We conclude that *only* the index-based            t/kmembugs_test.c:223:16
219       * And, KASAN compiled with clang 11 or lat           [13959.712994] index -5 is out of range for type 'char [20]'
220       */                                                   [13959.714762] CPU: 2 PID: 21538 Comm: run_tests Tainted: G    B D    0      5.10.60-dbg02-gcc #17
221      x = p[-3]; // invalid, OOB left read                  [13959.716692] Hardware name: innotek GmbH VirtualBox/VirtualBox, BIOS VirtualBox 12/01/2006
222                                                            [13959.718807] Call Trace:
223      y = local_arr[-5]; // invalid, not within            [13959.720696]  dump_stack+0xbd/0xfa
224      z = local_arr[5];  // valid, within bounds            [13959.722518]  ubsan_epilogue+0x9/0x45
225  } else if (mode == WRITE) {                               [13959.724358]  __ubsan_handle_out_of_bounds+0x70/0x80
226      /* Interesting: this OOB access isn't caught          [13959.726310]  global_mem_oob_left+0xdf/0x267 [test_kmembugs]
```

Figure 5.12 – Partial screenshot 2 of 3 showing UBSAN catching the left OOB read on a stack
local variable

Again, UBSAN even shows the source filename and line number where the buggy access was attempted!

It's more generic: UBSAN catches memory accesses when the variable in question *indexes the static memory array incorrectly* – when the index is out of bounds in any manner (left or right, underflow or overflow). It does appear, though, to miss buggy accesses made purely via pointers! KASAN has no issue with this and catches them all.

Just as we saw with KASAN (in the *Remaining tests with our custom buggy kernel module* section), UBSAN also cannot catch all memory defects. To prove this, we again run our custom buggy kernel module (in ch5/kmembugs_test), with pretty much identical results: *even on a UBSAN-enabled kernel, these three bugs – the UMR, UAR, and memory leakage bugs – aren't caught!* The following screenshot tells the story (to capture this, I (first) ran the run_tests script for the first three test cases with the --no-clear parameter, in order to preserve the kernel log content):

```
$ grep -w CONFIG_KASAN /boot/config-5.10.60-dbg02-gcc
CONFIG_KASAN=y
$ grep -w CONFIG_UBSAN /boot/config-5.10.60-dbg02-gcc
CONFIG_UBSAN=y
$ dmesg
[ 5147.233197] testcase to run: 1
[ 5147.233202] test_kmembugs:umr(): testcase 1: UMR (val=1039927376)
[ 5150.323534] testcase to run: 2
[ 5150.323541] test_kmembugs:uar(): testcase 2: UAR:
[ 5150.323546] testcase 2: UAR: res1 = "<whoops, it's NULL; UAR!>"
[ 5184.711447] testcase to run: 3.2
[ 5184.711455] test_kmembugs:leak_simple2(): testcase 3.2: simple memory leak testcase 2
[ 5184.711489]    res2 = "leaky!!"
$
```

Figure 5.13 – Screenshot 3 of 3: executing the first three – UMR, UAR, and leakage – test cases with our test module reveals that both KASAN and UBSAN (enabled in kernel) don't catch them

Also, don't forget: *UBSAN is quite adept at catching arithmetic-related UB too* – things such as overflowing or underflowing arithmetic calculations, the well-known **Integer OverFlow (IoF)** defect, and the divide-by-zero bugs being common and dangerous ones indeed! We mentioned the arithmetic UB that UBSAN can catch at the beginning of this section on UBSAN. We don't delve further into it as our topic is memory defects. To see more of UBSAN in action, you can always read the code of the UBSAN test module within the kernel (lib/test_ubsan.c) and try it out – I encourage you to do so. On a somewhat related note, understanding what *unaligned memory access* is, how it can cause issues, and how to avoid it is the topic of this kernel documentation page: *Unaligned Memory Accesses*: https://www.kernel.org/doc/html/latest/core-api/unaligned-memory-access.html#unaligned-memory-accesses.

Okay, let's tabulate the result of our experiments by running various test cases with UBSAN enabled within the kernel. Refer to the following table:

Test case # [1]	Memory defect type (below) / Infrastructure used (right)	Compiler warning? [2]	With UBSAN [3]
Defects not covered by the kernel's KUnit test_kasan.ko module			
1	Uninitialized Memory Read – UMR	Y [C1]	N
2	Use After Return – UAR	Y [C2]	N [SA]
3	Memory leakage [6]	N	N
Defects covered by the kernel's KUnit test_kasan.ko module			
4	OOB accesses on static global (compile-time) memory		
4.1	Read (right) overflow		
4.2	Write (right) overflow		
4.3	Read (left) underflow	N	
4.4	Write (left) underflow		Y [U1,U2]
4	OOB accesses on static global (compile-time) stack local memory		
4.1	Read (right) overflow		
4.2	Write (right) overflow		
4.3	Read (left) underflow		
4.4	Write (left) underflow	N	Y [U1,U2]
5	OOB accesses on dynamic (kmalloc-ed slab) memory		
5.1	Read (right) overflow		
5.2	Write (right) overflow		
5.3	Read (left) underflow	N	N
5.4	Write (left) underflow		
6	Use After Free - UAF	N	N
7	Double-free	N	N

8	Arithmetic UB (*via the kernel's test_ubsan.ko module*)		
8.1	Add overflow		Y
8.2	Sub(tract) overflow		N
8.3	Mul(tiply) overflow		N
8.4	Negate overflow		N
	Div by zero		Y
8.5	Bit shift OOB	N	Y [U3]
Other than arithmetic UB defects (copied from the kernel's KUnit test_ubsan.ko module)			
8.6	OOB		Y [U3]
8.7	Load invalid value		Y [U3]
8.8	Misaligned access	N	N
8.9	Object size mismatch		Y [U3]
9	OOB on copy_[to|from]_user*()	Y [C3]	N

Table 5.4 – Summary of memory defect and arithmetic UB test cases caught (or not) by UBSAN

The following is with respect to the numeric footnotes in the preceding table:

- [1] The test case number: do refer to the source of the test kernel module to see it
  – ch5/kmembugs_test/kmembugs_test.c, the debugfs entry creation and usage in debugfs_kmembugs.c, and the bash scripts load_testmod and run_tests, all within the same folder.

- [2] The compiler used here is GCC version 9.3.0 on x86_64 Ubuntu Linux. A later section covers using the **Clang 13** compiler.

- [3] To test with UBSAN, I booted via our custom production kernel (5.10.60-prod01) with CONFIG_UBSAN=y and CONFIG_UBSAN_SANITIZE_ALL=y.

- Test cases 4.1 through 4.4 work both upon static (compile-time allocated) global memory as well as stack local memory. That's why the test case numbers are 4.x in both.

The following section delves into the details. Don't miss out!

UBSAN – detailed notes on the tabulated results

The footnote notations in the preceding table (such as [U1], [U2], and so on) are explained in detail here. It's important to read through all the notes, as we've mentioned certain caveats and corner cases as well:

- [U1] UBSAN catches and reports the OOB access on global static memory:

  ```
  array-index-out-of-bounds in <C-source-pathname.
  c>:<line#>
  index <index> is out of range for type '<var-type>
  [<size>]'
  ```

- [U2] When relevant, UBSAN also reports an object size mismatch for [U1] as follows:

  ```
  object-size-mismatch in <C-source-pathname.c>:<line#>
  store to address <addr> with insufficient space for an
  object of type '<var-type>'
  ```

In the preceding cases, UBSAN also reports the actual violation in some detail along with the process context and kernel-mode stack call trace.

Note though, that with KASAN turned off (I rebuilt a test debug kernel with CONFIG_KASAN=n) and UBSAN turned on, the semantics seem a bit different: in this case, I got a segfault only, with, of course, the kernel log clearly showing the source of the bug (by looking up what the instruction pointer register, here, RIP, was pointing to at the time of the fault).

> **Note**
>
> As mentioned earlier, don't forget to look up *Table 6.4* in the following chapter, effectively, an all-results-in-one-place comparison table.

Great, now you're much better armed to catch memory bugs with both KASAN and UBSAN! I suggest you first take the time to absorb all this information, read the relevant detailed notes in the later *Catching memory defects in the kernel – comparisons and notes (Part 1)* section (pertaining to KASAN and UBSAN, at least for now), and practice trying out these test cases on your own. But wait: we saw that some OOB defects are only caught when compiled with Clang 11 or later. This is a key thing. So, let's now learn how to use the modern Clang compiler.

Building your kernel and modules with Clang

Low Level Virtual Machine (**LLVM**) is the original name given to this modular compiler tooling project. It now doesn't have much to do with traditional virtual machines and is instead a powerful backend for several compilers and toolchains.

Clang (the pronunciation rhymes with "slang") is a modern compiler frontend technology for C-type languages (includes support for C, C++, CUDA, Objective C/C++, and more) and is based on the LLVM compiler. It's considered a drop-in replacement for GCC. Clang currently seems to have a significant advantage over GCC – especially from our point of view – generating superior diagnostics as well as being able to intelligently generate code avoiding OOB accesses. This is critical. It paves the way to superior code. We saw (in the previous section on KASAN) that faulty left-OOB accesses on global memory, not reliably caught by GCC (versions 9.3, 10, and 11), are caught with Clang! The Android project is a key user of Clang, among many others.

Attempting to build your kernel module with Clang while the target kernel itself is compiled via GCC is simply not good enough! *You'll have to use the same compiler for both* – the underlying ABI needs to be completely consistent (this was one of the many things pointed out to me by Marco Elver when I was puzzled and asked why KASAN failed to catch certain test cases – again, the beauty of open source development). So, the upshot of it all is that we'll have to compile both our kernel and module with Clang 11.

Installing Clang and associated binaries in order to successfully compile your kernel module involves running the following command (on our Ubuntu 20.04 LTS guest):

```
sudo apt install clang-11 --install-suggests
```

Further, we seem to require setting up a soft link to `llvm-objdump-11` named `llvm-objdump` (this is likely as I have both Clang 10 and Clang 11 installed simultaneously):

```
sudo ln -s /usr/bin/llvm-objdump-11 /usr/bin/llvm-objdump
```

Hang on, a simpler approach follows...

Using Clang 13 on Ubuntu 21.10

For the purpose of using Clang on the kernel and module builds, instead of installing Clang 11 (or later) on Ubuntu 20.04 LTS, it might just be simpler to install Ubuntu 21.10 (I've done so as an x86_64 VM) *as it ships with Clang 13 preinstalled*. I then built the very same 5.10.60 kernel as a debug kernel, applying a similar debug config as was discussed back in *Chapter 1, A General Introduction to Debugging Software*, but this time with Clang.

Importantly, to specify using Clang (and not GCC) as the compiler, when building the kernel, set the CC variable to it:

```
$ time make -j8 CC=clang
   SYNC     include/config/auto.conf.cmd
*
* Restart config...
* Memory initialization
*
```

The first time you run this command, the *kbuild* system detects that with the Clang compiler, certain add-ons now become available and viable to use (that couldn't be used with GCC) and prompts us to configure it:

```
Initialize kernel stack variables at function entry
> 1. no automatic initialization (weakest) (INIT_STACK_NONE)
  2. 0xAA-init everything on the stack (strongest) (INIT_STACK_
ALL_PATTERN) (NEW)
  3. zero-init everything on the stack (strongest and safest)
(INIT_STACK_ALL_ZERO) (NEW)
choice[1-3?]:
```

Though it would be very useful to take advantage of this auto-initialization of kernel stack variables, I deliberately left it at the default (option 1) in order to check our tooling to catch the UMR defect. Similarly, the build asked the following. Here, I kept the defaults by simply pressing the *Enter* key. You could change them if you wish to:

```
Enable heap memory zeroing on allocation by default (INIT_ON_
ALLOC_DEFAULT_ON) [Y/n/?] y
Enable heap memory zeroing on free by default (INIT_ON_FREE_
DEFAULT_ON) [Y/n/?] y
*
* KASAN: runtime memory debugger
*
KASAN: runtime memory debugger (KASAN) [Y/n/?] y
  KASAN mode
  > 1. Generic mode (KASAN_GENERIC)
  choice[1]: 1
  [...]
  Back mappings in vmalloc space with real shadow memory
```

```
(KASAN_VMALLOC) [Y/n/?] y
  KUnit-compatible tests of KASAN bug detection capabilities
(KASAN_KUNIT_TEST) [M/n/?] m
  [...]
```

Once built, perform the usual remaining steps, not forgetting to add the `CC=clang` environment variable to the command line:

```
sudo make CC=clang modules_install && sudo make CC=clang
install
```

When done, reboot and ensure you boot into your spanking new Clang-built debug kernel! Verify with the following:

```
$ cat /proc/version
Linux version 5.10.60-dbg02 (letsdebug@letsdebug-VirtualBox)
(Ubuntu clang version 13.0.0-2, GNU ld (GNU Binutils for
Ubuntu) 2.37) #4 SMP PREEMPT Wed ...
```

Now, let's move on to building our kernel module with Clang:

```
cd <book_src>/ch5/kmembugs_test
make CC=clang
```

That's it – I've conditionally embedded this setting of the `CC` variable into our `load_testmod` bash script, based on which compiler was used to build the current kernel. Also, FYI, to distinguish between our custom debug kernel built with Clang and GCC, the former's `uname -r` output shows up as seen here, 5.10.60-dbg02, whereas the latter's name shows up as 5.10.60-dbg02-gcc.

Exercise

I'll leave it as an exercise to you to build both a (debug) kernel as well as our `test_kmembugs.ko` kernel module with *Clang* and run the test cases.

With this, we complete the first part of our detailed coverage on understanding and catching memory defects within the kernel! Great going. Let's complete this chapter with a kind of summarization of the many tools and techniques we've used so far.

Catching memory defects in the kernel – comparisons and notes (Part 1)

As we've already mentioned in this chapter, the following table tabulates our test case results for our test runs with all the tooling technologies/kernels – vanilla/distro kernel, compiler warnings, and with KASAN and UBSAN with our debug kernel – we employed in this chapter. In effect, *it's a compilation of all our findings so far in one place*, thus allowing you to make quick (and hopefully helpful) comparisons:

Testcase # [1]	Memory defect type (below) / Infrastructure used (right)	Distro kernel [2]	Compiler warning? [3]	With KASAN [4]	With UBSAN [5]
Defects not covered by the kernel's KUnit test_kasan.ko module					
1	Uninitialized Memory Read – UMR	N	Y [C1]	N	N
2	Use After Return – UAR	N	Y [C2]	N [SA]	N [SA]
3	Memory leakage [6]	N	N	N	N
Defects covered by the kernel's KUnit test_kasan.ko module					
4	OOB accesses on static global (compile-time) memory				
4.1	Read (right) overflow	N [V1]	N	Y [K1]	Y [U1,U2]
4.2	Write (right) overflow		N	Y [K1]	
4.3	Read (left) underflow		N	Y [K2]	
4.4	Write (left) underflow		N	Y [K2]	
4	OOB accesses on static global (compile-time) stack local memory				
4.1	Read (right) overflow	N [V1]	N	Y [K3]	Y [U1,U2]
4.2	Write (right) overflow		N	Y [K2]	
4.3	Read (left) underflow		N	Y [K2]	
4.4	Write (left) underflow				
5	OOB accesses on dynamic (kmalloc-ed slab) memory				
5.1	Read (right) overflow	N	N	Y [K4]	N
5.2	Write (right) overflow				
5.3	Read (left) underflow				
5.4	Write (left) underflow				
6	Use After Free - UAF	N	N	Y [K5]	N
7	Double-free	Y [V2]	N	Y [K6]	N

8	Arithmetic UB (*via the kernel's test_ubsan.ko module*)				
8.1	Add overflow				Y
8.2	Sub(tract) overflow				N
8.3	Mul(tiply) overflow				N
8.4	Negate overflow				N
	Div by zero	N	N	N	Y
8.5	Bit shift OOB	Y [U3]		Y [U3]	Y [U3]
Other than arithmetic UB defects (copied from the kernel's KUnit test_ubsan.ko module)					
8.6	OOB	Y [U3]		Y [U3]	Y [U3]
8.7	Load invalid value	Y [U3]	N	Y [U3]	Y [U3]
8.8	Misaligned access	N		N	N
8.9	Object size mismatch	Y [U3]		Y [U3]	Y [U3]
9	OOB on copy_[to|from]_user*()	N	Y [C3]	Y [K4]	N

Table 5.5 – Summary of various common memory defects and how various technologies react in catching them (or not)

Of course, the explanations of the footnotes within this table (such as [C1], [K1], [U1], and so on) can be found in the earlier relevant section.

So, here's a very brief summary:

- KASAN catches pretty much all OOB buggy memory accesses on global (static), stack local, and dynamic (slab) memory. UBSAN doesn't catch the dynamic slab memory OOB accesses (test cases 4.x and 5.x).

- KASAN does not catch the UB defects (test cases 8.x); UBSAN does catch (most of) them.

- Neither KASAN nor UBSAN catch the first three test cases – UMR, UAR, and leakage bugs, *but the compiler(s) generate warnings and static analyzers (cppcheck) can catch some of them*. (We in fact cover using a static analyzer to catch this tricky UAR bug in *Chapter 12, A few more kernel debugging approaches* in the *Examples – using cppcheck, checkpatch.pl for static analysis* section).

- The kernel **kmemleak** infrastructure catches kernel memory leaks allocated by any of k{m|z}alloc(), vmalloc(), or kmem_cache_alloc() (and friends) interfaces.

Regarding the preceding table, a few remaining notes now follow...

Miscellaneous notes

A few more points regarding *Table 5.5*:

- [V1]: The system could simply Oops, hang here, or even appear to remain unscathed, but that's not really the case... Once the kernel is buggy, the system is buggy.

- [V2]: Please see the explanation for this detailed in the following chapter, in the *Running SLUB debug test cases on a kernel with slub_debug turned off* section

A quick note on a KASAN alternative, especially for production systems, follows.

Introducing KFENCE – Kernel Electric-Fence

The Linux kernel has recent tooling named **Kernel Electric-Fence** (**KFENCE**). It's available from kernel version 5.12 onward (very recent, as of this writing).

KFENCE is described as a *low-overhead sampling-based memory safety error detector of heap use-after-free, invalid-free, and out-of-bounds access errors.*

It has recently added support for both x86 and ARM64 architectures with hooks to both the SLAB and SLUB memory allocators within the kernel. Why is KFENCE useful when we already have KASAN (which seems to overlap in function with it)? Here are a few points to help differentiate between them:

- KFENCE has been designed for use in *production* systems; KASAN's overhead would be too high for typical production systems and is suitable only on debug / development systems. KFENCE's performance overhead is minimal – close to zero.

- KFENCE works on a sampling-based design. It *trades precision for performance*, thus, with sufficiently lengthy uptime, KFENCE is almost certain to catch bugs! One way to have a really long total uptime is by deploying it across a fleet of machines.

- In effect, KASAN will catch all memory defects, but at a rather high performance cost. KFENCE also can catch all memory defects, at virtually no performance cost, but it takes time (very long uptimes are required, as it's a sampling-based approach). Thus, to catch memory defects on debug and development systems, use KASAN (and KFENCE, perhaps); to do the same on production systems, use KFENCE.

To enable KFENCE, set CONFIG_KFENCE=y (note, though, that as it's very recent, this config option *isn't* present in the 5.10 kernel series we work upon in this book). You can see more options and fine-tune them based on options present in the lib/Kconfig. kfence file.

We refer you to the details (including setup, tuning, interpreting error reports, internal implementation, and more) in the official kernel documentation page on KFENCE here: `https://www.kernel.org/doc/html/latest/dev-tools/kfence.html#kernel-electric-fence-kfence`.

A final point: with the 5.18 kernel (the latest stable one as of this writing), a new stricter `memcpy()` API family (covering the `memcpy()`, `memmove()` and `memset()` APIs), compile-time bounds checking kernel feature, has been introduced. It internally uses the compiler fortification feature (the kernel config is called `CONFIG_FORTIFY_SOURCE`). This being turned on helps catch a large class of typical buffer overflow defects within the kernel! Read more in LWN article here: *Strict memcpy() bounds checking for the kernel*: `https://lwn.net/Articles/864521/`.

Summary

With a non-managed programming language such as C, a trade-off exists: high power and the ability to code virtually anything you can imagine but at a significant cost. With memory being managed directly by the programmer, slipping in memory defects – bugs! – of all kinds, is rather easy to do, even for experienced folk.

In this chapter, we covered many tools, techniques, and approaches in this regard. First, you learned about the different (scary) types of memory defects. Then, we delved into how to use various tools and techniques to identify them and thus be able to fix them.

One of the most powerful tools in your arsenal for detecting memory bugs is KASAN. You learned how to configure and use it. We first learned how to use the kernel's built-in KUnit test framework to run memory test cases for KASAN to catch. We then developed our own custom module with test cases and even a neat way to test, via a debugfs pseudofile and custom scripts.

Catching UB with UBSAN came next. You learned how to configure it and leverage it to catch these kinds of defects, often overlooked, leading to not only buggy headaches but even security holes in production systems!

We learned that while GCC is solid and has been around for decades, a newer compiler, Clang, is in fact proving more adept at generating useful diagnostics (on our C code) and catching bugs that even GCC can miss! You saw how to use Clang to build the kernel and your modules, helping create more robust software, in effect.

As we covered these tools and frameworks, we tabulated the results, showing you the bugs a given tool can (or cannot) catch. To then summarize the whole thing, we built a larger table with columns covering all the test cases and all the tools – a quick and useful way for you to see and compare them (*Table 5.5*)! Note that we'll add to this table in the following chapter! Finally, we mentioned that the (very recent) KFENCE framework can (should) be used on production systems, in lieu of KASAN. The 5.18 kernel's `CONFIG_FORTIFY_SOURCE` config will likely be a big help as well.

So, congrats on completing this rather long – and really important – first chapter on catching memory bugs in kernel space! Do take the time to digest it and practice all you've learned. When set, I encourage you to move on to the next chapter where we'll complete our coverage on catching kernel memory defects.

Further reading

- Rust in the Linux kernel?

    - *Rust in the Linux kernel*, Apr 2021, Google security blog: `https://security.googleblog.com/2021/04/rust-in-linux-kernel.html`

    - *Let the Linux kernel Rust*, J Wallen, July 2021, TechRepublic: `https://www.techrepublic.com/article/let-the-linux-kernel-rust/`

    - *Linus Torvalds weighs in on Rust language in the Linux kernel*, ars technica, Mar 2021: `https://arstechnica.com/gadgets/2021/03/linus-torvalds-weighs-in-on-rust-language-in-the-linux-kernel/`

- Linux kernel security:

    - Several links and info here, from my *Linux Kernel Programming* book's *Further reading* section: `https://github.com/PacktPublishing/Linux-Kernel-Programming/blob/master/Further_Reading.md#kernel_sec`

    - *How a simple Linux kernel memory corruption bug can lead to complete system compromise*, Jann Horn, Project Zero, Oct 2021: `https://googleprojectzero.blogspot.com/2021/10/how-simple-linux-kernel-memory.html`

- Undefined Behavior (UB) – what is it?

  - Very comprehensive: A Guide to Undefined Behavior in C and C++, Part 1, John Regehr, July 2010: `https://blog.regehr.org/archives/213`

  - What Every C Programmer Should Know About Undefined Behavior #1/3, LLVM blog, May 2011: `http://blog.llvm.org/2011/05/what-every-c-programmer-should-know.html`

- KASAN – the Kernel Address Sanitizer:

  - Official kernel documentation: *The Kernel Address Sanitizer (KASAN)*: `https://www.kernel.org/doc/html/latest/dev-tools/kasan.html#the-kernel-address-sanitizer-kasan`

  - [K]ASAN internal working: `https://github.com/google/sanitizers/wiki/AddressSanitizerAlgorithm`

  - The ARM64 memory tagging extension in Linux, Jon Corbet, LWN, Oct 2020: `https://lwn.net/Articles/834289/`

  - How to use KASAN to debug memory corruption in an OpenStack environment: `https://www.slideshare.net/GavinGuo3/how-to-use-kasan-to-debug-memory-corruption-in-openstack-environment-2`

  - Android AOSP: Building a pixel kernel with KASAN+KCOV: `https://source.android.com/devices/tech/debug/kasan-kcov`

  - FYI, the original V2 KASAN patch post: *[RFC/PATCH v2 00/10] Kernel address sainitzer (KASan) - dynamic memory error deetector.*, LWN, Sept 2014: `https://lwn.net/Articles/611410/`

- UBSAN:

  - *The Undefined Behavior Sanitizer – UBSAN*: `https://www.kernel.org/doc/html/latest/dev-tools/ubsan.html#the-undefined-behavior-sanitizer-ubsan`

  - Improving Application Security with UndefinedBehaviorSanitizer (UBSan) and GCC, Meirowitz, May 2021: `https://blogs.oracle.com/linux/post/improving-application-security-with-undefinedbehaviorsanitizer-ubsan-and-gcc`

- Clang 13 documentation: UndefinedBehaviorSanitizer: `https://clang.llvm.org/docs/UndefinedBehaviorSanitizer.html`

- Android AOSP: *Integer Overflow Sanitization*: `https://source.android.com/devices/tech/debug/intsan`

- Kernel built-in test frameworks:

  - KUnit – Unit Testing for the Linux Kernel: `https://www.kernel.org/doc/html/latest/dev-tools/kunit/index.html#kunit-unit-testing-for-the-linux-kernel`

  - Linux Kernel Selftests: `https://www.kernel.org/doc/html/latest/dev-tools/kselftest.html#linux-kernel-selftests`

- KFENCE: official kernel documentation (only from ver 5.12): `https://www.kernel.org/doc/html/latest/dev-tools/kfence.html#kernel-electric-fence-kfence`

  With regard to the 5.18 mainline kernel: *Strict memcpy() bounds checking for the kernel*, Jon Corbet, July 2021: `https://lwn.net/Articles/864521/`

- Though it's with respect to userspace, useful: *Memory error checking in C and C++: Comparing Sanitizers and Valgrind*, Red Hat Developer, May 2021: `https://developers.redhat.com/blog/2021/05/05/memory-error-checking-in-c-and-c-comparing-sanitizers-and-valgrind`.

6
Debugging Kernel Memory Issues – Part 2

Welcome to the second portion of our detailed discussions on a really key topic – understanding and learning how to detect kernel memory corruption defects. In the preceding chapter, we introduced the reason why memory bugs are common and challenging and went on to cover some really important tools and technologies to help catch and defeat them – KASAN and UBSAN (along the way, covering the usage of the newer Clang compiler).

In this chapter, we continue this discussion. Here, we will focus on the following main topics:

- Detecting slab memory corruption via SLUB debug
- Finding memory leakage issues with kmemleak
- Catching memory defects in the kernel – comparison and notes (Part 2)

Technical requirements

The technical requirements and workspace remain identical to what's described in *Chapter 1, A General Introduction to Debugging Software*. The code examples can be found within the book's GitHub repository here: `https://github.com/PacktPublishing/Linux-Kernel-Debugging`.

Detecting slab memory corruption via SLUB debug

Memory corruption can occur due to various bugs or defects: **Uninitialized Memory Reads (UMR)**, **Use After Free (UAF)**, **Use After Return (UAR)**, **double-free**, **memory leakage**, or illegal **Out Of Bounds (OOB)** accesses that attempt to work upon (read/write/execute) illegal memory regions. They're unfortunately a very common root cause of bugs. Being able to debug them is a key skill. Having already checked out a few ways to catch them (the detailed coverage of setting up and using KASAN and UBSAN in the previous chapter), let's now leverage the kernel's built-in SLUB debug features to catch these bugs!

As you will know, memory is dynamically allocated and freed via the kernel's engine – the *page (or Buddy System) allocator*. To mitigate serious wastage (internal fragmentation) issues that it can face, the slab allocator (or slab cache) is layered upon it, serving two primary tasks – providing fragments of pages efficiently (within the kernel, allocation requests for small pieces of memory, from a few bytes to a couple of kilobytes, tend to be very common), and serving as a cache for commonly used data structures.

Current Linux kernels typically have three mutually exclusive implementations of the slab layer – the original SLAB, the newer and superior SLUB implementation, and the seldom-used SLOB implementation. It's key to realize that the following discussion is with respect to *only* the SLUB (unqueued allocator) implementation of the slab layer. It's typically the default in most Linux installations (the config option is named `CONFIG_SLUB`. It's found in the `menuconfig` UI here: `General setup|Choose SLAB allocator`).

> **Tip**
> Basic knowledge of the kernel memory management system, the page, slab allocator, and the various APIs to actually allocate (and free) kernel memory are a prerequisite for these materials. I've covered this (and much more) in the *Linux Kernel Programming* book (published by Packt in March 2021).

Let's quickly check out configuring the kernel for SLUB debug.

Configuring the kernel for SLUB debug

The kernel provides a good deal of support to help debug slab (SLUB, really) memory corruption issues. Within the kernel config UI, you'll find the following:

- `General Setup|Enable SLUB debugging support` (`CONFIG_SLUB_DEBUG`):

  - Turning this on buys you plenty of built-in SLUB debug support, the ability to view all slab caches via `/sys/slab`, and runtime cache validation support.

  - This config is automatically turned on (auto-selected) when Generic KASAN is on.

- `Memory Debugging|SLUB debugging on by default` (`CONFIG_SLUB_DEBUG_ON` is explained later).

Let's look up the kernel config for SLUB on my x86_64 Ubuntu guest running our custom debug kernel:

```
$ grep SLUB_DEBUG /boot/config-5.10.60-dbg02
CONFIG_SLUB_DEBUG=y
# CONFIG_SLUB_DEBUG_ON is not set
```

This config implies that SLUB debugging is available but disabled by default (as `CONFIG_SLUB_DEBUG_ON` is off). While always enabling it is perhaps useful for catching memory corruption, it can have quite a large (and adverse) performance impact. To mitigate this, you can – should, really – configure your debug kernel with `CONFIG_SLUB_DEBUG_ON` turned off by default (as seen here) and use the kernel command-line parameter `slub_debug` to fine-tune SLUB debugging as and when required.

The official kernel documentation here covers the usage of `slub_debug` in detail: `https://www.kernel.org/doc/html/latest/vm/slub.html`. We'll summarize it along with some examples to demonstrate how to use this powerful feature.

Leveraging SLUB debug features via the slub_debug kernel parameter

So, you'd like to leverage the `slub_debug` kernel command-line parameter! To do so, let's first understand the various option flags you can pass via it at boot time:

Flag to slub_debug= on kernel cmdline	Meaning	In more detail...
null (pass nothing after the =)	Switch all SLUB debugging on	All checks as specified by all the flags in this table are turned on.
F	Sanity checks on (enables SLAB_DEBUG_ CONSISTENCY_ CHECKS)	Performs (expensive) sanity checks at memory allocation and freeing times; minimally enables the double-free check. Corresponding sysfs pseudofile: `/sys/kernel/slab/<slabname>/sanity_checks` **Tip**: This option by itself can be useful even on production systems when a memory corruption bug must be identified. Look at the official kernel doc for more: `https://www.kernel.org/doc/html/latest/vm/slub.html#emergency-operations`.
Z	Red zoning	Cache objects will be red zoned (enabling OOB access checks). Corresponding sysfs pseudofile: `/sys/kernel/slab/<slabname>/red_zone`
P	Poisoning (object and padding)	Corresponding sysfs pseudofile: `/sys/kernel/slab/<slabname>/poison` See the *Understanding the SLUB layer's poison flags* section.
U	User tracking (free and alloc)	Stores last owner; useful for catching bugs! Corresponding sysfs pseudofile: `/sys/kernel/slab/<slabname>/store_user`
T	Trace (overhead; should only use on single slabs)	Traces allocs and frees of objects belonging to this slab cache. Corresponding sysfs pseudofile: `/sys/kernel/slab/<slabname>/trace`
A	Enable failslab filter mark for the cache	For fault injection purposes.
O	Switch SLUB debugging off	Applies to caches that would have caused higher minimum slab orders.
-	Switch all SLUB debugging off	Can be very useful (to remove checks and thus gain performance) when the kernel is configured with `CONFIG_SLUB_DEBUG_ON=y`.

Table 6.1 – The slub_debug=<NNN> flags and corresponding sysfs entries if any

A brief description of pretty much every (pseudo) file under `/sys/kernel/`
`slab/<slabname>` can be found in the kernel documentation here (a word of caution:
it seems to be quite aged): `https://www.kernel.org/doc/Documentation/`
`ABI/testing/sysfs-kernel-slab`.

Understanding the SLUB layer's poison flags

The poison flags defined by the kernel are defined as follows:

```
// include/linux/poison.h
#define POISON_INUSE 0x5a /* for use-uninitialised poisoning */
#define POISON_FREE 0x6b /* for use-after-free poisoning */
#define POISON_END  0xa5 /* end-byte of poisoning */
```

Here's the nitty-gritty on these poison values:

- When you use the `SLAB_POISON` flag when creating a slab cache (typically via
  the `kmem_cache_create()` kernel API) or set poisoning to on via the kernel
  parameter `slub_debug=P`, the slab memory gets auto-initialized to the value
  `0x6b` (which is ASCII k, corresponding to the `POISON_FREE` macro). In effect,
  when this flag is enabled, this (`0x6b`) is the value that *valid but uninitialized* slab
  memory regions are set to on creation.

- The `POISON_INUSE` value (`0x5a` equals ASCII z) is used to denote padding
  zones, before or after red zones.

- The last legal byte of the slab memory object is set to `POISON_END`, `0xa5`.

(You'll come across a nice example of seeing these poison values in action a bit later in this
section, in *Figure 6.4.*)

Our `ch5/kmembugs_test.c` code has the function `umr_slub()`. It employs the
`kmalloc()` API to dynamically allocate 32 bytes and then reads the just allocated
memory to test the **UMR** defect on slab (SLUB) memory. Here's the output when we run
this test case (`10 UMR on slab (SLUB) memory`) on a regular kernel with no slub
debug flags enabled:

```
[ 6845.100813] testcase to run: 10
[ 6845.101126] test_kmembugs:umr_slub(): testcase 10: simple
UMR on slab memory
[ 6845.101771] test_kmembugs:umr_slub(): q[3] is 0x0
```

```
[ 6845.102203] q: 00000000: 00 00 00 00 00 00 00 00 00 00 00 00
00 00 00 00  ...............
[ 6845.102946] q: 00000010: 00 00 00 00 00 00 00 00 00 00 00 00
00 00 00 00  ...............
```

With no slub debug flags or features enabled, the uninitialized memory region (all 32 bytes of it) shows up as the value 0x0. You'll soon find that, when run with slub_debug turned ON, though no error report is generated, a dump of the memory region shows the poison value 0x6b, denoting that its a valid but uninitialized memory region!

Passing the SLUB debug flags

Any and all SLUB debug flags – those seen in the earlier table – can be passed to a kernel configured with CONFIG_SLUB_DEBUG=y (as ours are, both the custom production and debug kernels) via the slub_debug kernel parameter. The format is as follows:

```
slub_debug=<flag1flag2...>,<slab1>,<slab2>,...
```

As can be seen, you can pass various slub debug flags. Don't leave any spaces between them; just concatenate them together. To set all flags on, set slub_debug to NULL; to turn all off, set it to -. Any combination is possible; for example, passing the kernel parameter slub_debug=FZPU enables, for all slab cache memory, the following SLUB features:

- Sanity checks (F)
- Red zoning (Z)
- Poisoning (P)
- User tracking (U)

Confirm this after boot with the following:

```
$ cat /proc/cmdline
BOOT_IMAGE=/boot/vmlinuz-5.10.60-dbg02-gcc root=UUID=<...> ro
quiet splash 3 slub_debug=FZPU
```

This is also reflected within the sysfs entry for the slab cache(s). Let's look up the slab cache `kmalloc-32`, which of course provides generic 32-byte memory fragments to any requester, as an example:

```
$ export SLAB=/sys/kernel/slab/kmalloc-32
$ sudo cat ${SLAB}/sanity_checks ${SLAB}/red_zone ${SLAB}/
poison ${SLAB}/store_user
1
1
1
1
$
```

They're all set to 1, indicating they're all on (the default is typically 0 – off).

All right, no dawdling. Let's run our (relevant) test cases to see where the kernel's SLUB debug infrastructure can help us.

Running and tabulating the SLUB debug test cases

All test cases are in the (same) module here: `ch5/kmembugs_test/kmembugs_test.c` (as well as the companion `debugfs_kmembugs.c`). Here, as we're testing SLUB debug, we only run the test cases that pertain to slab memory. We'll test on our custom production kernel as well as the distro kernel itself. Why? This is because most distros (including the one I'm using here, Ubuntu 20.04 LTS) configure the kernel with `CONFIG_SLUB_DEBUG=y`. This is also the default choice within the `init/Kconfig` file where it's defined. (Another reason we don't test with our debug kernel is obvious – with KASAN and UBSAN turned on, they tend to catch bugs first.)

Importantly, to test, we'll boot the system by passing the kernel parameter as follows:

- `slub_debug=-`, implying it's off.
- `slub_debug=FZPU`, implying these four flags and the SLUB debug features pertaining to them are turned on for *all slabs* on the system.

Then run the relevant test cases via the `test_kmembugs.ko` kernel module and associated `run_tests` script for each of these scenarios. The following table summarizes the results.

Test case #	Memory defect type (below) / Infrastructure used (right)	Production kernel with slub_debug=- (off)	Production kernel with slub_debug=FZPU
5	OOB accesses on dynamic kmalloc-ed slab (SLUB) memory		
5.1	Read (right) overflow		N
5.2	Write (right) overflow	N [V1]	Y [V4]
5.3	Read (left) underflow		N
5.4	Write (left) underflow		Y [V4]
Other memory corruption test cases			
6	Use After Free – UAF	N	Y [V5]
7	Double-free	N [V2]	Y [V6]
9	OOB on `copy_ [to	from]_ user*()`	N	N
10	Uninitialized Memory Read – UMR – on slab (SLUB) memory	N [V3]	N [V7]

Table 6.2 – Summary of findings when running relevant memory defect test cases against both our production kernel without slub_debug features and with slub_debug=FZPU

As mentioned earlier, don't forget to also look at *Table 6.4*, effectively, an all-in-one-place comparison table.

Okay, now let's dive into the details of running our test cases seen in the previous table.

Test environment: x86_64 guest running Ubuntu 20.04 LTS with our custom `5.10.60-prod01` production kernel (configured as mentioned earlier).

Running SLUB debug test cases on a kernel with slub_debug turned off

First, let's look at what occurs when we run our test cases without SLUB debug features enabled (corresponding to column 3 and points [V1], [V2], and [V3] in *Table 6.2*):

- No memory bugs are caught when `slub_debug=-`, that is, is off (FYI, our first three test cases – the UMR, UAR, and memory leakage – fail to be detected as well).

- [V1]: The system could simply Oops or hang here or even appear to remain unscathed, but that's not really the case... Once the kernel is buggy, the system is buggy.

- [V2]: A segfault might occur on the double-free defect. The vanilla or production 5.x kernel indicates it like this:

```
kernel BUG at mm/slub.c:305!
```

And the instruction pointer register (RIP on the x86_64) will be pointing at the `kfree()` API. The report, of course, has the usual details – the process context the bug occurred in and the kernel call trace.

Interestingly, what's at line 305 in mm/slub.c? I checked on the mainline kernel version 5.10.60, here: https://elixir.bootlin.com/linux/v5.10.60/source/mm/slub.c.

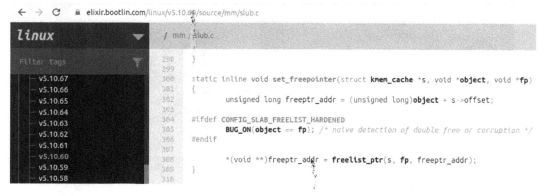

Figure 6.1 – Partial screenshot of the excellent Bootlin kernel source browser

You can see we're bang on target: line 305 is what triggered the double-free bug. The vanilla kernel has the intelligence to detect this (naïve) case of a double-free, which is a form of memory corruption.

- [V3]: The UMR on slab memory with our production kernel and `slub_debug` flags set to - - implying it's off – isn't caught. The kmalloc-ed memory region appears to be initialized to `0x0`.

Okay, let's move on now to testing with the kernel SLUB debug feature(s) turned on.

Running SLUB debug test cases on a kernel with slub_debug turned on

Now, let's rerun our test cases, this time with the kernel's SLUB debug features enabled by passing along the `slub_debug=FZPU` kernel parameter. Here's a screenshot showing setting the `slub_debug=FZPU` kernel parameter in the GRUB bootloader on the production kernel (as seen on VirtualBox):

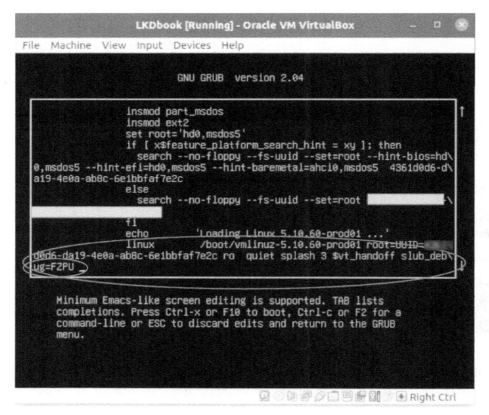

Figure 6.2 – The GRUB menu for editing the distro kernel parameters with the slub_debug=FZPU
kernel parameter added

Verify that the kernel command line we edited via the bootloader has made it intact:

```
$ dmesg |grep "Kernel command line"
[    0.094445] Kernel command line: BOOT_IMAGE=/boot/vmlinuz-
5.11.0-40-generic root=UUID=<...> ro quiet splash 3 slub_
debug=FZPU
```

It's fine; running `cat /proc/cmdline` will reveal the same. We run our test cases again, this time with SLUB debug enabled. The results are as seen in *Table 6.2*, in the fourth column.

Refer back to *Table 6.2*, noticing the places marked with [V4]. SLUB debug catches both the write over - and underflow (right and left) OOB accesses on slab memory. However, as we saw with UBSAN, it only seems able to catch it when the buggy access is via incorrect indices to the memory region, *not* when the OOB access is via a pointer! Also, the OOB reads do *not* seem to be caught.

Let's now learn a key skill: how to interpret the SLUB debug error report in detail.

Interpreting the kernel's SLUB debug error report

Let's look in detail at catching a few of our buggy test cases. We load up and use our `run_tests` script to execute them.

Interpreting the right OOB write overflow on slab memory

We begin with test case #5.2. The right OOB access, here the write overflow (right) on the slab object (marked as [V4] in the table), has the kernel's SLUB debug framework leap into action and complain quite loudly as follows:

```
[ 620.764707] testcase to run: 5.2
[ 620.764760] =============================================================
[ 620.764955] BUG kmalloc-32 (Tainted: G         OE    ): Right Redzone overwritten
[ 620.765116] -------------------------------------------------------------

[ 620.765370] Disabling lock debugging due to kernel taint
[ 620.765378] INFO: 0x00000000d0d6c75b-0x000000001b94c58a @offset=4640. First byte 0x78 instead of 0xcc
[ 620.765529] INFO: Allocated in dynamic_mem_oob_right+0x39/0x9c [test_kmembugs] age=0 cpu=5 pid=1697
[ 620.765659]   __slab_alloc.isra.0+0x8b/0xf0
[ 620.765723]   kmem_cache_alloc_trace+0x40b/0x450
[ 620.765791]   dynamic_mem_oob_right+0x39/0x9c [test_kmembugs]
[ 620.765873]   dbgfs_run_testcase+0x4d9/0x59a [test_kmembugs]
```

Figure 6.3 – Partial screenshot, 1 of 3, showing SLUB debug catching the right OOB while writing

First off, following the word `BUG` is the name of the affected slab cache (here, it's the `kmalloc-32` one, as our test case code performed a dynamic memory allocation of, in fact, exactly 32 bytes).

Next, the kernel taint flags are followed by the issue at hand – the OOB access defect that caused the SLUB debug code to say `Right Redzone overwritten`. This is pretty self-explanatory – it's what actually did occur. Within our `kmembugs_test.c:dynamic_mem_oob_right()` test case function, we did just this: performed a write at byte 32 (the legal range is bytes 0 to 31, of course).

Next, the first `INFO` line spits out the start and end of the corrupted memory region. Note that these kernel virtual addresses are hashed here, for security, preventing info leaks. Recall that we ran this test case on our production kernel, after all.

Next, the second `INFO` line shows where the buggy access took place in the code – via the usual `<func>+0xoff_from_func/0xlen_of_func [modname]` notation. (Here, it happens to be `dynamic_mem_oob_right+0x39/0x9c [test_kmembugs]`.) This implies that the defect occurred in the function named `dynamic_mem_oob_right()` at an offset of `0x39` bytes from the start of this function, and the kernel estimates the function's length to be `0x9c` bytes. (In the next chapter, we'll see how we can leverage this key information!)

Further, the process context running – its PID and the CPU core it ran upon – is displayed to the right.

This is followed by a (kernel-mode) stack trace:

```
kmem_cache_alloc_trace+0x40b/0x450
dynamic_mem_oob_right+0x39/0x9c [test_kmembugs]
dbgfs_run_testcase+0x4d9/0x59a [test_kmembugs]
full_proxy_write+0x5c/0x90
vfs_write+0xca/0x2c0
    [...]
```

We haven't shown the full stack call trace here. Read it bottom-up, ignoring any lines that begin with a `?` . So, here, it's quite clear – the `dynamic_mem_oob_right()` function, located in the kernel module `test_kmembugs`, is where the trouble seems to be...

Next, the third `INFO` line provides information on which task performed the *free*. This can be useful, helping us identify the culprit as, typically, the task that frees the slab is the one that allocated it in the first place:

```
INFO:Freed in kvfree+0x28/0x30 [...]
```

The call stack leading to the free is displayed under this `INFO:` line as well. Note that this *who-freed* information may not always be available or accurate though.

More information follows: a couple more `INFO` lines that display a few statistics on the slab and the particular object within it that got corrupted, the content of the left and right red zones, any padding, and the actual memory region content.

```
INFO: Slab 0x00000000d91ecea2 objects=19 used=5 fp=0x000000004fa4eb9d flags=0xffffffc0010201
INFO: Object 0x000000006489b63a @offset=4608 fp=0x0000000000000000

Redzone  000000003f2fee70: cc cc cc cc cc cc cc cc cc cc cc cc cc cc cc cc  ................
Redzone  00000000da09c2a2: cc cc cc cc cc cc cc cc cc cc cc cc cc cc cc cc  ................
Object   000000006489b63a: 6b 6b 6b 6b 6b 6b 6b 6b 6b 6b 6b 6b 6b 6b 6b 6b  kkkkkkkkkkkkkkkk
Object   00000000bb2f628f: 6b 6b 6b 6b 6b 6b 6b 6b 6b 6b 6b 6b 6b 6b 6b a5  kkkkkkkkkkkkkkk.
Redzone  00000000d0d6c75b: (78) cc cc (78) cc cc cc cc                          x..x....
Padding  00000000088b49804: 5a 5a 5a 5a 5a 5a 5a 5a 5a 5a 5a 5a 5a 5a 5a 5a  ZZZZZZZZZZZZZZZZ
Padding  00000000a984ce1: 5a 5a 5a 5a 5a 5a 5a 5a 5a 5a 5a 5a 5a 5a 5a 5a  ZZZZZZZZZZZZZZZZ
```

Figure 6.4 – Partial screenshot, 2 of 3, of the SLUB debug interpretation of corrupted slab memory, the red zones, object memory, and padding with faulty writes circled

Take a look at a snippet of our buggy test case code, the one that ran here:

```
// ch5/kmembugs_test.c
int dynamic_mem_oob_right(int mode)
{
volatile char *kptr, ch = 0;
char *volatile ptr;
size_t sz = 32;
kptr = (char *)kmalloc(sz, GFP_KERNEL);
    [...]
ptr = (char *)kptr + sz + 3; // right OOB
    [...]
} else if (mode == WRITE) {
/* Interesting: this OOB access isn't caught by UBSAN but is
caught by KASAN! */
*(volatile char *)ptr = 'x'; // invalid, OOB right write
/* ... but these below OOB accesses are caught by KASAN/UBSAN.
We conclude that only the index-based accesses are caught by
UBSAN. */
kptr[sz] = 'x'; // invalid, OOB right write
    }
```

As you can see highlighted, in two places, we (deliberately) perform an invalid right OOB access – writing the character x. Both are caught by the kernel SLUB debug infrastructure!

Do notice in *Figure 6.4* the value `0x78` is our `x` character being (wrongly) written by the test case code – I've circled the incorrect writes in the figure! Next, the poison values are used if the poison flag (`P`) is set for the slab, as is the case here. Here, the poison value `0x6b` denotes the value that's used to initialize the valid slab memory region, `0xa5` denotes the end poisoning marker byte, and `0x5a` denotes use-uninitialized poisoning – useful indeed.

Further, the typical output when most kinds of kernel bugs occur follows: a detailed `Call Trace` (the kernel-mode stack being unwound – read it bottom-up, ignoring lines that begin with `?`), and some of the CPU registers and their values:

```
CPU: 5 PID: 1697 Comm: run_tests Tainted: G    B      OE      5.10.60-prod01 #6
Hardware name: innotek GmbH VirtualBox/VirtualBox, BIOS VirtualBox 12/01/2006
Call Trace:
 dump_stack+0x76/0x94
 print_trailer+0x1de/0x1eb
 check_bytes_and_report.cold+0x6c/0x8c
 check_object+0x1c4/0x280
 free_debug_processing+0x165/0x2a0
 ? dynamic_mem_oob_right+0x63/0x9c [test_kmembugs]
 __slab_free+0x2e3/0x4a0
 ? vprintk_func+0x61/0x1b0
 ? _raw_spin_unlock_irqrestore+0x24/0x40
 kfree+0x4d8/0x500
 ? kmem_cache_alloc_trace+0x40b/0x450
 ? dynamic_mem_oob_right+0x63/0x9c [test_kmembugs]
 dynamic_mem_oob_right+0x63/0x9c [test_kmembugs]
 dbgfs_run_testcase+0x4d9/0x59a [test_kmembugs]
 full_proxy_write+0x5c/0x90
 vfs_write+0xca/0x2c0
 ksys_write+0x67/0xe0
 __x64_sys_write+0x1a/0x20
 do_syscall_64+0x38/0x90
 entry_SYSCALL_64_after_hwframe+0x44/0xa9
RIP: 0033:0x72d33f4d31e7
Code: 64 89 02 48 c7 c0 ff ff ff ff eb bb 0f 1f 80 00 00 00 00 f3 0f 1e fa 64 8b 04 25 18 00
 10 b8 01 00 00 00 0f 05 <48> 3d 00 f0 ff ff 77 51 c3 48 83 ec 28 48 89 54 24 18 48 89 74 24
RSP: 002b:00007ffdc666efd8 EFLAGS: 00000246 ORIG_RAX: 0000000000000001
RAX: ffffffffffffffda RBX: 0000000000000004 RCX: 000072d33f4d31e7
RDX: 0000000000000004 RSI: 0000558b9cde24e0 RDI: 0000000000000001
RBP: 0000558b9cde24e0 R08: 000000000000000a R09: 0000000000000003
R10: 0000558b9b2c4017 R11: 0000000000000246 R12: 0000000000000004
R13: 000072d33f5ae6a0 R14: 000072d33f5af4a0 R15: 000072d33f5ae8a0
FIX kmalloc-32: Restoring 0x00000000d0d6c75b-0x000000001b94c58a=0xcc

FIX kmalloc-32: Object at 0x000000006489b63a not freed
```

Figure 6.5 – Partial screenshot, 3 of 3, of the SLUB debug error report (continued) showing the process context, hardware, kernel stack trace, CPU register values, and FIX info

Finally, the kernel SLUB debug framework even informs us what it restores and what's to fix – see the last two lines of the preceding screenshot (beginning with `FIX kmalloc-32:`). With the `F` flag – the SLUB *sanity checks* feature – enabled, the kernel attempts to clean up the mess and restore the slab object state to what would be deemed the correct form. Of course, this may not always be possible to do. Also, this SLUB debug error report is generated before the slab object in question has been freed, hence the `... not freed` message (our code does free it).

For more info, do refer to the official kernel documentation here: *SLUB Debug output*: `https://www.kernel.org/doc/html/latest/vm/slub.html#slub-debug-output`.

Interpreting the UAF bug on slab memory

Now let's interpret the **Use After Free** (**UAF**) bug being caught (marked as [V5] in *Table 6.2*). The UAF bug is caught by the slub debug framework. The error report (within the syslog) looks like this:

```
BUG kmalloc-32 (Tainted: G    B       OE    ): Poison
overwritten
[ 3747.701588] ------------------------------------------------
[ 3747.707061] INFO: 0x00000000d969b0bf-0x00000000d969b0bf @
offset=872. First byte 0x79 instead of 0x6b
[ 3747.710110] INFO: Allocated in uaf+0x20/0x47 [test_kmembugs]
age=5 cpu=5 pid=2306
```

The format remains the same as described earlier. This time, the UAF defect caused the SLUB debug code to say `Poison overwritten`. Why? In our `uaf()` test case, we did just this, freed the slab object and then performed a write to a byte within it!

Next, the `INFO` line spits out the start and end of the corrupted memory region. Note that these kernel virtual addresses are hashed here, for security, preventing info-leaks. Recall that we ran this test case on our production kernel, after all.

The PID of the task performing the allocation (the process context) is seen – along with the kernel module name in square brackets, if applicable, as well as the function within it where the allocation took place.

This is followed by a stack trace (we don't show this here) and then information on which task performed the free. This can be useful, helping us identify the culprit as, typically, the task that frees the slab is the one that allocated it in the first place.

```
INFO: Freed in uaf+0x34/0x47 [test_kmembugs] age=5 cpu=5
pid=2306
```

Let's move along to the next test case, the double-free...

Interpreting the double-free on slab memory

Finally, a quick note on the double-free defect, successfully caught again (marked as [V6] in *Table 6.2*). Here, the kernel reports it as follows:

```
BUG kmalloc-32 (Tainted: G    B       OE    ): Object already
free
[ 3997.543154] -------------------------------------------------
[ 3997.544129] INFO: Allocated in double_free+0x20/0x4b [test_
kmembugs] age=1 cpu=5 pid=2330
```

The very same template as described earlier follows this output... Do try it out for yourself, both reading and interpreting it.

By the way, we've already seen how the SLUB debug framework deals with uninitialized memory reads (the UMR defect) on slab cache memory (test case #10, marked as [V3] and [V7] in *Table 6.2*). When run with slub_debug on, though no error report is generated, a dump of the memory region shows the poison value 0x6b, denoting the fact that this memory region is in an uninitialized state.

So, going by our experiments, while the kernel SLUB debug framework seems to catch most of the memory corruption issues on slab memory, it doesn't seem to catch the *read* OOB accesses on slab memory. Note that KASAN does (see *Table 6.4*)!

Learning how to use the slabinfo and related utilities

Another utility program that can prove to be very useful for understanding and helping debug the slab caches is one named slabinfo. Though a user-mode app, its code is in fact part of the kernel source tree, here:

```
tools/vm/slabinfo.c
```

Build it by simply changing directory to the `tools/vm` folder within your kernel source tree and typing `make`. Running it does require root access. Once built, for convenience, I like to create a soft or symbolic link to the binary executable named `/usr/bin/slabinfo`. Here it is on my system:

```
$ ls -l $(which slabinfo)
lrwxrwxrwx 1 root root 71 Nov 20 16:26 /usr/bin/slabinfo ->
<...>/linux-5.10.60/tools/vm/slabinfo
```

Display its help screen by passing the `-h` (or `--help`) option switch. This reveals a large number of possible option switches. The following screenshot shows them all (for the 5.10.60 kernel):

```
$ sudo slabinfo --help
slabinfo 4/15/2011. (c) 2007 sgi/(c) 2011 Linux Foundation.

slabinfo [-aABDefhilLnoPrsStTUvXz1] [N=K] [-dafzput] [slab-regexp]
-a|--aliases          Show aliases
-A|--activity         Most active slabs first
-B|--Bytes            Show size in bytes
-D|--display-active   Switch line format to activity
-e|--empty            Show empty slabs
-f|--first-alias      Show first alias
-h|--help             Show usage information
-i|--inverted         Inverted list
-l|--slabs            Show slabs
-L|--Loss             Sort by loss
-n|--numa             Show NUMA information
-N|--lines=K          Show the first K slabs
-o|--ops              Show kmem_cache_ops
-P|--partial           Sort by number of partial slabs
-r|--report           Detailed report on single slabs
-s|--shrink           Shrink slabs
-S|--Size             Sort by size
-t|--tracking         Show alloc/free information
-T|--Totals           Show summary information
-U|--Unreclaim        Show unreclaimable slabs only
-v|--validate         Validate slabs
-X|--Xtotals          Show extended summary information
-z|--zero             Include empty slabs
-1|--1ref             Single reference

-d  | --debug         Switch off all debug options
-da | --debug=a       Switch on all debug options (--debug=FZPU)

-d[afzput] | --debug=[afzput]
     f | F            Sanity Checks (SLAB_CONSISTENCY_CHECKS)
     z | Z            Redzoning
     p | P            Poisoning
     u | U            Tracking
     t | T            Tracing

Sorting options (--Loss, --Size, --Partial) are mutually exclusive
$
```

Figure 6.6 – The help screen of the kernel slabinfo utility

A few things to note when running `slabinfo`:

- By default, this tool will only display slabs that have data within them (the same as with the `-1` switch, in effect). You can change this by running `slabinfo -e`. It displays only the empty caches – there can be quite a few!

- All options may not work straight away. Most require that the kernel is compiled with SLUB debug on (`CONFIG_SLUB_DEBUG=y`). Typically, this is the case, even for distro kernels. Some options require the SLUB flags to be passed on the kernel command line (via the usual `slub_debug` parameter).

- You need to run it as root.

Let's begin by doing a quick run (with no parameters), seeing the header line and a line of sample output, that of the `kmalloc-32` slab cache – I've spaced the lines to fit):

```
$ sudo slabinfo |head -n1
Name Objects Objsize Space Slabs/Part/Cpu O/S O %Fr %Ef Flg
$ sudo slabinfo | grep "^kmalloc-32"
kmalloc-32 35072 32   1.1M   224/0/50      128 0   0 100
$
```

A quick roundup of the header line columns is as follows:

- The name of the slab cache (`Name`).

- The number of objects currently allocated (`Objects`).

- The size of each object is then shown (here, its 32 bytes, of course) (`Objsize`).

- The total space taken up in kernel memory by these objects (essentially, it's `Objects * Objsize`) (`Space`).

- The slab cache memory distribution for this cache: number of full slabs, partial slabs, and per-CPU slabs (`Slabs/Part/Cpu`).

- The number of objects per slab (`O/S`).

- The order of the page allocator from where memory is carved out for this cache (the order is from 0 to MAX_ORDER). Typically, MAX_ORDER is the value 11, to give us a total of 12 orders. `2^order` is the size of free memory chunks (in pages) on the page allocator free list for that order (`O`).

- The amount of cache memory free in percentage terms (`%Fr`).

- The effective memory usage as a percentage (`%Ef`).

- The slab flags (can be empty) (`Flg`).

Want to see the actual `printf()` that emits these slab cache stats? It's right here: https://elixir.bootlin.com/linux/v5.10.60/source/tools/vm/slabinfo.c#L640.

All possible values for the slab flags and their meaning are as follows (pertain to the column named `Flg` on the extreme right of `slabinfo` normal output):

- `*`: Aliases present
- `d`: For DMA memory slabs
- `A`: Hardware cache line (hwcache) aligned
- `P`: Slab is poisoned
- `a`: Reclaim accounting active
- `Z`: Slab is red zoned
- `F`: Slab has sanity checking on
- `U`: Slab stores user
- `T`: Slab is being traced

Note that the columns change when the `-D` (display active) option switch is passed.

An FAQ, perhaps: *Of the many slab caches that are currently allocated (and have some data content), which takes up the most kernel memory?* This is easily answered by `slabinfo`: one way is to run it with the `-B` switch, to display the space taken in bytes, allowing you to easily sort on this column. Even simpler, the `-S` option switch has `slabinfo` sort the slab caches by size (largest first) with nice, human-readable size units displayed. The following screenshot shows us doing so for the top 10 highest kernel memory consuming slab caches:

```
$ sudo slabinfo -S | head
Name                 Objects Objsize    Space Slabs/Part/Cpu  O/S  O  %Fr %Ef Flg
inode_cache            24726     600    15.5M     912/0/39     26  2    0  95 a
buffer_head           132015     104    13.8M    3369/0/16     39  0    0  99 a
ext4_inode_cache        7074    1176     8.5M      252/0/10     27  3    0  96 a
dentry                 40572     192     7.9M     1883/0/49     21  0    0  98 a
kmalloc-4k              1591    4096     6.5M      189/8/12      8  3    3  98
radix_tree_node         8603     576     5.0M      298/7/13     28  2    2  97 a
kernfs_node_cache      30144     128     3.8M      899/0/43     32  0    0 100
kmalloc-512             5040     512     2.5M      282/0/33     16  1    0 100
filp                    8816     256     2.2M      501/0/51     16  0    0  99 A
$ _
```

Figure 6.7 – The top 10 slab caches sorted by total kernel memory space taken (fourth column)

Interestingly (as often happens with software), the `-U` `'Show unreclaimable slabs only'` option of `slabinfo` came into being due to a system getting panicked. This occurred when the *unreclaimable* slab memory usage went too close to 100% and the **Out Of Memory (OOM)** killer was unable to find any candidate to kill! The patch has the utility – as well as the OOM kill code paths – display all unreclaimable slabs, to help with troubleshooting. This patch got mainlined in the 4.15 kernel. Here's the commit – do take a peek at it: `https://github.com/torvalds/linux/commit/7ad3f188aac15772c97523dc4ca3e8e5b6294b9c`. Along with the `-U` switch, the `-S` option (sort by size), makes troubleshooting these corner cases easier!

The *sort-by-loss* (`-L`) option switch has `slabinfo` sort the slab caches by the amount of kernel memory lost. A better word than *lost*, perhaps, is *wasted*. This is the usual well-known *internal fragmentation* issue: when memory is allocated via the slab layer, it internally does so via a *best-fit* model. This often results in a (hopefully small) amount of memory being wasted or lost. For example, attempting to allocate 100 bytes via the `kmalloc()` API will have the kernel actually allocate memory from the `kmalloc-128` slab cache (as it can't possibly give you less via the `kmalloc-96` cache), with the result that your slab object actually consumes 128 bytes of kernel memory. Thus, the loss or wastage in this case is 28 bytes. Running `sudo slabinfo -L |head` will quickly show you (in descending order) the slabs with maximum wastage (or loss – look at the fourth column, labeled `Loss`).

Once you've identified a slab cache that you'd like to further investigate, the `-r` (*report*) option will have `slabinfo` emit detailed statistics. By default, this is on all slabs. You can always pass a regular expression specifying which slabs you're interested in! For example, `sudo slabinfo -r vm.*` will display details on all slabs matching the regex pattern `vm.*`. With the SLUB debug flags enabled, it even shows the origin (and number) of allocs and frees for each cache; it can be useful!

At times, you might see a slab cache with a name that's unfamiliar. Trying the `-a` (or `--aliases`) option to show aliases can be useful to reveal what kernel object(s) it's being used to cache.

The `-T` option has `slabinfo` display *overall totals*, a summary snapshot of all slab caches. This is useful to get a quick overview of how many slab caches exist, how many are active, how much kernel memory in all is being used, and so on. This kind of information is extended when you use the `-X` option switch. It now shows even more detail. The following screenshot is an example of running `sudo slabinfo -X` on my x86_64 Ubuntu guest:

```
$ sudo slabinfo -X
[sudo] password for letsdebug:
Slabcache Totals
----------------
Slabcaches :          216   Aliases  :        0->0   Active:    133
Memory used:     90710016   # Loss   :     2548968   MRatio:     2%
# Objects  :       401015   # PartObj:        1444   ORatio:     0%

Per Cache        Average          Min          Max          Total
-----------------------------------------------------------------------
#Objects            3015           10       132132          401015
#Slabs                88            1         3388           11833
#PartSlab              0            0           31             101
%PartSlab             0%           0%          38%              0%
PartObjs               0            0          670            1444
% PartObj             0%           0%          23%              0%
Memory            682030         4096     15581184        90710016
Used              662865         3072     14835600        88161048
Loss               19165            0       745584         2548968

Per Object       Average          Min          Max
----------------------------------------------------------
Memory               221            8         8192
User                 219            8         8192
Loss                   1            0           64

Slabs sorted by size
--------------------
Name                   Objects Objsize              Space Slabs/Part/Cpu  O/S O %Fr %Ef Flg
inode_cache              24726     600           15581184      912/0/39   26 2   0  95 a

Slabs sorted by loss
--------------------
Name                   Objects Objsize               Loss Slabs/Part/Cpu  O/S O %Fr %Ef Flg
inode_cache              24726     600             745584      912/0/39   26 2   0  95 a

Slabs sorted by number of partial slabs
-----------------------------------------
Name                   Objects Objsize              Space Slabs/Part/Cpu  O/S O %Fr %Ef Flg
anon_vma                  2970      80             331776      40/31/41   46 0  38  71

$
```

Figure 6.8 – Screenshot showing extended summary information via slabinfo -X

These can serve as useful diagnostics when troubleshooting a system (you'll find more in a similar vein in the later section entitled *Practical stuff – what's eating my memory?*).

Running slabinfo with the -z (zero) option switch has it show all slab caches, both the ones with data as well as the empty ones.

Debug-related options for slabinfo

For debug purposes, slabinfo has a -d and a -v option switch, allowing you to pass *debug flags* and *validate slabs*, respectively. Note that both these option switches will only work when the system is booted with the slub_debug kernel parameter set to some non-null value.

It's interesting to see: when booted with `slub_debug=FZPU`, all the slab caches show up with (at least) these flags set!

```
$ sudo slabinfo -S |head
Name                   Objects Objsize       Space Slabs/Part/Cpu  O/S O  %Fr %Ef Flg
inode_cache              24162     600       23.3M     1425/4/0    17 2    0  62 PaZFU
kmalloc-4k                1255    4096       20.7M      634/13/0    2 3    2  24 PZFU
dentry                   35230     192       18.6M     1137/1/0    31 2    0  36 PaZFU
kernfs_node_cache        26301     128       12.7M     1558/25/0   17 1    1  26 PZFU
ext4_inode_cache          4949    1176        7.7M      237/4/0    21 3    1  74 PaZFU
kmalloc-32               16029      32        7.0M      856/78/0   19 1    9   7 PZFU
radix_tree_node           5192     576        5.0M      307/4/0    17 2    1  59 PaZFU
buffer_head              10231     104        4.6M      570/6/0    18 1    1  22 PaZFU
kmalloc-1k                1262    1024        4.2M      130/21/0   10 3   16  30 PZFU
$
```

Figure 6.9 – Partial screenshot – the focus is on the SLUB debug flags being set as we booted with slub_debug=FZPU

Notice how, for all slabs, the flags minimally contain `PZFU`.

Regarding the `-d` option switch, passing it by itself turns debugging off. (Quite non-intuitively, right? Then again, this is consistent with the way the kernel parameter `slub_debug` behaves.) When you want the kernel's SLUB debugging options on, pass along the usual SLUB debug flags, like so: `--debug=<flag1flag2...>`. The `slabinfo` help screen shows all of this clearly. Look at *Figure 6.6*, specifically the last few lines – the ones that describe the `--debug` option switch.

What really happens under the hood when you do pass, say, `--debug=fzput` (or, if 'a' is passed, *all* these SLUB debug flags are set), as a parameter to `slabinfo` is this: the utility opens (as root, of course) the underlying `/sys/kernel/slab/<slabname>` pseudofile for that slab cache (if you passed one or more of them as a parameter), else for all slabs, and arranges to set these to 1, meaning enabled:

- If `f | F` is passed in `--debug=<...>`, `/sys/kernel/slab/<slabname>/sanity_checks` is set to 1.

- If `z | Z` is passed in `--debug=<...>`, `/sys/kernel/slab/<slabname>/red_zone` is set to 1.

- And similarly for the rest...

FYI, the code that does this is here: `https://elixir.bootlin.com/linux/v5.10.60/source/tools/vm/slabinfo.c#L717`.

The -v option switch to slabinfo can again be useful for debugging: it *validates all slabs*, and on any errors being detected, it spews out diagnostics/error reports to the kernel log. The format of the report is in fact identical to the error report format that the kernel's SLUB debug infrastructure produces (we covered this in detail here: *Interpreting the kernel's SLUB debug error report*).

As with the debug option switch, the -v causes slabinfo to write 1 into the pseudofile /sys/kernel/slab/<slabcache>/validate. The kernel documents this as follows: *Writing to the validate file causes SLUB to traverse all of its cache's objects and check the validity of metadata. All slab objects will be checked. The detailed output is written to the kernel log.* This can be useful when troubleshooting a live system that you suspect might suffer from slab (SLUB) memory corruption.

Finally, we'll just mention the fact that there's even a utility script, slabinfo-gnuplot. sh, to plot graphs to help visualize slab (SLUB) functioning over time! I'll leave it to you to browse the kernel documentation that explains how to leverage it, here: https://www. kernel.org/doc/Documentation/vm/slub.txt, in the section named *Extended slabinfo mode and plotting*.

The /proc/slabinfo pseudofile

Also, the kernel of course exposes all this useful information on live slabs on the system via procfs, particularly, the pseudofile /proc/slabinfo (again, you'll require root access to view it). Here's a sampling of the large data available (internally, for each slab, it breaks the data into three types: statistics, tunables, slabdata). First, the header shows the version number and the columns:

```
$ sudo head -n2 /proc/slabinfo
slabinfo - version: 2.1
# name            <active_objs> <num_objs> <objsize>
<objperslab> <pagesperslab> : tunables <limit> <batchcount>
<sharedfactor> : slabdata <active_slabs> <num_slabs>
<sharedavail>
```

And here's some data from it:

```
$ sudo grep -C2 "^kmalloc-128" /proc/slabinfo
kmalloc-256     1982   2448   512   16   2 : tunables    0    0    0 : slabdata    153   153      0
kmalloc-192     3424   3424   256   16   1 : tunables    0    0    0 : slabdata    214   214      0
kmalloc-128     1968   1968   256   16   1 : tunables    0    0    0 : slabdata    123   123      0
kmalloc-96      1956   2368   128   32   1 : tunables    0    0    0 : slabdata     74    74      0
kmalloc-64      6907   8096   128   32   1 : tunables    0    0    0 : slabdata    253   253      0
$
```

Figure 6.10 – A screenshot showing some data from /proc/slabinfo the slabdata columns being pertinent

The man page on `slabinfo(5)` covers interpreting this data, in effect, all slab caches exposed via `/proc/slabinfo`. We'll leave it to you to check it out. (Unfortunately, it seems a bit dated in the sense that both the `statistics` and `tunables` information – the first two columns – pertain to only the older SLAB implementation.)

By the way, the `vmstat` utility also has the ability to display the kernel slab caches and some statistics (via its `-m` option switch). It essentially does so by reading `/proc/slabinfo`, and thus implies you must run it as root. Try this on your box:

```
sudo vmstat -m
```

Check out the man page on `vmstat(8)` for more information.

The slabtop utility

As with `top` (and the more modern `htop` and the cool `btop` utilities), to see who's consuming CPU in real time, we have the `slabtop(1)` utility to see *live, real-time kernel slab cache usage*, sorted by the maximum number of slab objects by default (the `sort` field can always be changed via the `-s` or `--sort` option switch). It too is based on data obtained from `/proc/slabinfo` and thus, as usual, you'll require root access to run it. Using `slabtop`, you can see for yourself how, besides the caches for specific kernel data structures, the small-sized generic caches (typically the ones named `kmalloc-*`) are often the ones being employed the most. Do try it and check out its man page for details.

eBPF's slabratetop utility

Finally, and a recent addition, is the *eBPF* `slabratetop` utility (it could be named `slabratetop-bpfcc`, as is the case on my system). It displays, in real time, the kernel's SLAB/SLUB memory cache allocation rate in a manner like the `top` utility does, refreshing it once a second by default. It internally tracks the `kmem_cache_alloc()` API to track the rate and total bytes allocated via this commonly used interface to allocate slab objects within the kernel. Via option switches, you can control the output interval (in seconds) and the number of times to show it (along with a couple of other switches).

So, entering the following will have the utility display the active caches (allocation rate and the number of bytes allocated) in 5-second interval summaries, thrice:

```
sudo slabratetop-bpfcc 5 3
```

Do refer to its man page and/or pass the `-h` option switch to see a brief help screen.

Practical stuff – what's eating my memory?

So, knowing about these utilities, how can they practically help? Well, *one common case is needing to know what's eating up memory (RAM) and from where exactly it is being eaten up.*

The first question is very wide-ranging. In terms of user-mode processes and threads, utilities such as smem, ps, and so on, can help. On a raw level, peeking at memory statistics under procfs can really help as well. For example, you can track memory usage of all threads by looking through procfs with something such as the following:

```
grep "^Vm.*:" /proc/*/status
```

Within this output, the VmRSS number is a reasonable measure of physical memory usage (with the unit being kilobytes). So, doing the following can quickly show you the PIDs of the top 10 processes or threads consuming the most RAM:

```
grep -r "^VmRSS" /proc/*/status |sed 's/kB$//'|sort -t: -k3n
|tail
```

Here, rather than userspace, you're perhaps more interested in who or what is eating kernel dynamic – slab cache – memory, right? Here's one investigative scenario:

- First, use slabratetop (or slabratetop-bpfcc) to figure out which slab cache, among the many present within the kernel, is being consumed the most.

- Second, use *dynamic kprobes* to look up the kernel-mode stack in real time to see who or what within the kernel is eating into this cache!

We can easily achieve the first step like this:

```
sudo slabratetop-bpfcc
[...]
CACHE                        ALLOCS      BYTES
names_cache                      18      78336
vm_area_struct                  176      46464
...
```

Okay, based on this sample output, there's a slab cache on my (guest) system named names_cache that is consuming the greatest number of bytes. Also, you can see that the vm_area_struct slab cache is currently seeing the greatest number of allocations, all within the given time interval (a second by default).

Let's say we want to dig deeper and investigate what code paths within the kernel are allocating memory from the vm_area_struct slab cache (pretty often, 176 times per second, as of right now). In other words, how can you figure out who or what within the kernel is performing these allocations? Okay, let's see: we know that most specific slab cache objects are allocated via the kmem_cache_alloc() kernel interface. Thus, seeing the (kernel) stack of kmem_cache_alloc() in real time will help you pinpoint who the allocation is being performed by, or where from!

So how do we do that? That's the second part. Recall what you learned in the chapter on Kprobes, specifically using dynamic kprobes (we covered this in *Chapter 4, Debugging via Instrumentation – Kprobes*, in the *Setting up a dynamic kprobe (via kprobe events) on any function*) section. Let's leverage that knowledge and look deeper into this. We'll begin by using the kprobe[-perf] command (Bash script, really) to probe all running instances of the kmem_cache_alloc() API in real time and reveal the internal kernel mode stack (by passing along the -s option switch):

```
sudo kprobe-perf -s 'p:kmem_cache_alloc
name=+0(+96(%di)):string'
```

Also, do recall from *Chapter 4, Debugging via Instrumentation – Kprobes*, in the *Understanding the basics of the Application Binary Interface (ABI) section*, that on x86_64, the RDI register holds the first parameter. Here, for the kmem_cache_alloc() API, the first parameter is a pointer to struct kmem_cache. Within this structure, at an offset of 96 bytes, is the thing we're after, the member named name – the name of the slab cache being allocated from!

On to the next point. The preceding command will probe for and show you *all* the slab cache allocations currently being performed by the popular kmem_cache_alloc() API. Let's filter its output to see only the one of interest to us right now, the one for the vm_area_struct slab cache:

```
sudo kprobe-perf -s 'p:kmem_cache_alloc
name=+0(+96(%di)):string' | grep -A10
"name=.*vm_area_struct"
```

A small portion of the output is seen in the following screenshot:

```
         <...>-3154    [001] ...1 13294.620610: kmem_cache_alloc: (kmem_cache_alloc+0x0/0x8d0)
name="vm_area_struct"
         <...>-3154    [001] ...1 13294.620616: <stack trace>
=> kmem_cache_alloc
=> do_brk_flags
=> __x64_sys_brk
=> do_syscall_64
=> entry_SYSCALL_64_after_hwframe
```

Figure 6.11 – Partial screenshot showing output from the kprobe-perf script with the kernel-mode stack showing the lead up to the kmem_cache_alloc() API for the VMA structure alloc

You might need to adjust the grep -An (I've kept n as 10 here) parameter to show a certain number of lines after the match). Quite clearly, this particular call chain shows us that the kmem_cache_alloc() API has been invoked via a system call, sys_brk(). FYI, this system call (named brk() in userspace) is typically the one issued when a memory region of a process needs to be created, or an existing one grown or shrunk.

Now, internally, the kernel manages the memory regions (technically, the mappings) of a process via the **Virtual Memory Area** (**VMA**) metadata structure. Thus, when creating a new mapping of a process – as is the case here – the VMA object will naturally need to be allocated. As the VMA is a frequently used kernel structure, it's kept on a custom slab cache and allocated from it – *the one named* vm_area_struct! This is the call chain from before that allocates a VMA object from this very slab cache:

```
sys_brk() --> do_brk() --> do_brk_flags() --> vm_area_alloc()
--> kmem_cache_alloc()
```

Here's the actual line of code where the vm_area_alloc() routine is invoked: https://elixir.bootlin.com/linux/v5.10.60/source/mm/mmap.c#L3110, which in turn issues the kmem_cache_alloc() API, allocating an instance of a VMA object from its slab cache and then initializing it. Interesting.

> **Security Tip**
>
> Though unrelated to this topic, I think security is important. To guarantee that slab memory is always wiped, both at the time of allocation and freeing, pass these on the kernel command line: init_on_alloc=1 init_on_free=1. Of course, doing this can result in performance impact; do test and ascertain whether to use one, both, or none on your project. More information in a similar vein can be found here: https://kernsec.org/wiki/index.php/Kernel_Self_Protection_Project/Recommended_Settings.

Good job on covering this section on SLUB debug and its offshoots. Now let's finish this large topic on kernel memory debugging by – finally! – learning how to catch those dangerous leakage bugs. Read on!

Finding memory leakage issues with kmemleak

What is a memory leak and why does it matter? A memory leak is a situation where you have allocated memory dynamically but failed to free it. Well, you *think* you have succeeded in freeing it but the reality is that it hasn't been freed. The classic pedagogical case is a simple one like this (let's just make it a userspace example for simplicity):

static foo(void) {

```
    char *ptr = malloc(1024);
    /* ... work with it ... */
    // forget to free it
}
```

Now, once you return from the function foo(), it's basically impossible to free the memory pointed to by the variable ptr. Why? You know it, ptr is a local variable and is out of scope once you've returned from foo(). Now, the 1,024 bytes of allocated memory is, in effect, locked up, inaccessible, wasted – we call this a **leak**. Of course, once the user process dies, it's freed back to the system.

Catching Memory Leaks in Userspace Apps

This book focuses purely on kernel debugging. Userspace app memory issues and debugging them are covered in a lot of detail in my earlier book, *Hands-On System Programming with Linux*, Packt, October 2018 (see chapters 5 and 6). In a word, using the newer Sanitizer toolset (especially the all-powerful **Address Sanitizer (ASAN)**, as well as the older Valgrind suite of tools will certainly help in userspace debug.

This can certainly happen within the kernel too (just substitute the preceding user code for a kernel module and the malloc() function with a kmalloc() or a similar API!). Even a small leakage of a few bytes can become a huge issue when the code that causes the leak runs often (in a loop, perhaps). Not just that, unlike userspace, the kernel isn't expected to die... the leaked memory is thus lost forever (yes, even when a module has a leak and is unloaded, the dynamic kernel memory allocated isn't within the module; it's typically slab or page allocator memory!).

Now, you might say *Hey, I know, but come on, if I allocate memory, I'll certainly free it*. True, but when the project is large and complex, believe me, you can miss it. As a simple example, check out this pseudocode snippet:

```
static int kfoo_leaky(void) {
    char *ptr1 = kmalloc(GFP_KERNEL, 1024), *ptr2;
    /* ... work with ptr1 ... */
        // ...
        if (bar() < 0)
        return -EINVAL;
    // ...
    ptr2 = vmalloc(5120); [...]
    // ... work with ptr2 ...
    if (kbar() < 0)
        return -EIO;
    // ...
    vfree(ptr2);
    kfree(ptr1);
    return 0;
}
```

You see it, don't you? At both the error return call sites before the final return, we've returned *without freeing* the memory buffers previously allocated – classic memory leaks! This kind of thing is in fact a pretty common pattern, so much so that the kernel community has evolved a set of helpful *coding style guidelines*, one of which will certainly have helped avoid a disaster of this sort: using the (controversial) goto to perform cleanup (like freeing memory buffers!) before returning. Don't knock it till you try it – it works really well when used correctly. (This technique is formally called *Centralized exiting of functions* – read all about it here: https://www.kernel.org/doc/html/latest/process/coding-style.html#centralized-exiting-of-functions).

Here's a fix (via the *centralized exiting of functions* route):

```
static int kfoo(void) {
    char *ptr1 = kmalloc(GFP_KERNEL, 1024), *ptr2;
    int ret = 0;
    /* ... work with ptr1 ... */
    // ...
```

```
    if (bar() < 0) {
        ret = -EINVAL;
        goto bar_failed;
    }
    // ...
    ptr2 = vmalloc(5120); [...]
        // ... work with ptr2 ...
    if (kbar() < 0){
        ret = -EIO;
        goto kbar_failed;
    }
    // ...
kbar_failed:
    vfree(ptr2);
bar_failed:
    kfree(ptr1);
    return ret;
}
```

Quite elegant, right? The later you fail, the higher/the earlier (in the code) you jump, so as to perform all the required cleanup. So here, if the function `kbar()` fails, it goes to the `kbar_failed` label, performing the required free via the `vfree()` API, *and then neatly falls through* performing the next required free via the `kfree()` API as well. I'm betting you've seen code like this all over the kernel; it's a very common and useful technique. (Again, a random example: see the code of a function belonging to the Cadence MACB/GEM Ethernet Controller network driver here: `https://elixir.bootlin.com/linux/v5.10.60/source/drivers/net/ethernet/cadence/macb_main.c#L3578`).

Here's another common root cause of leakage bugs: an interface is designed in such a way that it allocates memory, usually to a pointer passed by reference to it as a parameter. It's often deliberately *designed such that the caller is responsible for freeing the memory buffer* (after using it). But what happens when the caller fails to do so? A leakage bug, of course! Typically, this will be well documented, but then who reads documentation... (hey, that means read it!).

Can I see some real kernel memory leakage bugs?

Sure. First, head over to the kernel.org Bugzilla site, `https://bugzilla.kernel.org/`. Go to the **Search | Advanced Search** tab there. Fill in some search criteria – in the **Summary** tab, perhaps you'd want to type in something (such as `memory leak`). You can filter down by **Product** (or subsystem), **Component** (within **Product**), and even **Status** and **Resolution**! Then, click on the **Search** button. Here's a screenshot of the search screen for an example search I did, for your reference:

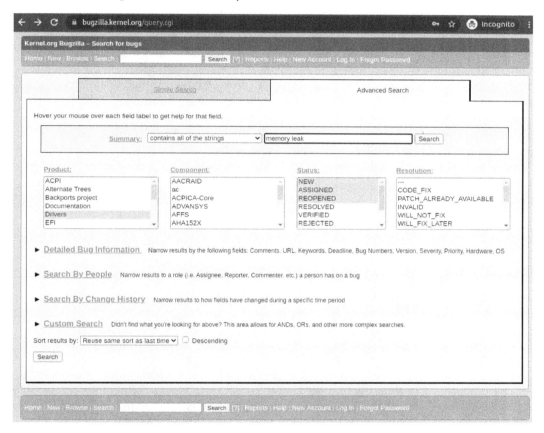

Figure 6.12 – kernel.org Bugzilla, searching for "memory leak" within bugs submitted under Drivers

(This particular search, at the time I made it, yielded 11 results.)

Again, as an example, here's one memory leak bug report from the kernel *Bugzilla: backport-iwlwifi: memory leak when changing channels* (`https://bugzilla. kernel.org/show_bug.cgi?id=206811`). Look at it and download the attachment labeled `dmesg.log after reboot`. The kernel log output shows that none other than **kmemleak** detected the leakage! We cover the interpretation of the report in the *Interpreting kmemleak's report* section.

The real challenge is this: without explicitly employing powerful tools – such as **kmemleak** for the kernel (for userspace, there's ASAN, MSAN, Valgrind's memcheck, and so on) – memory leaks often go unnoticed in development and even in testing and in the field. But when they strike one fine day, the symptoms can appear random – the system might run perfectly for a long while (even several months), when all of a sudden, it experiences random failures or even abruptly crashes. Debugging such situations can, at times, be next to impossible – the team blames power variances/outages, lightning, anything convenient to explain away the inexplicable random crash! How often have you heard support say *Just reboot and try it again – it'll likely work*? And the unfortunate thing is that it often does, thus, the real issue is glossed over. Unfortunately, this will simply not do on mission-critical projects or products; it will eventually cause customer confidence to deteriorate and could ultimately result in failure.

It's a serious problem. Don't let it become one on your project! Take the trouble to perform long-spanning coverage testing (and for long durations – a week or more).

Configuring the kernel for kmemleak

Before going further, do realize that *kmemleak* is, again, not a magic bullet: it's designed to track and catch memory leakage for dynamic kernel memory allocations performed via the `kmalloc()`, `vmalloc()`, `kmem_cache_alloc()`, and friends APIs *only*. These being the interfaces via which memory is typically allocated, it does serve an extremely useful purpose.

The key kernel config option, the one we need to enable, is `CONFIG_DEBUG_ KMEMLEAK=y` (several related ones are mentioned here as well). Of course, there's the usual trade-off: being able to catch leakage bugs is a tremendous thing, but can cause a pretty high overhead on memory allocations and freeing. Thus, of course, we recommend setting this up in your custom *debug* kernel (and/or in your production kernel during intense testing, where performance isn't what matters and catching defects does).

The usual `make menuconfig` UI can be used to set up the kernel config; the relevant menu is here: **Kernel hacking | Memory Debugging | Kernel memory leak detector**. Turn it on.

> **Tip**
>
> You can edit the kernel config file non-interactively by leveraging the kernel's built-in config script here: `scripts/config` (it's a Bash script). Just run it and it will display a help screen.

Once configured for kmemleak, a quick grep for `DEBUG_KMEMLEAK` on our debug kernel's config file (`/boot/config-5.10.60-dbg02` on my system) reveals that it's indeed all set and ready to go; see the following screenshot:

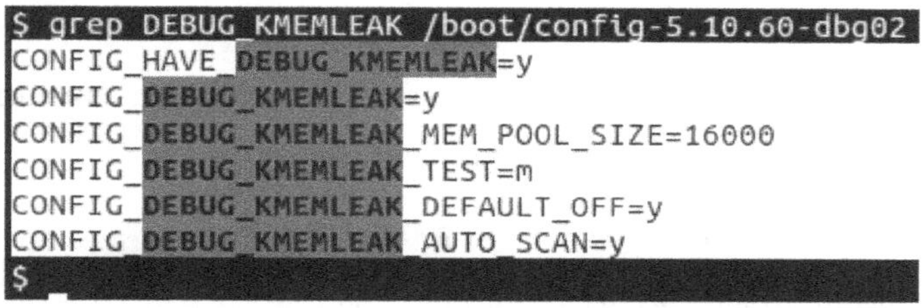

```
$ grep DEBUG_KMEMLEAK /boot/config-5.10.60-dbg02
CONFIG_HAVE_DEBUG_KMEMLEAK=y
CONFIG_DEBUG_KMEMLEAK=y
CONFIG_DEBUG_KMEMLEAK_MEM_POOL_SIZE=16000
CONFIG_DEBUG_KMEMLEAK_TEST=m
CONFIG_DEBUG_KMEMLEAK_DEFAULT_OFF=y
CONFIG_DEBUG_KMEMLEAK_AUTO_SCAN=y
$
```

Figure 6.13 – Our debug kernel configured for kmemleak

A one-liner regarding each of the configs seen in the preceding screenshot is shown here:

- Does the architecture support kmemleak? Yes, as `CONFIG_HAVE_DEBUG_KMEMLEAK=y` (implies that kmemleak is supported on this CPU).
- Is the kmemleak config on? Yes, as `CONFIG_DEBUG_KMEMLEAK=y`.
- For allocations that might occur before kmemleak is fully initialized, a memory pool of this size (in bytes) is used to hold metadata: `CONFIG_DEBUG_KMEMLEAK_MEM_POOL_SIZE=16000`
- Build a module for testing kmemleak? Yes, as `CONFIG_DEBUG_KMEMLEAK_TEST=m`.
- Is kmemleak disabled by default? Yes, as `CONFIG_DEBUG_KMEMLEAK_DEFAULT_OFF=y`. Enable it by passing `kmemleak=on` on the kernel command line.
- Is scanning memory for leaks enabled (every 10 minutes) by default? Yes, as `CONFIG_DEBUG_KMEMLEAK_AUTO_SCAN=y`. This is considered reasonable for most systems, except perhaps low-end embedded systems.

Better chances of finding leaks with kmemleak are gained by enabling SLUB debug features on the kernel (`CONFIG_SLUB_DEBUG=y`). This is mainly due to the poisoning of slabs, which helps the leak detector as well. (As you'll realize from the previous section, this config option is typically on by default in any case!)

Using kmemleak

Using kmemleak is straightforward. Here's a basic five-step checklist and steps to follow:

1. First, verify the following:

    A. The debug filesystem (debugfs) is mounted and visible: we'll assume it is and mounted in the usual location, `/sys/kernel/debug`.

    B. Kmemleak's enabled and running: for now, we assume it's fine. What if it isn't? More on this in the *Addressing the issue – unable to write to the kmemleak pseudofile* section that follows...

2. Run your (possibly buggy) code or test cases, or just let the system run...

3. Initiate a memory scan. As root, do the following:

    ```
    echo scan > /sys/kernel/debug/kmemleak
    ```

    This kicks a kernel thread (no prizes for guessing that it's named `kmemleak`) into action, actively scanning memory for leaks... Once done, if a leak (or the suspicion of one) is found, a message in this format is sent to the kernel log:

    ```
    kmemleak: 1 new suspected memory leaks (see /sys/kernel/
    debug/kmemleak)
    ```

4. View the result of the scan by looking up the `kmemleak` debugfs pseudofile:

    ```
    cat /sys/kernel/debug/kmemleak
    ```

5. (Optional) Clear all the current memory leak results. As root, do the following:

    ```
    echo clear > /sys/kernel/debug/kmemleak
    ```

Note that as long as kmemleak memory scanning is active (it is by default), new leaks could come up. They can be seen by again simply reading the `kmemleak` debugfs pseudofile.

Before going any further, it's quite possible that *Step 1.B* will come up – kmemleak may not be enabled in the first place! The following section will help you troubleshoot and figure it out. Once you've read it, we'll move on to trying out kmemleak with our leaky test cases!

Addressing the issue – unable to write to the kmemleak pseudofile

A common issue could turn up: when attempting to write to the `kmemleak` debugfs file, often, you get an error like this:

```
# echo scan > /sys/kernel/debug/kmemleak
bash: echo: write error: Operation not permitted
```

Search the kernel log (via `dmesg` or `journalctl -k`) for a message like this:

```
kmemleak: Kernel memory leak detector disabled
```

If it does show up, it obviously shows that kmemleak, though configured, is still disabled at runtime. How come? *It usually implies that kmemleak hasn't been correctly or fully enabled yet.*

Here's a simple yet interesting way to debug what went wrong at boot (we certainly mentioned this technique in *Chapter 3, Debug via Instrumentation – printk and Friends*): at boot, ensure you pass the `debug` and `initcall_debug` parameters on the kernel command line, in order to enable debug printks and see details of all kernel init hooks. Now, once booted and running, do this:

```
$ cat /proc/cmdline
BOOT_IMAGE=/boot/vmlinuz-5.10.60-dbg02-gcc root=UUID=<...> ro
quiet splash 3 debug initcall_debug
```

Look up the kernel log, searching for `kmemleak`:

```
$ journalctl --output=short-unix -k |grep -iC2 "kmemleak"
1637844902.306232 dbg-LKD kernel: random: get_random_u64 called from __kmem_cache_create+0x2f/0x500 wit
h crng_init=0
1637844902.306303 dbg-LKD kernel: SLUB: HWalign=64, Order=0-3, MinObjects=0, CPUs=6, Nodes=1
1637844902.306367 dbg-LKD kernel: kmemleak: Kernel memory leak detector disabled
1637844902.306441 dbg-LKD kernel: Kernel/User page tables isolation: enabled
1637844902.306506 dbg-LKD kernel: ftrace: allocating 44433 entries in 174 pages

1637844902.629853 dbg-LKD kernel: calling  split_huge_pages_debugfs+0x0/0x29 @ 1
1637844902.629942 dbg-LKD kernel: initcall split_huge_pages_debugfs+0x0/0x29 returned 0 after 23 usecs
1637844902.630024 dbg-LKD kernel: calling  kmemleak_late_init+0x0/0xa1 @ 1
1637844902.630093 dbg-LKD kernel: initcall kmemleak_late_init+0x0/0xa1 returned -12 after 30 usecs
1637844902.630159 dbg-LKD kernel: calling  check_early_ioremap_leak+0x0/0x9e @ 1
1637844902.630225 dbg-LKD kernel: initcall check_early_ioremap_leak+0x0/0x9e returned 0 after 872 usecs
```

Figure 6.14 – Screenshot showing how kmemleak failed at boot

We can see the following:

- The init function `kmemleak_late_init()` failed, returning the value `-12`. This is the negative `errno` value, of course (recall the kernel's `0`/`-E` return convention).

- `errno` value `12` is ENOMEM, implying it failed as it ran out of memory.

- The error perhaps occurred here, in the initialization code of kmemleak:

```
// mm/kmemleak.c
static int __init kmemleak_late_init(void)
{
    kmemleak_initialized = 1;
```

```
debugfs_create_file("kmemleak", 0644, NULL, NULL,
&kmemleak_fops);
if (kmemleak_error) {
/* Some error occurred and kmemleak was disabled.
 *There is a [...] */
    schedule_work(&cleanup_work);
    return -ENOMEM;
} [...]
```

One possible reason that the `kmemleak_error` variable gets set is that the early log buffer used by kmemleak at boot isn't quite large enough. The size is a kernel config, `CONFIG_DEBUG_KMEMLEAK_MEM_POOL_SIZE`, and typically defaults to 16,000 bytes (as you can see above). So, let's try changing it to a larger value and retry (by the way, this config was earlier named `CONFIG_DEBUG_KMEMLEAK_EARLY_LOG_SIZE`; it was renamed to `CONFIG_DEBUG_KMEMLEAK_MEM_POOL_SIZE` in the 5.5 kernel):

```
$ scripts/config -s CONFIG_DEBUG_KMEMLEAK_MEM_POOL_SIZE
16000
$ scripts/config --set-val CONFIG_DEBUG_KMEMLEAK_MEM_POOL_SIZE
32000
$ scripts/config -s CONFIG_DEBUG_KMEMLEAK_MEM_POOL_SIZE
32000
```

(The `-s` option switch to the `config` script is to show the current value of the kernel config supplied as a parameter; the `--set-val` switch is to set a kernel config). We now build the reconfigured (debug) kernel, reboot, and test.

Guess what? We get the very same error: `dmesg` again shows that kmemleak is disabled!

A dollop of thought will reveal the actual – and rather silly – issue: to enable kmemleak, we *must* pass `kmemleak=on` via the kernel command line. In fact, we already mentioned this very point in the section on configuring kmemleak: *Is kmemleak disabled by default? Yes, as CONFIG_DEBUG_KMEMLEAK_DEFAULT_OFF=y).*

Important

Want to use kmemleak to detect kernel-space memory leaks? Then, after configuring the kernel for it, you have to explicitly enable it by passing `kmemleak=on` on the kernel command line.

Once this is done, all seems well (to verify, I even set the value of CONFIG_DEBUG_
KMEMLEAK_MEM_POOL_SIZE back to its default of 16,000, rebuilt the kernel,
and rebooted):

```
$ dmesg |grep "kmemleak"
[    0.000000] Command line: BOOT_IMAGE=/boot/vmlinuz-5.10.60-
dbg02-gcc root=UUID=<...> ro quiet splash 3 kmemleak=on
[...]
[    6.743927] kmemleak: Kernel memory leak detector
initialized (mem pool available: 14090)
[    6.743956] kmemleak: Automatic memory scanning thread
started
```

All good! Here's the kmemleak kernel thread:

```
$ ps -e|grep kmemleak
 144 ?        00:00:07 kmemleak
```

It, in fact, is deliberately run at a lower priority (a nice value of 10), thus only running
when most other threads yield (recall that on Linux, the nice value ranges from -20 to
+19, with -20 being the highest priority). By the way, you can check the nice value by
running ps -el instead of just ps -e.

The Nature of Debugging

So, this particular debug session turned out to be a bit of a non-event. That's
okay – we eventually figured it out and got kmemleak enabled, and that's what
matters. It also shows a truth about the nature of debugging – quite often, we'll
chase down a path (or several) that really doesn't lead anywhere (so-called **red
herrings**). Worry not, it's all part of the experience! In fact, it helps – you will
always learn something!

For the curious: passing kmemleak=on as a kernel parameter caused the mm/
kmemleak.c:kmemleak_boot_config() function to set the kmemleak_skip_
disable variable to 1, which skips disabling it at boot, which is what occurs otherwise.

Running our test cases and catching leakage defects

Now that kmemleak's enabled and running, let's get to the interesting bit – running
our buggy, leaky as heck test cases! We have three of them: without further ado, let's get
started with the first one.

Running test case 3.1 – simple memory leakage

The code of our first memory leakage test case is as follows:

```
// ch5/kmembugs_test/kmembugs_test.c
void leak_simple1(void)
{
    volatile char *p = NULL;
    pr_info("testcase 3.1: simple memory leak testcase 1\n");
    p = kzalloc(1520, GFP_KERNEL);
    if (unlikely(!p))
        return;
    pr_info("kzalloc(1520) = 0x%px\n", p);
    if (0)              // test: ensure it isn't freed
        kfree((char *)p);
#ifndef CONFIG_MODULES
    pr_info("kmem_cache_alloc(task_struct) = 0x%px\n",
    kmem_cache_alloc(task_struct, GFP_KERNEL));
#endif
    pr_info("vmalloc(5*1024) = 0x%px\n", vmalloc(5*1024));
}
```

Clearly, this code has three memory leaks – the allocation of 1,520 bytes via the kzalloc() API, the allocation of a task_struct object from its slab cache via the kmem_cache_alloc() API, and the allocation of 5 kilobytes via the vmalloc() API. Notice though, that due to CONFIG_MODULES being set, the second case doesn't actually run, leaving us with two leaks. (Here, I haven't shown explicit code to check for the failure case for the latter two allocation APIs; you should check, of course.)

As mentioned at the beginning of this section, *Using kmemleak*, let's now perform steps 1 to 5 (1 to 4 actually, as step 5 is optional – it's shown later):

1. Verify kmemleak's enabled, running, and its kthread is alive and well.

2. Run the test case:

    ```
    cd <booksrc>/ch5/kmembugs_test
    ./load_testmod
    [...]
    sudo ./run_tests --no-clear
    ```

```
--no_clear: will not clear kernel log buffer after
running a testcase
Debugfs file: /sys/kernel/debug/test_kmembugs/lkd_dbgfs_
run_testcase
Generic KASAN: enabled
UBSAN: enabled
KMEMLEAK: enabled


Select testcase to run:
1  Uninitialized Memory Read - UMR
[...]
Memory leakage
3.1  simple memory leakage testcase1
3.2  simple memory leakage testcase2 - caller to free
memory
3.3  simple memory leakage testcase3 - memleak in
interrupt ctx
[...]
(Type in the testcase number to run):
3.1
Running testcase "3.1" via test module now...
[...]
[ 4053.909155] testcase to run: 3.1
[ 4053.909169] test_kmembugs:leak_simple1(): testcase
3.1: simple memory leak testcase 1
[ 4053.909212] test_kmembugs:leak_simple1():
kzalloc(1520) = 0xffff888003f17000
[ 4053.909390] test_kmembugs:leak_simple1():
vmalloc(5*1024) = 0xffffc9000005c000
```

You can see from the output that the run_tests bash script first does a few quick config checks and determines that KASAN, UBSAN, and KMEMLEAK are all enabled (do browse through the script's code on the book's GitHub repo). It then displays the menu of available test cases and has us select one. We type in 3.1. The debugfs write hook, upon seeing this, invokes the test function – in this case, leak_simple1(). It executes and you can see its printk output. Of course, it's buggy, leaking memory twice, as expected.

3. Here's the key part! Initiate a kmemleak memory scan, as root:

```
sudo sh -c "echo scan > /sys/kernel/debug/kmemleak"
```

Hang on tight, the memory scan can take some time (on my x86_64 Ubuntu VM running our custom debug kernel, it takes approximately 8 to 9 seconds)...

4. We read the content within the kmemleak pseudofile (in the section immediately following this). Once done, and a potential leak(s) is found, the kernel log will show something like this:

```
dmesg | tail -n1
kmemleak: 2 new suspected memory leaks (see /sys/kernel/
debug/kmemleak)
```

It even prompts you to *now* look up the kmemleak pseudofile, /sys/kernel/ debug/kmemleak. (You could always rig up a script to poll for a line like this within the kernel log and only then read the scan report; I'll leave stuff like this to you as an exercise.)

Interpreting kmemleak's report: So, let's look up the details, as kmemleak's urging us to:

```
$ sudo cat /sys/kernel/debug/kmemleak
unreferenced object 0xffff8880127f8000 (size 2048):
  comm "run_tests", pid 5498, jiffies 4296684850 (age 84.737s)
  hex dump (first 32 bytes):
    00 00 00 00 00 00 00 00 00 00 00 00 00 00 00 00  ................
    00 00 00 00 00 00 00 00 00 00 00 00 00 00 00 00  ................
  backtrace:
    [<00000000c0b84cb6>] slab_post_alloc_hook+0x78/0x5b0
    [<00000000f76c1d8d>] kmem_cache_alloc_trace+0x16b/0x370
    [<00000000896eb2a4>] leak_simple1+0x45/0x90 [test_kmembugs]
    [<000000000fca301f>] dbgfs_run_testcase+0x1c7/0x51a [test_kmembugs]
    [<00000000f0fd1df8>] full_proxy_write+0xaf/0xe0
    [<000000000d54f8ef>] vfs_write+0x148/0x500
    [<000000007f738be9>] ksys_write+0xd9/0x180
    [<000000001fce737f>] __x64_sys_write+0x43/0x50
    [<000000001a646102>] do_syscall_64+0x38/0x90
    [<0000000024b0a009>] entry_SYSCALL_64_after_hwframe+0x44/0xa9
unreferenced object 0xffffc90000065000 (size 8192):
  comm "run_tests", pid 5498, jiffies 4296684851 (age 84.734s)
  hex dump (first 32 bytes):
    00 00 00 00 00 00 00 00 00 00 00 00 00 00 00 00  ................
    00 00 00 00 00 00 00 00 00 00 00 00 00 00 00 00  ................
  backtrace:
    [<000000001fb65f64>] __vmalloc_node_range+0x476/0x4f0
    [<00000000c80cce1d>] __vmalloc_node+0xa7/0xd0
    [<000000001fd83f6a>] vmalloc+0x21/0x30
    [<000000005e2eaf52>] leak_simple1+0x71/0x90 [test_kmembugs]
    [<000000000fca301f>] dbgfs_run_testcase+0x1c7/0x51a [test_kmembugs]
    [<00000000f0fd1df8>] full_proxy_write+0xaf/0xe0
    [<000000000d54f8ef>] vfs_write+0x148/0x500
    [<000000007f738be9>] ksys_write+0xd9/0x180
    [<000000001fce737f>] __x64_sys_write+0x43/0x50
    [<000000001a646102>] do_syscall_64+0x38/0x90
    [<0000000024b0a009>] entry_SYSCALL_64_after_hwframe+0x44/0xa9
$
```

Figure 6.15 – kmemleak showing the memory leakage report for our test case #3.1

Aha, the screenshot shows that both leakage bugs have indeed been caught! Bravo.

Let's interpret in detail the first of kmemleak's reports:

- `unreferenced object 0xffff8880127f8000 (size 2048)::` The **KVA** – the **kernel virtual address** – of the unreferenced object, the orphaned memory chunk, the one that was allocated but not freed, is displayed, followed by it's size in bytes.

- Our test code issued a call: `kzalloc(1520, GFP_KERNEL);` asking for 1,520 bytes whereas the report shows the allocated size as 2,048 bytes. We know why this is: the slab layer allocates memory on a best-fit basis; the closest slab cache greater than or equal to the size we want is the `kmalloc-2k` one, thus the size shows up as 2,048 bytes.

- `comm "run_tests", pid 5498, jiffies 4296684850 (age 84.737s):` This line shows the (process) context in which the leak occurred, the value of the `jiffies` variable when it occurred, and the `age` – this is the time elapsed from when the process context (the one that ran the leaky kernel code) ran to now... (doing `sudo cat /sys/kernel/debug/kmemleak` a little later will show that the age has increased!). Keeping an eye on the `age` field can be useful: it allows you to see if a detected leak is an old one (you can then clear the list by writing `clear` to the `kmemleak` pseudofile).

- Next, a hex dump of the first 32 bytes of the affected memory chunk is displayed (as we issued the `kzalloc()` API here, the memory is initialized to all zeros).

- This is followed by the crucial information – a stack backtrace, which of course you read bottom-up. In this particular first leak test case, you can see from the trace that a write system call was issued (showing up as the frame `__x64_sys_write` here; this originates from the `echo` command we issued, of course). It, as expected, ended up in our debugfs write routine.

- Next, we see `dbgfs_run_testcase+0x1c7/0x51a [test_kmembugs]`. As mentioned earlier, this implies that the code that ran was at an offset of `0x1c7` bytes from the start of this function and the length of the function is `0x51a` bytes. The module name is in square brackets, showing that this function lives within that module.

  - It called the function `leak_simple1()`, again, within our module.

  - This function, as we know, issued the `kzalloc()` API, which is a simple wrapper around the `kmalloc()` API, which works by allocating memory from an existing slab cache (one of the `kmalloc-*()` slab caches – as mentioned earlier, it will be the one named `kmalloc-2k`).

- This allocation is internally done via the kmem_cache_alloc() API, which kmemleak tracks (thus it shows up in the stack backtrace as kmem_cache_ alloc_trace()).

So, there we are – we can see that our test case indeed caused the leak! The interpretation of the second leak is completely analogous. This time, the stack backtrace clearly shows that the leak_simple1() function in the test_ kmembugs module allocated a memory buffer by invoking the vmalloc() API, which we quite deliberately didn't subsequently free, causing the leak.

5. Optionally (and step 5 of this procedure), we can clear (as root) all the current memory leak results:

```
$ sudo sh -c "echo clear > /sys/kernel/debug/kmemleak"
$ sudo cat /sys/kernel/debug/kmemleak
```

Done. Clearing the previous results is useful, allowing you to de-clutter the report output; this is especially true when running development code (or test cases) over and over.

Running test case 3.2 – the "caller-must-free" case

Our second memory leakage test case is interesting: here, we invoke a function (named leak_simple2()) that allocates a small 8-byte piece of memory via the kmalloc() API and sets it content to the string literal, leaky!!. It then returns the pointer to this memory object to the caller. This is fine. The caller then collects the result in another pointer and prints its value – it's as expected. Here's the code of the *caller*:

```
// ch5/kmembugs_test/debugfs_kmembugs.c
[...]
else if (!strncmp(udata, "3.2", 4)) {
    res2 = (char *)leak_simple2();
    // caller's expected to free the memory!
    pr_info(" res2 = \"%s\"\n", res2 == NULL ? "<whoops, it's
NULL>" : (char *)res2);
    if (0) /* test: ensure it isn't freed by us, the caller */
        kfree((char *)res2);
}
```

Run it (via our run_tests script), then perform the usual:

```
$ sudo sh -c "echo scan > /sys/kernel/debug/kmemleak"
$ sudo cat /sys/kernel/debug/kmemleak
```

```
unreferenced object 0xffff8880074b5d20 (size 8):
comm "run_tests", pid 5779, jiffies 4298012622 (age 181.044s)
hex dump (first 8 bytes):
6c 65 61 6b 79 21 21 00                          leaky!!.
backtrace:
[<00000000c0b84cb6>] slab_post_alloc_hook+0x78/0x5b0
[<00000000f76c1d8d>] kmem_cache_alloc_trace+0x16b/0x370
[<000000009f614545>] leak_simple2+0xc0/0x19b [test_kmembugs]
[<00000000747f9f09>] dbgfs_run_testcase+0x1e6/0x51a [test_
kmembugs]
    [...]
```

Great, it's caught. Interestingly, the first time I ran the scan, nothing seemed to be detected. Running it again after a minute or so yielded the expected result – it reported one new suspected memory leak. Also, having the address of the unreferenced (or orphaned) memory buffer lets you investigate more about it via the kmemleak dump command. We cover it and related stuff in the upcoming section, *Controlling the kmemleak scanner.*

Running test case 3.3 – memory leak in interrupt context

Until now, we've been pretty much exclusively running our test cases by having a process run through our (buggy) kernel module code. This, of course, implies that the kernel code was run in *process context*. The other context in which kernel code can possibly run is *interrupt context*, literally, within the context of an interrupt.

Types of Interrupt Contexts

More precisely, within interrupt context, we can have a hardirq (the actual hardware interrupt handler), the so-called softirq, and the tasklet (the common way in which bottom halves are implemented – the tasklet is in fact a type of softirq). These details (and a lot more!) are covered in depth in my earlier book *Linux Kernel Programming – Part 2* (hey, it's freely downloadable too).

So, what if we have a memory leak in code that runs in interrupt context? Will kmemleak detect it? The only way to know is to try – the empirical approach!

The code of our third – interrupt-context – memory leakage test case is as follows:

```
// ch5/kmembugs_test/kmembugs_test.c
void leak_simple3(void)
{
```

```
    pr_info("testcase 3.3: simple memory leak testcase 3\n");
    irq_work_queue(&irqwork);
}
```

To achieve running in interrupt context without an actual device that generates interrupts, we make use of a kernel feature – the `irq_work*` functionality. It allows the ability to run code in interrupt (hardirq) context. Without going into details, to set this up, in the init code of our module, we called the `init_irq_work()` API. It registers the fact that our function named `irq_work_leaky()` will be invoked in hardirq context. But when will this happen? Whenever the `irq_work_queue()` function triggers it! This is the code of the actual interrupt context function:

```
/* This function runs in (hardirq) interrupt context */
void irq_work_leaky(struct irq_work *irqwk)
{
    int want_sleep_in_atomic_bug = 0;
    PRINT_CTX();
    if (want_sleep_in_atomic_bug == 1)
        pr_debug("kzalloc(129) = 0x%px\n",
            kzalloc(129, GFP_KERNEL));
    else
        pr_debug("kzalloc(129) = 0x%px\n",
            kzalloc(129, GFP_ATOMIC));
}
```

The leakage bug is obvious. Did you see the sneaky bug that we can cause to surface (if you set the variable `want_sleep_in_atomic_bug` to 1?) This leads to an allocation with the `GFP_KERNEL` flag in an *atomic* context – a bug! Okay, we'll ignore that for now as it won't trigger as the variable is set to 0 (try it out and see, though).

We execute the test case (via our trusty `run_tests` wrapper script). To be safe, let's clear the kmemleak internal state first:

```
sudo sh -c "echo clear > /sys/kernel/debug/kmemleak"
```

Then, run the relevant test case:

```
$ sudo ./run_tests
[...]
(Type in the testcase number to run):
3.3
[...]
```

Now, let's have kmemleak scan kernel memory for any suspected leaks, and dump its report:

```
$ sudo sh -c "echo scan > /sys/kernel/debug/kmemleak" ; dmesg |tail
[34619.682989] test_kmembugs:kmembugs_test_init(): KASAN configured
[34619.684794] test_kmembugs:kmembugs_test_init(): CONFIG_UBSAN configured
[34619.686614] test_kmembugs:kmembugs_test_init(): CONFIG_DEBUG_KMEMLEAK configured
[34619.688443] debugfs file 1 <debugfs_mountpt>/test_kmembugs/lkd_dbgfs_run_testcase created
[34619.690270] debugfs entry initialized
[35412.528017] testcase to run: 3.3
[35412.530040] test_kmembugs:leak_simple3(): testcase 3.3: simple memory leak testcase 3
[35412.532750] test_kmembugs:irq_work_leaky(): 001)  run_tests :11781   |  d.h1   /* irq_work_leaky() */
[35412.537365] test_kmembugs:irq_work_leaky(): kzalloc(129) = 0xffff88803c1e0e00
[35438.671971] kmemleak: 1 new suspected memory leaks (see /sys/kernel/debug/kmemleak)
$
$ sudo cat /sys/kernel/debug/kmemleak
unreferenced object 0xffff88803c1e0e00 (size 192):
  comm "hardirq", pid 0, jiffies 4305500943 (age 34.834s)
  hex dump (first 32 bytes):
    00 00 00 00 00 00 00 00 00 00 00 00 00 00 00 00  ................
    00 00 00 00 00 00 00 00 00 00 00 00 00 00 00 00  ................
  backtrace:
    [<00000000c0b84cb6>] slab_post_alloc_hook+0x78/0x5b0
    [<00000000f76c1d8d>] kmem_cache_alloc_trace+0x16b/0x370
    [<0000000002912ff8c>] irq_work_leaky+0x1f3/0x226 [test_kmembugs]
    [<00000000b094c375>] irq_work_single+0x8f/0xf0
    [<000000005a10cafa>] irq_work_run_list+0x52/0x70
    [<00000000e07f0913>] irq_work_run+0x6b/0x110
    [<000000006d70efc1>] __sysvec_irq_work+0x75/0x2b0
    [<0000000038851639>] asm_call_irq_on_stack+0x12/0x20
    [<000000006e1838aa>] sysvec_irq_work+0xc3/0xe0
    [<0000000043c320fa>] asm_sysvec_irq_work+0x12/0x20
    [<000000007864aefa>] native_write_msr+0x6/0x30
    [<0000000041cbb6ac>] x2apic_send_IPI_self+0x3c/0x50
    [<00000000b30d6970>] arch_irq_work_raise+0x5d/0x90
    [<00000000848d8ab3>] __irq_work_queue_local+0xf8/0x170
    [<00000000a3bb972c>] irq_work_queue+0x32/0x50
    [<000000005b977e7a>] leak_simple3+0x2f/0x31 [test_kmembugs]
$
```

Figure 6.16 – kmemleak catches the leak in interrupt context

Notice the following:

- Test case 3.3 runs. The custom `convenient.h:PRINT_CTX()` macro shows the context: you can see the `d.h1` token within, showing that the `irq_work_leaky()` function ran in the hardirq interrupt context (we covered interpreting the `PRINT_CTX()` macro's output in *Chapter 4, Debug via Instrumentation – Kprobes*, in the *Interpreting the PRINT_CTX() macro's output* section).

- The top line shows we ran the kmemleak `scan` command, getting it to check for any leakage.

- The read of the `kmemleak` pseudofile tells the story: the orphaned or unreferenced object, the memory buffer we allocated but didn't free. This time, the context is `"hardirq"` – perfect, the leak did indeed occur in an interrupt, not process, context. This is followed by the hex dump of the first 32 bytes and then the stack backtrace (whose output verifies the situation).

Next, let's check out the kernel's built-in kmemleak test module.

The kernel's kmemleak test module

While configuring the kernel for kmemleak, we set `CONFIG_DEBUG_KMEMLEAK_TEST=m`. This has the build generate the kmemleak test kernel module, for my (guest) system, here: `/lib/modules/$(uname -r)/kernel/samples/kmemleak/kmemleak-test.ko`.

This module's code is present within the `samples` folder of the kernel source tree, here: `samples/kmemleak/kmemleak-test.c`. Please do take a peek; though short and sweet (and full of leaks), it quite comprehensively runs memory leakage tests! I inserted it into kernel memory with the command:

```
sudo modprobe kmemleak-test
```

The `dmesg` output is seen below. I also did the usual kmemleak scan and dumped it's report; it caught all 13 memory leaks (!). The first of its reports (catching the first leak) is visible here as well:

```
[ 8825.985116] kmemleak: Kmemleak testing
[ 8825.985147] kmemleak: kmalloc(32) = 00000000cab708dd
[ 8825.985172] kmemleak: kmalloc(32) = 000000008d5c540a
[ 8825.985196] kmemleak: kmalloc(1024) = 000000006d719a53
[ 8825.985221] kmemleak: kmalloc(1024) = 00000000de599e5e
[ 8825.985247] kmemleak: kmalloc(2048) = 00000000b5e60406
[ 8825.985272] kmemleak: kmalloc(2048) = 000000000309c294
[ 8825.985299] kmemleak: kmalloc(4096) = 000000009200f455
[ 8825.985324] kmemleak: kmalloc(4096) = 000000001cfde96d
[ 8825.985555] kmemleak: vmalloc(64) = 00000000b7894b61
[ 8825.985672] kmemleak: vmalloc(64) = 00000000bbb401d6
[ 8825.985796] kmemleak: vmalloc(64) = 000000009c4e811f
[ 8825.985893] kmemleak: vmalloc(64) = 000000001e8fcc4a
[ 8825.985999] kmemleak: vmalloc(64) = 000000007f7b580a
[ 8825.986025] kmemleak: kzalloc(sizeof(*elem)) = 00000000d68f3627
[ 8825.986048] kmemleak: kzalloc(sizeof(*elem)) = 000000008bcc71cd
[ 8825.986070] kmemleak: kzalloc(sizeof(*elem)) = 00000000d90adbf5
[ 8825.986092] kmemleak: kzalloc(sizeof(*elem)) = 000000004c07e127
[ 8825.986115] kmemleak: kzalloc(sizeof(*elem)) = 00000000226b752f
[ 8825.986141] kmemleak: kzalloc(sizeof(*elem)) = 00000000d7eaeed8
[ 8825.986164] kmemleak: kzalloc(sizeof(*elem)) = 000000006ed69561
[ 8825.986187] kmemleak: kzalloc(sizeof(*elem)) = 00000000a79442e4
[ 8825.986209] kmemleak: kzalloc(sizeof(*elem)) = 0000000083a42752
[ 8825.986231] kmemleak: kzalloc(sizeof(*elem)) = 00000000412c4a56
[ 8825.986259] kmemleak: kmalloc(129) = 000000005c48a002
[ 8825.986281] kmemleak: kmalloc(129) = 000000000700d3c9
[ 8825.986304] kmemleak: kmalloc(129) = 0000000000e572f9
[ 8825.986327] kmemleak: kmalloc(129) = 000000002943f11c
[ 8825.986351] kmemleak: kmalloc(129) = 00000000f9236807
[ 8825.986372] kmemleak: kmalloc(129) = 00000000b9efae8e
$ time sudo sh -c "echo scan > /sys/kernel/debug/kmemleak"

real    0m8.950s
user    0m0.000s
sys     0m8.947s
$ dmesg |tail -n1
[ 8860.390327] kmemleak: 13 new suspected memory leaks (see /sys/kernel/debug/kmemleak)
$ sudo cat /sys/kernel/debug/kmemleak
unreferenced object 0xffff88800df30540 (size 32):
  comm "modprobe", pid 5647, jiffies 4297524992 (age 866.434s)
  hex dump (first 32 bytes):
    00 00 00 00 00 00 00 00 00 00 00 00 00 00 00 00  ................
    00 00 00 00 00 00 00 00 00 00 00 00 00 00 00 00  ................
  backtrace:
    [<00000000c0b84cb6>] slab_post_alloc_hook+0x78/0x5b0
    [<00000000f76c1d8d>] kmem_cache_alloc_trace+0x16b/0x370
    [<00000000e1aa9887>] 0xffffffffc080f058
    [<00000000deb5ae43>] do_one_initcall+0xcb/0x430
    [<00000000fc291604>] do_init_module+0x10f/0x3b0
    [<00000000977ca321>] load_module+0x3f49/0x4570
    [<0000000040c61d85>] __do_sys_finit_module+0x12a/0x1b0
    [<00000000d87c4816>] __x64_sys_finit_module+0x43/0x50
    [<000000001a646102>] do_syscall_64+0x38/0x90
    [<0000000024b0a009>] entry_SYSCALL_64_after_hwframe+0x44/0xa9
```

Figure 6.17 – Output from trying out the kernel's kmemleak-test module; the first kmemleak report is seen at the bottom

Notice that the addresses of the allocated memory buffers are printed with the `%p` specifier, leading to the kernel hashing it (info leak prevention, security), but the kmemleak report shows the actual kernel virtual address. Do try it out yourself and read the full report.

Controlling the kmemleak scanner

This is the `kmemleak` debugfs pseudofile, our means to work with kmemleak:

```
$ sudo ls -l /sys/kernel/debug/kmemleak
rw-r—r-- 1 root 0 Nov 26 11:34 /sys/kernel/debug/kmemleak
```

As you know by now, reading from it has the underlying kernel callback display the last memory leakage report, if any. We've also seen a few values that can be written to it, in order to control and modify kmemleak's actions at runtime. There are a few more; we summarize all values you can write (you'll need root access, of course) in the following table:

String to write to /sys/kernel/debug/kmemleak (as root)	Effect
clear	Clears all existing memory leak suspects from its internal list; if disabled, frees all kmemleak meta objects
dump=<kva>	Dump information on the object (memory chunk) at the given KVA. For example, # echo dump=0xffff88800df30540 > /sys/kernel/debug/kmemleak dumps information about the particular memory object at that address (look up the address from the kmemleak report, of course).
off	Disables kmemleak; note that this action is irreversible. Internal objects might still be kept (and can sometimes occupy significant amounts of RAM). To free them, write clear to the kmemleak pseudofile. Can also disable kmemleak at boot by passing kmemleak=off as a kernel parameter.
scan	Kick-starts a memory scan; kmemleak searches for orphaned (leaked) memory. (View its report by reading the kmemleak pseudofile as root).
scan=on	Start the automatic scan of memory via the kmemleak kernel thread.
scan=<seconds>	Sets the automated memory scan (with the kmemleak kthread) interval in seconds. The default is 600; 0 disables automatic scanning.
scan=off	Stop the automatic scan of memory via the kmemleak kernel thread.
stack=on	Enable task stack scanning (default).
stack=off	Disable task stack scanning.

Table 6.3 – Values to write to the kmemleak pseudofile to control it

The code that governs the action to be taken on these writes can be seen here: mm/kmemleak.c:kmemleak_write().

A quick tip: If you need to test something specific and want a clean slate, as such, it's easy to: first clean the kmemleak internal list, run your module or test(s), and run the `scan` command, followed by the read to the `kmemleak` pseudofile. So, something like this:

```
# echo clean > /sys/kernel/debug/kmemleak
//... run your module / test cases(s) / kernel code / ...
// wait a bit ...
# echo scan > /sys/kernel/debug/kmemleak
// check dmesg last line to see if new leak(s) have been found
by kmemleak
// If so, get the report
# cat /sys/kernel/debug/kmemleak
```

As with any such tool, the possibility of false positives is present. The official kernel documentation provides some tips on how you could deal with it (if required): `https://www.kernel.org/doc/html/latest/dev-tools/kmemleak.html#dealing-with-false-positives-negatives`. This document also covers some details about the internal algorithm used by kmemleak to detect memory leakage. Do check it out (`https://www.kernel.org/doc/html/latest/dev-tools/kmemleak.html#basic-algorithm`).

A few tips for developers regarding dynamic kernel memory allocation

Though not directly related to debugging, we feel it's well worth mentioning a few tips with respect to dynamic kernel memory allocation and freeing, for a modern driver or module author. This is in line with the age-old principle that prevention is better than cure!

Preventing leakage with the modern devres memory allocation APIs

Modern driver authors should definitely exploit the kernel's **resource-managed** (or **devres**) devm_k{m,z}alloc() APIs. The key point: *they allow you to allocate memory and not worry about freeing it!* Though there are several (they all are of the form devm_*()), let's focus on the common case, the following dynamic memory allocation APIs for you, the typical driver author:

```
void *devm_kmalloc(struct device *dev, size_t size, gfp_t gfp);
void *devm_kzalloc(struct device *dev, size_t size, gfp_t gfp);
```

Why do we stress that only driver authors are to use them? Simple: the first, mandatory parameter is a pointer to the device structure, typical in all kinds of device drivers.

The reason why these resource-managed APIs are useful is that there is no need for the developer to explicitly free the memory allocated by them. The kernel resource management framework guarantees that it will automatically free the memory buffer upon driver detach, and/or, if a kernel module, when the module is removed (or the device is detached, whichever occurs first).

As you'll surely realize, this feature immediately enhances code robustness. Why? Simple, we're all human and make mistakes. Leaking memory (especially on error code paths) is indeed a fairly common bug!

A few relevant points regarding the usage of these devres APIs:

- A key point – don't attempt to blindly replace k[m|z]alloc() with the corresponding devm_k[m|z]alloc() APIs! These resource-managed allocations are really designed to be used only in the init and/or probe() methods of a device driver (all drivers that work with the kernel's unified device model will typically supply the probe() and remove() (or disconnect()) methods. We will not delve into these aspects here.

- devm_kzalloc() is usually preferred as it initializes the buffer as well, thus eclipsing the, again all too common, **uninitialized memory read** (**UMR**) types of defects. Internally (as with kzalloc()), it is merely a thin wrapper over the devm_kmalloc() API. (It's popular: the 5.10.60 kernel has the devm_kzalloc() function being invoked well over 5,000 times.)

- The second and third parameters are the usual ones, as with the k[m|z]alloc() APIs – the number of bytes to allocate and the **Get Free Page** (**GFP**) flags to use. The first parameter, though, is a pointer to struct device. Quite obviously, it represents the device that your driver is driving.

- As the memory allocated by these APIs is auto-freed (on driver detach or module removal), you don't have to do anything after allocation. It can, though, be freed via the devm_kfree() API. Doing this, however, is usually an indication that the managed APIs are the wrong ones to use...

- The managed APIs are exported (and thus available) only to modules licensed under the GNU **General Public License** (**GPL**) (ah, the sweet revenge of the kernel community).

A few more tips on memory-related issues for developers follow...

Other (more developer-biased) common memory-related bugs

Studies have shown that <*insert anything you'd like here*>. Okay, jokes aside, there's evidence to suggest that preventing bugs during the development cycle (and/or early unit testing) itself causes the least impact on the product (both cost-wise and otherwise). Good, solid coding practices are skills one continually hones as a developer. As we've seen, when it comes to working directly with memory, a non-managed language such as C can be a nightmare, both bug- and security-wise. Thus, here's a hopefully useful quick list with regard to common development-time defects that can occur when using the kernel's memory allocation/free slab APIs:

- Performing a kernel slab allocation with the wrong GFP flag(s); for example, with GFP_KERNEL when in an atomic context (such as an interrupt context of any sort, or when holding a spinlock), here, you should use the GFP_ATOMIC flag, of course!

  For instance, here's the patch to one such bug: https://lore.kernel.org/lkml/1420845382-25815-1-git-send-email-khoroshilov@ispras.ru/.

- Memory allocated via k{m|z}alloc() but freed with vfree(), and vice versa.

- Not checking the failure case (NULL, of course, for memory allocations). This might seem pedantic, but it can and does happen! Using the if (unlikely(!p)) { [...] kind of semantic is fine.

- Doing things such as the following:

```
if (p)
    kfree(p);
```

  It's not required but quite harmless – still, don't. The reverse isn't: only performing some action after a free conditionally, if the pointer is NULL. In other words, assuming that the free interface sets the pointer variable to NULL! It does not (though that would be quite intuitive).

- Failing to realize the wastage (internal fragmentation) that can occur when allocating memory via the slab layer. Use the ksize() API to see the *actual* number of bytes allocated. For example, in this pseudocode, p = kmalloc(4097); n = ksize(p);, you'll find the value of n – the actual memory allocated – is 8,192, implying a wastage of *8,192-4,097 = 4,095* bytes, or almost 100%! Ask yourself: could I not redesign to allocate 4,096 bytes via kmalloc(), instead of 4,097? Also, recall, you can leverage using slabinfo with the -L option switch to see losses/wastage in all slab caches.

With this, we complete our detailed coverage on understanding and catching dangerous memory leakage defects within the kernel! Great going. Let's complete this long chapter with a kind of summarization of the many tools and techniques we've used.

Catching memory defects in the kernel – comparisons and notes (Part 2)

The table that follows is an extension of the one in the previous chapter (*Table 5.5*), adding the rightmost column, that of employing the kernel's SLUB debug framework. Here, we tabulate and hence summarize our test case results for our test runs with all the tooling technologies/kernels – vanilla/distro kernel, compiler warnings, with KASAN, with UBSAN, and with SLUB debug with our debug kernel – we employed in the preceding and this chapter. In effect, *it's a compilation of all the findings in one place*, thus allowing you to make quick (and hopefully helpful) comparisons.

Testcase # [1]	Memory defect type (below) / Infrastructure used (right)	Distro kernel [2]	Compiler warning? [3]	With KASAN [4]	With UBSAN [5]	With slub_debug=FZPU
Defects not covered by the kernel's KUnit test_kasan.ko module						
1	Uninitialized Memory Read – UMR	N	Y [C1]	N	N	N
2	Use After Return – UAR	N	Y [C2]	N [SA]	N [SA]	N
3	Memory leakage [6]	N	N	N	N	N
Defects covered by the kernel's KUnit test_kasan.ko module						
4	OOB accesses on static global (compile-time) memory					
4.1	Read (right) overflow	N [V1]	N	Y [K1]	Y [U1,U2]	N
4.2	Write (right) overflow		N	Y [K1]		
4.3	Read (left) underflow		N	Y [K2]		
4.4	Write (left) underflow		N	Y [K2]		
4	OOB accesses on static global (compile-time) stack local memory					

4.1	Read (right) overflow	N [V1]	N	Y [K3]	Y [U1,U2]	N
4.2	Write (right) overflow		N	Y [K2]		
4.3	Read (left) underflow		N	Y [K2]		
4.4	Write (left) underflow					
5	OOB accesses on dynamic (kmalloc-ed slab) memory					
5.1	Read (right) overflow	N	N	Y [K4]	N	N
5.2	Write (right) overflow					Y [S1]
5.3	Read (left) underflow					N
5.4	Write (left) underflow					Y [S1]
6	Use After Free – UAF	N	N	Y [K5]	N	Y
7	Double-free	Y [V2]	N	Y [K6]	N	Y
8	Arithmetic UB (via the kernel's test_ubsan.ko module)					
8.1	Add overflow				Y	
8.2	Sub(tract) overflow				N	
8.3	Mul(tiply) overflow				N	
8.4	Negate overflow				N	
	Div by zero	N	N	N	Y	N
8.5	Bit shift OOB	Y [U3]		Y [U3]	Y [U3]	
Other than arithmetic UB defects (copied from the kernel's KUnit test_ubsan.ko module)						
8.6	OOB	Y [U3]	N	Y [U3]	Y [U3]	
8.7	Load invalid value	Y [U3]		Y [U3]	Y [U3]	
8.8	Misaligned access	N		N	N	
8.9	Object size mismatch	Y [U3]		Y [U3]	Y [U3]	N
9	OOB on copy_[to|from]_user*()	N	Y [C3]	Y [K4]	N	N

Table 6.4 – Summary of various common memory defects and how various technologies react in catching them (or not)

As mentioned in the previous chapter, an explanation of the footnotes within this table (such as [C1], [K1], [U1], and so on) can be found in earlier relevant section. (For the footnotes within the third to sixth columns, refer to the previous chapter).

So, again, a very brief summary:

- KASAN catches pretty much all OOB buggy memory accesses on global (static), stack local, and dynamic (slab) memory. UBSAN doesn't catch the dynamic slab memory OOB accesses (test cases 4.x, 5.x).

- KASAN does not catch the UB defects (test cases 8.x). UBSAN does catch (most of) them.

- Neither KASAN nor UBSAN catch the first three test cases – UMR, UAR, and leakage bugs, *but the compiler(s) generate warnings and static analyzers (cppcheck, others) can catch some of them.*

- The kernel's SLUB debug framework is adept at catching most of the slab memory corruption defects, but no others.

- The kernel **kmemleak** infrastructure catches kernel memory leaks allocated by any of the k{m|z}alloc(), vmalloc(), or kmem_cache_alloc() (and friends) interfaces.

Miscellaneous notes

Again, a few more points on the footnotes regarding *Table 6.4*:

- [V1]: The system could simply Oops or hang here or even appear to remain unscathed, but that's not really the case... Once the kernel is buggy, the system is buggy.

- [V2]: Please see the explanation of this detailed note in the section *Running SLUB debug test cases on a kernel with slub_debug turned off.*

- [S1]: The kernel's SLUB debug infrastructure – when slub_debug=FZPU is passed as a kernel parameter – catches both write over and underflow (right and left) OOB accesses on slab memory. However, just as we saw with UBSAN, it only seems able to catch it when the buggy access is via incorrect indices to the memory region, *not* when the OOB access is via a pointer! Also, the OOB reads do not seem to be caught, only the writes.

So, there we are! We (finally) went through our single summary table (*Table 6.4*) for pretty much all the common memory defects and how they're caught, or not, by the tooling we've discussed in some depth.

Summary

Most dynamic memory allocation (and freeing) in the kernel is done via the kernel's powerful slab (internally, SLUB) interfaces. To debug them, the kernel provides a strong SLUB debug framework and several associated utilities (`slabtop`, `slabratetop` `[-bpfcc]`, `vmstat`, and so on). Here, you learned how to catch SLUB bugs via the kernel's SLUB debug framework as well as how to leverage these utilities.

Among memory bugs, the very mention of the leakage defect raises dread and fear, even in very experienced developers! It's a deadly one indeed, as we (hopefully) showed you in the *Can I see some real kernel memory leakage bugs?* section! The kernel's powerful kmemleak framework can catch these dangerous leakage bugs. Be sure to test your product (for long durations) with it running!

As we covered these tools and frameworks, we tabulated the results, showing you the bugs a given tool can (or cannot) catch. To then summarize the whole thing, we built a larger table with columns covering all the test cases and all the tools (*Table 6.4*) – a quick and useful way for you to see and compare tooling and the memory defects they do and do not catch (this table's a superset of a similar table, *Table 5.5*, in the previous chapter)!

Good job! You've now completed the long but really important chapters on catching memory bugs in kernel space! Whew, plenty to chew on, right?! I'd definitely recommend you take the time to think about and digest these topics, practicing as you go (please do the few exercises suggested as well!). Then, when you've done so, take a break and let's meet in the next, very interesting chapter, where we'll tackle head-on the topic of what a kernel *Oops* is and how we diagnose it. See you there!

Further reading

- SLUB debug:

  - Kernel documentation: Short users guide for SLUB: `https://www.kernel.` `org/doc/html/latest/vm/slub.html#short-users-guide-for-` `slub`

  - `slub_debug`: Detect kernel heap memory corruption, TechVolve, Mar 2014: `http://techvolve.blogspot.com/2014/04/slubdebug-detect-` `kernel-heap-memory.html`

  - `slabratetop` example by Brendan Gregg: `https://github.com/` `iovisor/bcc/blob/master/tools/slabratetop_example.txt`

- Interesting: Network Jitter: An In-Depth Case Study, Alibaba Cloud, Jan 2020, Medium: `https://alibaba-cloud.medium.com/network-jitter-an-in-depth-case-study-cb42102aa928`

- LLVM/Clang: LLVM FAQs, omnisci: `https://www.omnisci.com/technical-glossary/llvm`

- Kmemleak: *Kernel Memory Leak Detector*: `https://www.kernel.org/doc/html/latest/dev-tools/kmemleak.html#kernel-memory-leak-detector`

- Linux Kernel Memory Leak Detection, Catalin Marinas, 2011: `https://events.static.linuxfound.org/images/stories/pdf/lceu11_marinas.pdf`

- GRUB bootloader:

  - *How To Configure GRUB2 Boot Loader Settings In Ubuntu*, Sk, Sept 2019: `https://ostechnix.com/configure-grub-2-boot-loader-settings-ubuntu-16-04/`

  - *GRUB: How do I change the default boot kernel*: `https://askubuntu.com/questions/216398/set-older-kernel-as-default-grub-entry`

- The Heartbleed OpenSSL (TLS) vulnerability

  - `https://heartbleed.com/`

  - `https://xkcd.com/1354/` (Brilliantly illustrated here)

7
Oops! Interpreting the Kernel Bug Diagnostic

Kernel code is supposed to be perfect. It mustn't ever crash. But, of course, it does on occasion... Welcome to the real world.

When userspace code hits a (typical) bug – an invalid memory access, say – the processor's **Memory Management Unit (MMU)**, upon failing to translate the invalid userspace virtual address to a physical one (via the process context's paging tables), raises a fault. The fault handler within the kernel then takes control. It ultimately (and typically) results in a fatal signal (often, SIGSEGV) being sent to the faulting process (or thread). This, of course, has the process possibly handle the signal and terminate.

Now take exactly the same case – except that this time, the invalid memory access occurs in kernel space (in kernel mode)! Hey, that's not supposed to happen, right? True, but bugs do happen, within kernel space too. This time, the kernel fault handler, on realizing that it's kernel-mode code that triggered the fault, runs code to generate an **Oops** – *a kernel diagnostic that details what happened*. The unfortunate process context can die as well, as a side effect.

Here, you will learn about a key topic – what exactly a kernel Oops diagnostic message is, and more importantly, how to interpret it in detail. Along the way, you will generate a simple kernel Oops and understand exactly how to interpret it. Further along, several tools and techniques to help with this task will be shown. Getting to the bottom of the Oops often helps pinpoint the root cause of the kernel bug! To help you understand more – and better spot – typical issues, a few actual kernel Oopses will also be discussed and/or pointed to.

In this chapter, we will focus on the following main topics:

- Generating a simple kernel bug and Oops
- A kernel Oops and what it signifies
- The devil is in the details – decoding the Oops
- Tools and techniques to help determine the location of the Oops
- An Oops on an ARM Linux system and using netconsole
- A few actual Oops

Technical requirements

The technical requirements and workspace remain identical to what's described in *Chapter 1, A General Introduction to Debugging Software*. The code examples can be found within the book's GitHub repository here: `https://github.com/PacktPublishing/Linux-Kernel-Debugging`. The only thing new is we'll show you how to clone and use the useful `procmap` utility as well.

Generating a simple kernel bug and Oops

You've heard the quote *It takes a thief to catch a thief*. So, let's first learn how to generate a kernel bug (it shouldn't be too much of a challenge).

As you'll know, the classic pedagogical bug is the (in)famous NULL pointer dereference (the upcoming section, *What's this NULL trap page anyway?* elaborates on it). So, here's the plan:

- We'll first write a very simple kernel module that performs the cardinal sin of dereferencing the NULL pointer (the address `0x0`). We'll call it our version 1 `oops_tryv1` module.

- Once you try it out, we'll move on to a slightly more sophisticated version 2 `oops_tryv2` module. Within it, we'll provide three distinct ways to generate an Oops!

Before embarking on our *generate-an-Oops* quest, let's better understand what the `procmap` utility does and what the NULL trap page is. First, let's go with the utility.

The procmap utility

Being able to *visualize* the complete memory map of the kernel **Virtual Address Space** (**VAS**) as well as any given process's user VAS is what the `procmap` utility is designed to do. (Full disclosure: I'm the original author.)

The description on its GitHub page (`https://github.com/kaiwan/procmap`) sums it up:

> procmap is designed to be a console/CLI utility to visualize the complete memory map of a Linux process, in effect, to visualize the memory mappings of both the kernel and usermode Virtual Address Spaces (VAS).

> It outputs a simple visualization of the complete memory map of a given process in a vertically-tiled format ordered by descending virtual address (see screenshots below). The script has the intelligence to show kernel and userspace mappings as well as calculate and show the sparse memory regions that will be present. Also, each segment or mapping is (very approximately) scaled by relative size and color-coded for readability. On 64-bit systems, it also shows the so-called non-canonical sparse region or 'hole' (typically close to a whopping 16,384 PB on the x86_64).

The utility includes options to see only kernel space or userspace, verbose and debug modes, the ability to export its output in convenient CSV format to a specified file, as well as other options. It has a kernel component as well (a module) and currently works on (auto-detects) x86_64, AArch32, and AArch64 CPUs.

Do note, though, that it's not complete in any real sense; development is ongoing. There are several caveats. Feedback and contributions are most appreciated!

Download/clone it from here: `https://github.com/kaiwan/procmap`.

What's this NULL trap page anyway?

On all Linux-based systems (indeed, pretty much on all modern virtual memory-based operating systems), the kernel splits the virtual memory region available to a process into two portions – user and kernel VAS (we call it the *VM split* – there's a very detailed discussion in the *Linux Kernel Programming*, *Packt* book, in *Chapter 7*, *Memory Management Internals - Essentials*).

On the x86_64, the size of the complete VAS per process is of course 2^64 bytes. Now, that's a phenomenally huge number. It's 16 EB (EB stands for exabytes – 1 exabyte = 1,024 petabytes = 1 million terabytes = 1 billion gigabytes!). The VAS is simply far too large. So, the kernel, by default on the x86_64, is designed to split it like this:

- Kernel VAS of size 128 TB anchored to the top of the VAS (from the **kernel virtual address (KVA)** `0xffffffffffffffff` at the very top of the VAS to KVA `0xffff800000000000`)

- User VAS of size 128 TB anchored to the bottom of the VAS (from the **user virtual address (UVA)** `0x00007fffffffffff` to UVA `0x0` at the very bottom of the VAS)

> **Think about This**
>
> The 64-bit VAS is so big that, in this case, we end up using just a tiny fraction of the available address space. 16 EB is 16,384 PB. Of that – on the x86_64 - we're using *128 TB + 128 TB = 256 TB* (which is *256/1024 = 0.25 PB*). This implies that about 0.0015% of the available VAS is being used.

Now, to the point of interest here: at the low end of the user VAS, the very first virtual page – from byte 0 to byte 4095 – is called the **NULL trap page**. Let's quickly run the `procmap` utility (we assume you've installed it by now) on our shell process (which happens to have PID 1076) to see it display the NULL trap page:

```
$ </path/to/>procmap
 --pid=1076
[...]
```

We can see the NULL trap page in the following screenshot:

```
+------------------------------------------------------------+ 000057826c590000
|       /usr/bin/bash   [   36 KB,rw-,p,0x118000]            |
+------------------------------------------------------------+ 000057826c587000
|       /usr/bin/bash   [   16 KB,r--,p,0x114000]            |
+------------------------------------------------------------+ 000057826c583000
|       /usr/bin/bash   [  220 KB,r--,p,0xde000]             |
|                                                            |
+------------------------------------------------------------+ 000057826c54c000
|       /usr/bin/bash   [  708 KB,r-x,p,0x2d000]             |
|                                                            |
+------------------------------------------------------------+ 000057826c49b000
|       /usr/bin/bash   [  180 KB,r--,p,0x0]                 |
|                                                            |
+------------------------------------------------------------+ 000057826c46e000
|<... Sparse Region ...> [   87.50 TB,---,-,0x0]             |
|                                                            |
|                                                            |
|                                                            |
|                                                            |
|                                                            |
|                                                            |
|                                                            |
|                                                            |
~     .        .        .         .        .         .       ~
|                                                            |
|                                                            |
|                                                            |
|                                                            |
|                                                            |
+------------------------------------------------------------+ 0000000000001000
|     < NULL trap >   [    4 KB,---,-,0x0]                   |
+------------------    U S E R   V A S  start uva  ----------+ 0000000000000000
VAS mappings:  name    [ size,perms,u:maptype,u:0xfile-offset]
```

Figure 7.1 – Partial screenshot of the lower portion of the user VAS from the procmap utility

You can spot the NULL trap page right at the bottom of the preceding screenshot (some mappings of the `bash` process are visible higher up). The NULL trap page works by having all permissions – `rwx` – set to `- - -` so that no process (or thread) can read, write, or execute anything therein! This is why, when a process attempts to read or write the NULL byte at address `0x0`, it doesn't work. Briefly, what actually happens is this:

- A process attempts to access (read/write/execute) or dereference the NULL byte.

- In fact, accessing *any byte* within this page will lead to this same sequence of events, as the `- - -` mode applies to all bytes within the page, which is why it's called the NULL trap page! It traps access to any bytes within it.

- The permissions for all bytes in the page are zero: no read, no write, no execute. Now (unless cached), all virtual addresses end up at the MMU. The MMU does its checks and then performs runtime address translation, translating the virtual address to a physical one. Here, the MMU detects the fact that all bytes in the page have no permissions and thus raises a fault (typically on x86, a general protection fault).

- The OS has fault (and trap/exception) handlers preinstalled. Control is passed on to the appropriate fault handling function.

- This function – the fault handler – runs in the process context of the process that caused the fault. It, via a rather elaborate algorithm, figures out what the issue is.

- Here, the fault handler will conclude that a process executing in user mode attempted a buggy access. It thus sends it a fatal signal (`SIGSEGV`) to it. This is what can ultimately lead to the process dying and the `Segmentation fault [(core dumped)]` message showing on the console. Of course, the process could install a signal handler to handle this signal. Ultimately though, after cleanup, it must terminate.

Now that you understand what exactly the NULL trap page is and its workings, let's do what we're not supposed to: try and read/write the NULL address *in kernel mode*, causing a kernel bug!

A simple Oops v1 – dereferencing the NULL pointer

In this, our first simple version of a buggy kernel module, we simply read or write the NULL address. As you just learned in the previous section, any access – read, write, or execute – on any byte within the NULL trap page will cause the MMU to jump up and trigger a fault. This remains true in kernel mode as well.

Here's the relevant code snippet for the buggy module (please do clone this book's GitHub repo, browse through, and try things yourself!):

```
// ch7/oops_tryv1/oops_tryv1.c
[...]
static bool try_reading;
module_param(try_reading, bool, 0644);
MODULE_PARM_DESC(try_reading,
"Trigger an Oops-generating bug when reading from NULL; else,
do so by writing to NULL");
```

We keep a Boolean module parameter named try_reading. It's 0 (or off) by default. If set to 1 (or the value yes), the module code will attempt to read the content of the NULL address. If left as 0, the code, detecting this, will instead try to write a byte ('x') to the NULL address. Here's the code of the initialization function where this is done:

```
static int __init try_oops_init(void)
{
    size_t val = 0x0;
    pr_info("Lets Oops!\nNow attempting to %s something
            %s the NULL address 0x%p\n",
        !!try_reading ? "read" : "write",
        !!try_reading ? "from" : "to",  // pedantic, huh
        NULL);
    if (!!try_reading) {
        val = *(int *)0x0;
        /* Interesting! If we leave the code at this, the
compiler actually optimizes it away, as we're not working with
the result of the read. This makes it appear that the read does
NOT cause an Oops; this ISN'T the case, it does, of course. So,
to prove it, we try and printk the variable, thus forcing the
compiler to generate the code, and voila, we're rewarded with a
nice Oops ! */
        pr_info("val = 0x%lx\n", val);
    } else // try writing to NULL
        *(int *)val = 'x';
    return 0;       /* success */
}
```

It's pretty straightforward. Do read the detailed comment above regarding compiler optimization in the read case and how we can sidestep this.

The key point here, of course, is that both the read and write accesses are buggy – as described in detail in the previous section, *What's this NULL trap page anyway?*. Any attempt to read/write/execute any byte in the NULL trap page is disallowed and results in a fault! Here and now, the kernel module code, running in the process context of the insmod process, will perform the buggy access.

Now, think about this: the kernel isn't a process. The fault handler code, upon detecting that a buggy access was made **in kernel mode** (yes, it can and does happen!), realizes that something's dramatically wrong – the kernel is buggy. It thus triggers an Oops! (We also mention that there are exceptions to this; handling vmalloc faults and interrupts are some of them.)

> **What's with the !!<boolean> Syntax?**
>
> It's one of the C coding features being taken advantage of: using !!<boolean_expression> guarantees the expression evaluates to either 0 or 1, no matter what value is passed (for example, passing 5 makes it !!(5). Now, !5 is 0 and !0 is 1). Clever.

The following partial screenshot shows just the initial portion of the Oops messages written to the kernel log. Worry not – we'll definitely cover the rest and learn how to interpret it in detail. For now, just take a look at it. Here, in the example shown, we attempted to write into the NULL byte, triggering the Oops:

```
[  302.546331] oops_tryv1:try_oops_init():37: Lets Oops!
               Now attempting to write something to the NULL address 0x0000000000000000
[  302.546351] BUG: kernel NULL pointer dereference, address: 0000000000000000
[  302.546374] #PF: supervisor write access in kernel mode
[  302.546388] #PF: error_code(0x0002) - not-present page
[  302.546402] PGD 0 P4D 0
[  302.546411] Oops: 0002 [#1] PREEMPT SMP PTI
[  302.546424] CPU: 5 PID: 2903 Comm: insmod Tainted: G           OE     5.10.60-prod01 #6
[  302.546466] Hardware name: innotek GmbH VirtualBox/VirtualBox, BIOS VirtualBox 12/01/2006
[  302.546489] RIP: 0010:try_oops_init+0xdb/0x1000 [oops_tryv1]
```

Figure 7.2 – A partial screenshot showing the classic Oops that results when attempting to write to the NULL address (via our oops_tryv1 module)

The Oops: at the beginning of a line in the kernel log messages (spot it above) denotes the kernel printks that follow are the Oops diagnostic message.

A perhaps useful (silly) workaround to rebooting

Did you notice, once buggy, the kernel module can't be unloaded (via `rmmod`) as the reference count is non-zero? `lsmod` verifies this:

```
$ lsmod |grep oops
oops_tryv1                 16384  1
```

This is typically as the Oops occurred prior to the process context (`insmod`, in our case) exiting and thus having the module reference count being decremented down to 0. The 1 at the extreme right of the preceding output shows that the current module reference count is 1, preventing the unloading of this module.

Now, if you can't unload the module, you can't load it up again (to retry it after editing the source file(s)). The correct approach is to reboot the box and start over. A very silly workaround to this annoying problem is to simply clean up (`make clean`), rename the source file, edit the `Makefile` to use the new name, and build it. Now it will load up, under the new name! Very silly, but effective when you're in development and in a hurry to try things out.

Doing a bit more of an Oops – our buggy module v2

As mentioned at the beginning of this chapter, in our version 2 buggy module, we'll do a few more slightly (and hopefully) more realistic things to trigger a kernel Oops. This module has three distinct ways to trigger an Oops:

- One, by writing to a randomly generated KVA within the NULL trap page.

- Two, by allowing the user to pass a (random) invalid KVA and attempting to write something there (you can leverage the `procmap` utility to find an invalid KVA).

- Three, we spin up a simple workqueue function. This will have a kernel worker thread run its code when it's scheduled. Within the workqueue function, we'll trigger an Oops by attempting to write something to a member of a structure where the structure pointer is NULL (as this scenario can be quite realistic, we'll make it a use case pretty much throughout this chapter).

Let's begin by using the first approach mentioned above to trigger an Oops!

Case 1 – Oops by writing to a random location within the NULL trap page

Being very similar to the first v1 module, I won't delve much into this. It suffices to say that we use a kernel interface (the `get_random_bytes()` API) to generate a random number and scale it down to numbers between 0 and 4,095 (by using the modulo operator). The relevant code in the module's init function is seen here:

```
// ch7/oops_tryv2/oops_tryv2.c
[...]
static int __init try_oops_init(void)
{
    unsigned int page0_randptr = 0x0;
    [...]
} else { // no module param passed, write to random kva in NULL
trap
        pr_info("Generating Oops by attempting to write to a
random invalid kernel address in NULL trap page\n");
        get_random_bytes(&page0_randptr, sizeof(unsigned
int));
    bad_kva = (page0_randptr %= PAGE_SIZE);
    }
    pr_info("bad_kva = 0x%lx; now writing to it...\n", bad_
kva);
    *(unsigned long *)bad_kva = 0xdead;
    [...]
```

The last line seen here is where we attempt to write into this *bad* KVA. This, of course, triggers an Oops. To try this out, simply `insmod` the module without passing any parameters. This will have the code go to this use case (I'll leave it to you to try it out for yourself and see the kernel log).

Case 2 – Oops by writing to an invalid unmapped location within the kernel VAS

For this second use case, we have a module parameter named mp_randaddr. To run this case, you're to pass it to the module the usual way, setting it to an invalid kernel address (or KVA):

```
// ch7/oops_tryv2/oops_tryv2.c
[...]
static unsigned long mp_randaddr;
module_param(mp_randaddr, ulong, 0644);
MODULE_PARM_DESC(mp_randaddr, "Random non-zero kernel virtual
address; deliberately invalid, to cause an Oops!");
```

Now, when the module's init function detects that you've passed a non-zero value in this parameter, it invokes the following code:

```
} else if (mp_randaddr) {
        pr_info("Generating Oops by attempting to write to the
invalid kernel address passed\n");
        bad_kva = mp_randaddr;
    } else {
        [... << code of the first case above >> ...]
    }
    pr_info("bad_kva = 0x%lx; now writing to it...\n", bad_
kva);
    *(unsigned long *)bad_kva = 0xdead;
```

The approach is pretty much identical to the first case; what makes it interesting is this: how will I know which kernel address (or KVA) to pass? How will I know it's an invalid (or unmapped) location in the kernel VAS?

Ah, this is where the procmap utility comes into play! Simply run procmap (passing any PID and specifying the --only-kernel option switch, as we're not interested in the user VAS now). Here's how I invoked it, for example, on my x86_64 guest VM (you will need to update the PATH environment variable to include the directory where you installed procmap):

```
$ procmap --pid=1 --only-kernel
...
```

Here's a partial screenshot of the output it displays, focused on the upper portion of the kernel VAS:

```
[=================---    P R O C M A P    ---=================]
Process Virtual Address Space (VAS) Visualization utility
https://github.com/kaiwan/procmap

Wed Dec 15 14:55:03 IST 2021
[=====---  Start memory map for 1:systemd  ---=====]
[Pathname: /usr/lib/systemd/systemd ]
+-----------------  K E R N E L   V A S    end kva  -----------------+ ffffffffffffffff
|<... K sparse region ...> [   8.00 MB,--- ]                         |
|                                                                    |
|                                                                    |
+--------------------------------------------------------------------+ fffffffffff7ff000
|       fixmap region [   2.52 MB,r-- ]                              |
|                                                                    |
|                                                                    |
+--------------------------------------------------------------------+ fffffffff579000  <-- FIXADDR_START
|<... K sparse region ...> [   5.47 MB,--- ]                         |
|                                                                    |
|                                                                    |
+--------------------------------------------------------------------+ fffffffff000000  <-- MODULES_END
|       module region [1008.00 MB,rwx ]                             |
|                                                                    |
|                                                                    |
|                                                                    |
|                                                                    |
|                                                                    |
+--------------------------------------------------------------------+ ffffffffc0000000  <-- MODULES_VADDR
|<... K sparse region ...> [   37.78 TB,--- ]                        |
|                                                                    |
|                                                                    |
|                                                                    |
|                                                                    |
|                                                                    |
|                                                                    |
~                                                                    ~
|                                                                    |
|                                                                    |
|                                                                    |
|                                                                    |
+--------------------------------------------------------------------+ ffffda377fffffff  <-- VMALLOC_END
|       vmalloc region [  31.99 TB,rw- ]                            |
```

Figure 7.3 – Partial screenshot showing the procmap utility's output focused on the upper portion of the kernel VAS; some sparse (unmapped) regions are clearly visible

Okay, look carefully at the preceding screenshot. The regions marked `<... K sparse region ...>` are empty holes in the kernel VAS. There's nothing mapped here. This is quite common. Memory like this is often referred to as a *sparse region* or a *hole* in the address space.

The point is this: sparse regions are unmapped regions, thus, if you attempt to access any of these locations in any manner – read, write, or execute – it's a bug! So, let's pick a KVA within a sparse region. I'll pick one between the module region (where kernel modules live) and the kernel vmalloc region (where the `vmalloc()` allocates memory from), that is, any address between `0xffffffffc0000000` and `0xffffda377fffffff`. So, I'll take the KVA `0xffffffffc000dead` as the value for my invalid kernel address and run with it.

Right, ensure you've built the `oops_tryv2` module, then load it up passing the parameter as just discussed:

```
$ modinfo -p ./oops_tryv2.ko
mp_randaddr:Random non-zero kernel virtual address;
deliberately invalid, to cause an Oops! (ulong)
bug_in_workq:Trigger an Oops-generating bug in our workqueue
function (bool)
$
```

We use the `modinfo` utility to show that our module accepts two parameters (please ignore the second one for now – it's our next topic). Let's (finally!) get going:

```
$ sudo insmod ./oops_tryv2.ko  mp_randaddr=0xffffffffc000dead
Killed
$
```

Aha! Our module (deliberately) attempting to write to the invalid kernel address `0xffffffffc000dead` (passed via the module parameter) has it run headlong into a buggy ending. We got what we wanted – an Oops has hit. The following (partial) screenshot shows you a good portion of it:

```
[49132.584848] oops_tryv2:try_oops_init():92: Generating Oops by attempting to write to the invalid kernel a
ddress passed
[49132.585606] oops_tryv2:try_oops_init():100: bad_kva = 0xffffffffc000dead; now writing to it...
[49132.586023] BUG: unable to handle page fault for address: ffffffffc000dead
[49132.586450] #PF: supervisor write access in kernel mode
[49132.586961] #PF: error_code(0x0002) - not-present page
[49132.587417] PGD 33c15067 P4D 33c15067 PUD 33c17067 PMD 182d067 PTE 0
[49132.587875] Oops: 0002 [#2] PREEMPT SMP PTI
[49132.588296] CPU: 5 PID: 15255 Comm: insmod Tainted: G      D    OE     5.10.60-prod01 #6
[49132.588727] Hardware name: innotek GmbH VirtualBox/VirtualBox, BIOS VirtualBox 12/01/2006
[49132.589134] RIP: 0010:try_oops_init+0xf4/0x1000 [oops_tryv2]
[49132.589543] Code: 42 64 0d 00 b9 64 00 00 00 48 c7 c2 d0 93 6d c0 48 c7 c6 3c 90 6d c0 48 c7 c7 90 92 6d
c0 e8 98 35 a8 c6 48 8b 05 1c 64 0d 00 <48> c7 00 ad de 00 00 e9 78 ff ff ff b9 5f 00 00 00 48 c7 c2 d0 93
[49132.590928] RSP: 0018:ffffba3783dffc20 EFLAGS: 00010246
[49132.591423] RAX: ffffffffc000dead RBX: 0000000000000000 RCX: 0000000000000000
[49132.591954] RDX: 0000000000000000 RSI: 0000000000000027 RDI: 00000000ffffffff
[49132.592398] RBP: ffffba3783dffc38 R08: 0000000000000000 R09: ffffba3780e9f020
[49132.592864] R10: 0000000000000001 R11: 00000000ffffffff R12: ffffffffc0604000
[49132.593322] R13: ffff8f90766f6530 R14: ffffba3783dffe70 R15: ffffffffc06da158
[49132.593769] FS:  0000785ef7e11540(0000) GS:ffff8f90bdd40000(0000) knlGS:0000000000000000
[49132.594258] CS:  0010 DS: 0000 ES: 0000 CR0: 0000000080050033
[49132.594711] CR2: ffffffffc000dead CR3: 000000005a5e4001 CR4: 00000000000706e0
[49132.595199] Call Trace:
[49132.595739]  do_one_initcall+0x48/0x210
[49132.596217]  ? kmem_cache_alloc_trace+0x3ae/0x450
[49132.596666]  do_init_module+0x62/0x240
[49132.597119]  load_module+0x2a04/0x3080
[49132.597596]  ? security_kernel_post_read_file+0x5c/0x70
[49132.598078]  __do_sys_finit_module+0xc2/0x120
[49132.598648]  ? __do_sys_finit_module+0xc2/0x120
[49132.599087]  __x64_sys_finit_module+0x1a/0x20
[49132.599549]  do_syscall_64+0x38/0x90
[49132.600066]  entry_SYSCALL_64_after_hwframe+0x44/0xa9
[49132.600557] RIP: 0033:0x785ef7f5689d
[49132.600987] Code: 00 c3 66 2e 0f 1f 84 00 00 00 00 00 90 f3 0f 1e fa 48 89 f8 48 89 f7 48 89 d6 48 89 ca
4d 89 c2 4d 89 c8 4c 8b 4c 24 08 0f 05 <48> 3d 01 f0 ff ff 73 01 c3 48 8b 0d c3 f5 0c 00 f7 d8 64 89 01 48
```

Figure 7.4 – Screenshot showing the Oops generated by attempting to write to an invalid/unmapped kernel address

I hope the key point is clear: of course, we're going to have a bug – an Oops. We wrote to an invalid unmapped kernel address within a sparse region of the kernel VAS, which `procmap` literally helped us see.

Why does attempting to access an invalid address cause the Oops? The answer's very similar to what we discussed regarding the NULL trap page. Here's what essentially occurs:

1. The virtual address being worked upon (read, written, or executed) goes to the MMU.

2. The MMU, knowing where the current process context's paging tables are (for x86, the physical address of the base of the paging tables is in the CR3 register), proceeds to now translate this virtual address (KVA) to a physical address (here, we'll ignore hardware optimizations such as the CPU caches and the **Translation Lookaside Buffers** (**TLBs**) that might already hold the physical address, thus short-circuiting the lengthy translation and providing a speed-up).

3. Normally, it will find a mapping and perform the translation, placing the physical address on the bus. The CPU takes over and the work gets done. In this case, though, the kernel address passed along is invalid (deliberately so) – it's literally part of a hole in the kernel VAS! So, the address translation fails. The MMU, being hardware, does the best it can: it informs the OS that something's wrong by raising a (page) fault.

4. The OS's page fault handler takes over (running in the context of the process that caused the fault – here, it's insmod of course). It figures that *an invalid write was attempted while in kernel mode* – it thus triggers an Oops!

What about understanding and interpreting this messy Oops thing in detail? That's precisely what we do in the coming section, *The devil is in the details – decoding the Oops*. Hang tight – we'll get there!

Case 3 – Oops by writing to a structure member when the structure pointer's NULL

This use (or test) case is a bit more involved, helping make it a bit more realistic as well. The end result's the same as the prior two cases though – we get the kernel to trigger an Oops.

This time, we'd like the buggy code path to not run in the insmod process context. To arrange for this to happen, we initialize a (kernel default) workqueue and schedule it, having its code execute. The execution of the kernel default workqueue is done in the context of a kernel worker thread. We arrange for the work function to have a bug – a write to an invalid memory location, a pointer (to a structure) that hasn't been assigned any memory. This of course causes an Oops to trigger. Here are the relevant code snippets (as usual, I urge you to browse the full code and try these things out yourself as well):

```
// ch7/oops_tryv2/oops_tryv2.c
[...]
static bool bug_in_workq;
module_param(bug_in_workq, bool, 0644);
MODULE_PARM_DESC(bug_in_workq, "Trigger an Oops-generating bug
in our workqueue function");
```

This time we have a module parameter named bug_in_workq, with data type Boolean. It's false by default. Set it to 1 (or yes) to have this use case get underway:

```
static struct st_ctx {
    int x, y, z;
    struct work_struct work;
    u8 data;
} *gctx, *oopsie; /* careful, pointers have no memory! */
```

The preceding is the structure we use – notice the pointers to it. In our module's init function, if the bug_in_workq parameter is set, we call the function setup_work(), which sets up some work on the kernel-default workqueue:

```
if (!!bug_in_workq) {
[...]
    setup_work();
    return 0;
}
```

The function allocates memory to the gctx pointer, calls the INIT_WORK() macro to set up work – the function do_the_work() – on the kernel's default (or events) workqueue (it's the default workqueue):

```
static int setup_work(void)
{
    gctx = kzalloc(sizeof(struct st_ctx), GFP_KERNEL);
    [...]
    gctx->data = 'C';
    /* Initialize our workqueue */
    INIT_WORK(&gctx->work, do_the_work);
```

Next, we call `schedule_work()` on our workqueue to have the kernel actually run the code of our work function:

```
    // Do it!
    schedule_work(&gctx->work); [...]
}
```

Finally, here's the actual workqueue function that's run (by a kernel worker thread) when the `schedule_work()` API triggers. It's buggy, of course (quick – spot the bug!):

```
static void do_the_work(struct work_struct *work)
{
    struct st_ctx *priv = container_of(
                        work, struct st_ctx, work);
    [...]
    if (!!bug_in_workq) {
        pr_info("Generating Oops by attempting to
                write to an invalid kernel
                memory pointer\n");
        oopsie->data = 'x';
    }
    kfree(gctx);
}
```

Well, it's obvious in retrospect: the pointer to our structure named `oopsie` (appropriate, huh?) has no memory (its value is NULL as it's a global static variable within our module). Yet, we attempt to try and write into a member of the structure via it. This triggers the Oops. Here's how I invoke it:

```
sudo insmod ./oops_tryv2.ko bug_in_workq=yes
```

Did you notice? This time the `Killed` message does not appear. This is because the `insmod` process isn't killed. Instead, the kernel worker thread that consumes our workqueue function will suffer the consequences of the bug.

Here's a partial screenshot:

```
[  448.049270] oops_tryv2:try_oops_init():87: Generating Oops via kernel bug in workqueue function
[  448.049408] oops_tryv2:do_the_work():57: In our workq function: data=67
[  448.049409] oops_tryv2:do_the_work():59: delta: 137891 ns
[  448.049410] oops_tryv2:do_the_work():59:  137 us
[  448.049411] oops_tryv2:do_the_work():61: Generating Oops by attempting to write to an invalid kernel memo
ry pointer
[  448.049414] BUG: kernel NULL pointer dereference, address: 0000000000000030
[  448.049435] #PF: supervisor write access in kernel mode
[  448.049449] #PF: error code(0x0002) - not-present page
[  448.049462] PGD 0 P4D 0
[  448.049471] Oops: 0002 [#1] PREEMPT SMP PTI
[  448.049483] CPU: 0 PID: 16 Comm: kworker/0:1 Tainted: G           OE     5.10.60-prod01 #6
[  448.049504] Hardware name: innotek GmbH VirtualBox/VirtualBox, BIOS VirtualBox 12/01/2006
[  448.049547] Workqueue: events do_the_work [oops_tryv2]
[  448.049562] RIP: 0010:do_the_work+0x124/0x15e [oops_tryv2]
[  448.049578] Code: c0 e8 d0 1d ad df f6 c3 01 74 27 b9 3d 00 00 00 48 c7 c2 c0 63 5a c0 48 c7 c6 3c 60 5a
c0 48 c7 c7 18 61 5a c0 e8 61 25 0e e0 <c6> 04 25 30 00 00 00 78 48 8b 3d cd 23 00 00 e8 a8 aa 79 df 5b 41
[  448.049686] RSP: 0018:ffffb6e1c008be48 EFLAGS: 00010246
[  448.049704] RAX: 0000000000000067 RBX: 0000000000000001 RCX: 0000000000000000
[  448.049734] RDX: 0000000000000000 RSI: 0000000000000027 RDI: 00000000ffffffff
[  448.049775] RBP: ffffb6e1c008be58 R08: 0000000000000000 R09: ffffffffffffc9c88
[  448.049801] R10: ffffffffa10c3820 R11: 3fffffffffffffff R12: 0000000000021aa3
[  448.049827] R13: ffff9ddffdc31700 R14: 0000000000000000 R15: ffff9ddffdc2b9c0
[  448.049853] FS:  0000000000000000(0000) GS:ffff9ddffdc00000(0000) knlGS:0000000000000000
[  448.049882] CS:  0010 DS: 0000 ES: 0000 CR0: 0000000080050033
[  448.049904] CR2: 0000000000000030 CR3: 000000005f410003 CR4: 00000000000706f0
[  448.049934] Call Trace:
[  448.049949]  process_one_work+0x1b8/0x3b0
[  448.049967]  worker_thread+0x50/0x3a0
[  448.049984]  ? process_one_work+0x3b0/0x3b0
[  448.050002]  kthread+0x154/0x180
[  448.050018]  ? kthread_unpark+0xa0/0xa0
[  448.050034]  ret_from_fork+0x22/0x30
[  448.050050] Modules linked in: oops_tryv2(OE) intel_rapl_msr snd_intel8x0 snd_ac97_codec intel_rapl_commo
```

Figure 7.5 – Partial screenshot of the Oops triggered by our oops_tryv2 module due to a bug in our workqueue function

By the way, if required, you can read the details of setting up and using kernel workqueues, timers, and kernel threads in *Chapter 5, Working with Kernel Timers, Threads, and Workqueues* of my earlier *Linux Kernel Programming, Part 2* book (the e-book is freely downloadable).

Of course, here we always assume that you do have access to the kernel log (via dmesg, journalctl, on a safe place on a flash chip, and so on). What if you don't know where the Oops message is in the first place? Well, the kernel community has documented what you can do about this here: *Where is the Oops message located?* (https://www.kernel.org/doc/html/latest/admin-guide/bug-hunting.html#where-is-the-oops-message-is-located). Also, we shall cover some of the techniques mentioned therein later. Netconsole is covered in the section, *An Oops on an ARM Linux system and using netconsole*, and kdump/crash is briefly covered in *Chapter 12, A Few More Kernel Debug Approaches*.

Okay, you now know how to trigger a kernel bug, an Oops, in several ways! A quick look at what a kernel Oops is and isn't follows.

A kernel Oops and what it signifies

Here are a quick few things to realize regarding a kernel Oops.

First off, an Oops is not the same as a segfault – a segmentation fault... It might, as a side effect, cause a segfault to occur, and thus the process context might receive the fatal SIGSEGV signal. This, of course, has the poor process caught in the crossfire.

Next, an Oops is not the same thing as a full-fledged kernel panic. A panic implies the system is in an unusable state. It might lead up to this, especially on production systems (we cover kernel panic in *Chapter 10, Kernel Panic, Lockups and Hangs*). Note though, that the kernel provides several sysctl tunables (editable by root, of course) regarding what circumstances can lead to the kernel panicking. We can check them out – on my x86_64 Ubuntu 20.04 guest running our custom production kernel, here they are:

```
$ cd /proc/sys/kernel/
$ ls panic_on_*
panic_on_io_nmi    panic_on_oops    panic_on_rcu_stall    panic_on_
unrecovered_nmi    panic_on_warn
```

And, as you can see, if you cat them, all of their values are zero by default, implying that a kernel panic will *not* be triggered. It also shows us that setting, for example, the panic_on_oops tunable to 1 will cause the kernel to panic on any Oops, no matter how trivial it might seem.

It's important to understand that this can be the right thing to do on many installations. When a system suffers an Oops, we usually want a bright red flag to show up – a showstopper, a way to understand that the system's in (or was in) an unhealthy state! This does depend on the nature of the project or product: a deeply embedded system might not afford to remain down due to a kernel panic. There, a *watchdog* will typically detect that the system's in an unhealthy state and reboot it. We shall cover using watchdogs and whatnot in *Chapter 10, Kernel Panic, Lockups and Hangs*).

Even though an Oops isn't a kernel panic, depending on the circumstances and the severity of the bug, the kernel can be rendered unresponsive, unstable, or both. Or it might continue to work as though nothing alarming has occurred! Whatever the case, *an Oops is, ultimately, a kernel-level bug; it must be detected, interpreted, and fixed!*

Right, let's get to the juicy bit: learning how to interpret, in detail, the Oops kernel output. Let's go!

The devil is in the details – decoding the Oops

We'll use the third scenario (or use/test case), covered in the section, *Case 3 – Oops by writing to a structure member when the structure pointer's NULL*. To quickly recap, this is what we did to trigger this particular kernel Oops (case #3):

```
cd ch7/oops_tryv2
make
sudo insmod ./oops_tryv2.ko bug_in_workq=yes
```

As seen earlier, it triggers an Oops. Now we get to the interesting part – deciphering the Oops, step by step, line by line.

Before starting, it's important to realize that the detailed discussion below is necessarily arch-specific, here and now pertaining to the x86_64 platform (as portions of the Oops output are, of course, very arch-specific). We shall also show how a typical Oops appears on the ARM platform in a later section.

Line-by-line interpretation of an Oops

The initial, and really key, portion of the Oops we get is seen in *Figure 7.5*. Now, to help refer to it line by line, here's an annotated diagram of the same screenshot (zoomed in a bit more, for clarity):

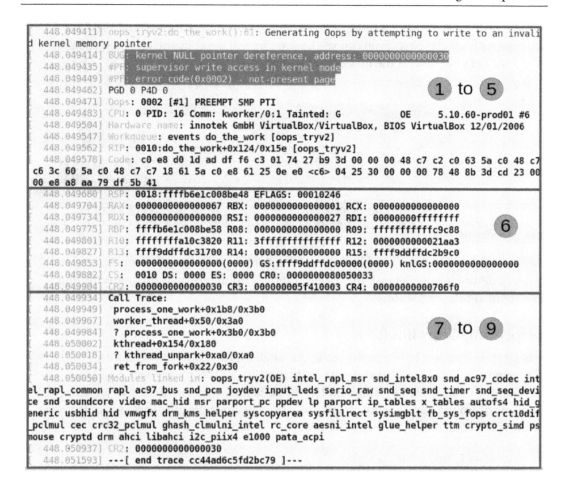

```
 448.049411] oops_tryv2:do_the_work():61: Generating Oops by attempting to write to an invali
d kernel memory pointer
 448.049414] BUG: kernel NULL pointer dereference, address: 0000000000000030
 448.049435] #PF: supervisor write access in kernel mode
 448.049449] #PF: error code(0x0002) - not-present page
 448.049462] PGD 0 P4D 0
 448.049471] Oops: 0002 [#1] PREEMPT SMP PTI
 448.049483] CPU: 0 PID: 16 Comm: kworker/0:1 Tainted: G          OE      5.10.60-prod01 #6
 448.049504] Hardware name: innotek GmbH VirtualBox/VirtualBox, BIOS VirtualBox 12/01/2006
 448.049547] Workqueue: events do_the_work [oops_tryv2]
 448.049562] RIP: 0010:do_the_work+0x124/0x15e [oops_tryv2]
 448.049578] Code: c0 e8 d0 1d ad df f6 c3 01 74 27 b9 3d 00 00 00 48 c7 c2 c0 63 5a c0 48 c7
c6 3c 60 5a c0 48 c7 c7 18 61 5a c0 e8 61 25 0e e0 <c6> 04 25 30 00 00 00 78 48 8b 3d cd 23 00
00 e8 a8 aa 79 df 5b 41
 448.049680] RSP: 0018:ffffb6e1c008be48 EFLAGS: 00010246
 448.049704] RAX: 0000000000000067 RBX: 0000000000000001 RCX: 0000000000000000
 448.049734] RDX: 0000000000000000 RSI: 0000000000000027 RDI: 00000000ffffffff
 448.049775] RBP: ffffb6e1c008be58 R08: 0000000000000000 R09: ffffffffffc9c88
 448.049801] R10: ffffffffa10c3820 R11: 3fffffffffffffff R12: 0000000000021aa3
 448.049827] R13: ffff9ddffdc31700 R14: 0000000000000000 R15: ffff9ddffdc2b9c0
 448.049853] FS:  0000000000000000(0000) GS:ffff9ddffdc00000(0000) knlGS:0000000000000000
 448.049882] CS:  0010 DS: 0000 ES: 0000 CR0: 0000000080050033
 448.049904] CR2: 0000000000000030 CR3: 000000005f410003 CR4: 00000000000706f0
 448.049934] Call Trace:
 448.049949]  process_one_work+0x1b8/0x3b0
 448.049967]  worker_thread+0x50/0x3a0
 448.049984]  ? process_one_work+0x3b0/0x3b0
 448.050002]  kthread+0x154/0x180
 448.050018]  ? kthread_unpark+0xa0/0xa0
 448.050034]  ret_from_fork+0x22/0x30
 448.050050] Modules linked in: oops_tryv2(OE) intel_rapl_msr snd_intel8x0 snd_ac97_codec int
el_rapl_common rapl ac97_bus snd_pcm joydev input_leds serio_raw snd_seq snd_timer snd_seq_devi
ce snd soundcore video mac_hid msr parport_pc ppdev lp parport ip_tables x_tables autofs4 hid_g
eneric usbhid hid vmwgfx drm_kms_helper syscopyarea sysfillrect sysimgblt fb_sys_fops crct10dif
_pclmul cec crc32_pclmul ghash_clmulni_intel rc_core aesni_intel glue_helper ttm crypto_simd ps
mouse cryptd drm ahci libahci i2c_piix4 e1000 pata_acpi
 448.050937] CR2: 0000000000000030
 448.051593] ---[ end trace cc44ad6c5fd2bc79 ]---
```

Figure 7.6 – Annotated (full) screenshot of the Oops output from our oops_tryv2 workqueue (case 3) function bug

Now, for clarity and to make it more approachable, we'll split up this diagram and the discussion into several parts (you can see how we intend the splitting via the rectangles in *Figure 7.6*). Let's begin with the first of them – here it is:

```
[  448.049411] oops_tryv2:do_the_work():51: Generating Oops by attempting to write to an invali
d kernel memory pointer
[  448.049414] BUG: kernel NULL pointer dereference, address: 0000000000000038
[  448.049435] #PF: supervisor write access in kernel mode
[  448.049449] #PF: error code(0x0002) - not-present page                          1
[  448.049462] PGD 0 P4D 0
[  448.049471] Oops: 0002 [#1] PREEMPT SMP PTI                            2
[  448.049483] CPU: 0 PID: 16 Comm: kworker/0:1 Tainted: G        OE    5.10.60-prod01 #6  3
[  448.049504] Hardware name: innotek GmbH VirtualBox/VirtualBox, BIOS VirtualBox 12/01/2006
[  448.049547] Workqueue: events do_the_work [oops_tryv2]                       4
[  448.049562] RIP: 0010:do_the_work+0x124/0x15e [oops_tryv2]
[  448.049578] Code: c0 e8 d0 1d ad df f6 c3 01 74 27 b9 3d 00 00 00 48 c7 c2 c0 63 5a c0 48 c7  5
c6 3c 60 5a c0 48 c7 c7 18 61 5a c0 e8 61 25 0e e0 <c6> 04 25 30 00 00 00 78 48 8b 3d cd 23 00
00 e8 a8 aa 79 df 5b 41
```

Figure 7.7 – Annotated screenshot 1 of 3: Oops output from our oops_tryv2 workqueue function bug

Okay, let's delve into the details! The material shown from here on is arch-specific and applies *only* to the x86 platform.

Interpreting Oops line(s) 1

The brightly visible lines (within the rectangle labeled 1 in *Figure 7.7*) with a red background emanate from this code (it's arch-specific – this is for the x86_64). It's a portion of the code of the OS fault handler, the code that, when it's detected, an abnormal condition within the kernel, a bug, begins to write the Oops diagnostic message. Here's a portion of the actual x86 fault handling code:

```
// arch/x86/mm/fault.c
static void
show_fault_oops(struct pt_regs *regs,
                unsigned long error_code,
                unsigned long address)
{
[...]
if (address < PAGE_SIZE && !user_mode(regs))
        pr_alert("BUG: kernel NULL pointer
                dereference, address: %px\n",
                (void *)address);

    else
        pr_alert("BUG: unable to handle page fault
                for address: %px\n", (void *)address);
```

Check out the preceding `if` condition – it's now amply clear why we got this output:

BUG: kernel NULL pointer dereference, address: 0000000000000030

It's emitted when the faulting address is within the first page and we're running in kernel mode. Recall that the very first page of the user VAS is the NULL trap page, one where all addresses are within `PAGE_SIZE` (typically 4,096 bytes). If the condition is false, the kernel prints an alternate message. Further, here, the address that caused the fault – the one within the first NULL trap page – is then printed (it's always in hexadecimal). Here, it's the value `0x30`.

Now, this too is important: why `0x30` and not `0x0`? Think back to the code that generated this particular Oops (you can refer back to the section, *Case 3 – Oops by writing to a structure member when the structure pointer's NULL* to see this). The buggy line of code is here:

```
ch7/oops_tryv2/oops_tryv2.c:do_the_work():oopsie->data = 'x';
```

Now, `oopsie` is a pointer to the `st_ctx` structure in our code, but its value is NULL (recall, it was never allocated). So, the value of `0x30` is the offset from the beginning of the structure to the member being referenced! A key point we thus learn is that *when the faulting address shows up as a small integer value within the size of a page (as is the case here), it's very likely that the structure (or other) pointer was NULL and the number displayed is the offset from the beginning of the structure (or possibly array, or whatever) to the member being referenced.*

The next line of output in the Oops is this:

#PF: supervisor write access in kernel mode

This is generated from the code that continues, in the same `show_fault_oops()` function (by the way, PF stands for `Page Fault`):

```
pr_alert("#PF: %s %s in %s mode\n",
        (error_code & X86_PF_USER)   ?
                        "user" : "supervisor",
        (error_code & X86_PF_INSTR) ?
                        "instruction fetch" :
        (error_code & X86_PF_WRITE) ? "write access" :
                        "read access",
            user_mode(regs) ? "user" : "kernel");
```

Take the trouble to read the code and match it with the output we obtained. It clearly shows us that the kernel figured a lot out: the code was executed in *supervisor* mode (which means kernel mode), and there was a *write* attempt, again executing in *kernel* mode.

Here's the line after that:

```
#PF: error_code(0x0002) - not-present page
```

We'll cover what exactly this means shortly. This is from the code within the show_ fault_oops() function that immediately follows the preceding code:

```
        pr_alert("#PF: error_code(0x%041x) -
                %s\n", error_code,
  !(error_code & X86_PF_PROT) ? "not-present page" :
  (error_code & X86_PF_RSVD)  ? "reserved bit violation" :
  (error_code & X86_PF_PK)    ? "protection keys violation" :
 "permissions violation");
```

So, there, we figured out how each of these three lines of output came to be. How did the color show up with a red background? Ah, that's easy: dmesg interprets the pr_ alert() log level and colors it accordingly.

We'll skip the details on the **Page Global Directory** (**PGD**) and *P4D* (a level inserted between the PGD and **Page Upper Directory** (**PUD**) in 4.11 Linux). These are references to the paging tables of the process context the process was running in. See the code of the dump_pagetable() kernel function if you're interested.

Interpreting Oops line(s) 2

For your convenience, the following portion of the screenshot is duplicated from *Figure 7.7*. The next line of output (line 2) in the Oops is this:

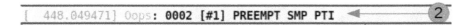

Figure 7.8 – Line(s) 2 of the Oops output from our buggy oops_tryv2 module, test case 3

Clearly, this tells us that an Oops has occurred (it's not so ridiculous – grepping for the string Oops: can be useful). The number immediately following this string – here, it's 0002 – is important. It's the arch-specific Oops bitmask. Learning how to interpret it will definitely help.

Interpreting the (arch-specific) Oops bitmask

The overall function responsible for displaying the Oops content from here on is named `arch/x86/kernel/dumpstack.c:__die()`. It's split into two portions – the `__die_header()` and the `__die_body()` functions. The Oops bitmask and remaining tokens on that line (such as PREEMPT SMP ...) are displayed from the header function. To give you a sampling of how the actual Oops-diagnostic-display kernel code works, here's a screenshot of the `__die_header()` function (for the 5.10.60 kernel). It's within the source file `arch/x86/kernel/dumpstack.c`:

```c
static void __die_header(const char *str, struct pt_regs *regs, long err)
{
    const char *pr = "";

    /* Save the regs of the first oops for the executive summary later. */
    if (!die_counter)
        exec_summary_regs = *regs;

    if (IS_ENABLED(CONFIG_PREEMPTION))
        pr = IS_ENABLED(CONFIG_PREEMPT_RT) ? " PREEMPT_RT" : " PREEMPT";

    printk(KERN_DEFAULT
           "%s: %04lx [#%d]%s%s%s%s%s\n", str, err & 0xffff, ++die_counter,
           pr,
           IS_ENABLED(CONFIG_SMP)          ? " SMP"            : "",
           debug_pagealloc_enabled()       ? " DEBUG_PAGEALLOC" : "",
           IS_ENABLED(CONFIG_KASAN)        ? " KASAN"          : "",
           IS_ENABLED(CONFIG_PAGE_TABLE_ISOLATION) ?
           (boot_cpu_has(X86_FEATURE_PTI) ? " PTI" : " NOPTI") : "");
}
NOKPROBE_SYMBOL(__die_header);
```

Figure 7.9 – Screenshot of a part of the kernel's Oops output functionality on the x86_64

As mentioned above, the **arch-specific Oops bitmask** is actually very meaningful, further clueing us in as to why the kernel bug occurred!

You can see it being emitted from the preceding `printk()` (it's the second part of the `printk` format string `%04lx`, corresponding to the `err & 0xffff` code). How do we interpret this bitmask? Here's how – but, again remember: it's arch-specific – this interpretation applies *only* to the x86 platform.

The MMU sets up a page fault error as an encoded value. On the x86 platform, this is how the encoding of the page fault error code bits is done:

```
bit 0 == 0: no page found          1: protection fault
bit 1 == 0: read access            1: write access
bit 2 == 0: kernel-mode access     1: user-mode access
bit 3 == 1: use of reserved bit detected
bit 4 == 1: fault was an instruction fetch
```

This information in fact used to be a comment in the code base in earlier kernel versions.

A nicer way, perhaps, to more easily visualize and thus interpret the particular error – the reason why the Oops occurred – is to examine the five **Least Significant Bit (LSB)** bits of the page fault error code in a tabular format:

Value of bit	Bit 4	Bit 3	Bit 2	Bit 1	Bit 0
0	-na-	-na-	kernel mode	Read attempt	No page found
1	Instruction fetch fault	Reserved bit used	user mode	Write attempt	Protection fault

Table 7.1 – The meaning of the LSB 5 bits of the page fault error code on the x86 platform

So, now it's easy! We got the Oops bitmask as 0002 (it's in hexadecimal – see the printk: the format specifier is %041x). This translates to 00010 in binary. According to the preceding table, this implies the following (here, as bits 3 and 4 are zero, they don't matter):

```
Bit 2 is 0 : kernel mode
Bit 1 is 1 : write attempt
Bit 0 is 0 : no page found
```

Well, well, no surprise there – this is *exactly* what the Oops diagnostic tells us (point 1 in *Figure 7.7*):

```
#PF: supervisor write access in kernel mode
#PF: error_code(0x0002) - not-present page
```

The remainder of the line is as follows:

```
...  [#1] PREEMPT SMP PTI
```

This line is easy to interpret:

- [#1]: This is the number of the Oops that has occurred during this system session. [#1] tells us it's the first Oops (it's a session value – a power cycle resets it).

- PREEMPT: The code was running on a kernel configured for preemption (CONFIG_PREEMPT=y).

- SMP: The kernel has **Symmetric Multi-Processing (SMP)** enabled. It supports multicore.

- PTI: **PTI** is short for **Page Table Isolation**. The **Meltdown/Spectre** hardware bugs circa early 2018 had the kernel community build a protection mechanism against this serious vulnerability called PTI (see the *Further reading* section for more on this).

Let's move on to interpreting the following two lines within the Oops diagnostic.

Interpreting Oops line(s) 3

For your convenience, this portion of the screenshot is duplicated from *Figure 7.7*:

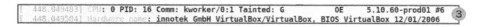

Figure 7.10 – Line(s) 3 of the Oops output from our buggy oops_tryv2 module, test case 3

The first of these lines essentially informs us of the process context – the process or thread that executed the buggy code in kernel mode that caused the fault (along with a few other details). We'll take it a token at a time:

- CPU: Denotes the CPU core the code was running on at the time of the Oops (here it's CPU 0).

- PID: The PID of the process or thread that was executing the code at the time of the Oops.

- Comm: The name of the process or thread that was executing the code at the time of the Oops.

- Tainted: The kernel tainted flags bitmask – we cover this shortly.

- The output of uname -r, the kernel release, followed by a number prefixed with the # symbol. This is the number of times this kernel has been built (here, it's 6).

It's important to note, though, that when the Oops is triggered from an interrupt context, parts of the data seen here become suspect (see more on this in the upcoming section, *Leveraging the console device to get the kernel log after Oopsing in IRQ context* where we examine an Oops triggered in interrupt context).

Interpreting the kernel tainted flags

The Linux kernel community likes to know whether the running kernel is clean or dirty. A *dirtied* kernel or a *tainted kernel* (the word *tainted* means polluted) is one that isn't in a pristine state. This state information is kept as bits within a bitmask. The entire bitmask, currently consisting of a total of 18 bits or flags, is called the **tainted flags**.

In our particular use case, the tainted flags show up like so, following the string `Tainted:` (highlighted here):

```
CPU: 0 PID: 16 Comm: kworker/0:1 Tainted: G          OE
5.10.60-prod01 #6
```

You can interpret the letters – the tainted flags, as they're called – as shown here:

Bit #	Logged as: When bit is Cleared (_) or Set (X)	Meaning if bit is set (1)
0	G or P	A proprietary module was loaded (P) or only GPL'ed modules loaded (G).
1	_ or F	Forced loading of one or more modules.
2	_ or S	Out of specification system.
3	_ or R	Forced unloading of one or more modules.
4	_ or M	**Machine Check Exception** (MCE) reported by a CPU core.
5	_ or B	Bad page referenced or unexpected page flags.
6	_ or U	User space app requested taint to be set.
7	_ or D	The kernel died recently (due to an Oops or `BUG()`).
8	_ or A	ACPI table overridden by user.
9	_ or W	A warning was issued (via one of the `WARN*()` macros) by the kernel.
10	_ or C	A staging (experimental) driver was loaded.
11	_ or I	Platform firmware bug, workaround applied.

12	_ or O	Out-of-tree / externally built module was loaded.
13	_ or E	An unsigned module was loaded.
14	_ or L	A soft lockup occurred.
15	_ or K	Live patch applied on the kernel.
16	_ or X	Distros use this flag, called auxiliary taint.
17	_ or T	The kernel was built with structure randomization enabled.

Table 7.2 – Interpreting the kernel tainted flags

The _ symbol in the second column implies a blank, a space to indicate that that particular taint bit is cleared (notice how the tainted flags are printed in the Oops with blank spaces as required, to show that a particular bit(s) is/are unset: `Tainted:  G          OE`).

So, in our use case, the G | O | E tainted flags imply all GPL'ed modules were loaded (G), one or more externally-built (*out-of-tree*) modules were loaded (O), and (one or more) unsigned modules were loaded (E). Indeed, our `oops_tryv2` module has a dual license that includes GPL, is an out-of-tree one, and is unsigned.

So, it's easy to look up the table and figure out what the tainted flags imply. It's even easier to employ a helper script that does the work for you! – that's exactly what the `tools/debugging/kernel-chktaint` script (within the kernel source tree) is designed to do. We cover using this script in the section, *Are we clean? The kernel-chktaint script*.

The official kernel documentation covers these flags (along with more in-depth details) here: `https://www.kernel.org/doc/html/latest/admin-guide/tainted-kernels.html`.

The second line (as seen in *Figure 7.10*) is simply some hardware platform details – useful.

Interpreting Oops line(s) 4

For your convenience, this portion of the screenshot is duplicated from *Figure 7.7*:

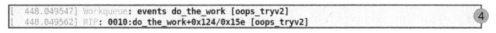

Figure 7.11 – Line4 of the Oops output from our buggy oops_tryv2 module, test case 3

In this particular case, let's begin with interpreting the second of the two preceding lines first – the one beginning with RIP:.

Finding the code where the Oops occurred

Perhaps the key line in the Oops output is this:

```
RIP:  0010:do_the_work+0x124/0x15e [oops_tryv2]
```

Let's interpret it a token at a time. Again, I remind you, much of this is very arch-specific (here, for the x86_64 platform/CPU):

- `RIP::` This, of course, is the name of the CPU register that holds the address of the text (code) to execute. On the x86 platform, it's called the **Instruction Pointer** register. On the x86_64, it's a 64-bit register (in order to hold 64-bit virtual addresses) and is named `RIP`. So, what is it holding? Read on!

- `0010::` On the x86, the notion of a hardware segment still exists (it's mostly historical and continues in a vestigial form even on the modern 64-bit x86 today). This value, `0010`, represents the code segment. This makes sense – the CPU core's `RIP` register should be pointing to code.

- `do_the_work+0x124/0x15e [oops_tryv2]`: Perhaps the *most important* thing here, this is the format being employed above:

```
function_name+off_from_func/size_of_func [module-name]
```

It identifies which function was being executed on the processor at the time of the Oops. Following it is a + sign followed by two numeric (hexadecimal) values (so, in the form +x/y):

  - The first number (x) is the offset (in bytes) from the beginning of the function. In other words, it references the actual byte of machine code that was being executed at the time of the Oops!

  - The second number (y) is (the kernel's best guess of) the size of the function (in bytes).

Next, if this string – `funcname+x/y` – is followed by a name in square brackets (of the form `[modulename]`), that is the kernel module where this function (`funcname`) resides! If not, the function is part of the kernel image itself.

So, for our particular Oops, we can now conclude that the bug (likely) occurred in the function `do_the_work()`, which belongs to the module `oops_tryv2`. Further, the instruction pointer was `0x124` (decimal 292) bytes from the start of this function. The size of the function, as estimated by the kernel (and it's typically dead right), is `0x15e` (decimal 350) bytes.

This information – the precise location of the CPU Instructor Pointer register – can be, and often is, critical in figuring out exactly where in the code the bug occurred!

So, now that we know more or less the exact location in the code where the Oops occurred, how exactly do we leverage this information – 292 bytes from the start of `oops_tryv2:do_the_work()` – to find the C source line where the issue, and likely the root cause, lies?

Ah, that's the crux of it, isn't it?! We cover precisely this in depth in the upcoming section, *Tools and techniques to help determine the location of the Oops*. It's very important to read the detailed discussion there on the usage of various tools, technologies, and indeed, kernel helper scripts to literally pinpoint the buggy code's location in the (module or kernel) source.

Is this code location (the `funcname+x/y`) always guaranteed to be the root cause of the defect, the bug? No. There are no guarantees! With the more difficult, subtle bugs, the root cause can be miles away. The symptoms have shown up here perhaps. Debugging these is where you really earn your money. With the really wide breadth of tools, techniques, and internal details that this book covers, you'll be in a much better position to do so.

In fact, this use case (our `oops_tryv3` case 3 bug) is itself a bit involved. The bug occurred not in the context of the `insmod` process but rather, within the kernel's default `events` workqueue's **kthread (kernel thread)** worker. This information is in fact revealed by the first line in *Figure 7.11*:

```
Workqueue: events do_the_work [oops_tryv2]
```

The word `events` in the preceding line is significant. It's the name of the workqueue: `events` is the name of the kernel's default workqueue. And `do_the_work()` is our workqueue function that was consumed by the kthread worker.

By the way, the kernel has `printk` format specifiers to print function pointer symbols in this common and useful format: `func+x/y`. The format specifiers include `%pF`, `%pS [R]`, and `%pB` (do refer to the kernel documentation here for the gory details: `https://www.kernel.org/doc/Documentation/printk-formats.txt`).

Let's continue with the line-by-line exploration of the Oops output.

Interpreting Oops line(s) 5

For your convenience, this portion of the screenshot is duplicated from *Figure 7.7*:

```
[  448.049578] Code: c0 e8 d0 1d ad df f6 c3 01 74 27 b9 3d 00 00 00 48 c7 c2 c0 63 5a c0 48 c7
c6 3c 60 5a c0 48 c7 c7 18 61 5a c0 e8 61 25 0e e0 <c6> 04 25 30 00 00 00 78 48 8b 3d cd 23 00  5
00 e8 a8 aa 79 df 5b 41
```

Figure 7.12 – Line(s) 5 of the Oops output from our buggy oops_tryv2 module, test case 3

The bytes following the string `Code:` are indeed just that – the machine code that was running on the CPU core at the time of the Oops! Decoding this machine code is now automated via kernel helper scripts (both `scripts/decodecode` and `scripts/decode_stacktrace.sh`). Please refer to our coverage in the section, *Interpreting machine code with the decodecode script*.

We now move on to the next portion of our Oops test case interpretation.

Interpreting Oops line(s) 6

The lines of `printk` output at position 6 (see *Figure 7.6*) are the processor registers and their runtime values at the time of the Oops.

CPU registers and the Oops

The Oops diagnostic includes the value of all general-purpose CPU registers on the processor core that ran the buggy code at the time the Oops occurred.

How is this useful? Recall our quite detailed discussion on the processor **Application Binary Interface** (**ABI**) (if you're hazy on it, do refer back to *Chapter 4, Debug via Instrumentation – Kprobes*, in the section, *Understanding the basics of the Application Binary Interface (ABI)*). This can be crucial to understanding what happened at the level of bare metal.

For your convenience, this portion of the screenshot is duplicated from *Figure 7.7*:

```
[  448.049680] RSP: 0018:ffffb6e1c008be48 EFLAGS: 00010246
[  448.049704] RAX: 0000000000000067 RBX: 0000000000000001 RCX: 0000000000000000
[  448.049734] RDX: 0000000000000000 RSI: 0000000000000027 RDI: 00000000ffffffff
[  448.049775] RBP: ffffb6e1c008be58 R08: 0000000000000000 R09: fffffffffffc9c88
[  448.049801] R10: ffffffffa10c3820 R11: 3fffffffffffffff R12: 0000000000021aa3     6
[  448.049827] R13: ffff9ddffdc31700 R14: 0000000000000000 R15: ffff9ddffdc2b9c0
[  448.049853] FS:  0000000000000000(0000) GS:ffff9ddffdc00000(0000) knlGS:0000000000000000
[  448.049882] CS:  0010 DS: 0000 ES: 0000 CR0: 0000000080050033
[  448.049904] CR2: 0000000000000030 CR3: 000000005f410003 CR4: 00000000007006f0
```

Figure 7.13 – Line6 of the Oops output from our buggy oops_tryv2 module, case 3, showing the x86_64 CPU register values (in hex) at the time of the Oops

The first register to be shown is `RSP` – obviously, it's the stack pointer register. Being able to access the (kernel) stack and interpret the frames therein is crucial, again, to understanding why and where the Oops occurred. Here, the top of the kernel-mode stack happens to be the kernel virtual address `0xffffb6e1c008be48`. Remember, pretty much all architectures (CPUs) follow the *stack-grows-down* semantic, so the top of the stack is really the lowest legal virtual address on it.

The `EFLAGS` register content follows. It contains the value of various state flags on the processor, for example, the sign flag, carry flag, irq-enable flag, and so on. See a link in the *Further reading* section for a detailed look at the x86 registers. Also, look at the value of the **Code Segment (CS)** register – it's `0x0010`. As we saw, this is the value prefixed to the instruction pointer.

As an example, for our particular Oops, some of the x86 **control registers** are (perhaps) useful to look up. Let's check them out.

The control registers on the x86_64

The x86_64 has 16 control registers, named `CR0` through `CR15`. Of these, 11 are reserved (`CR1`, `CR5`-`CR7`, `CR9`-`CR15`). We mention the few that the x86_64 kernel Oops diagnostic is designed to reveal (and are thus meaningful here):

- `CR0`: Can be programmed (only in kernel mode). Contains control bits such as protected mode enable, emulation, write protect, alignment mask, cache disable, paging, and so on.

- `CR2`: Contains the KVA, which when accessed, caused the MMU to raise the page fault that led to the Oops. Hence, the `CR2` content is a key value! Look it up in *Figure 7.13*. Here, `CR2` contains the value `0x30`, which, as you'll recall, is the offset from the beginning of the structure we looked up – in the line of code that caused this Oops: `oopsie->data = 'x';`.

- `CR3`: A bitmask; holds (among other stuff) the physical address of the base of paging tables (called **PML4**) for the process context running at the time of the Oops. In effect, it tells the MMU how to get to the paging tables for the running context.

- `CR4`: Various control bits (for example, the V86 mode extension, debug extensions, page size extension, **Physical Address Extension (PAE)** bit, **Performance Monitoring Counter Enable (PCE)** bit, security feature bits such as **Supervisor Mode Executions Protection Enable (SMEP)**, **Supervisor Mode Access Protection Enable (SMAP)**, and so on).

What about CPU register values earlier in the execution path? Read the following callout for more on this...

CPU Register Values on Other Call Frames and Other Details

So, think about this: The CPU register values we saw above (*Figure 7.13*) are those that were on the CPU at the time the Oops occurred. But what about the CPU register values on all the other call frames (on the kernel-mode stack) that led up to the crash or Oops? Aren't they important? Yes, indeed. In fact, their values might be just what's required to decipher the really deep details of how the bug occurred, the root cause. But the Oops diagnostic doesn't reveal them – it's limited to the top of the stack: the function and register values in place at the instant the Oops occurred. So, is there any way to get to the deeper details, to be able to examine all the call frames and register values? Yes: one way is by leveraging the **kdump** kernel feature, which when enabled, allows saving all pertinent data (in fact, it lets you access a snapshot of the entire kernel memory segment taken at the time of the crash!). Along with the powerful userspace **crash** app to investigate the dump, you're in business! We briefly cover kdump/crash in *Chapter 12, A few more kernel debugging approaches*. Another powerful way to get details leading up to the crash (or panic) is to be able to trace the kernel in depth – ftrace provides just such functionality. We cover using ftrace (and its many frontends) in detail in *Chapter 9, Tracing the Kernel Flow*.

Right, let's move on to interpreting the last portion of the Oops diagnostic.

Interpreting Oops lines 7, 8, and 9

For your convenience, this portion of the screenshot is duplicated from *Figure 7.6*:

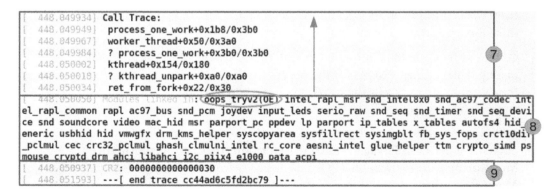

Figure 7.14 – Lines 7, 8, and 9 of the Oops output from our buggy oops_tryv2 module, case 3, showing the call (stack) trace, modules, and CR2 value

The really key thing to see and interpret here is the call (or stack) trace. Let's get to it!

Interpreting the call stack within the Oops

The lines that are below the string `Call Trace:` and slightly indented to the right represent the call or stack trace. This, of course, is the kernel mode stack of the process context (though it could also be that of an interrupt – more on this follows shortly...) that caused the Oops.

This call trace is very valuable to the development team, to you, debugging an Oops. Why? It literally shows you the history – how we got to where we did – of the bug. So, how do you interpret it? A few key points follow:

- Firstly, each (slightly right indented) line below `Call Trace:` represents a function in the call path, abstracted by a call frame.

- Each frame has the same notation as we saw earlier, `funcname+x/y`, where `x` is the distance from the start of the function in bytes (the start offset), and `y` is the size (length) of the function in bytes.

- Read the call trace bottom-up, of course (the vertical arrow in *Figure 7.14* points up to show this). Recall that pretty much all modern processors follow the *stack-grows-down* semantic.

- Ignore any line that begins with a question mark symbol (`?`). This implies the call frame is very likely invalid, a stale leftover from earlier stack memory usage, perhaps. The kernel's stack unwind algorithm (there are several – they're even configurable!) is smart enough to pretty much guarantee getting this right, so trust it.

Putting together what we learned, this is the call sequence that led to our Oops:

```
ret_from_fork() --> kthread() --> worker_thread() --> process_
one_work()
```

Now, of course, understanding this properly requires at least a basic understanding of what our code was doing when the Oops occurred. We do know that the Oops was actually triggered in the function `do_the_work()`. Refer back to our notes in the section, *Finding the code where the Oops occurred*. This function was our custom workqueue routine. Now how did it get invoked? Ah, indirectly, when our module invoked the `schedule_work()` API passing the pointer to our work structure (glance back at the code, `ch7/oops_tryv2/oops_tryv2.c`, if you need to). This work function was serviced by the kernel's default `events` workqueue. The servicing – which really means the consuming or execution of our work function – is done by spinning up (or using an existing) kernel worker thread (that belongs to the kernel's default `events` workqueue).

The `kthread()` routine – that you see in the call sequence just above – is an internal kernel interface that performs this task, spinning up a kernel thread. It invokes the kernel workqueue function `worker_thread()` whose job it is to process all work items on the workqueue. This is in turn done by calling each work item in a loop, by within it calling the function `process_one_work()`, whose sole job is to process the one work item it's given – ours!

So there we are – the kernel stack does indeed reveal exactly how we got to our worker routine `do_the_work()`. Whew – it can be painstaking work! But, then again, you're the forensic detective at work. No one said it's easy!

> **Tip – Seeing the Kernel-Mode Stack of All CPU Cores**
>
> The kernel has a tunable, `/proc/sys/kernel/oops_all_cpu_backtrace`, whose default value is 0, off. Turning it on (by writing 1 to it as root), will have the call stacks for all CPU cores on the system displayed as part of the Oops diagnostic. This can be very useful in a deep debug scenario. (Internally, the call tracing on other CPU cores is done via the **Non-Maskable Interrupt** (**NMI**) backtrace facility.)

The line following the call trace – the one starting with the string `Modules linked in:` – displays all modules loaded in the kernel at the time of the Oops. Why? Modules are very often third-party code (for device drivers, custom network firewall rules, custom filesystems, and so on), hence they are highly suspect when the kernel encounters a bug! Hence the list of all modules. Indeed, our very own buggy module, `oops_trv2`, stands proudly first in this list, as it's the last one loaded. Also, the tainted flags `OE` denote that it's an out-of-tree unsigned module.

Finally, the last line in the Oops diagnostic is what the kernel developers call the *executive summary* – the value of the CR2 register! By now you should realize why: it's the faulting (virtual) address, access to which caused the Oops.

> **Tip – Forcing a Pause to Read the Oops Diagnostic**
>
> Printing the detailed Oops diagnostic when an Oops occurs is all well and good, but what if it scrolls off the console window? Or, there are additional secondary Oopses that follow, causing the primary one's content to scroll away? For such cases, pass the `pause_on_oops=n` kernel parameter. It will have all CPUs halt for n seconds after the first kernel Oops is printed.

Finally, we're done interpreting our Oops in great detail, literally line by line, learning aplenty along the journey (hey, it's the journey that matters, not the destination).

We'd like to also mention (in passing) that there are several frameworks to help capture a kernel Oops and report it back to the vendor (or distributor). Among these is the `kerneloops(8)` program (man page: `https://linux.die.net/man/8/kerneloops`). Many modern distributions use the **kdump** feature to collect the entire kernel memory image when an Oops or panic occurs, for later analysis (typically via the `crash` app).

> **Exercise**
>
> Run this same Oops-ing test case (`ch7/try_oopsv2`) by passing the module parameter `bug_in_workq=yes` on your custom *debug* kernel and see what happens. On mine, with KASAN enabled, the kernel does Oops but not before KASAN catches the bug!

Great – now, it's important to continue on to the following section, where we cover how exactly you can use various tools and techniques to uncover the buggy line(s) of code!

Tools and techniques to help determine the location of the Oops

While analyzing a kernel Oops, we can certainly use all the help we can get, right?! There are several tools and helper scripts that can be leveraged. Among them, and part of the (cross) toolchain, are the `objdump`, the **GNU DeBugger (GDB)**, and `addr2line` programs. Besides them, a few kernel helper scripts (found within the kernel source tree) can prove very useful as well.

In this section, we'll start learning how to exploit these tools to help interpret an Oops.

> **Tip – Getting the Unstripped vmlinux Kernel Image with Debug Symbols**
>
> Many, if not most, of the tools and techniques to help debug kernel issues do *depend on your having an unstripped uncompressed vmlinux kernel image with debug symbols*. Now, if you've built both a debug and production kernel, as we've recommended from literally the outset of this book, you'll of course have the debug vmlinux kernel image file (which fulfills this requirement).
>
> And if not? Well, pretty much all enterprise (and desktop) Linux distros provide a package – separate ones for the commonly used kernel versions that get integrated into the distro – which will provide it (and more). Often, the package is named `linux-devel*` or `linux-headers*`. It's essentially just a compressed archive that contains the kernel headers, the unstripped `vmlinux` with debug symbols, and possibly more goodies within it. Download the package, install it, and see for yourself.
>
> For example, a note on the `kernel-*` RPMs for Red Hat RHEL 8 can be found here: `https://access.redhat.com/documentation/en-us/red_hat_enterprise_linux/8/html/managing_monitoring_and_updating_the_kernel/the-linux-kernel-rpm_managing-monitoring-and-updating-the-kernel#the-linux-kernel-rpm-package-overview_the-linux-kernel-rpm`. Also, as another example, here's a link to download RPM packages for the `kernel-devel` Linux packages for the following Linux distros – AlmaLinux, ALT Linux, CentOS, Fedora, Mageia, OpenMandriva, openSUSE, PCLinuxOS, and Rocky Linux: `https://pkgs.org/download/kernel-devel`.

Next, it's important to realize that, when debugging an Oops on a non-native arch (for example, ARM, ARM64, PowerPC, and so on), *you'll need to run the cross-toolchain version* of the tools we examine and use below. As a concrete example, if you're debugging an Oops that occurred on an ARM-32, compiled with the `arm-linux-gnueabihf-` toolchain prefix (this would be the value of the environment variable `CROSS_COMPILE`), then you'll need to run not `objdump` but `${CROSS_COMPILE}objdump`, that is, `arm-linux-gnueabihf-objdump`. The same goes for the other tools in the toolchain (such as GDB, `readelf`, `addr2line`, and so on).

Okay, let's begin using these tools!

Using objdump to help pinpoint the Oops code location

First off, you'll gain the maximum benefit from these tools by (re)building your buggy kernel, or kernel module, with CONFIG_DEBUG_INFO=y. In other words, boot from a debug kernel and build your module(s) there.

In our so-called better Makefile, set the following variable:

```
MYDEBUG := y
```

It's left as n by default. Now rebuild the module. Due to debug symbols and information embedded within it, its size is typically (much) larger than for the non-debug or production variant.

The objdump utility has the intelligence to deeply examine and interpret an **Executable and Linker Format** (**ELF**) object file (which includes the kernel vmlinux uncompressed image as well as a kernel module image, the .ko file). We'll typically run objdump with the -dS option switches to disassemble the module and intermix source code with assembly wherever possible (which is possible when it's compiled with -g)! If you're running live (that is, the module is still loaded in kernel memory), you can try this:

```
$ grep oops_tryv2 /proc/modules
oops_tryv2 16384 0 - Live 0x0000000000000000 (OE)
```

Due to security reasons, to prevent info leaks, the address isn't revealed. Run as root to see it:

```
$ sudo grep oops_tryv2 /proc/modules
oops_tryv2 16384 0 - Live 0xffffffffc0604000 (OE)
$ objdump -dS --adjust-vma=0xffffffffc0604000 ./oops_tryv2.ko >
oops_tryv2.disas
```

With the correct VMA object address of our module specified, objdump is able to display the full kernel virtual address on the left extreme of its output.

The objdump option switches we use are as follows:

- -d or --disassemble: Display assembler content of executable sections.

- -S or --source: Display a source code assembly intermix wherever possible (implies -d).

Of course, there are many other options. Do have a look at the man page of objdump. Here's some sample output from the objdump command just issued:

```
static void do_the_work(struct work_struct *work)
{
ffffffffc0604000: e8 00 00 00 00   callq   ffffffffc0604005 <do_
the_work+0x5>
ffffffffc0604005: 55                push    %rbp
[...]
```

Here's the analysis procedure:

1. In the preceding output from objdump, you can see that the start address of our do_the_work() function – the function where the Oops occurred (via the RIP value – see *Figure 7.6*) – is 0xffffffffc0604000.

2. Add the offset as shown by the first RIP line in the Oops diagnostic message:

    ```
    0xffffffffc0604000 + 0x124 = 0xffffffffc0604124
    ```

3. Look for the closest match to this address within the objdump output:

```
ffffffffc0604103:   74 27                   je      ffffffffc060412c <do_the_work+0x12c>
        pr_info("Generating Oops by attempting to write to an invalid kernel memory pointer\n");
ffffffffc0604105:   b9 3d 00 00 00          mov     $0x3d,%ecx
ffffffffc060410a:   48 c7 c2 00 00 00 00    mov     $0x0,%rdx
ffffffffc0604111:   48 c7 c6 00 00 00 00    mov     $0x0,%rsi
ffffffffc0604118:   48 c7 c7 00 00 00 00    mov     $0x0,%rdi
ffffffffc060411f:   e8 00 00 00 00          callq   ffffffffc0604124 <do_the_work+0x124>
        oopsie->data = 'x';
ffffffffc0604124:   c6 04 25 30 00 00 00    movb    $0x78,0x30
ffffffffc060412b:   78
        }
    kfree(gctx);
```

Figure 7.15 – Screenshot showing the output of objdump -dS -adjust-vma=… for our buggy module

We've annotated the same screenshot here, highlighting it so that you can see the relevant address – 0xffffffffc0604124 – highlighted by the blue rectangle and the relevant code region by the red rectangle:

```
ffffffffc0604103:   74 27                       je      ffffffffc060412c <do_the_work+0x12c>
         pr_info("Generating Oops by attempting to write to an invalid kernel memory pointer\n");
ffffffffc0604105:   b9 3d 00 00 00              mov     $0x3d,%ecx
ffffffffc060410a:   48 c7 c2 00 00 00 00        mov     $0x0,%rdx
ffffffffc0604111:   48 c7 c6 00 00 00 00        mov     $0x0,%rsi
ffffffffc0604118:   48 c7 c7 00 00 00 00        mov     $0x0,%rdi
ffffffffc060411f:   e8 00 00 00 00              callq   ffffffffc0604124 <do_the_work+0x124>
         oopsie->data = 'x';
ffffffffc0604124:   c6 04 25 30 00 00 00        movb    $0x78,0x30
ffffffffc060412b:   78
         }
    kfree(gctx);
```

Figure 7.16 – The same screenshot annotated to show the portion where the module code caused the Oops!

The C source line is just before the machine and assembly, showing that this is precisely where the fault – and thus the Oops – occurred!

If you aren't doing a live run, no matter, just run objdump in the same way, leaving out the --adjust-vma= parameter (see an example of this in the upcoming section, *On ARM with objdump*).

Also, objdump can be very useful when analyzing an Oops where the culprit is likely in-kernel code (as opposed to module code). In these cases, you'll require to generate the kernel code with source code and assembly intermixed, using the uncompressed vmlinux kernel image file as input to objdump:

```
${CROSS_COMPILE}objdump -dS <path/to/kernel-src/>/vmlinux >
vmlinux.disas
```

We assume the vmlinux image used here has been compiled with debug symbols. In fact, this is a one-time thing to do (unless the kernel itself is updated, of course).

Using GDB to help debug the Oops

The powerful GDB debugger can also be exploited to help with pinpointing the line of source code that triggered the Oops. GDB's `list` command can really help here. To use it though, you'll need to (re)build your module with debug symbols (that is, with the `-g` and other useful option switches). You can see, within our better `Makefile`, if we set the variable `MYDEBUG` to `y` (the default being `n`), how it employs compiler option switches that are important for debugging purposes:

```
MYDEBUG := y
ifeq (${MYDEBUG}, y)
# EXTRA_CFLAGS deprecated; use ccflags-y
        ccflags-y   += -DDEBUG -g -ggdb -gdwarf-4 -Og -Wall
-fno-omit-frame-pointer -fvar-tracking-assignments
```

After building it for debug, let's have GDB have a go:

```
$ gdb -q ./oops_tryv2.ko
Reading symbols from ./oops_tryv2.ko...
(gdb) list *do_the_work+0x124
0x160 is in do_the_work (<...>/ch7/oops_tryv2/oops_tryv2.c:62).
  [...]
61              pr_info("Generating Oops by attempting to write
to an invalid kernel memory pointer\n");
62              oopsie->data = 'x';
63          }
64          kfree(gctx);
(gdb)
```

Check it out: line 62 in the source is precisely where our bug is – GDB is bang on target!

More on using GDB for kernel/module debug is available on the official kernel documentation site: `https://www.kernel.org/doc/html/latest/admin-guide/bug-hunting.html#gdb`.

Using addr2line to help pinpoint the Oops code location

The addr2line utility has the ability to translate (virtual) addresses into their corresponding pathname(s) and line number(s)! It can be immensely useful to quickly pinpoint the place in the source code where the bug was triggered.

The address (or even multiple addresses) can be specified to addr2line via the -e (executable) option switch. (The utility takes several other optional parameters as well – do a quick addr2line -h to see them.)

For our module, let's invoke addr2line as follows, passing the offset from the start of the function reported by the Oops diagnostic as the RIP register value (the value 0x124 – see *Figure 7.6*):

```
$ addr2line -e ./oops_tryv2.o -p -f 0x124
do_the_work at <...>/ch7/oops_tryv2/oops_tryv2.c:62
```

Again, perfect and so easy to use! The optional parameter -p pretty prints the output, while -f displays the function name as well.

The addr2line utility is also useful when you have a kernel (not module) crash and the kernel virtual address. In these cases, supply the uncompressed vmlinux file (which has all debug symbols) to addr2line via the -e option switch:

```
addr2line -e </path/to/>vmlinux -p -f  <faulting_kernel_
address>
```

Note that addr2line will *not* work correctly on addresses generated on a system with **Kernel Address Space Layout Randomization (KASLR)**, a kernel security/hardening feature, configured (from kernel version 3.14). In this case (and it's usually the case due to security), use the faddr2line script instead (we cover it in the following section). Alternatively, you can disable KASLR at boot by passing the kernel command-line option nokaslr via the bootloader. For more on KASLR, see the *Further reading* section.

Let's now check out some kernel helper scripts.

Taking advantage of kernel scripts to help debug kernel issues

The modern Linux kernel has many helper scripts, helping you to debug kernel bugs. Here's a quick table summarizing them. A bit of a more detailed take on how to practically use them follows:

Script	Purpose
`scripts/ checkstack.pl`	Estimates the stack size used by functions within the kernel (or module), in descending order by size.
`scripts/ decode_ stacktrace.sh`	A script that tries to convert all kernel (virtual) addresses passed to it (usually by redirecting standard input from the `dmesg` output or piped to it), to source filenames with line numbers.
`scripts/ decodecode`	A script that attempts to add useful information by parsing the `Code: <...machine code bytes...>` line in a typical Oops report; figures out and specifies the instruction where the fault actually occurred and shows it as `<-- trapping instruction` to the right of that line.
`scripts/ faddr2line`	The same as `addr2line` but appropriate for use on systems using KASLR (for security) and for interpreting stack dumps from kernel modules. (Note – use the most recent version, as earlier versions were flawed; the upcoming *Exploiting the faddr2line script on KASLR systems* section has the details.)
`tools/debugging/ kernel-chktaint`	Interprets the kernel tainted flags. Can pass the tainted bitmask as a parameter. If not passed, it looks up the current system taint state and prints its report.
`scripts/ get_maintainer.pl`	With the `-f file|directory` parameter, this script identifies and prints details on the maintainer(s), the mailing list, and so on. It's useful to quickly find who maintains a given piece of code in the kernel!

Table 7.3 – A summary table of several useful kernel helper scripts

The list isn't exhaustive but is plenty to work with. Let's get going!

Using the checkstack.pl script

As you should be aware, every user-mode thread alive has two stacks – a user-mode stack and a kernel-mode stack. The former is dynamic and can grow pretty large (up to 8 MB by default on your typical Linux); the latter is used when execution enters kernel space. The **kernel-mode stack is fixed in size and small**, typically two pages on 32-bit and four pages on 64-bit systems – in effect, just 8 KB or 16 KB, with a typical 4 KB page size (also, don't assume the page size that the MMU's using – the kernel macro PAGE_SIZE has the correct value).

Overflowing the kernel-mode stack is thus relatively easy to do. The kernel maintainers consider it worthwhile to have a Perl script handy. It parses the kernel functions and reports the largest possible stack size required by it. This script outputs the stack size used by functions passed to it (often via objdump across a pipe, as seen here), in descending order by size:

```
$ objdump -d <...>/linux-5.10.60/vmlinux | <...>/linux-5.10.60/
scripts/checkstack.pl
0xffffffff810002100 sev_verify_cbit [vmlinux]:          4096
0xffffffff81a554300 od_set_powersave_bias [vmlinux]:   2064
0xffffffff817b24100 update_balloon_stats [vmlinux]:    1776
[...]
```

There's no reason you can't run it on a kernel module, for example:

```
$ objdump -d /lib/modules/5.10.60-dbg02-gcc/kernel/drivers/net/
netconsole.ko | <...>/scripts/checkstack.pl
0x00000000000013800 enabled_store [netconsole]:        224
0x00000000000000000 init_module [netconsole]:          224
0x0000000000000c300 remote_ip_store [netconsole]:      208
[...]
```

Overflowing the stack, especially a kernel-mode stack, is no joking matter. The resulting stack corruption *can result in an abrupt and dramatic system hang*, with no real way to debug what exactly occurred. It's better to be prepared than sorry!

A quick kernel internals note: precisely because kernel stack overflows can simply and immediately hang the system, recent kernels have shifted to enabling an arch-specific kernel config called CONFIG_VMAP_STACK (the x86_64 has it enabled from the 4.9 kernel and ARM64 from 4.14). Very briefly, the vmalloc() interface is used to allocate (task and IRQ) stack memory. This ensures that a later bad page fault (perhaps from an overflow) on these pages can be handled by the kernel's fault/Oops handling code.

Leveraging the decode_stacktrace.sh script

A raw (kernel) stack dump is of limited usefulness when the text – the function names – is all that's seen. The `decode_stacktrace.sh` script attempts to remedy this situation by showing, for each function name on the stack call trace, its source code location and line number within the kernel and/or module!

In effect, it's a kind of combination of the raw stack trace along with the information that the `addr2line` utility provides (just that it's done for *every* call frame on the stack. In fact, this script is at some level a wrapper over the `addr2line` utility, internally invoking `${CROSS_COMPILE}addr2line`). Its usage is as follows:

```
$ </path/to/>/linux-5.10.60/scripts/decode_stacktrace.sh
Usage:
<...>/linux-5.10.60/scripts/decode_stacktrace.sh -r <release> |
<vmlinux> [base path] [modules path]
```

You can see that, to be effective, this script requires the pathname of the uncompressed kernel `vmlinux` image with debug symbols. Next, the `base-path` parameter is the pathname of the directory where this file is available. We provide the path to our `vmlinux` image file within the 5.10.60 kernel source tree (where we built it). Alternately, you can specify the kernel release via the `-r` option switch. The script will attempt to retrieve the `vmlinux` image (with debug symbols) based on this value. The `modules path` parameter is the location where the defective kernel module lives. Here, it's the current directory (as we're working from there), so we specify it as `./`.

Again using our usual bug-in-workqueue module example (on an x86_64 guest system), we've saved the `dmesg` output into a file named `dmesg_oops_buginworkq.txt`. Just as the `vmlinux` file with debug symbols must be passed, you should ensure you (re) compile the module with debug flags on (set the `MYDEBUG` variable in the `Makefile` to `y` and rebuild it).

Here's what the `decode_stacktrace.sh` script buys us when we run it through the saved `dmesg` output:

```
$ ~/lkd_kernels/productionk/linux-5.10.60/scripts/decode_
stacktrace.sh ~/lkd_kernels/debugk/linux-5.10.60/vmlinux ~/lkd_
kernels/debugk/linux-5.10.60 ./ < dmesg_oops_buginworkq.txt
[...]
[  448.049414] BUG: kernel NULL pointer dereference, address:
0000000000000030
[...]
[  448.049547] Workqueue: events do_the_work [oops_tryv2]
```

```
[  448.049562] RIP: 0010:do_the_work (/home/letsdebug/Linux-
Kernel-Debugging/ch7/oops_tryv2/oops_tryv2.c:62) oops_tryv2
<< ... output of the decodecode script ... >>
[...]
[  448.049934] Call Trace:
[  448.049949] process_one_work (kernel/workqueue.c:1031
(discriminator 19) kernel/workqueue.c:2194 (discriminator 19))
[  448.049967] worker_thread (./arch/x86/include/asm/
current.h:15 kernel/workqueue.c:979 kernel/workqueue.c:1815
kernel/workqueue.c:2381)
[  448.049984] ? process_one_work (kernel/workqueue.c:2222)
[  448.050002] kthread (kernel/kthread.c:277)
[...]
[  448.050937] CR2: 0000000000000030
[  448.051593] ---[ end trace cc44ad6c5fd2bc79 ]---
```

This script also interprets the machine code running on the processor at the time of the fault – this work is actually performed by yet another helper script, the `scripts/decodecode` one (which we show next). Here, you can see the bash function that invokes it:

```
$ cat <...>/linux-5.10.60/scripts/decode_stacktrace.sh
[...]
decode_code() {
local scripts=`dirname "${BASH_SOURCE[0]}"`
echo "$1" | $scripts/decodecode
}
```

The original commit of the `decode_stacktrace.sh` script into the kernel source tree (in version 3.16), is interesting to browse through. Check it out here: `https://github.com/torvalds/linux/commit/dbd1abb209715544bf37ffa0a3798108e140e3ec`.

Interpreting machine code with the decodecode script

As with the `decode_stacktrace.sh` script, this script reads from standard input, thus, passing it the `dmesg` Oops output via a file or across a pipe is typical. It attempts to decode the machine code that was running on the processor core at the time of the bug (or crash) and displays useful output. It's even able to identify the particular instruction that caused the fault (or trap) to be raised by the MMU and shows it by printing `<--trapping instruction` to the right of that line. Check out the example following screenshot showing running this script on our `oops_tryv2` module's printks when it triggered an Oops:

```
$ ~/lkd_kernels/productionk/linux-5.10.60/scripts/decodecode < dmesg_oops_buginworkq.txt
[ 53.695794] Code: c0 e8 d0 2d 47 c6 f6 c3 01 74 27 b9 3d 00 00 00 48 c7 c2 c0 53 60 c0 48
c7 c6 3c 50 60 c0 48 c7 c7 18 51 60 c0 e8 61 35 a8 c6 <c6> 04 25 30 00 00 00 78 48 8b 3d cd
23 00 00 e8 a8 ba 13 c6 5b 41
All code
========
   0:   c0 e8 d0                shr    $0xd0,%al
   3:   2d 47 c6 f6 c3          sub    $0xc3f6c647,%eax
   8:   01 74 27 b9             add    %esi,-0x47(%rdi,%riz,1)
   c:   3d 00 00 00 48          cmp    $0x48000000,%eax
  11:   c7 c2 c0 53 60 c0       mov    $0xc06053c0,%edx
  17:   48 c7 c6 3c 50 60 c0    mov    $0xffffffffc060503c,%rsi
  1e:   48 c7 c7 18 51 60 c0    mov    $0xffffffffc0605118,%rdi
  25:   e8 61 35 a8 c6          callq  0xffffffffc6a8358b
  2a:*  c6 04 25 30 00 00 00    movb   $0x78,0x30                <-- trapping instruction
  31:   78
  32:   48 8b 3d cd 23 00 00    mov    0x23cd(%rip),%rdi        # 0x2406
  39:   e8 a8 ba 13 c6          callq  0xffffffffc613bae6
  3e:   5b                      pop    %rbx
  3f:   41                      rex.B

Code starting with the faulting instruction
===========================================
   0:   c6 04 25 30 00 00 00    movb   $0x78,0x30
   7:   78
   8:   48 8b 3d cd 23 00 00    mov    0x23cd(%rip),%rdi        # 0x23dc
   f:   e8 a8 ba 13 c6          callq  0xffffffffc613babc
  14:   5b                      pop    %rbx
  15:   41                      rex.B
$
```

Figure 7.17 – Screenshot showing the output of the decodecode script

It has indeed shown the trap at the precise place in the machine code/assembly line where it occurred (notice the operand of `0x30` – the offset we're working with – to the `movb` machine instruction where the trap occurred).

The original commit of this script into the kernel source tree happened many years back (in July 2007, kernel version 2.6.23) and can be found here: `https://github.com/torvalds/linux/commit/dcecc6c70013e3a5fa81b3081480c03e10670a23`.

As mentioned already, this script is itself invoked by the `decode_stacktrace.sh` script to better interpret the machine code bytes, thus the `decode_stacktrace.sh` script is a superset of this one.

Exploiting the faddr2line script on KASLR systems

Are you running on a KASLR-enabled kernel? Let's check:

```
$ grep CONFIG_RANDOMIZE_BASE /boot/config-$(uname -r)
CONFIG_RANDOMIZE_BASE=y
```

Yes – here, we are. In cases like this (as mentioned already), the `addr2line` script may not work as expected. In such cases, use the `faddr2line` script instead (as you'll guess, it is a wrapper over the `addr2line` utility):

```
<...>/linux-5.10.60/scripts/faddr2line
usage: faddr2line [--list] <object file> <func+offset>
<func+offset>...
```

So, let's appropriately invoke the `faddr2line` script:

```
$ ~/lkd_kernels/productionk/linux-5.10.60/scripts/faddr2line ./
oops_tryv2.ko do_the_work+0x124
bad symbol size: base: 0x0000000000000000 end:
0x0000000000000000
```

Hey, that's really not what we expected!

> **Tip – Patch the faddr2line Script or Use a Newer Fixed Version**
>
> Upon encountering this issue with `faddr2line`, I reported it to the maintainer, Josh Poimboeuf (link: `https://lkml.org/lkml/2022/1/16/305`). By May 2022, Josh had fixed it (the underlying issue was that the nm utility wasn't good enough; he switched to using `readelf` – you'll find the details in the patch: `https://lore.kernel.org/lkml/29ff99f86e3da965b6e46c1cc2d72ce6 528c17c3.1652382321.git.jpoimboe@kernel.org/`). So, until this fix hits the upcoming mainline kernel (it will, and I am hoping it happens soon – as of this writing, the process is just getting started), you'll have to manually apply this patch to the existing `scripts/faddr2line` script. (The fixed faddr2line should make it into the 5.19 kernel.)

Once the patch (mentioned just above) is applied, or, you have a fixed version of the `faddr2line` script from a later kernel source tree (this should definitely be the case soon enough), let's retry:

```
$ <...>/scripts/faddr2line  ./oops_tryv2.ko do_the_
work+0x124/0x15e
do_the_work at <...>/Linux-Kernel-Debugging/ch7/oops_tryv2/
oops_tryv2.c:62
```

Ah, that's perfect! Line 62 (`oopsie->data = 'x';`) is indeed the buggy one.

Are we clean? The kernel-chktaint script

We covered the interpretation of the kernel's *tainted flags* in the previous section, *Interpreting the kernel tainted flags*.

The `kernel-chktaint` script is a simple helper script that interprets the kernel's tainted bitmask and prints its report. Here's an example when I ran it on my x86_64 Ubuntu guest that our `oops_tryv2` buggy module caused an Oops upon:

```
$ tools/debugging/kernel-chktaint $(cat /proc/sys/kernel/tainted)
Kernel is "tainted" for the following reasons:
 * kernel died recently, i.e. there was an OOPS or BUG (#7)
 * externally-built ('out-of-tree') module was loaded  (#12)
 * unsigned module was loaded (#13)
For a more detailed explanation of the various taint flags see
 Documentation/admin-guide/tainted-kernels.rst in the the Linux kernel sources
 or https://kernel.org/doc/html/latest/admin-guide/tainted-kernels.html
Raw taint value as int/string: 12416/'G      D    OE    '
$
```

Figure 7.18 – Screenshot showing output from the kernel-chktaint helper script

If you don't pass a parameter, the script looks up the proc pseudofile, `/proc/sys/kernel/tainted`, and interprets its value. Note that this script lives in the `tools/debugging` directory under the kernel source tree, not the `scripts/` one.

Locating our saviors – the get_maintainer.pl script

Heard this one? When you fail, try, try again; if you fail non-stop, deny you ever tried. Ha ha, very funny. We prefer this: when all else fails, contact the maintainer(s)!

It's easy to do with the `scripts/get_maintainer.pl` script. Typically, provide the file or directory via the `-f` option switch and the details are revealed (do see its pretty verbose help screen though for more options).

Here's an example: say you're having trouble with the kernel's **Kernel GDB (KGDB)** feature and want to ask someone pertinent questions regarding your troubles. Who do you ask? The maintainers, and/or the mailing list if there's one, of course. Well, who maintains it? The following screenshot shows how this question is easily answered, via the `get_maintainer.pl` Perl script:

```
$ cd ~/lkd_kernels/productionk/linux-5.10.60/
$ scripts/get_maintainer.pl
scripts/get_maintainer.pl: missing patchfile or -f file - use --help if necessary
$ scripts/get_maintainer.pl -f kernel/debug/
scripts/get_maintainer.pl: No supported VCS found.  Add --nogit to options?
Using a git repository produces better results.
Try Linus Torvalds' latest git repository using:
git clone git://git.kernel.org/pub/scm/linux/kernel/git/torvalds/linux.git
Jason Wessel <jason.wessel@windriver.com> (maintainer:KGDB / KDB /debug_core)
Daniel Thompson <daniel.thompson@linaro.org> (maintainer:KGDB / KDB /debug_core)
Douglas Anderson <dianders@chromium.org> (reviewer:KGDB / KDB /debug_core)
kgdb-bugreport@lists.sourceforge.net (open list:KGDB / KDB /debug_core)
linux-kernel@vger.kernel.org (open list)
$
```

Figure 7.19 – Screenshot showing output from the kernel get_maintainer.pl helper script

The last few lines provide the key portion of the answer – the KGDB maintainers, their email, and more importantly, the email address of the KGDB mailing list. Note that you need to run it from the root of the kernel source tree. Also, running this script within a Git-based kernel source tree can produce superior results.

Do realize that this script searches the MAINTAINERS file in the root of the kernel source tree to provide its results. There's no reason you can't do the same with a simple `grep`:

```
$ grep -A15 -w "KGDB" MAINTAINERS
KGDB / KDB /debug_core
M:      Jason Wessel <jason.wessel@windriver.com>
M:      Daniel Thompson <daniel.thompson@linaro.org>
R:      Douglas Anderson <dianders@chromium.org>
L:      kgdb-bugreport@lists.sourceforge.net
S:      Maintained
W:      http://kgdb.wiki.kernel.org/
T:      git git://git.kernel.org/pub/scm/linux/kernel/git/jwessel/
kgdb.git
F:      Documentation/dev-tools/kgdb.rst
[...]
F:      kernel/debug/
```

In the preceding, L denotes the mailing list. So, fire off your well-thought-out email to the mailing list!

To report the bug I found in the `faddr2line` script (see the section, *Exploiting the faddr2line script on KASLR systems*), I used this technique to find the maintainer:

```
5.10.60 $ grep -i -w -A1 faddr2line MAINTAINERS
FADDR2LINE
M:    Josh Poimboeuf <jpoimboe@redhat.com>
--
F:    scripts/faddr2line
```

Then, you ask, what happens when you (and/or your team) are the maintainer(s)? You got it – keep reading this book and learning (tongue-in-cheek grin).

With that, we'll close this section. Any other kernel helper scripts? Oh, there are plenty. Here's a sampling of the `check*` ones (from the root of the kernel source tree):

```
ls scripts/check*
scripts/check-sysctl-docs   scripts/checkkconfigsymbols.
py  scripts/checksyscalls.sh    scripts/check_extable.sh
scripts/checkpatch.pl       scripts/checkversion.pl  scripts/
checkincludes.pl    scripts/checkstack.pl
```

We've seen a couple of these already (recall, the `checkpatch.pl` Perl script is invoked by our better `Makefile`!).

On this note, there's a helper script, `scripts/extract-vmlinux`. It's used to extract an uncompressed `vmlinux` image from an existing kernel image file. In a similar vein, the `kdress` utility attempts to extract an uncompressed debug-symbols `vmlinux` file from an existing `vmlinuz` image (and `/proc/kcore`) – see `kdress` here: https://github.com/elfmaster/kdress. Of course, with these, it's often a case of **Your Mileage May Vary** (**YMMV**), though!

We'll leave it to you now, intrepid explorer, to check them out!

Leveraging the console device to get the kernel log after Oopsing in IRQ context

When you tried out the first two versions of our trivial Oops-generating buggy modules (`ch7/oops_tryv1` and `ch7/oops_tryv2`), you'd have typically found that, though the kernel has a bug and generated an Oops diagnostic, the system is still usable (of course, no guarantees on this!).

These two modules generated the Oops while running kernel (module) code in process context (often, it's the `insmod` process, but our workqueue test case had the Oops occur in the context of a kernel worker thread). Now, what if we do the same thing – generate an Oops by, say, attempting to read an address within the NULL trap page (as we did earlier), *but this time while running in interrupt context*!

Well, our `ch7/oops_inirqv3` module does precisely this: it sets up a function that will run in (hard) interrupt context by leveraging the kernel's `irq_work*` functionality (we in fact used this same feature for running one of the memory leakage test cases in interrupt context in the previous chapter). Here's the relevant code snippet – the interrupt context work function that generates a simple Oops:

```
// ch7/oops_inirqv3/oops_inirqv3.c
void irq_work(struct irq_work *irqwk)
{
    int want_oops = 1;
    PRINT_CTX();
    if (!!want_oops) // okay, let's Oops in irq context!
        // a fatal hang can happen here!
        *(int *)0x100 = 'x';
}
```

Simple enough. Try running this module on your system. In my case at least, running my Ubuntu 20.04 guest VM (with our custom production 5.10.60-prod01 kernel) on Oracle VirtualBox 6.1, it has the VM simply freeze! No printks are seen, the login shell is unusable, and the system appears to be hung.

Now what? How do you debug when even dmesg can't be run?

Setting up Oracle VirtualBox with a virtual serial port

Ah, welcome to the real world. For now, here's what we'll do: leveraging the kernel console device concept, *we can set up an additional serial console* (pseudo) device on our guest VM that actually backs onto a log file on our host system. (Do realize that this case is very specific to using an x86_64 guest with Oracle VirtualBox as the hypervisor app. The general concept, though, is applicable pretty much everywhere.)

Follow these steps to set up your hypervisor and guest system to log all kernel printks from guest to host in a file on the host:

1. If already running, shut down your x86_64 guest Linux VM.

2. On your host system, go to your Oracle VirtualBox app GUI, select your guest VM (typically seen in the column on the left side) and open your guest VM settings (by clicking on the settings gearwheel).

3. The **Settings** dialog box opens. Here's a screenshot as it appears on my host system:

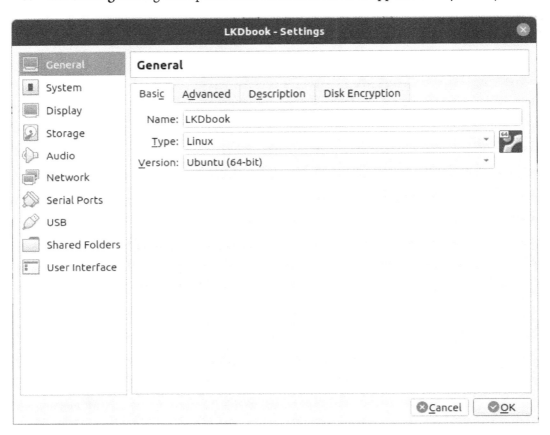

Figure 7.20 – Screenshot of the Oracle VirtualBox Settings dialog for our guest VM

4. Now navigate to the **Serial Ports** option. Enable **Port 1** (by clicking its toggle button). Set **Port Number** to **COM1** (equivalent to `ttyS0` on Linux), **Port Mode** to **Raw File** (via the drop-down menu), and within the **Path/Address** textbox, enter the pathname to the console log file (the path is with respect to your host system):

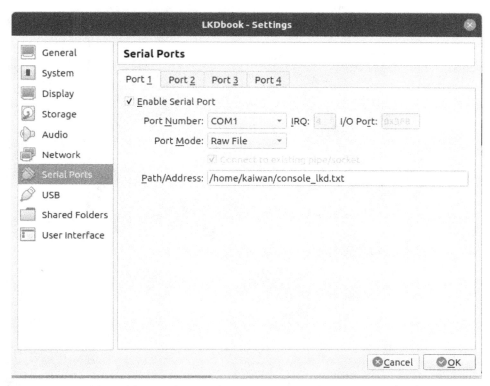

Figure 7.21 – Screenshot of the Oracle VirtualBox Settings / Serial Port setup dialog for our guest VM, setting up a serial console to log to a file on the host

(My host system is Linux too, hence the pathname follows the usual Unix/Linux conventions. On a Windows/Mac host, provide the pathname to the file on your host in accordance with its naming conventions.) Click the **OK** button when done.

5. Start the guest system. Press a key (usually *Shift*) to interrupt the GRUB bootloader and enter its menu interface screen. Navigate to the **Advanced options for Ubuntu** menu. Within it, highlight the appropriate kernel (I'm selecting our custom 5.10.60-prod01 kernel) via the up/down arrow keys.

6. Type e to edit this kernel entry. Scroll down to the line beginning with `linux /boot/vmlinux-5.10.60-prod01 root=UUID=... ro quiet splash ....`

Most importantly, edit the kernel command line, adding the new serial console(s). Take the cursor to the end of this entry and type `console=ttyS0 console=tty0 ignore_loglevel`:

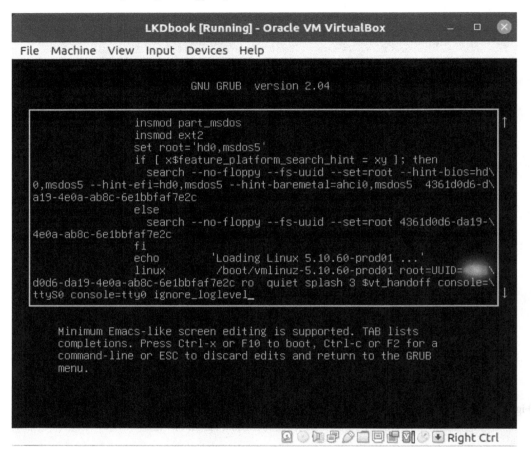

Figure 7.22 – Screenshot of the GRUB CLI, showing how we've edited the kernel command line to add additional serial console(s)

On boot, this has the target kernel understand there are additional serial console devices to send kernel printks to. The `ignore_loglevel` directive (as we saw before), has the kernel send all printks to the consoles, irrespective of their log level (good for debugging scenarios). When done editing, press Ctrl + X to boot.

Retrying oops_inirqv3 with the serial console enabled

All right, by now, I assume you've followed the preceding detailed steps and are logged into your guest VM with the new serial console(s) enabled. Check this with the following:

```
$ cat /proc/cmdline
BOOT_IMAGE=/boot/vmlinuz-5.10.60-prod01 root=UUID=<...> ro
quiet splash 3 console=ttyS0 console=tty0 ignore_loglevel
```

As mentioned back in *Chapter 3, Debug via Instrumentation – printk and Friends*, in the section, *Emitting a printk from userspace*, a quick test to see if the (pseudo) serial console works as expected is to do this, on the guest, as root:

```
# echo "testing serial console 1 2 3" > /dev/kmsg
```

On the host, look up your console file (by the pathname you provided earlier). The message sent above should be appended to it. Here it is on my Linux host:

```
$ tail -n1 ~/console_lkd.txt
[  646.403129] testing serial console 1 2 3
```

It's fine. Let's proceed, running our buggy-in-irq module again:

```
cd <...>/ch7/oops_inirqv3
make
sudo insmod ./oops_inirqv3.ko ; sudo dmesg
[... <hangs> ...]
```

Again, the system seems to be hung. But this time, the Oops diagnostic printks would have made it across the (pseudo) serial line to the file on the host! Check it. On my host, this is what I see within the serial console file:

```
[  770.407919] BUG: kernel NULL pointer dereference, address: 0000000000000100
[  770.408580] #PF: supervisor write access in kernel mode
[  770.409050] #PF: error_code(0x0002) - not-present page
[  770.409521] PGD 0 P4D 0
[  770.409757] Oops: 0002 [#1] PREEMPT SMP PTI
[  770.410143] CPU: 1 PID: 1699 Comm: insmod Tainted: G           OE     5.10.60-prod01 #6
[  770.410869] Hardware name: innotek GmbH VirtualBox/VirtualBox, BIOS VirtualBox 12/01/2006
[  770.411615] RIP: 0010:irq_work+0x36/0x150 [oops_inirqv3]
[  770.412100] Code: 05 6f fb 9a 3f a9 00 01 ff 00 74 19 a9 00 00 0f 00 0f 84 f5 00 00 00 ba 48 00 00 00 f6
 c4 ff 0f 84 e7 00 00 00 0f 1f 44 00 00 <c7> 04 25 00 01 00 00 78 00 00 00 c3 55 65 4c 8b 04 25 c0 7b 01 00
[  770.413793] RSP: 0018:ffff9cc4800f0f78 EFLAGS: 00010006
[  770.414274] RAX: 0000000080010001 RBX: ffffffffc066b480 RCX: 000000000000080b
[  770.414922] RDX: 0000000000000068 RSI: ffffffffb4f27968 RDI: ffffffffc066b480
[  770.415546] RBP: ffff9cc4800f0f98 R08: 0000000000000000 R09: 0000000000000000
[  770.416168] R10: 0000000000000000 R11: 0000000000000000 R12: 0000000000000022
[  770.416787] R13: 0000000000000020 R14: 0000000000000000 R15: 0000000000000000
[  770.417397] FS:  00007f514a975540(0000) GS:ffff903c7dc40000(0000) knlGS:0000000000000000
[  770.418136] CS:  0010 DS: 0000 ES: 0000 CR0: 0000000080050033
[  770.418662] CR2: 0000000000000100 CR3: 00000000265fe006 CR4: 00000000000706e0
[  770.419283] Call Trace:
[  770.419500]  <IRQ>
[  770.419684]  ? irq_work_single+0x34/0x50
[  770.420032]  irq_work_run_list+0x31/0x50
[  770.420398]  irq_work_run+0x5a/0xf0
[  770.420711]  __sysvec_irq_work+0x30/0xd0
[  770.421085]  asm_call_irq_on_stack+0x12/0x20
[  770.421479]  </IRQ>
[  770.421677]  sysvec_irq_work+0x9f/0xc0
[  770.422478]  asm_sysvec_irq_work+0x12/0x20
[  770.423215] RIP: 0010:native_write_msr+0x6/0x30
[  770.423990] Code: 0f 1f 40 00 0f 1f 44 00 00 55 48 89 e5 0f 0b 48 c7 c7 60 07 07 b5 e8 51 70 c1 00 66 0f
 1f 84 00 00 00 00 89 f9 89 f0 0f 30 <0f> 1f 44 00 00 c3 55 48 c1 e2 20 89 f6 48 09 d6 31 d2 48 89 e5 e8
[  770.426876] RSP: 0018:ffff9cc482603bb0 EFLAGS: 00000206
[  770.427732] RAX: 00000000000000f6 RBX: 0000000000000010 RCX: 000000000000083f
[  770.429039] RDX: 0000000000000000 RSI: 00000000000000f0 RDI: 000000000000083f
[  770.430134] RBP: ffff9cc482603bb8 R08: 0000000000000010 R09: ffff903c137550a0
[  770.431195] R10: ffff903c01ae5410 R11: 0000000000000000 R12: ffffffffc066b480
[  770.432229] R13: 00000000000288a8 R14: 0000000000000001 R15: ffffffffc066b0d8
[  770.433264]  ? native_apic_msr_write+0x2b/0x30
[  770.434066]  x2apic_send_IPI_self+0x20/0x30
[  770.434859]  arch_irq_work_raise+0x2a/0x40
[  770.435620]  __irq_work_queue_local+0xbf/0x130
[  770.436398]  irq_work_queue+0x32/0x50
[  770.437091]  ? 0xffffffffc0662000
[  770.437751]  try_oops_init+0x2a/0x1000 [oops_inirqv3]
[  770.438559]  do_one_initcall+0x48/0x210
[  770.439353]  ? kmem_cache_alloc_trace+0x3ae/0x450
[  770.440431]  do_init_module+0x62/0x240
[  770.441121]  load_module+0x2a04/0x3080
[  770.441814]  ? security_kernel_post_read_file+0x5c/0x70
[  770.442651]  __do_sys_finit_module+0xc2/0x120
[  770.443390]  ? __do_sys_finit_module+0xc2/0x120
[  770.444141]  __x64_sys_finit_module+0x1a/0x20
```

Figure 7.23 – Partial screenshot showing the content of the serial console raw file on the host; we can clearly see (and thus analyze) the Oops diagnostic!

Fantastic – this time we can see the Oops diagnostic in all its glory and therefore analyze it.

Take a closer look – it's interesting: here, we know that the bug occurred in an interrupt context. The `RIP` register points to the correct IRQ work function (`irq_work()`). Further verifying this, the call (stack) trace now has two distinct parts – the IRQ stack (delimited by the `<IRQ>` ... `</IRQ>` tokens) and, further down, the non-IRQ process-mode kernel stack trace. Together, reading them bottom-up (ignoring the frames prefixed with `?`), they paint a pretty clear picture of what happened.

Several Stacks

The reality is that there can be several stacks in existence simultaneously; it depends on the arch (CPU) and on what happened, on the code paths taken. On modern architectures, for example, interrupt processing occurs on a separate stack, the IRQ stack. On the x86_64, there can be regular stacks – the user-mode and the kernel-mode stack (which are called the task stacks), an IRQ stack, a hardware exception stack (for handling double faults, NMI, debug, and the **machine-check exception (mce)**), and an entry stack. Starting with the current stack pointer, the stacks' frames are unwound. Each stack has a pointer to the next one.

Of course, the line at the top also clearly indicates the NULL pointer page dereference that is the root cause of this Oops (where it occurred is what the call trace(s) reveal):

```
BUG: kernel NULL pointer dereference, address: 0000000000000100
```

The Oops output contains the process context in play at the time of the Oops:

```
CPU: 1 PID: 1699 Comm: insmod Tainted: G          OE
5.10.60-prod01 #6
```

Now, this might have you think that the `insmod` process was the one executing kernel code at the time of the crash and that it's the culprit. Well, not this time! This is because – as you'll recall – this time, the bug occurred while the kernel was executing our IRQ work function, *in interrupt context*. The presence of the IRQ call stack gives us that information as well...

So, an important thing to realize is the `insmod` process was merely the value that the `current` macro happened to point to at the time the Oops occurred. In other words, the `insmod` process was actually interrupted by the interrupt (a work IRQ here)! That's why it shows up; it doesn't necessarily mean it was running the code that triggered the Oops.

Quite often, when analyzing Oopses, you might find the process context is `swapper` (`PID 0`). That, of course, is the kernel thread that runs on the processor when that core is idle (it's the so-called *idle thread* – there's one per CPU core, of the form `swapper/n`, where n is the core number starting with 0). So, the point here is that when this shows up as the process context, it's perhaps more likely that it was interrupted by some interrupt (or softirq) that is actually the one that ran the buggy code.

The last line of output (not seen in the preceding truncated screenshot) shows why the system hung – it's considered a fatal error, the kernel did actually panic:

```
[  770.483105] ---[ end Kernel panic - not syncing: Fatal
exception in interrupt ]---
```

A word of caution: unfortunately, the serial console log file seemed to get truncated. (This seems to be a known issue with VirtualBox – see the link in the *Further reading* section.)

This general approach here has us specifying a serial console device to log kernel printks to. Extending this approach, the **netconsole** facility allows you to log kernel printks across a network! In effect, *it's a remote printk facility*. We shall shortly cover the basic usage of netconsole.

An Oops on an ARM Linux system and using netconsole

To get the most out of this section, I'd definitely recommend you be at least a little conversant with the processor ABI conventions (especially stuff such as function calls, parameter passing, and return values) for the processor your code's running upon. So, here, it's for the ARM32. Again, do review the basics that we covered in *Chapter 4, Debug via Instrumentation – Kprobes*, in the section, *Understanding the basics of the Application Binary Interface (ABI)*.

Here, our test environment is a Raspberry Pi 0W running Raspbian 10 (Buster) with the standard 5.10.17+ kernel. This popular prototyping (and product) board has the Broadcom BCM2835 **System on Chip** (**SoC**), which internally sports a single ARM32 CPU core for this board.

We cross-compile and run our usual test case on the device: our `oops_tryv2` module case 3 (passing the `bug_in_workq=yes` parameter). It does cause an Oops, of course, but the bug was severe enough to hang the board entirely! So how do we get to see the kernel log?

Ah, the kernel's netconsole facility turns out to be the answer. Well, at least one way. Another way is to use the Raspberry Pi's serial UART along with a USB-to-serial converter and to see the console output on a terminal – minicom / Hyperterminal – window! Of course, we assume you have the target kernel – in our case, the (embedded ARM) target system kernel configured for netconsole. The kernel config CONFIG_NETCONSOLE should be set to either y or m (here, we assume it's built as a module, the typical case).

When loading the netconsole driver as a module, this is the format of the key parameter, named netconsole, the means to specify the sender system's source address and the receiver system's destination addresses:

```
netconsole=[+][src-port]@[src-ip]/[<dev>],[tgt-port]@<tgt-ip>/
[tgt-macaddr]
```

Please read the official kernel doc for completeness here: https://www.kernel.org/doc/Documentation/networking/netconsole.txt. We leave the source and destination ports as the defaults.

Netconsole works by configuring a sender and sending all kernel printks to a receiver system across a network. Quite obviously, the sender is the one that sends data – the kernel printk content – across the network (over UDP) to the receiver system. Very briefly, I set up netconsole as follows:

- Set the Raspberry Pi 0W as the *sender* system. Strictly speaking, the sender should have a static IP address so that the receiver can reliably specify it. Here, to test, we don't bother setting it up. It does work (of course, the IP addresses and interface names that follow are examples, change them to suit your systems' addresses – do type this on one line):

    ```
    sudo modprobe netconsole netconsole=@192.168.1.24/
    wlan0,@192.168.1.101/
    ```

- Set the Linux host system (it could be your x86_64 guest VM) as the *receiver* (you can also set up a Win/Mac host as the receiver system, but I won't delve into that here). On the receiver system, you don't actually require the netconsole module to be installed. Simply run the netcat utility as follows to capture the incoming network stream (from the sender system) and log the data it receives to a file:

    ```
    netcat -d -u -l 6666 | tee -a dmesg_arm.txt
    ```

The options passed to `netcat` are as follows:

- `-d`: Won't attempt to read from standard input.

- `-u`: Use UDP (not TCP) as the transport layer protocol.

- `-l 6666`: Listens for an incoming connection on port number `6666` (the default target port for netconsole).

Now run the buggy module on the embedded board, while running `netcat` on the host receiver system (to capture the kernel log being sent by netconsole from the embedded system).

A Practical Consideration – ARM (Cross) Compiler Fails

I find that, quite often, when building the module for ARM (using the x86_64-to-ARM32 cross compiler, `arm-linux-gnueabihf-gcc`), it fails with an error of this sort:

`ERROR: modpost: "__aeabi_ldivmod" [<...>/ch7/oops_tryv2/oops_tryv2.ko] undefined!`

This seems to be an issue with the way division is carried out for ARM. A silly workaround is to just comment out code that performs division. Here, the `convenient.h:SHOW_DELTA()` macro. Once removed, it compiles correctly.

The following screenshot captures these steps being carried out (the top window with the light background is the embedded sender system, where we load netconsole and then our buggy module; the lower window with the darker background is the receiver host system where netcat is running):

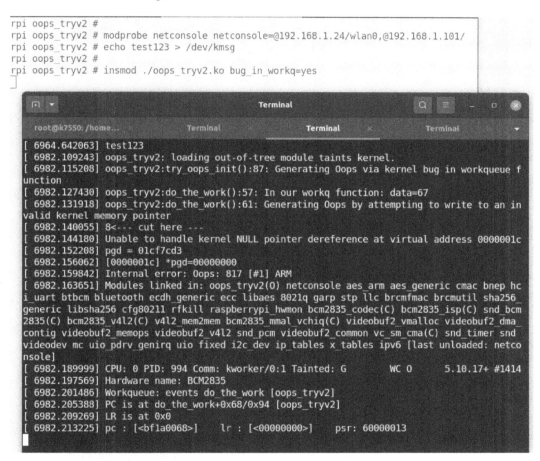

Figure 7.24 – Screenshot showing netconsole in action; the left, lighter window is the embedded sender; the right, darker window is the host running netcat

Voila. We capture the kernel log on the receiver system, and can now analyze it at leisure.

Though we don't delve into the details here, I'll point this out: the Oops bitmask here, the value `817` (it's in the line `Internal error: Oops: 817 [#1] ARM` – in hexadecimal), is definitely *not* interpreted the same way as we did for the x86. So how do you interpret it? You'll need to refer to the **Technical Reference Manual** (**TRM**) of the particular ARM core in question – here it's the core on the BCM2835, the ARM1176JZF-S. Interpreting its **Fault Status Register** (**FSR**) encoding is what needs to be done to see what it means. Here, it's a bit pedantic as the line `PC is at do_the_work+0x68/0x94 [oops_tryv2]` is the giveaway; moreover, using `[f]addr2line` will pinpoint the source line as well! This is what follows.

Figuring out the actual buggy code location (on ARM)

The key lines in the Oops diagnostic (from the ARM system's kernel, sent over netconsole to our receiver system) are the following:

```
Workqueue: events do_the_work [oops_tryv2]
PC is at do_the_work+0x68/0x94 [oops_tryv2]
LR is at irq_work_queue+0x6c/0x90
```

In this particular case, the system seems to have hard-hung before the call stack can even be shown! No matter: the line with **Program Counter** (**PC**) – the equivalent of the `RIP` register on the x86_64 – clearly tells us what occurred: the bug seems to have been triggered at the function `do_the_work()`, within the `oops_tryv2` module, at an offset of 0x68 (decimal 104) bytes from the start of this function; the length of this function is 0x94 (decimal 148) bytes. Interestingly, the **Link Register** (**LR**) on ARM specifies the return address. In effect, we can tell that the function `irq_work_queue()` called our workqueue routine `do_the_work()`! (Note that this info doesn't show up in *Figure 7.22* though. Also, as with the x86_64, the `Workqueue:` line tells us that the kernel's default `events` workqueue functionality was invoked to run our work routine.)

Okay, let's make good use of what we learned in the previous section and use some of the tooling at our disposal to actually identify the buggy line of code that triggered this Oops! We shall do so using three of these tools – `addr2line`, GDB, and `objdump`. Read on to see the magic!

On ARM with addr2line

On the device, let's run the powerful `addr2line` utility on our module, providing the start offset that the Oops reported (the value `0x68`, as you saw above) and, hey, the `rpi oops_tryv2 $` prefix is simply our shell prompt, as the environment variable `PS1=rpi \W $`:

```
rpi oops_tryv2 $ addr2line -e ./oops_tryv2.ko 0x68
</path/
to/>Linux-Kernel-Debugging/ch7/oops_tryv2/oops_tryv2.c:62
```

Here's a relevant snippet of the source file with line numbers prefixed:

```
61              pr_info("Generating Oops by attempting to write to
an invalid kernel memory pointer\n");
62              oopsie->data = 'x';
63      }
```

The offending buggy line is highlighted. The `addr2line` utility (this time running on ARM) has nailed it!

On ARM with GDB

On the device, let's now run GDB on our module, using its powerful `list` command to get the job done:

```
rpi oops_tryv2 $ gdb -q ./oops_tryv2.ko
Reading symbols from ./oops_tryv2.ko...done.
(gdb) list *do_the_work+0x68
0x68 is in try_oops_init (/home/pi/Linux-Kernel-Debugging/ch8/oops_tryv2/oops_tryv2.
c:62).
57              pr_info("In our workq function: data=%d\n", priv->data);
58              t2 = ktime_get_real_ns();
59      //      SHOW_DELTA(t2, t1);
60              if (!!bug_in_workq) {
61                      pr_info("Generating Oops by attempting to write to an invali
d kernel memory pointer\n");
62                      oopsie->data = 'x';
63              }
64              kfree(gctx);
65      }
66
(gdb) ▌
```

Figure 7.25 – Screenshot showing how ARM GDB's list command caught the offending source line

Look carefully at the output of the `list *do_the_work+0x68` command in the preceding screenshot. Again, exactly right!

On ARM with objdump

On the device, let's now run the `objdump` utility on our module (I've removed some empty lines from the following output):

```
rpi oops_tryv2 $ objdump -dS ./oops_tryv2.ko
./oops_tryv2.ko:    file format elf32-littlearm
Disassembly of section .text:
00000000 <do_the_work>:
/* Our workqueue callback function */
static void do_the_work(struct work_struct *work)
{
    0: e1a0c00d  mov   ip, sp
    4: e92dd800  push  {fp, ip, lr, pc}
    8: e24cb004  sub   fp, ip, #4
[...]
```

Scroll down to the nearest match to the start offset (as seen in the usual `funcname+x/y` format). Here, the x (offset) value is `0x68`:

```
   5c: ebffffe  bl    0 <printk>
oopsie->data = 'x';
   60: e3a03000   mov   r3, #0
   64: e3a02078   mov   r2, #120  ; 0x78
   68: e5c3201c   strb  r2, [r3, #28]
}
kfree(gctx);
```

Clearly, the offending source line that generated this (ARM) assembler is highlighted. Again, bang on target!

As mentioned earlier, a practical consideration is *what if you can't run these tools on the embedded device?* Then, you'll need to have the following:

- The unstripped debug kernel (`vmlinux` kernel image) with debug symbols available

- The cross toolchain with all tools/utilities used to generate the bootloader, kernel, and root filesystem images for your target

Again, this is why we highly recommend you always build and keep a debug kernel in addition to the production one.

When running other kernel helper scripts on an Oops generated on a non-x86 platform, don't forget to set the ARCH and CROSS_COMPILE environment variables appropriately (just as we do when cross-compiling). For example, to run the decode_stacktrace. sh script for Oops output generated on an ARM machine, do something like this:

```
ARCH=arm CROSS_COMPILE=arm-linux-gnueabihf-  scripts/decode_
stacktrace.sh < oops_from_arm.txt
```

One last thing: for a bit of variety, here's the same bug (and Oops) being triggered on another popular ARM development/prototyping board – the BeagleBone Black (from Texas Instruments)!

```
[20178.051346] oops_tryv2:try_oops_init():87: Generating Oops via kernel bug in workqueue function
[20178.064694] oops_tryv2:do_the_work():57: In our workq function: data=67
[20178.075333] oops_tryv2:do_the_work():61: Generating Oops by attempting to write to an invalid kernel me
mory pointer
[20178.097847] Unable to handle kernel NULL pointer dereference at virtual address 0000001c
[20178.107986] pgd = 7f31d0d1
[20178.110727] [0000001c] *pgd=00000000
[20178.116429] Internal error: Oops: 805 [#2] PREEMPT SMP ARM
[20178.121959] Modules linked in: oops_tryv2(O) oops_tryv1(O+) usb_f_acm u_serial usb_f_ecm usb_f_mass_sto
rage usb_f_rndis u_ether libcomposite wkup_m3_rproc pm33xx wkup_m3_ipc uio_pdrv_genirq uio pruss_soc_bus p
ru_rproc pruss irq_pruss_intc remoteproc virtio virtio_ring spidev
[20178.146455] CPU: 0 PID: 3912 Comm: kworker/0:1 Tainted: G      D    O     4.19.94-ti-r42 #1buster
[20178.155452] Hardware name: Generic AM33XX (Flattened Device Tree)
[20178.161596] Workqueue: events do_the_work [oops_tryv2]
[20178.166763] PC is at do_the_work+0x84/0xa0 [oops_tryv2]
[20178.172025] LR is at wake_up_klogd+0x7c/0xa8
[20178.176313] pc : [<bf10e084>]    lr : [<c01ac370>]    psr: 600f0013
[20178.182606] sp : dae09ee8  ip : dae09e10  fp : dae09efc
[20178.187853] r10: 00000000  r9 : dc761b10  r8 : 00000000
[20178.193100] r7 : df900a00  r6 : df8fd700  r5 : dc121200  r4 : dc761b0c
[20178.199654] r3 : 00000000  r2 : 00000078  r1 : c10ed348  r0 : 00000067
[20178.206212] Flags: nZCv  IRQs on  FIQs on  Mode SVC_32  ISA ARM  Segment none
[20178.213379] Control: 10c5387d  Table: 9c4cc019  DAC: 00000051
[20178.219155] Process kworker/0:1 (pid: 3912, stack limit = 0x77671b58)
[20178.225625] Stack: (0xdae09ee8 to 0xdae0a000)
[20178.230005] 9ee0:                   00000043 c0169a40 dae09f34 dae09f00 c0159b20 bf10e00c
[20178.238223] 9f00: df8fd700 df8fd700 df8fd700 dc121200 dc121214 df8fd700 00000008 df8fd718
[20178.246440] 9f20: c1504d00 df8fd700 dae09f74 dae09f38 c015aa84 c0159978 c0d3d4c8 c10e1598
[20178.254658] 9f40: c15dd636 ffffe000 c015ffb0 d9ed6cc0 d9ed65c0 00000000 dae08000 dc121200
[20178.262874] 9f60: c015aa24 d9a09e74 dae09fac dae09f78 c01604c0 c015aa30 d9ed6cdc d9ed6cdc
[20178.271091] 9f80: 00000000 d9ed65c0 c0160354 00000000 00000000 00000000 00000000 00000000
[20178.279308] 9fa0: 00000000 dae09fb0 c01010e8 c0160360 00000000 00000000 00000000 00000000
[20178.287524] 9fc0: 00000000 00000000 00000000 00000000 00000000 00000000 00000000 00000000
[20178.295741] 9fe0: 00000000 00000000 00000000 00000000 00000013 00000000 00000000 00000000
[20178.303997] [<bf10e084>] (do_the_work [oops_tryv2]) from [<c0159b20>] (process_one_work+0x1b4/0x504)
[20178.313180] [<c0159b20>] (process_one_work) from [<c015aa84>] (worker_thread+0x60/0x508)
[20178.321312] [<c015aa84>] (worker_thread) from [<c01604c0>] (kthread+0x16c/0x174)
[20178.328747] [<c01604c0>] (kthread) from [<c01010e8>] (ret_from_fork+0x14/0x2c)
[20178.335998] Exception stack(0xdae09fb0 to 0xdae09ff8)
[20178.341072] 9fa0:                   00000000 00000000 00000000 00000000
[20178.349289] 9fc0: 00000000 00000000 00000000 00000000 00000000 00000000 00000000 00000000
[20178.357504] 9fe0: 00000000 00000000 00000000 00000000 00000013 00000000
[20178.364154] Code: e34b0f10 eb427ace e3a03000 e3a02078 (e5c3201c)
[20178.387196] ---[ end trace 1218b813e308db06 ]---
```

Figure 7.26 – Screenshot showing the same Oops on a TI BeagleBone Black running Debian Linux with a custom 4.19.94 kernel

I'll leave it an exercise for you to carefully browse this screenshot and spot all the key points that we've spent so many pages discussing!

A tip: as mentioned earlier, interpreting the Oops bitmask on a system other than the x86 involves looking up the TRM for that processor. Here, the Oops bitmask shows up in this line (highlighted):

```
Internal error: Oops: 805 [#2] PREEMPT SMP ARM
```

So, we need to look up the TRM for the TI Sitara AM335x SoC (it's a Cortex-A8 core – that's the one this board's running on). Within it, the **Memory Protection Fault Status Register** (**MPFSR**) holds the error code, which shows up as the Oops bitmask! The relevant TRM document is available here as a PDF document: `https://www.ti.com/lit/ug/spruh73q/spruh73q.pdf`. The details on the internal encoding of the MPFSR register are available in section 11.4.1.87 (as of this writing, the relevant MPFSR register details can be found on page 1735 of this PDF document).

A few actual Oopses

Here are a few actual Oopses (that I've very arbitrarily searched for via the keyword `Oops`, from the **Linux Kernel Mailing List** (**LKML**) archives):

- A kernel NULL pointer dereference (on 4.14-rc2): `https://groups.google.com/g/linux.kernel/c/rG2uYWdoteo/m/6RacvsJ6BwAJ?hl=en`

- An Oops on 4.9.33 (read the email): `https://groups.google.com/g/linux.kernel/c/t4IRjnxo2Kc/m/7Me5AEVIBwAJ`

- An Oops flagged by Intel's superb kernel test robot (this has been reproduced on a QEMU-based x86-32, so look for the `EIP` register, not `RIP`!): `https://lkml.org/lkml/2020/8/10/1390`

- A recent one (as of this writing) – an Oops on 5.14.19: `https://lkml.org/lkml/2021/11/18/1116`

- An Oops on ARM64 while booting 5.8.0-rc5 (read the analysis): `https://lkml.org/lkml/2020/7/20/139`

Then there's the interesting **Linux Driver Verification** (**LDV**) project. They have a set of rules that are validated via their static and dynamic analysis frameworks, as well as other tooling. As far as kernel bugs go, this project has found several. They're documented here, under the heading *Problems in Linux Kernel*: `http://linuxtesting.org/results/ldv`. Do take a look!

Of course, you can always simply search the kernel Bugzilla site (`https://bugzilla.kernel.org/query.cgi`) for Oopses. Do note, though, that the kernel community really wants you to write your bug report directly to the appropriate mailing list (recall the `get_maintainer.pl` script) and copied to the LKML, not this site.

A key concern, of course, is *being able to obtain the kernel log* in the first place, else, how can you analyze and interpret the Oops or panic that possibly occurred. With a severe-enough bug, the saving of the kernel log to disk (or flash memory) may be compromised. For this reason, there are alternatives. Here's a short list:

- **Serial console**: The kernel printks are saved on another system across a physical serial console; it can also be a virtual serial console, as we saw in the previous section.

- **Netconsole**: A facility to enable the transfer of kernel printks across a network.

- Employing persistent RAM to save the kernel log buffer; for example, the kernel *Ramoops* framework has the kernel continually save kernel printks into a circular buffer in a persistent memory region (allowing the content to be accessed after a reboot). See the details in the official kernel doc here: `https://www.kernel.org/doc/html/latest/admin-guide/ramoops.html`.

- The kernel's elaborate kdump framework to capture the entire kernel memory image. This, along with the `crash` app to analyze it can be very powerful. We'll provide an introduction in *Chapter 12, A few more kernel debugging approaches*.

Many similar (to the ones mentioned above) independent implementations have been done by both individuals and organizations. You may come across some while working on projects or products.

Summary

Awesome! Great job on completing this really important chapter!

Here, you first learned what a kernel Oops is. You can perhaps think of it as the equivalent to a user-mode segfault, but as it's the kernel that's buggy, all guarantees are off. We began by showing you how to generate a simple NULL pointer dereference bug, triggering an Oops (though it may sound silly and obvious, these bugs still do occur – the last portion of this chapter points you to some actual Oopses, some of which are NULL pointer dereference bugs). We then went a bit further, triggering bugs in the NULL trap page and then in a random sparse region of kernel VAS (recall the useful `procmap` utility, which allows you to see the entire memory map of any process). Still further, more realistically, we used the kernel's default `events` workqueue to have a kernel worker thread illegally access an invalid pointer, causing an Oops (case 3)! We used this as a useful test case throughout the remainder of the chapter.

The meat of this topic – actually interpreting the detailed Oops diagnostic – was then covered in a lot of detail, with many screenshots to show you how it looks. Of course, being arch-specific, we covered it mainly from the viewpoint of the x86_64. We also covered generating and interpreting an Oops on ARM (32-bit) systems, using the Raspberry Pi 0W (and a quick look at the BeagleBone Black Oops screenshot) as a test board. Learning how to use various toolchain utilities and kernel helper scripts to help you debug the Oops was critical learning here. We even covered using the powerful netconsole facility along the way.

The chapter closed by pointing you to a few actual Oopses that are interesting to see. Importantly, we mentioned a few techniques to help capture the kernel log in situations where it can get problematic.

Needless to say (but I'll say it!), please do take the trouble to look at (and even generate!) actual Oopses, and practice using the various available tools and techniques to try and interpret them.

With this behind you, I'll see you in the following chapter where we'll look at another key topic – figuring out locking bugs.

Further reading

- My earlier book: *Linux Kernel Programming, Part 2 – Char Device Drivers and Kernel Synchronization* is freely downloadable as an e-book here: `https://github.com/PacktPublishing/Linux-Kernel-Programming/blob/master/Linux-Kernel-Programming-(Part-2)/Linux%20Kernel%20Programming%20Part%202%20-%20Char%20Device%20Drivers%20and%20Kernel%20Synchronization_eBook.pdf`

- VirtualBox serial console log file getting truncated? This Q&A seems related: *[Solved] How to create unique path for serial port log file*: `https://forums.virtualbox.org/viewtopic.php?f=1&t=86254`

- CPU Registers on the x86_64: `https://wiki.osdev.org/CPU_Registers_x86-64`

- Official kernel documentation: *Bug hunting*: `https://www.kernel.org/doc/html/latest/admin-guide/bug-hunting.html`

- KASLR:

  - Kernel address space layout randomization, LWN, Oct 2013: `https://lwn.net/Articles/569635/`

  - A brief description of ASLR and KASLR, Sep 2019: `https://dev.to/satorutakeuchi/a-brief-description-of-aslr-and-kaslr-2bbp`

- The Meltdown/Spectre hardware bugs:

  - Meltdown and Spectre: `https://meltdownattack.com/`

  - Spectre and Meltdown explained: A comprehensive guide for professionals, Tech Republic, May 2019: `https://www.techrepublic.com/article/spectre-and-meltdown-explained-a-comprehensive-guide-for-professionals/`

  - KPTI/KAISER Meltdown Initial Performance Regressions, B Gregg, Feb 2018: `https://www.brendangregg.com/blog/2018-02-09/kpti-kaiser-meltdown-performance.html`

- Netconsole:

  - Official kernel doc: `https://www.kernel.org/doc/Documentation/networking/netconsole.txt`

  - Also briefly covered here: Debugging by printing, eLinux: `https://elinux.org/Debugging_by_printing#NetConsole`

- Article: Much ado about NULL: Exploiting a kernel NULL dereference, Oracle Linux Blog, Apr 2010: `https://blogs.oracle.com/linux/post/much-ado-about-null-exploiting-a-kernel-null-dereference`

8
Lock Debugging

Imagine this: two threads, T1 and T2, running on different CPU cores, concurrently work upon a shared (global) writable data item. If one (or both) of these memory accesses is a write (a *store*), then congratulations, you've just witnessed a wily difficult-to-spot-and-catch bug or defect: a *data race*. This can happen in both user as well as kernel space. In the latter, the possibility of racing with both process (thread) and interrupt contexts arises as well.

A data race is a bug of course. What's worse, it's often a clue, or symptom, to the fact that there's often a higher-level issue or defect (like the proverbial tip of the iceberg). Untangling buggy code, finding the data race, fixing it (and finding any higher-level root defect) is necessary! As will be covered in detail, data races occur when a critical section in the code path is left unprotected. So how do you protect the critical section? **Locking** is one common way to do so (the Linux kernel provides several locking primitives: the mutex, the spinlock, and atomic and refcount-based locking on integers). Locking does cause performance issues – bottlenecks. Thus, the kernel also provides **lock-free** mechanisms to help overcome this. These include the usage of per-CPU data, lock-free data structures by design, and the powerful **Read Copy Update (RCU)** mechanism.

The Linux kernel is a complex beast indeed. With literally thousands of shared writable data items, the possibility of data races is very real. Correctly employing locking (or lock-free mechanisms) will ensure correctness. But of course, human programmers being, well, human, mistakes can and do occur (with somewhat alarming frequency at times). To be fair, concurrency is a really complex topic for our human minds. It's very hard to fully comprehend all its possible side effects. Marco Elver (currently the maintainer of **Kernel Concurrency Sanitizer (KCSAN)**), shows us in *Data-race detection in the Linux kernel*: `https://linuxplumbersconf.org/event/7/contributions/647/attachments/549/972/LPC2020-KCSAN.pdf` how several commits to the Linux kernel are to do with fixing or avoiding data races. Within the root of a recent (here, 5.15.0) Git-based kernel source tree, do the following:

```
git log-format=oneline v5.3..v5.15 |grep -iE '(Fix|avoid) .*[
-]race[ -]'|wc -l
197
```

Not just that, reliably reproducing, or even realizing, that a bug's root cause is a concurrency issue – a data race – can be very hard to do. They tend to be delicate timing coincidences. Further, these classes of defects are often called **Heisenbugs**: they subtly change or even disappear when being observed! (The name is inspired, of course, by the well-known **Heisenberg uncertainty principle** in quantum mechanics, the classic case being that the better an observer can predict an electron's position, the less they can figure out its momentum, and vice versa.) Adding instrumentation to a concurrency issue can cause Heisenbugs – the result: confusion in the mind of a developer new to this.

On a similar note, do realize that in the long run, for the product or project you're working on, detecting and fixing kernel data races is not sufficient; even userspace data races have to be detected and fixed! Tooling exists to do so: `helgrind`, **Thread Sanitizer (TSAN)**, and even **lockdep**. Though we don't cover them here (we do point you to lockdep's coverage), I'd urge you to familiarize yourself with their usage!

In this chapter, we shall focus on and cover the following main topics:

- Locking and lock debugging
- Locking – a quick summarization of key points
- Catching concurrency bugs with KCSAN
- A few actual use cases of kernel bugs due to locking defects

Technical requirements

The technical requirements and workspace remain identical to what's described in *Chapter 1, A General Introduction to Debugging Software*. The code examples can be found within the book's GitHub repository here: `https://github.com/PacktPublishing/Linux-Kernel-Debugging`.

We also make reference to the last two chapters of my earlier (free!) eBook, *Linux Kernel Programming – Part 2*. Its GitHub repository is available here: `https://github.com/PacktPublishing/Linux-Kernel-Programming-Part-2`.

Locking and lock debugging

As this book is explicitly meant for the subject matter of Linux kernel debugging, we don't even attempt to cover the basics of locking, why it's required, and the various kernel technologies that provide locking (which includes the mutex lock, the spinlock, atomic and refcount-based locking for integers, lock-free technologies such as per-CPU variables, RCU, and so on). Much, in fact, pretty much most, of this content, is covered in my earlier book (referred to below).

Further, a lot of material regarding the debugging of kernel-level locking issues, typically deadlock (of different types), and the tools to catch them (including **lockdep**, one of the most powerful!), is also covered in detail in this earlier book. If you're new to these topics, I urge you to refer to the *Linux Kernel Programming – Part 2* book (it's free to download as an eBook), for our purposes here, particularly these sections:

- *Chapter 6, Kernel Synchronization – Part 1*, covers the basics regarding locking in a lot of detail: what a critical section is, locking concepts and terminology, concurrency concerns in kernel space, how to actually use the mutex lock and spinlock APIs, and more (including using locks in both process and interrupt contexts). Also, I provide a summarization of important points in the next *Locking – a quick summarization of key points* section.

- *Chapter 7, Kernel Synchronization – Part 2*, in the *Lock debugging within the kernel* section.

- This chapter also has coverage of more advanced locking, including lock-free techniques.

The cool thing is that the LKP-2 e-book is *free to download*! Here's the link: `https://github.com/PacktPublishing/Linux-Kernel-Programming/blob/master/Linux-Kernel-Programming-(Part-2)/Linux%20Kernel%20Programming%20Part%202%20-%20Char%20Device%20Drivers%20and%20Kernel%20Synchronization_eBook.pdf`. You can also download the Kindle edition from Amazon for free. Quick, grab it!

Locking – a quick summarization of key points

As mentioned just above, do refer to the *Linux Kernel Programming – Part 2* book to brush up on the basics of locking within the Linux kernel (if you need to) – more importantly, on kernel-level debug techniques as well as guidelines on preventing and detecting dangerous locking bugs such as the deadly deadlock.

Nevertheless, I'd like to summarize some really key points with respect to locking here as well. Here they are:

- A **critical section** is a code path that can execute in parallel and that works on (reads and/or writes) shared writeable data (also known as **shared state**).

- Because it works on shared writable data, *the critical section requires protection* from the following:

  - Parallelism (that is, it must run alone, serialized, in a mutually exclusive fashion)

  - When running in an atomic (for example, interrupt) non-blocking context, it *must run atomically*: indivisibly, to completion, without interruption. To do so, every critical section in the code path needs to be first identified and then protected from concurrent access; how? The point that follows covers this. Every critical section in the code base must be identified and protected:

    - Identifying critical sections is critical! Carefully review your code and make sure you don't miss them.

    - Protecting them can be achieved via various technologies. One very common technique is **locking** (there's also a more efficient technology, lock-free programming).

    - A common mistake is only protecting critical sections that *write* to global writeable data. You are required to also protect critical sections that read global writeable data; otherwise, you risk *a torn or dirty read*!

- Another deadly mistake is not using the same lock to protect a given data item. Alternatively, using the incorrect lock variable, in effect, the wrong lock.

- Failing to protect critical sections leads to a **data race**, a situation where the outcome – the actual value of the data being read/written – is *racy*, which means it varies, depending on runtime circumstances and timing. This is a bug – a bug that, once in "the field," is extremely difficult to see, reproduce, determine its root cause, and fix. Also, there's more to a data race; do see the section that follows, *What exactly is a data race?*.

- **Exceptions**: You are safe (implicitly, without explicit protection) in the following situations:

  - When you are working on local variables. They're allocated on the private (kernel) stack of the process/thread context (or, in the interrupt context, on the local IRQ stack) and are thus, by definition, safe.

  - When you are working on shared writeable data in code that cannot possibly run in another context; that is, it's serialized by nature. In our context, the init and cleanup methods of an LKM qualify (they run exactly once, serially, on `insmod` (or `modprobe`) and `rmmod` only).

  - When you are working on shared data that is truly constant and read-only (don't let C's `const` keyword fool you, though!).

  - Places where the use of plain C memory accesses are fine are documented here: *MARKING SHARED-MEMORY ACCESSES* in the *Use of Plain C-Language Accesses section*: `https://git.kernel.org/pub/scm/linux/kernel/git/torvalds/linux.git/tree/tools/memory-model/Documentation/access-marking.txt`.

Locking is inherently complex. You must very carefully think about, design, and implement your code to avoid deadlocks. Again, do refer to the *Linux Kernel Programming – Part 2* book, *Chapter 6, Kernel Synchronization – Part 1*, in the *Locking guidelines and deadlocks* section, for details. Also, do check out some useful links on locking in the *Further reading* section of this chapter.

Understanding data races – delving deeper

As you dig deeper into this complex topic, you'll discover there's a lot more to learn!

What exactly is a data race?

The Linux kernel has its own memory model called the **Linux Kernel's Memory (Consistency) Model (LKMM)**. With this model, there's a more precise way to define a **data race** – it's a situation where two (or more) memory accesses occur so that the following applies:

- Both access the same memory location (address).

- Both occur concurrently (in parallel).

- (At least) one access is a write (or store) operation.

- (At least) one access is a plain C-language access.

- They run on different CPU cores (or within different threads on the same core).

The screenshot that follows makes it clear:

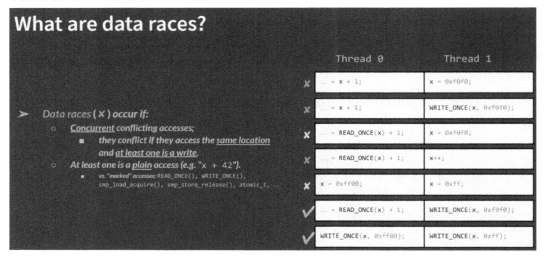

Figure 8.2 – Screenshot from M Elver's presentation on data race detection in the kernel

(Credit: the above slide is from Marco Elver's presentation *Data-race detection in the Linux kernel*, at the Linux Plumbers Conference, August 2020: https://linuxplumbersconf.org/event/7/contributions/647/attachments/549/972/LPC2020-KCSAN.pdf.)

In the preceding screenshot, an X implies a (red color) data race and a (green) check mark implies there's none. Here's a highly abbreviated set of key points:

- In *Figure 8.2*, the first row in the table is an obvious data race: **Thread 0** is reading the shared writable variable x while **Thread 1** is concurrently writing to (updating) it.

- A normal C statement accessing a variable is termed a **plain access**. Memory accesses wrapped via the READ_ONCE(), WRITE_ONCE(), and similar macros are termed **marked accesses**. Such accesses are designed to be inherently atomic (thus safe from concurrency) and follow the LKMM. See the *Further reading* section for links to more on this. (In fact, it's not just these macros; the atomic_*(), refcount_*(), smp_load_acquire(), and similar macros fall into the class of marked accesses.)

- See the second row; it's a data race as – as already explained – at least one of the threads is performing a plain access at the same location, in parallel, and one of them is a write access.

- Rows 3, 4, and 5 have a similar explanation to row 2.

- Rows 6 and 7 (the last two) are fine (no data races) as they use only marked accesses, resulting in code that is atomic with memory-ordering guarantees.

A Lot to Read and Learn!

These details should have you thinking harder about concurrency and its myriad (and complex!) related areas: concurrency general concepts, memory ordering and memory barriers, the LKMM, marked accesses, lock-free programming and how it works, and so on. We have neither the scope nor space to cover these in detail here; some have been covered in my earlier book *Linux Kernel Programming – Part 2* (a freely downloadable eBook). Do refer to the *Further reading* section in this chapter for links to articles to help sate your thirst for knowledge in these areas!

With all this complexity, we could surely do with some help! The next section covers a powerful kernel framework that does indeed help.

Catching concurrency bugs with KCSAN

The **Kernel Concurrency Sanitizer** (**KCSAN**) is a powerful kernel framework for helping catch data races within the Linux kernel (and modules). It was merged into the kernel in the 5.8 series (Aug 2020). It currently works on the x86_64 platform with support for ARM64 being very recent (the 5.17 kernel, March 2022).

What KCSAN does, in a nutshell

KCSAN figures out data races (if you haven't already, please first read this section: *What exactly is a data race?*) and reports them. In a nutshell, KCSAN treats all aligned writes up to the processor word size as atomic (regardless of whether they're plain or marked accesses). In effect, KCSAN works by checking for unmarked (or plain) reads that race with these writes (that is, any write to the same address where the unmarked read occurred)!

KCSAN is essentially a robot that (with the help of a syzbot instance), continuously scans the kernel's main branches, setting up watchpoints on memory locations that are accessed, teasing out patterns that will result in a data race and reporting them to the kernel log. Given that it scans pretty much all kernel memory locations, **it uses a statistical approach** to this (see more on this below, as well as in the *Enabling KCSAN section*, especially the tunable named CONFIG_KCSAN_SKIP_WATCH), in the hope that it will catch many, if not most, data races. (As a rule of thumb, KCSAN will by default only check a memory access once every 2,000 times it occurs; else, the system would be effectively unusable.) The default config tuning does work very well indeed (see the section just mentioned for some details) and thus, so does KCSAN. This scanning has been going on since October 2019! Here's where you can peek at the results being generated: https://syzkaller. appspot.com/upstream?manager=ci2-upstream-kcsan-gce.

> **What the Heck is a Syzbot?**
>
> A **syzbot** (short for **syzkaller robot**) is essentially a fuzzing technology that continually fuzzes the main branches of the Linux kernel, teasing out bugs and reporting them to a dashboard (a web interface). They're also copied to a syzkaller-bugs mailing list. Details here: https://github.com/ google/syzkaller/blob/master/docs/syzbot.md.

(We cover a bit more on fuzzing in *Chapter 12, A few more kernel debugging approaches,* in the *What is fuzzing?* section).

Concurrency bugs are by definition hard to observe, as they depend on delicate timing coincidences. In order for KCSAN to detect them, it has to introduce deliberate (small) delays into code paths and set up various watchpoints (via compiler instrumentation and so-called soft watchpoints). It's like this: KCSAN sets up a (soft) watchpoint on a given memory address. Then, when memory accesses are made to this address, they're deliberately stalled for a short tunable duration. This delay is the value of the kernel config KCSAN_UDELAY_TASK for task delays and defaults to 80 microseconds (whereas interrupt delays are just 20 microseconds by default).

Now, if two threads (or interrupt contexts, or one of each) access the same memory location, both read/write watchpoints are triggered, and we have a race! KCSAN then checks, and if the conditions for a data race are fulfilled (see the *What exactly is a data race?* section), it reports it as a defect. If the content at that address changes, it reports the old and the new data values. It also shows a stack trace of both the racing threads or interrupt contexts (to help you see how they landed up here). If the memory accesses are marked, no watchpoint is set up. The explanation below on plain and marked accesses is deliberately kept brief; you can learn more about the inner workings here: `https://www.kernel.org/doc/html/latest/dev-tools/kcsan.html#implementation-details`. The article *Finding race conditions with KCSAN*, Jonathan Corbet, LWN, 14 Oct 2019: `https://lwn.net/Articles/802128/`, does a stellar job of explaining how KCSAN works – do look it up!

As mentioned above, KCSAN works by checking for unmarked (plain) reads that race with any write (marked or plain) to the same address. A marked access is a racy access being marked as legal. This isn't right according to the strict LKMM definition: there, *any* unmarked write that's concurrent with *any* read to the same address constitutes a race. If this is what's required (tip: it usually isn't; it's a very strict way to use KCSAN), then set the following like this in your kernel config (worry not, we cover configuring KCSAN in the very next section):

```
CONFIG_KCSAN_ASSUME_PLAIN_WRITES_ATOMIC=n
CONFIG_KCSAN_REPORT_VALUE_CHANGE_ONLY=n
CONFIG_KCSAN_INTERRUPT_WATCHER=y
```

KCSAN does incur significant overhead; the most significant tunable that affects the performance overhead is `CONFIG_KCSAN_SKIP_WATCH`. It specifies the number of per-CPU memory operations to skip before it sets up another watchpoint; the default is 4,000. The smaller this value is, the more precise and aggressive KCSAN will be in catching data races. This will be at the cost of larger system overhead (there's always a trade-off, right?).

Also realize that KCSAN, precisely because it works by using a statistical approach (that is, it only actually checks memory accesses once in a while), can indeed miss many data races. This is why you should run your test cases with KCSAN enabled over a long period of time, thus increasing the chances of races being caught.

Configuring the kernel for KCSAN

To enable KCSAN, simply set the kernel config `CONFIG_KCSAN=y`. This config has non-trivial dependencies though, which need to be fulfilled.

Dependencies for enabling KCSAN

The dependencies for enabling KCSAN are summarized here:

- Arch: currently supported only on x86_64, with support for ARM64 coming in very recently (the 5.17 kernel, March 2022.)

- Kernel version: x86_64: 5.8 (Aug 2020) or later; ARM64: 5.17 or later.

- Compiler-wise, KCSAN requires GCC or Clang version 11 or later. The `CONFIG_HAVE_KCSAN_COMPILER` config directive encodes these requirements by checking for specific features being supported (this is within the `lib/Kconfig.kcsan` file).

- Further, you'll need to turn on kernel debug (via the `CONFIG_DEBUG_KERNEL` option). Note though, that `CONFIG_DEBUG_KERNEL=y` merely makes kernel debug features – within the `Kernel Hacking` menu – available for config; it doesn't automatically enable anything by itself. So, in your custom debug kernel, ensure you enable KCSAN.

- The KCSAN dependencies can be seen via its config file, `lib/Kconfig.kcsan`:

```
menuconfig KCSAN
bool "KCSAN: dynamic data race detector"
depends on HAVE_ARCH_KCSAN && HAVE_KCSAN_COMPILER
depends on DEBUG_KERNEL && !KASAN
depends on !KCSAN_KCOV_BROKEN
select STACKTRACE
```

Interestingly, KASAN (we covered KASAN in depth in *Chapter 5, Debugging Kernel Memory Issues – Part 1*) and KCSAN don't get along; *you can enable either or neither but not both at the same time.*

The config directive `CONFIG_KCSAN_KCOV_BROKEN` tells us that Clang can support either KCSAN or **KCOV** (the **kernel coverage** tool) but not both together.

Finally, selecting KCSAN also turns on `CONFIG_STACKTRACE` (via the `select STACKTRACE` directive), enabling detailed call traces (as part of its data race reports).

Right, assuming your x64 system has all these dependencies fulfilled, let's enable it.

Enabling KCSAN

To enable KCSAN with the usual `make menuconfig` UI, look for it here: `Kernel hacking|Generic Kernel Debugging Instruments|KCSAN: dynamic data race detector`. Note that if KCSAN doesn't even show up in the menu, it's likely that your system doesn't fulfill all the dependencies (see the *Further reading* section for a link on installing GCC-11 on Ubuntu). Quick tip: to ensure the basic dependencies are fulfilled, I simply worked on an x86_64 Ubuntu 21.10 VM.

Click (press *Enter*) on it; a screenshot of the x86_64 KCSAN sub-menu (with its config values all set to defaults) follows:

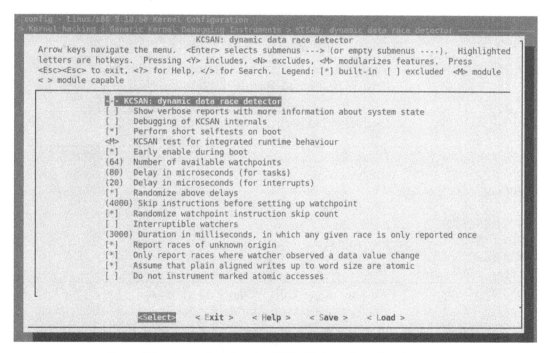

Figure 8.3 – Screenshot showing KCSAN's kernel config sub-menu

The default values for KCSAN tunables are deliberately set to be conservative. A few (four) of them can be overridden by updating them at runtime as they're treated as kernel module parameters as well, here: `/sys/module/kcsan/parameters/`. We show them in the following table by mentioning the module parameter name in the leftmost column in parentheses (as `kcsan.<foo>`). You can find these tunables and their details in the kernel config file `lib/Kconfig.kcsan`. The following table summarizes some of them:

Kernel config (or kernel module parameter)	Meaning	Default value
`CONFIG_KCSAN_VERBOSE`	Show more in the report (including locks held and IRQ trace events); can cause instability.	`n`
`CONFIG_KCSAN_SELFTEST`	Runs KCSAN self-tests at boot time; kernel panics on failure.	`y`
`CONFIG_KCSAN_TEST`	Various KCSAN test cases (internally uses the kernel's KUnit and Torture test frameworks); can be built as a module by specifying m here.	`n`
`CONFIG_KCSAN_EARLY_ENABLE`	Enables KCSAN during early boot.	`y`
`CONFIG_KCSAN_UDELAY_TASK (or kcsan.udelay_task)`	Delay after setting up a watchpoint, for tasks (microseconds).	`80`
`CONFIG_KCSAN_UDELAY_INTERRUPT (or kcsan.udelay_interrupt)`	Delay after setting up a watchpoint, for interrupts (microseconds).	`20`
`CONFIG_KCSAN_SKIP_WATCH (or kcsan.skip_watch)`	Number of per-CPU memory operations to skip before it sets up another watchpoint; this tunable has the most impact on system performance and detecting data races. A smaller number implies better race detection with more degradation in system performance (and vice versa).	`4000`
`CONFIG_KCSAN_INTERRUPT_WATCHER (or kcsan.interrupt_watcher)`	When enabled, a task that set up a watchpoint can be interrupted while delayed allowing detection of races in this situation. Disabled by default (safer, else can generate false positives).	`n`

CONFIG_KCSAN_ REPORT_ONCE_IN_MS	Rate limiting of data race reports (set to a duration of 3 s by default) to avoid flooding the console and log buffers with reports; set to 0 to disable rate limiting.	3000
CONFIG_KCSAN_ REPORT_RACE_ UNKNOWN_ORIGIN	If yes (default), report races where only one access is known (the others are unknown); only reported if the data value changed while delayed.	y
CONFIG_KCSAN_ REPORT_VALUE_ CHANGE_ONLY	Only report a data race when the data value changed (implying that, if a conflicting write was seen but the data value remained unchanged, don't report it).	y
CONFIG_KCSAN_ ASSUME_PLAIN_ WRITES_ATOMIC	If yes (the default), assume that plain aligned writes up to the processor word size are atomic; turning this off results in more reports (stricter mode). You'll find that this needs to be changed to n to test a simple two-plain-integer-writes data race.	y
CONFIG_KCSAN_ IGNORE_ATOMICS	Don't check marked atomic accesses; has implications on what is reported as a data race (for example, this, along with CONFIG_ KCSAN_REPORT_RACE_UNKNOWN_ ORIGIN=n, implies that races where at least one access is marked atomic never get reported).	n

Table 8.1 – Summary of some of the KCSAN kernel config tunables

---

Getting the Details on KCSAN Internal Configs

More details (in increasing verbosity) on each of the KCSAN configs (mentioned in the preceding table) can be found here:

`lib/Kconfig.kcsan`

KCSAN official kernel doc: `https://www.kernel.org/doc/html/ latest/dev-tools/kcsan.html`

Concurrency bugs should fear the big bad data-race detector (part 1), LWN, Apr 2020: `https://lwn.net/Articles/816850/`

Once configured, build and boot from your shiny new KCSAN-enabled kernel in the usual manner. (I've configured, built, and booted from a debug 5.10.60 kernel with KCSAN enabled. To ensure am using GCC >= 11, I did this on my x86_64 Ubuntu 21.10 VM.) Now, let's get going using KCSAN to catch those naughty data races.

Using KCSAN

Once you're all set up to run KCSAN, you can start catching these dangerous data races (bugs!) with KCSAN. I'll assume that by now you're running a debug kernel with KCSAN enabled.

A simple data race test case

We write a simple test case within a module (it's based on our earlier ch7/oops_tryv2 code, in fact). You'll find the code here: ch8/kcsan_datarace. To set up a data race, we'll need (at least) two contexts that do actually race on a piece of global data. So, this time, we have our setup_work() function initialize and schedule work on two (kernel-default) events work queues. Thus, we'll have two kernel worker threads that consume the work functions, do_the_work1() and do_the_work2(). Within these functions, provided a Boolean module parameter (named race_2plain_w) is set to y, we race on a global data variable! Next, we parametrize the number of times we loop in each of the work queue functions, operating upon the shared global gctx->data, via the module parameters iter1 and iter2 respectively. The reason is so that we can test with different values to see when KCSAN – with its statistical approach, setting up watchpoints after every CONFIG_KCSAN_SKIP_WATCH per-CPU memory accesses – actually catches our data race!

Here's the relevant code, showing the actual data race:

```
// ch8/kcsan_datarace.c
[...]
static void do_the_work1(struct work_struct *work1)
{
    int i; u64 bogus = 32000;
    PRINT_CTX();
    if (race_2plain_w) {
        pr_info("data race: 2 plain writes:\n");
        for (i=0; i<iter1; i++)
            gctx->data = (u64)bogus + i;
            /* unprotected plain write on global */
```

```
    }
}
static void do_the_work2(struct work_struct *work2)
{
    int i; u64 bogus = 98000;
    PRINT_CTX();
    if (race_2plain_w) {
        pr_info("data race: 2 plain writes:\n");
        for (i=0; i<iter2; i++)
            gctx->data = (u64)gctx->y + i;
                /* unprotected plain write on global */
    }
}
```

As you can see, each work function performs a C language plain write. It races as they're concurrent, unprotected, and the plain write is upon the same address.

But guess what? **KCSAN doesn't catch this (pretty glaring) data race**. How come? The kernel config CONFIG_KCSAN_ASSUME_PLAIN_WRITES_ATOMIC is set to y by default (see *Table 8.1*). It turns out that this is the issue. Here's a (truncated) screenshot of its help screen. Do scan it carefully (you'll find this help text is in lib/Kconfig.kcsan):

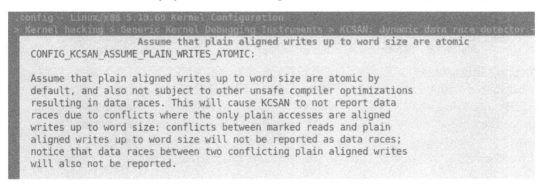

Figure 8.4 – Truncated screenshot of the Help screen for the CONFIG_KCSAN_ASSUME_PLAIN_
WRITES_ATOMIC kernel config option

So, if this is the case for you, go back to your debug kernel source tree, use make menuconfig and toggle the KCSAN_ASSUME_PLAIN_WRITES_ATOMIC config to *off*. Now, *KCSAN no longer assumes that plain writes up to word size are atomic.*

Rebuild and reboot; `insmod` our test module (passing the module parameters as appropriate) and *now* KCSAN indeed catches our simple data race! The following screenshot shows its report in all its glory:

```
kcsan_datarace $ sudo rmmod kcsan_datarace 2>/dev/null; sudo dmesg -C; sudo insmod ./kcsan_datarace.ko race
_2plain_w=y iter1=50000 iter2=30000; dmesg
[ 6441.048400] kcsan_datarace:kcsan_datarace_init():109: Setting up a deliberate data race via our workqueu
e functions:
[ 6441.048409] kcsan_datarace:kcsan_datarace_init():111: 2 plain writes; #loops in workfunc1:50000 workfunc
2:30000
[ 6441.048415] kcsan_datarace:setup_work():84: global data item address: 0xffff9fc3cc9e3238
[ 6441.048730] kcsan_datarace:do_the_work1():58: 005) [kworker/5:1]:69    |  ...0   /* do_the_work1() */
[ 6441.048792] kcsan_datarace:do_the_work1():60: data race: 2 plain writes:
[ 6441.052375] kcsan_datarace:do_the_work2():74: 001) [kworker/1:0]:5785   |  ...0   /* do_the_work2() */
[ 6441.052396] kcsan_datarace:do_the_work2():76: data race: 2 plain writes:
[ 6441.052448] =================================================================
[ 6441.056772] BUG: KCSAN: data-race in process_one_work / process_one_work

[ 6441.065308] write to 0xffff9fc3cc9e3238 of 8 bytes by task 69 on cpu 5:
[ 6441.069638]  process_one_work+0x4ee/0xa60
[ 6441.069643]  worker_thread+0x320/0x770
[ 6441.069647]  kthread+0x225/0x250
[ 6441.069653]  ret_from_fork+0x22/0x30

[ 6441.073846] write to 0xffff9fc3cc9e3238 of 8 bytes by task 5785 on cpu 1:
[ 6441.078131]  process_one_work+0x4ee/0xa60
[ 6441.078136]  worker_thread+0x320/0x770
[ 6441.078140]  kthread+0x225/0x250
[ 6441.078146]  ret_from_fork+0x22/0x30

[ 6441.082488] Reported by Kernel Concurrency Sanitizer on:
[ 6441.086869] CPU: 1 PID: 5785 Comm: kworker/1:0 Tainted: G        O      5.10.60-dbg02-kcsan #8
[ 6441.086873] Hardware name: innotek GmbH VirtualBox/VirtualBox, BIOS VirtualBox 12/01/2006
[ 6441.086882] Workqueue: events do_the_work2 [kcsan_datarace]
[ 6441.086887] =================================================================
kcsan_datarace $
```

Figure 8.5 – Screenshot showing KCSAN catching our 2 plain writes data race

The preceding screenshot shows the actual (kernel virtual) address of our chosen global shared writable data item (`gctx->data`; it happens to be the value `0xffff9fc3cc9e3238` here). This is confirmed by KCSAN's report as well. Also, you can see the output from our `PRINT_CTX()` macro showing the context when the kernel worker threads (belonging to the kernel-default `events` work queue) execute. This is followed by KCSAN's report. How do you interpret it? Read on!

Interpreting KCSAN's report

KCSAN's typical report format is as follows:

```
BUG: KCSAN: data-race in func_x / func_y
```

This line (in red) implies that the functions `func_x()` and `func_y()` are involved in a data race! Above, it's the `process_one_work()` function for both threads as this is the underlying kernel routine that has both our (`events` work queues') kernel worker threads consume their work function.

Next, you'll see lines in the format (and in red):

```
read/write [(marked)] to <kernel-virt-addr> of <n> bytes by
task <PID> on cpu <CPU#>
```

This is followed by the kernel stack trace for `func_x()`, with `func_x()` as the topmost frame, showing, in effect, how we got here:

```
[ ... kernel stack call frames for func_x() ]
```

This is followed by the same format for the other function involved in the data race (with the first line in red):

```
read/write [(marked)] to <kernel-virt-addr> of <n> bytes by
task <PID> on cpu <CPU#>
[ ... kernel stack call frames for func_y() ]
```

These lines – and the fact that in both conflicting threads it's a (plain) write, as well as the kernel stack frames leading up to the race – are clearly seen in *Figure 8.5*.

The report indicates whether it was a read or write operation that was involved; further, if the token `(marked)` is seen after the word `read` or `write`, it indicates a marked read or write access; else, it was a plain access. The location – kernel virtual address – where the data race occurred is then seen, along with the number of bytes actually being read or written, followed by the CPU core on which the racy accesses ran. Next, follows these lines:

```
Reported by Kernel Concurrency Sanitizer on:
[...]
```

This is essentially a summary paragraph where KCSAN shows a few more details: the process/interrupt context details as well as the hardware details along with any other relevant information (in our test case, the fact that it occurred on the kernel-default `events` work queue is recorded; again, see *Figure 8.5*). That's it for the KCSAN report – brief and to the point.

This isn't the only manner in which KCSAN reports data races. If the concerned data item's value change has been seen, KCSAN will report it with both the old and new values. Further, if the kernel config CONFIG_KCSAN_REPORT_RACE_UNKNOWN_ORIGIN is y (it is by default), then KCSAN reports data races even where it cannot determine one or both of the racing threads (or entities). Do refer to the official kernel documentation on KCSAN for more on these aspects: https://www.kernel.org/doc/html/v5.10/dev-tools/kcsan.html.

You can see reports of plenty of actual data races caught (and subsequently fixed) by KCSAN here: https://github.com/google/kernel-sanitizers/blob/master/KCSAN.md#upstream-fixes-of-data-races-found-by-kcsan.

Running our data race test case with a wrapper script

We write a simple wrapper script over the test case in order to test, in this particular case, how many loop iterations are required in the racy code paths before KCSAN catches the (two plain writes) data race. The bash script is located here: ch8/kcsan_datarace/tester.sh. Its code is simple, please check it out. These are the results I obtained when I ran the script in this manner:

```
sudo ./tester.sh 1 10000 5000
```

The parameters to this script are (in order) the number of trial runs to execute, the number of times to loop in work function 1, and the number of times to loop in work function 2 respectively (the latter two values become the value of the iter1 and iter2 module parameters respectively). Here are my (limited) findings:

# loops in workfunc 1	# loops in workfunc 2	KCSAN catches the data race?
10,000	5,000	No
20,000	10,000	Yes
75,000	50,000	Yes

Table 8.2 – Effect of differing loop iterations via our tester.sh script wrapper over the kcsan_datarace module

I deliberately ran it with only one trial due to how the config CONFIG_KCSAN_REPORT_ONCE_IN_MS works. It's set to the value 3000 by default; this is a rate-limiting construct. In effect, KCSAN reports data races once within a 3-second interval. This is why, when I run the module via my test script more than once in quick succession, KCSAN only reports the data race once, on the first trial run. To counter this, the script is programmed to perform a sleep of just over 3 seconds on every loop iteration.

So, in *Table 8.2*, the middle row represents the (approximate) minimum number of loop iterations before KCSAN catches this data race. You will realize that this isn't a general conclusion and is very particular to this specific test case. Its results will almost certainly vary on different systems. Don't read too much into this; our wrapper script is simply a means to let you more easily test, that's all. (*Quick tip*: if you're not getting the data race caught by KCSAN, try using large values for the 'number of loops' parameters).

Moving along, the kernel has a module to test KCSAN in depth; if you set `CONFIG_KCSAN_TEST=m`, the `kcsan-test.ko` module gets installed. It employs both the kernel KUnit and Torture test frameworks to conduct a large number of test cases (literally torturing the system with concurrency bugs). Running it can take a while (close to 7 minutes on my x86_64 VM) and you'll have plenty of KCSAN data race reports to look through. The source of this test module is here: `kernel/kcsan/kcsan-test.c`. I'll leave it to you to browse through its source and try it out on your KCSAN-enabled Linux system.

Runtime control via debugfs

KCSAN makes the `/sys/kernel/debug/kcsan` pseudofile available (under debugfs); reading or writing to it has an effect. The following table summarizes just this (you'll need root access):

Action on `/sys/kernel/debug/kcsan`	Effect
Reading it	Shows statistics regarding KCSAN runtime; includes the number of watchpoints, data races detected, blacklisted functions, and so on
Writing `on` / `off` to it	Toggles KCSAN on/off
Writing `!funcname` to it	Blacklists reporting any data race where the function `funcname` is one of the top stack frames in either function involved in the race
Writing `blacklist`	Stop reporting frequently occurring data races
Writing `whitelist`	Keep reporting frequently occurring data races; helpful for testing/reproducing data races

Table 8.3 – Summary of action and effect on the KCSAN debugfs pseudofile /sys/kernel/debug/kcsan

Let's complete our coverage of KCSAN with a really key point – learning more on how to, or rather, how not to, react to its data race reports!

Knee-jerk reactions to KCSAN reports – please don't!

A key point: *don't* react in a knee-jerk fashion to a KCSAN error report, blindly attempting to fix the issue by using the READ_ONCE(), WRITE_ONCE(), and/or data_race() macros – them making the racy accesses legal as they're now deemed marked accesses – in your code.

Why not? The premise is that reads and writes to shared variables are not supposed to race. If you mark every (or almost every) shared variable memory access with the READ_ONCE() or WRITE_ONCE() macros, this in effect *prevents* KCSAN from detecting buggy races that they may encounter! *Thus it's important that they not be protected via these macros and that the reads/writes they perform should be in plain C language.* Instead, we expect that your memory accesses are correctly protected by design and via your code-level implementation (perhaps by using a mutex, spinlock, or atomic_t / refcount_t primitive, a lock-free technique such as RCU or per-CPU variables, whatever). Moreover, KCSAN reporting a data race is often a precursor, a hint, to the fact that a (severe) logic bug exists in the code. Simply turning it off by using the READ_ONCE() or WRITE_ONCE() macros would do everybody a grave injustice.

Sometimes, on the other hand, you know there's a data race in the code but either it's benign or doesn't really matter (for example, statistics/diagnostic code paths performing racy reads on shared variables being referenced within a sysfs or procfs pseudofile). In cases like this, it's better that KCSAN is made aware of the fact and ignores them. This is achieved by using the data_race() macro, marking the racy code as being intentional (lockdep uses it in places as well). Here's an example of usage from the process/thread creation code path in the kernel:

```
// kernel/fork.c:
/* If multiple threads are within copy_process(), then this
check triggers too late. This doesn't hurt, the check is only
there to stop root fork bombs. */
retval = -EAGAIN;
if (data_race(nr_threads >= max_threads))
    goto bad_fork_cleanup_count;
```

A list of bugs found by KCSAN (link: `https://github.com/google/kernel-sanitizers/blob/master/kcsan/FOUND_BUGS.md`) includes several including the phrase *"annotate data race"* in their commit header. These tend to be done by using the techniques mentioned just above.

Another way is to mark an entire function as not being a candidate for race detection by KCSAN by prefixing the function with the `__no_kcsan` compiler attribute. There are several ways to selectively enable/disable code from KCSAN detection. Please refer to this link for more: `https://www.kernel.org/doc/html/v5.10/dev-tools/kcsan.html#selective-analysis`. Of course, you're expected to not abuse these features!

For completeness, I'd definitely recommend you read the detailed documentation in these LWN articles: *Concurrency bugs should fear the big bad data-race detector (part 1)*, LWN, Apr 2020: `https://lwn.net/Articles/816850/` (the section entitled *How to use KCSAN* in this first part shows an example of catching a deemed data race with the strict settings on). More in-depth details on various strategies to practically use KCSAN (for kernel maintainers and developers) are covered in the second part of this LWN article series: *Concurrency bugs should fear the big bad data-race detector (part 2)*, LWN, Apr 2020: `https://lwn.net/Articles/816854/`.

Another point: we all understand the usage of a lock to protect a critical section. It is key to realize, of course, that the entire safety aspect is based on the premise that the two (or more) parties (processes/threads/interrupt contexts) accessing the shared memory region concurrently actually perform the access while the very same lock is being held. This is obvious. But, what if one party does *not* hold the lock and performs the shared memory access? These locks are all *advisory* of course; the access would go through. The result is a data race! Now, traditional tooling that depends on dynamic runtime checking (such as lockdep) can't detect this. What about KCSAN? This is where KCSAN shines (as well): it can catch these too! How? Without delving into details, a class of macros, `ASSERT_EXCLUSIVE*()`, can determine whether an access actually occurs exclusively or not. For critical sections, *exclusive access* is just what the doctor ordered! It's what prevents data races, after all. This also has us understand that KCSAN itself works without using locks, via compiler instrumentation (like KASAN).

Great – with this, we complete our coverage of the really powerful kernel concurrency sanitizer, KCSAN. Let's move on to something interesting and practical: we'll explore a few actual kernel defects whose root causes lie in – guess what – locking issues.

A few actual use cases of kernel bugs due to locking defects

Looking up existing, fixed bugs helps us better understand their root cause, and thus helps when we design and implement code. Here are a few instances of actual kernel bugs related to locking. We don't delve into the details of each (just a few); that's left to you! Quite clearly, the kernel bugs identified here aren't in the least bit exhaustive, merely an attempt to get you started on exploring kernel bugs – caused primarily by locking defects – faced by others and how they were tackled.

Defects identified by KCSAN

As was just covered in detail in the previous section, from the 5.8 kernel (Aug 2020), we have a really powerful weapon to catch kernel concurrency issues – **KCSAN**. Bugs found by KCSAN include those seen here: `https://github.com/google/kernel-sanitizers/blob/master/kcsan/FOUND_BUGS.md`.

Identifying locking rules and bugs from the LDV project

The **Linux Driver Verification** (**LDV**) project is an interesting one; among other stuff, it encompasses a set of *rules* to be followed by developers working on Linux drivers (which of course apply to pretty much any kernel code actually). The relevant site link, showing the *LDV Rules*, is `http://linuxtesting.org/ldv/online?action=rules`. The following screenshot highlights the point of interest to us here – the rules that apply to locking!

Figure 8.6 – Screenshot of the LDV project's Rules section with the rules applying to locking highlighted

Let's quickly check out the LDV rules relevant to locking:

- **Rule** (corresponding to the **Mutex lock/unlock** link in *Figure 8.6*): *Locking a mutex twice or unlocking without prior locking* (link: http://linuxtesting. org/ldv/online?action=show_rule&rule_id=0032). It's straightforward – doing, or attempting to do, the following is *illegal* and results in a locking defect, a bug:

  - Attempting to lock a mutex twice; aka *double-locking*. Interestingly, the kernel simply disallows this type of recursive locking (as it often leads to problems). The userspace **POSIX Threads (Pthreads)** implementation, on the other hand, allows it via a special mutex type, PTHREAD_MUTEX_RECURSIVE. Within the kernel though, any attempt to double-lock causes a serious defect – (self) deadlock!

  - Attempting to unlock a mutex that you haven't locked; in other words, you can only unlock a mutex that you have locked, that you currently hold or "own."

- Exiting without unlocking your mutex(es).

An example of one of these bugs, an attempt to double-lock, and the subsequent fix (the commit) can be found here: `https://www.mail-archive.com/git-commits-head@vger.kernel.org/msg18392.html`. Here, the code first grabbed a mutex lock (see the function `edac_device_reset_delay_period()`) and then invoked another function, `edac_device_workq_teardown()`. The problem, the bug, is that this latter function *also* attempts to take the same mutex lock, resulting in a (self) deadlock defect! The fix essentially has the order of the functions reversed, so that the teardown function runs without the mutex lock held.

Let's move on to the next LDV rule with regard to locking.

- **Rule** (corresponding to the **Memory allocation inside spinlocks** link in *Figure 8.6*): *Using a blocking memory allocation when spinlock is held* (`http://linuxtesting.org/ldv/online?action=show_rule&rule_id=0043`). Again, it's simple (but can be quite easily forgotten in the heat of the moment!): when allocating memory while holding a spinlock, you must use the `GFP_ATOMIC` (not `GFP_KERNEL`) flag.

Again, here's an example of this very defect (within a wireless network driver) as well as the fix: `https://git.kernel.org/pub/scm/linux/kernel/git/torvalds/linux.git/commit/?id=5b0691508aa99d309101a49b4b084dc16b3d7019` (check out the call trace, context info, and so on, provided the warning that got triggered when a `kzalloc()` was performed with `GFP_KERNEL` within an atomic context).

The final LDV rule is here.

- **Rule** (corresponding to the **Spinlocks lock/unlock** link in *Figure 8.6*): *Usage of spinlock and unlock functions* (link: `http://linuxtesting.org/ldv/online?action=show_rule&rule_id=0039`). This rule essentially mirrors the first one above, which is with respect to the mutex lock; here, it's with respect to spinlocks. So, the following are defects/bugs:

  - Attempting to acquire the same spinlock more than once/double-locking (results in self deadlock).

  - Attempting to unlock a spinlock that you haven't locked; in other words, you can only unlock a spinlock that you have locked, that you currently hold or own.

  - Exiting without unlocking your spinlock(s).

Though basic, these rules are easy to break unless you're careful!

Local Locks

A new synchronization primitive was added to the 5.8 kernel from the **Real-Time Linux (RTL)** project (earlier called the PREEMPT_RT project), which deserves a quick mention: **local locks**. These locks simply enable a clear context when "locking" is done by disabling interrupts and/or preemption; a use case is per-CPU locking. Earlier, it wasn't clear what exactly was being protected; local locks solve this issue. In effect, a local lock is a wrapper over preemption and interrupt enabling/disabling primitives. It is a necessary construct, especially for debug kernels where using lockdep and static analysis is key to finding bugs. This LWN article has the details: `https://lwn.net/Articles/828477/`.

Right, let's move on to looking up a few locking bugs from the kernel Bugzilla.

Identifying locking bugs from the Linux kernel Bugzilla

Quite obviously, the kernel Bugzilla site tends to be quite a rich source of bugs, including some related to locking. One way to reveal some reported locking bugs is to search by the string emanated by the kernel when it feels there's an issue.

Lockdep

Do read up on using the powerful lockdep infrastructure within the kernel to catch kernel deadlocks and defects here: *Linux Kernel Programming – Part 2, Chapter 7, Kernel Synchronization – Part 2*, in the *The lock validator lockdep – catching locking issues early* section.

When lockdep detects a typical deadlock (or even a potential one), it emits a warning message containing the string `possible circular locking dependency detected`.

Figure 8.7 is a screenshot showing some results (of course, by the time you try this, it could change). Simply searching for the string `locking  bug` reveals a few as well...

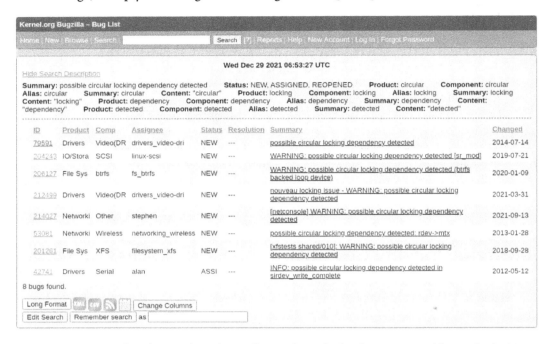

Figure 8.7 – Screenshot showing kernel Bugzilla search results for the string possible circular locking dependency detected

> **Tip – Catching Those Sleep-in-Atomic Defects**
>
> On a related note, turning on the kernel config CONFIG_DEBUG_ATOMIC_SLEEP as well as CONFIG_DEBUG_KERNEL (as the former depends on this one; in other words, running within a debug kernel) helps catch bugs where code sleeps in an atomic section (which isn't allowed of course).

Do note though, that you're expected to report Linux kernel bugs to the kernel mailing list (and associated subsystem maintainer(s)/subsystem lists).

Identifying some locking defects from various blog articles and the like

This section is an attempt to try and give you a broader view, taking others' experiences into account – people who've clearly learned from their experiences debugging kernel code and have taken the trouble to write good blog articles about it! Obviously, this isn't in the least exhaustive (just as the previous section on some actual kernel bugs wasn't), even with the fact that it's primarily about bugs due to locking defects; it's simply here to give you a broader perspective.

Exploiting an incorrect spinlock usage bug to gain complete control

Article: *How a simple Linux kernel memory corruption bug can lead to complete system compromise*, Jann Horn, Google Project Zero, Oct 2021: `https://googleprojectzero.blogspot.com/2021/10/how-simple-linux-kernel-memory.html`.

Without getting into too many details, Jann Horn, a security researcher at Google's Project Zero, found a bug within the kernel's pseudoterminal tty driver code (here: `drivers/tty/tty_jobctrl.c:tiocspgrp()`). Basically, the bug – which boiled down to the use of an incorrect spinlock – allowed one to set up data races between `struct pid` structures that can skew its reference count. By itself, this may not buy you much, but clever (white hat) security researchers such as Horn use it to set up a chain, an exploit, which ultimately leads to the complete compromising of a Debian Linux system (running the relatively recent 4.19 kernel), even getting a root shell, thus making it a **Privilege Escalation (privesc)** attack! Do read the details at the link provided. Once the exploit description is done, he also shows several possible defensive measures that can be taken.

We're concerned about correctly using locks and finding incorrect usage of them. The incorrect usage of a particular spinlock, a more or less humdrum bug, ultimately led to this exploit. The fix involves using the appropriate spinlock variable (instead of how it was being done before, using a spinlock belonging to a structure that could be arbitrarily specified). The following screenshot of the commit (the fix) shows this:

```
diff --git a/drivers/tty/tty_jobctrl.c b/drivers/tty/tty_jobctrl.c
index 28a23a0fef21c..baadeea4a289b 100644
--- a/drivers/tty/tty_jobctrl.c
+++ b/drivers/tty/tty_jobctrl.c
@@ -494,10 +494,10 @@ static int tiocspgrp(struct tty_struct *tty, struct tty_struct *real_tty, pid_t
        if (session_of_pgrp(pgrp) != task_session(current))
                goto out_unlock;
        retval = 0;
-       spin_lock_irq(&tty->ctrl_lock);
+       spin_lock_irq(&real_tty->ctrl_lock);
        put_pid(real_tty->pgrp);
        real_tty->pgrp = get_pid(pgrp);
-       spin_unlock_irq(&tty->ctrl_lock);
+       spin_unlock_irq(&real_tty->ctrl_lock);
 out_unlock:
        rcu_read_unlock();
        return retval;
```

Figure 8.7 – Screenshot showing the actual fix applied to the tty layer code

The preceding commit's clear: *the fix is to use the correct spinlock*! (Recall, we mentioned this in the *Locking – a quick summarization of key points.* section) This fix was in fact very recent (as of this writing), in kernel version 5.16.

Long delays due to interrupts being disabled while holding a lock

In general, a rule of thumb, a heuristic, to always follow is this: *disable interrupts only when absolutely essential, and then, for as short a time as is possible.*

Why is this so important? It's quite obvious: hardware interrupts are the means by which peripheral devices interrupt the processor and muscle their way onto the core, running their (allegedly) urgent code paths, getting stuff done.

Let's consider a (very simplified) case where hardware interrupt handling plays a critical role: think of a typical Ethernet adapter. When it detects a network packet with the MAC address being equal to its MAC ID, it pulls it into an internal buffer. When full, it interrupts the processor (run the command cat /proc/interrupts on the console to see all registered **Interrupt Requests (IRQs)**). We expect the system – the network driver, really – to then react. The kernel pretty much immediately redirects control to the network driver's interrupt handler code. It executes, pulling in the packet (often via DMA). The driver performs some basic processing on it and finally hands it off to higher layers of the network protocol stack to process. They send the packet up to its final destination, typically a userspace process.

Now, imagine disabling this network interrupt for a rather long time! The system literally suffers... network throughput drops. So, how do you disable hardware interrupts? Being in kernel mode, it's entirely possible. APIs such as `local_irq_disable()`, `local_irq_save()`, and `disable_[hard]irq()` can all achieve this (of course, their enabling counterparts are used to reenable interrupts). But, you say, why would I deliberately do that, especially for a long time?! Ah, this is where it gets interesting: a *spinlock internally disables interrupts (and kernel preemption) on the local processor while it's held*!

This is certainly the case with the pretty common `spin_lock_irq()` and `spin_lock_irqsave()` APIs. Check out the pseudo-code here:

```
spin_lock_irq[save](&mylock[, flags]); /* disables interrupts
*/
// time t1
/* ... do the work ... */
// time t2
spin_unlock_irq[restore](&mylock[, flags]); /* enables
interrupts */
```

The scenarios we're interested in are with regard to the time delta, the time spent running code while the spinlock's held, `(t2-t1)`. In effect, the time spent running code *while interrupts (and kernel preemption) are disabled*:

- If `(t2-t1)` is small (a few microseconds or up to single-digit milliseconds), it's generally okay (of course, this is a very generic statement; the actual affordable latency really depends on your project and its worst-case response time characteristics).

- If `(t2-t1)` is large (in the order of tens of milliseconds or more), it's generally *not okay* and can cause all kinds of latency issues, even livelock, on the system.

The latter case is of course the bad one, the dangerous one. This is at the crux of what happened in this actual use case (the following). Do read the article and see for yourself!

Article: *Network Jitter: An In-Depth Case Study*, Alibaba Cloud, Jan 2020: https://www.alibabacloud.com/blog/network-jitter-an-in-depth-case-study_595742.

Executive summary: Network jitter is caused here due to very long latencies in the system. Investigating it, engineers find it boils down to latencies introduced when slab statistics look up code calling `spin_lock_irq()` – which, as we know, internally disables hardware interrupts – for a pretty long while as this code runs in a loop! It locks up networking! Why does the loop run for a long while? It was iterating over `dentry` objects in a linked list. It so happened that there were a huge number of these objects, causing this (O(n) time-complexity, non-scalable code path) to take a long time...

This is a pretty typical issue: iterating over a list that's larger than expected, causing unexpected latencies! Keep a keen eye out!

You just learned that, because the `spin_lock_irq[save]()` APIs disable both hardware IRQs and kernel preemption, we must reenable them as soon as possible (by invoking the complementary APIs, `spin_unlock_irq[restore]()`). Anything more than a few tens of milliseconds spent with a spinlock held is considered too long. Okay, but how can I measure this?

One way to find critical sections that take a long time to complete is by leveraging the powerful **eBPF** infrastructure (we briefly covered eBPF in *Chapter 4, Debug via Instrumentation – Kprobes*, in the *Observability with eBPF tools – an introduction* section). Among the many eBPF tools available, for our purposes here, you can leverage the `criticalstat[-bpfcc]` one – it's used to measure the length of atomic critical sections – and you can filter based on duration. So, as an example, to measure which code paths where kernel preemption is disabled are taking longer than, say, 5 ms (the time unit for the tool is microseconds), run this command:

```
sudo criticalstat-bpfcc -p -d 5000 2>/dev/null
```

It even generates a stack trace showing the origins of lengthy critical sections – very useful (see its man page for details)! Another way to measure the duration for which hardware interrupts (and kernel preemption) is off is mentioned in *Chapter 9, Tracing the Kernel Flow*, in the *Ftrace - miscellaneous remaining points via FAQs* section.

How blocking while atomic and failing to take references can be your downfall

Working at the level of the kernel can certainly be more than you bargained for! The complexity can get pretty high. This article illustrates this really well. Though long, it's worth reading through fully.

Article: *My First Kernel Module: A Debugging Nightmare*, Ryan Eberhardt, Nov 2020: https://reberhardt.com/blog/2020/11/18/my-first-kernel-module.html.

Here's a very brief summary of the issues faced and lessons learned:

- Don't sleep (block) in any kind of atomic context, in critical sections! Relevant to this particular use case, here's one of the bugs (the comments are self-explanatory):

```
rcu_read_lock();   // begin RCU read critical section
[...]
msleep(10); /* bug! *sleep() helpers all block; if
you must, use the *delay() helpers instead (they're
non-blocking) */
rcu_read_unlock(); // end RCU read critical section
```

Recall our earlier advice: turning on the kernel configs CONFIG_DEBUG_ATOMIC_SLEEP and CONFIG_DEBUG (the former depends on this being enabled; in other words, testing with a debug kernel) helps catch bugs – like this one – where code sleeps in an atomic section!

- When using (even reading) several kernel global data structures, ensure you *take a reference to it* so that it's not freed "under" you! Release the reference when done. A couple of typical examples are the following:

  - For the task structure (struct task_struct):

    ```
    get_task_struct();
    /* ... use it ... */
    put_task_struct();
    ```

  - For the open file structure (struct file):

    ```
    get_file(file);
    /* ... use the file ... */
    fput(file);
    ```

  FYI, this is the commit (appropriately labeled *Fix race conditions in kernel module*) on the preceding project's GitHub repo where the racy bugs are fixed: https://github.com/reberhardt7/cplayground/commit/e14b9eb9d9ed616d9c030b8dd99c09b85349da28.

Interestingly, Ryan mentions how the on-the-surface-ridiculous technique of simply commenting out a bunch of code, running it and checking if it works, then uncommenting some statements, running it and checking if it works, over and over, until it doesn't, is actually the way he made real progress with these bugs! It reminds me of what was mentioned back in *Chapter 1, A General Introduction to Debugging Software*, in the *Debugging – a few quick tips*: *"Think small"* section (reread the section and see for yourself).

Summary

Good going! I think you'll agree with me that this chapter was a really key one. Locking and concurrency are inherently complex topics, with all kinds of bad side effects (such as unexplained hangs, deadlock, performance issues, and even livelock) when used incorrectly. In this chapter, you began by refreshing the basics on several key points with regard to locking.

We mentioned – and again emphasize – that detailed coverage on locking technologies within the kernel (mutex, spinlock, `atomic_t`, `refcount_t`, per-CPU, and so on) are covered in detail in my earlier *Linux Kernel Programming – Part 2* book's last two chapters. The eBook is freely downloadable (as both a PDF as well as a Kindle edition). The last chapter in the *Linux Kernel Programming – Part 2* book covers key information on lock debugging techniques (especially lockdep) that are important.

This chapter then delved into what actually constitutes a data race (as defined by the LKMM). You then learned what KCSAN is – a really powerful means to detect concurrency-related bugs, and how to enable and use it. Don't blindly attempt to paper over issues with indiscriminate use of the `{READ|WRITE}_ONCE()` and `data_race()` helpers! When KCSAN has detected a data race, unless it's intentional, it's your job to investigate and fix it.

The last section in this chapter showed you a few actual cases of concurrency/locking related bugs in the kernel (and drivers). It's very interesting and educational to learn from them!

So, do take the time to digest these key areas. With this, we complete part 2 of this book! In the last part of this book (part 3), we'll begin with the next chapter, covering how you can trace the flow of kernel (and driver) code – interesting and useful stuff to learn!

Further reading

- My earlier book: *Linux Kernel Programming – Part 2*, Kaiwan N Billimoria, Packt, Mar 2021. Freely downloadable as an eBook: `https://github.com/ PacktPublishing/Linux-Kernel-Programming/blob/master/ Linux-Kernel-Programming-(Part-2)/Linux%20Kernel%20 Programming%20Part%202%20-%20Char%20Device%20Drivers%20 and%20Kernel%20Synchronization_eBook.pdf`

  (Its last two chapters are relevant to the coverage in this chapter.)

- A number of very useful links (a few might get repeated below) from my earlier *Linux Kernel Programming* book:

  - *Chapter 12, Kernel Synchronization, Part 1 – Further reading*: `https://github. com/PacktPublishing/Linux-Kernel-Programming/blob/master/ Further_Reading.md#chapter-12-kernel-synchronization- part-1---further-reading`

  - *Chapter 13, Kernel Synchronization, Part 2 – Further reading*: `https://github. com/PacktPublishing/Linux-Kernel-Programming/blob/master/ Further_Reading.md#chapter-13-kernel-synchronization- part-2---further-reading`

- *What every systems programmer should know about concurrency*, Matt Kline, April 2020: `https://assets.bitbashing.io/papers/concurrency- primer.pdf`

- *An Introduction to Lock-Free Programming*, Preshing on Programming blog, June 2012: `https://preshing.com/20120612/an-introduction-to-lock- free-programming/`

- *Memory Barriers Are Like Source Control Operations*, Preshing on Programming blog, July 2012: `https://preshing.com/20120710/memory-barriers- are-like-source-control-operations/`

- The **Linux-Kernel Memory Consistency Model (LKMM)**:

  - *Explanation of the Linux-Kernel Memory Consistency Model*: `https://git. kernel.org/pub/scm/linux/kernel/git/torvalds/linux.git/ tree/tools/memory-model/Documentation/explanation.txt`

  - *Linux-Kernel Memory Model*, Paul E. McKenney, Apr 2015: `http://www.open- std.org/jtc1/sc22/wg21/docs/papers/2015/n4444.html`

- *Why kernel code should use READ_ONCE and WRITE_ONCE for shared memory accesses*, Andrey Konovalov, Google Sanitizers: `https://github.com/google/kernel-sanitizers/blob/master/other/READ_WRITE_ONCE.md`

- The **Kernel Concurrency Sanitizer (KCSAN)**:

  - Official kernel documentation: *The Kernel Concurrency Sanitizer (KCSAN)*: `https://www.kernel.org/doc/html/latest/dev-tools/kcsan.html#the-kernel-concurrency-sanitizer-kcsan`

  - *Finding race conditions with KCSAN*, Jonathan Corbet, LWN, 14 Oct 2019: `https://lwn.net/Articles/802128/`. Also explains how KCSAN works.

  - *Data-race detection in the Linux kernel*, Marco Elver, Linux Plumbers Conference, Aug 2020; PDF slides: `https://linuxplumbersconf.org/event/7/contributions/647/attachments/549/972/LPC2020-KCSAN.pdf`

  - LWN's "big bad" series:

    - *Who's afraid of a big bad optimizing compiler?* Jade Alglave, Paul E. McKenney, et al, LWN, July 2019: `https://lwn.net/Articles/793253/`

    - *Concurrency bugs should fear the big bad data-race detector (part 1)*, Marco Elver, Paul E. McKenney, et al, LWN, Apr 2020: `https://lwn.net/Articles/816850/`

    - *Concurrency bugs should fear the big bad data-race detector (part 2)*, Marco Elver, Paul E. McKenney, et al, LWN, Apr 2020: `https://lwn.net/Articles/816854/`

  - The KCSAN Google Wiki site: `https://github.com/google/kernel-sanitizers/blob/master/KCSAN.md`

  - Installing GCC-11 on Ubuntu: StackOverflow, Apr/May 2021: `https://stackoverflow.com/questions/67298443/when-gcc-11-will-appear-in-ubuntu-repositories`

- The **Android Open Source Project (AOSP)** uses the kernel **lockstat** to solve some performance issues by figuring out where exactly kernel lock contention is occurring. See the case study as well within this section: `https://source.android.com/devices/tech/debug/ftrace#lock_stat`.

- The blog articles covered in this chapter (in the *Identifying locking defects from various blog articles and the like section*):

  - *How a simple Linux kernel memory corruption bug can lead to complete system compromise*, Jann Horn, Google Project Zero, Oct 2021: `https://googleprojectzero.blogspot.com/2021/10/how-simple-linux-kernel-memory.html`

  - *Network Jitter: An In-Depth Case Study*, Alibaba Cloud, Jan 2020: `https://www.alibabacloud.com/blog/network-jitter-an-in-depth-case-study_595742`

- *My First Kernel Module: A Debugging Nightmare*, Ryan Eberhardt, Nov 2020: `https://reberhardt.com/blog/2020/11/18/my-first-kernel-module.html`. FYI, Ryan introduces the complex subject of **Read-Copy-Update (RCU)** lock-free synchronization concepts superbly in this (above-mentioned) article. I especially mention this as I didn't cover this key topic in the *Linux Kernel Programming – Part 2* book. A very brief introduction to RCU *is given* in *Chapter 10, Kernel Panic, Lockups, and Hangs*, in the *Conceptually understanding RCU in a nutshell section.*

Part 3: Additional Kernel Debugging Tools and Techniques

In this section, you will begin by learning about powerful technologies that allow you to trace the flow of kernel code in detail. Then, you'll move on to learning all about kernel panic and what you can do if it ever occurs! Next, you'll use KGDB within the kernel and modules to single-step through their source. The section – and the book – winds up with an introduction to even more approaches to debugging the Linux kernel.

The following chapters will be covered in this section:

- *Chapter 9, Tracing the Kernel Flow*
- *Chapter 10, Kernel Panic, Lockups, and Hangs*
- *Chapter 11, Using Kernel GDB (KGDB)*
- *Chapter 12, A Few More Kernel Debugging Approaches*

9
Tracing the Kernel Flow

Tracing is the ability to collect relevant details as code executes. Typically, data collected will include function names (and perhaps parameters and return values) of function calls made along the code path being followed, the context that issued the call, when the call was made (a timestamp), the duration of the function call, and so on. Tracing allows you to study and understand the detailed flow of a system or a component within it. It's akin to the black box in an aircraft – it simply collects data, allowing you to interpret and analyze it later. (You can also consider tracing to be loosely analogous to logging.)

Profiling is different from tracing in that it typically works by taking samples (of various interesting events/counters) at periodic points in time. It won't capture everything; it (usually) captures just enough to help with runtime performance analysis. A profile of code execution, a report, can usually be generated, allowing you to catch outliers. So, profiling is statistical by nature, while tracing isn't. It captures literally everything.

Tracing can be, and certainly often is, a debugging technique well worth understanding and using; profiling, on the other hand, is meant for performance monitoring and analysis. This book is about kernel debugging; hence, in this chapter, we keep the focus on a few tracing technologies (among the many available) and their frontends, which can prove useful. (To be honest, there will be overlap at times – some tools serve as both tracers as well as profilers, depending on how they're invoked.)

In this chapter, we're going to cover the following main topics:

- Kernel tracing technology – an overview
- Using the ftrace kernel tracer
- Using the trace-cmd, KernelShark, and perf-tools ftrace frontends
- An introduction to kernel tracing with LTTng and Trace Compass

Technical requirements

The technical requirements and workspace remain identical to what's described in *Chapter 1, A General Introduction to Debugging Software*. The code examples can be found within the book's GitHub repository here: `https://github.com/PacktPublishing/Linux-Kernel-Debugging`. The only new requirements are installing LTTng and Trace Compass for your Ubuntu 20.04 LTS system.

Kernel tracing technology – an overview

In order to trace or profile, a data source (or several) are required; the Linux kernel provides them, of course. **Tracepoints** are a primary data source within a kernel (in fact, we covered using the kernel's dynamic event tracing in *Chapter 4, Debug via Instrumentation – Kprobes* in the *The easier way – dynamic kprobes or kprobe-based event tracing* section). The kernel has several predefined tracepoints; you can see them here: `/sys/kernel/tracing/events/`. Many tracing tools rely on them. You can even set up tracepoints dynamically by writing to `/sys/kernel/tracing/kprobe_events` (we covered this too in *Chapter 4, Debug via Instrumentation – Kprobes* via dynamic kprobes, as just mentioned).

Other data sources include kprobes, uprobes (the equivalent of kprobes for userspace), USDT/dprobes and LTTng-ust (these latter two are for user mode tracing; also, LTTng has several kernel modules that it inserts into the kernel for kernel tracing – there will be more on LTTng later in this chapter).

A well-known look at the state of Linux tracing (and the many tools and technologies encompassed within it) is this blog article by Julia Evans (@b0rk): *Linux tracing systems & how they fit together*, Julia Evans, July 2017: https://jvns.ca/blog/2017/07/05/linux-tracing-systems/. Do check it out. Here, I use the same methodology to organize the rather large Linux tracing infrastructure; we divide it up into data sources (as mentioned previously), infrastructure technologies that collect or extract data from them, and finally, frontends, enabling you to use them more easily and effectively. The following diagram is my attempt to sum this up in one place:

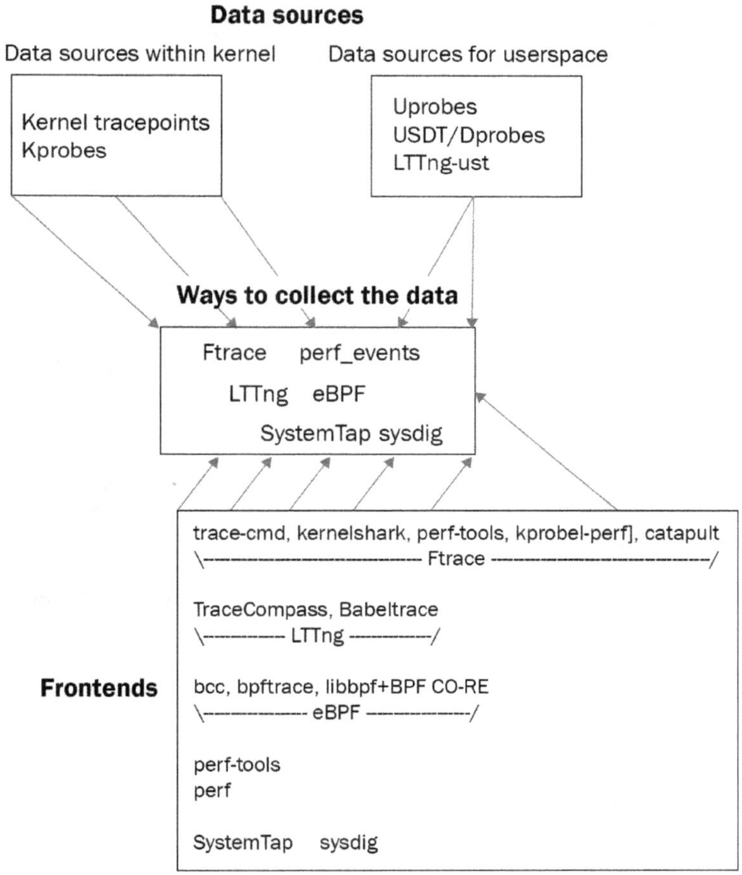

Data sources

Data sources within kernel   Data sources for userspace

Kernel tracepoints
Kprobes

Uprobes
USDT/Dprobes
LTTng-ust

Ways to collect the data

Ftrace   perf_events

LTTng   eBPF

SystemTap sysdig

trace-cmd, kernelshark, perf-tools, kprobel-perf], catapult
\\------------------------------ Ftrace -------------------------------/

TraceCompass, Babeltrace
\\----------- LTTng ------------/

Frontends

bcc, bpftrace, libbpf+BPF CO-RE
\\---------------- eBPF ----------------/

perf-tools
perf

SystemTap   sysdig

Figure 9.1 – The Linux tracing infrastructure

Steven Rostedt is the original developer of **ftrace** and, I dare say, is intimately familiar with much of Linux's vast tracing landscape. A slide from one of his many presentations on Linux tracing goes a long way toward summing up the state of the system (as of 2019 at least):

Commonality

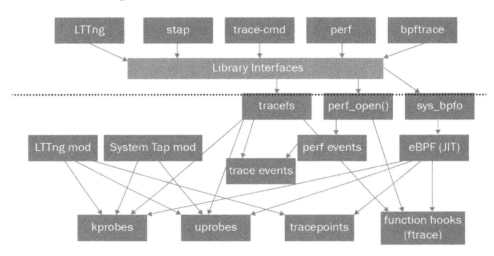

Figure 9.2 – The rich Linux tracing ecosystem, one where the underlying tech is shared

Here is the credit for the previous slide: *Unified Tracing Platform, Bringing tracing together*, Steven Rostedt, VMware 2019: `https://static.sched.com/hosted_files/osseu19/5f/unified-tracing-platform-oss-eu-2019.pdf`.

As you can see (*Figure 9.1* and *Figure 9.2*), there's a pretty vast plethora of technologies here; taking into account the fact that this book is about kernel debugging (and the space constraints within the book), we will only cover a few key kernel tracing technologies, focusing on their usage rather than on their internals. Again, the point Rostedt convincingly makes (which is echoed in this article: *Unifying kernel tracing*, Jack Edge, Oct 2019: `https://lwn.net/Articles/803347/`) is that in today's world, Linux tracing technologies aren't really competing with each other; rather, they're building off each other, as sharing ideas and code is not only allowed, it's encouraged! Thus, he visualizes a common userspace library that unifies all the disparate (yet powerful) kernel tracing technology in such a way that anyone can take advantage of all of it.

Do refer to the *Further reading* section for links to the other tracing technologies (as well as to what's covered here). Also, we've already covered kprobes and the related event tracing tooling, the usage of the `kprobe[-perf]` script, and the basics of using eBPF tooling back in *Chapter 4, Debug via Instrumentation – Kprobes*.

So, buckle up, and let's dive into kernel tracing with ftrace!

Using the ftrace kernel tracer

Ftrace is an inbuilt kernel feature; its code is deeply ingrained into that of the kernel itself. It provides developers (anyone with root access, really) with a way to look deep within the kernel, perform detailed traces to see exactly what's going on inside, and to even get help with performance/latency issues that may crop up.

A simple way to think about ftrace's functionality is this – if you'd like to see what a process is up to, performing `strace` on it can be very useful indeed; it will display every system call that the process invokes in a meaningful way, with parameters, return values, and so on. Thus, `strace` is useful and interesting, as it shows what occurs at the interesting system call point – the boundary between user and kernel space. But that's it; `strace` *cannot* show you anything beyond the system call; what does the system call code do within the kernel? What kernel APIs does it invoke, and thus which kernel subsystems does it touch? Does it move into a driver? Ftrace can answer these questions (and more)!

> **Tip**
>
> Don't knock the sheer usefulness of `strace` (and even `ltrace` – a library call tracer) in helping to understand and solve issues, obviously more so at the userspace layers. I highly recommend you learn to leverage them; read their man pages and search for tutorials.

Ftrace works essentially by setting up function hooks via compiler instrumentation (a bit simplistically, by enabling the compiler's -pg profiler option, which adds a special mcount call), ensuring that the kernel is aware of the entry (prologue) and possibly the exit/return (epilogue) of (almost) every single function in kernel space. (The reality is more complex; this would be too slow. For performance, there's a sophisticated *dynamic ftrace* kernel option – more on this is mentioned in the *Ftrace and system overhead* section.) In this way, ftrace is more like the **Kernel Address Sanitizer** (**KASAN**), which uses compiler instrumentation to check for memory issues, and less like the **Kernel Concurrency Sanitizer** (**KCSAN**), which works using a statistical sampling-based approach. (We covered KASAN in *Chapter 5, Debugging Kernel Memory Issues – Part 1*, and KCSAN in *Chapter 8, Lock Debugging*.) But by virtue of the dynamic ftrace option, it runs with native performance for the vast majority of the time, making it a superb debug tool on even production systems.

Accessing ftrace via the filesystem

Modern ftrace (kernel 4.1 onward) is implemented as a virtual (API) filesystem named tracefs; this is how you're expected to work with it. The default mount point is a leaf directory under the debugfs mount point named tracing; it's also made available under sysfs:

```
mount | grep "^tracefs"
tracefs on /sys/kernel/tracing type tracefs
(rw,nosuid,nodev,noexec,relatime)
tracefs on /sys/kernel/debug/tracing type tracefs
(rw,nosuid,nodev,noexec,relatime)
```

As mentioned earlier, it's possible – especially on a production kernel – that the CONFIG_DEBUG_FS_DISALLOW_MOUNT kernel config is set to y, implying that although debugfs is available, it isn't visible. In cases like this, having access to kernel tracepoints via sysfs (/sys/kernel/tracing) becomes important. Until Linux 4.1, only the traditional mount point – /sys/kernel/debug/tracing – was present. From 4.1 onward, mounting debugfs (typically done at boot, as systemd is configured to mount it) will also result in the sys/kernel/tracing mount being automatically set up. As the kernel version we typically work with here is a much later one (5.10), we'll assume from now on that you'll work within the /sys/kernel/tracing directory.

Configuring the kernel for ftrace

Most modern Linux distributions are preconfigured to support ftrace out of the box. The relevant config is CONFIG_FTRACE, which should be set to y. Using the familiar make menuconfig UI, you'll find ftrace (along with its sub-menus) here: Kernel hacking | Tracers.

The key dependency is the TRACING_SUPPORT config; it's arch-dependent and must be y. Realistically, most architectures (CPU types) will have this dependency satisfied. Here's a screenshot of the default sub-menu for ftrace on x86:

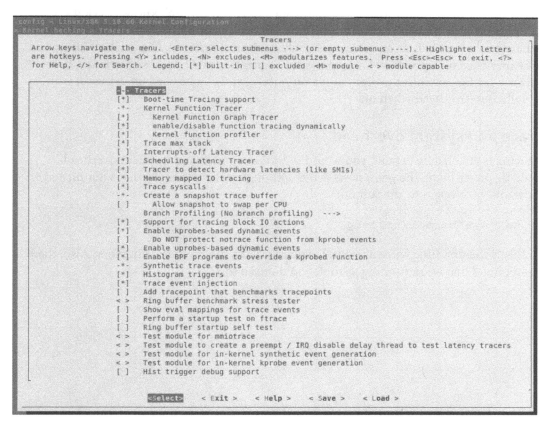

Figure 9.3 – A screenshot of the Tracers sub-menu (defaults) on x86_64 (5.10.60)

Also, if you want to enable the *Interrupts-off Latency Tracer* (CONFIG_IRQSOFF_ TRACER), it depends on TRACE_IRQFLAGS_SUPPORT=y (which is usually the case; we will briefly cover the **interrupt (IRQ)** and preempt-off latency tracers within the *Ftrace – miscellaneous remaining points via FAQs* section). In addition, several options depend upon the ability to figure out and display (kernel) stack traces (CONFIG_STACKTRACE_ SUPPORT=y), which, again, is typically on by default.

All the sub-menus and further configs belonging to ftrace are defined (and described) in the `Kconfig` file here: `kernel/trace/Kconfig`; you can look it up for details on any particular ftrace config directive.

So, practically speaking, on a common distro kernel (such as that of Ubuntu 20.04.3 LTS), is ftrace enabled or not? Let's check:

```
$ grep -w CONFIG_FTRACE /boot/config-5.11.0-46-generic
CONFIG_FTRACE=y
```

It is. What about the embedded system on your project? I wouldn't know; check it out (run grep on your project's kernel config file). On our custom 5.10.60 production kernel, we have ftrace enabled. As an aside, this book's technical reviewer, Chi Thahn Hoang, has vast experience on embedded Linux projects; he mentions that, in his experience, ftrace is always configured into a project, as it's very useful and can be used on demand, with virtually zero overhead when off.

Ftrace and system overhead

If tracing is enabled by default, you would indeed imagine that the system overheads would be pretty high. The good news is that although tracing is enabled, it's not turned on by default. Let's begin to check it out:

```
# cd /sys/kernel/tracing
```

(As this is the first time we're using it, I explicitly show the `cd` to the directory. Also, the # prompt hints that we're running as root; you do need to.)

The `tracefs` filesystem has many control knobs (pseudofiles, of course, just as with procfs, sysfs, and other API-based filesystems); one of them, named `tracing_on`, toggles the actual tracing functionality on or off; here it is:

```
# ls -l tracing_on
-rw-r--r-- 1 root root 0 Jan 19 19:00 tracing_on
```

Let's query its current value:

```
# cat tracing_on
1
```

It's quite intuitive – 0 means it's off, and 1 means it's on. So, ftrace is on by default? Isn't that risky performance-wise? No, it's actually/practically off – as the `current_tracer` pseudofile's value (soon explained) is nop, implying it's not tracing.

Back to the question of performance – even with the ability to toggle ftrace on/off, performance will still be an issue. Think about this – pretty much every function entry and return point (if not more fine-grained) will have to perform an `if` clause, something akin to the pseudocode: `if tracing's enabled, trace...`. This `if` clause itself constitutes far too much overhead; remember, this is the OS we're talking about; every nanosecond saved is a nanosecond earned!

The brilliant solution to this situation is to enable a config option called *dynamic ftrace* – `CONFIG_DYNAMIC_FTRACE`. When set to `y`, the kernel performs something amazing (and, in truth, scary); it can (and does!) modify kernel machine instructions on the fly in RAM, patching kernel functions to jump into ftrace or not, as required (this is often called a trampoline)! This config option is turned on by default, resulting in native performance of the kernel when tracing is disabled and near-native performance when tracing for only some functions is enabled.

Using ftrace to trace the flow of the kernel

By now, you'll begin to realize that, as with `tracing_on`, there are several pseudofiles under `tracefs` (that is, within the `/sys/kernel/tracing` directory) that are considered to be the "control knobs" of ftrace! (By the way, now that you know that the files under `tracefs` are pseudofiles, we'll mostly refer to them as simply files from now on.) The following screenshot shows us that there are indeed plenty of them:

```
# pwd
/sys/kernel/tracing
# ls
available_events                 max_graph_depth         stack_max_size
available_filter_functions       options/                stack_trace
available_tracers                per_cpu/                stack_trace_filter
buffer_percent                   printk_formats          synthetic_events
buffer_size_kb                   README                  timestamp_mode
buffer_total_size_kb             saved_cmdlines          trace
current_tracer                   saved_cmdlines_size     trace_clock
dynamic_events                   saved_tgids             trace_marker
dyn_ftrace_total_info            set_event               trace_marker_raw
enabled_functions                set_event_notrace_pid   trace_options
error_log                        set_event_pid           trace_pipe
events/                          set_ftrace_filter       trace_stat/
free_buffer                      set_ftrace_notrace      tracing_cpumask
function_profile_enabled         set_ftrace_notrace_pid  tracing_max_latency
hwlat_detector/                  set_ftrace_pid          tracing_on
instances/                       set_graph_function      tracing_thresh
kprobe_events                    set_graph_notrace       uprobe_events
kprobe_profile                   snapshot                uprobe_profile
#
```

Figure 9.4 – A screenshot showing the content within the tracefs pseudo-filesystem

Don't stress regarding the meaning of all of them; we'll consider just a few key ones for now. Understanding what they're for and how to use them will have you using ftrace in a jiffy.

Ftrace uses the notion of a **tracer** (sometimes referred to as a **plugin**) to determine the kind of tracing that will be done under the hood. The tracers need to be configured (enabled) within the kernel; several are by default, but not all of them. The ones that are already enabled can be easily seen (here we assume you're running as root within the `/sys/kernel/tracing` directory):

```
# cat available_tracers
hwlat blk function_graph wakeup_dl wakeup_rt wakeup function
nop
```

So, among the ones seen in the preceding snippet, which one will be used when tracing? Can you change it? The tracer (plugin) to be used is the content of the file named current_tracer; and yes, you can modify it as root. Let's look it up:

```
# cat current_tracer
nop
```

The default tracer is called nop; it's an abbreviation for **No Operation** (**No-Op**), implying that nothing will actually be done. Another tracer's named function; when tracing's enabled (toggled on) and the function tracer's selected, it will show every function that's executed within the kernel! Shouldn't we use it? Sure we can; there's an even better one though – function_graph. It will also have the trace display every kernel function executed during the trace session; in addition, it has the intelligence to indent the function name output in such a manner that it becomes almost like reading code – like a call graph!

> **Documentation on All the ftrace Tracers**
>
> As seen just previously, ftrace makes several tracers or plugins available. Here, we focus on using the function_graph tracer. Some of them are latency-related (hwlat, wakeup*, irqsoff, and preempt*off). How come the irqsoff and preempt[irqs]off tracers didn't show up previously? They're kernel configurables. The ones that show under available_tracers are the ones configured in; you'll have to reconfigure (enable) and rebuild the kernel to get any others! (*Tip* – execute grep "CONFIG_.*_TRACER" on your kernel config file to see which are enabled/disabled.)
>
> Do refer to the official kernel docs for details on all ftrace tracers: https://www.kernel.org/doc/html/v5.10/trace/ftrace.html#the-tracers.

The following table enumerates a few key files under tracefs; do check it out carefully:

Name of the file under /sys/kernel/tracing	What it's for	Default value
tracing_on	Toggles tracing on/off	1
current_tracer	The current tracer (or plugin) in effect	nop
available_tracers	A list of all configured (enabled) tracers within the kernel	<varies with config>
trace	Holds the trace report	<none>

Table 9.1 – A few key files under tracefs (/sys/kernel/tracing)

The file named `trace` is a key one; it contains the actual report – the output from the trace. We use this knowledge to save this output to a regular file.

Trying it out – a first trial (run one)

Here, we'll keep things as simple as possible – just getting ftrace enabled and running, tracing everything that occurs within the kernel for just a second then we turn tracing off. Ready? Let's do it; before that:

```
# cat tracing_on
1
```

Surprised? It appears that tracing's currently on (by default)! It's not actually tracing anything though; this is because the current tracer is set to the nop value by default, which of course means that it isn't really tracing anything:

```
# cat current_tracer
nop
```

So, we'll have to change the current tracer. What's available? Check by seeing the content of the `available_tracers` pseudofile:

```
# cat available_tracers
hwlat blk function_graph wakeup_dl wakeup_rt wakeup function
nop
#
```

You'll realize that the list of available tracers depends on your kernel configuration. Great, let's use one of them, the function graph tracer:

```
# echo function_graphix > current_tracer
bash: echo: write error: Invalid argument
```

Oops, we made a typo; it isn't accepted. That's good; `tracefs` internally validates the input passed to it. Now, let's do it correctly. Realize though that setting a valid tracer will have the kernel immediately start tracing! So, we first turn tracing off and then set up a valid tracer:

```
# echo 0 > tracing_on
# echo function_graph > current_tracer
```

Done. Now, let's begin tracing the kernel – for just a second – and then turn it off:

```
# echo 1 > tracing_on ; sleep 1; echo 0 > tracing_on
#
```

Tracing done, for 1 second. Where's the report? It's within the file named `trace`:

```
# ls -l trace
-rw-r--r-- 1 root root 0 Jan 19 17:25 trace
```

The `trace` file is empty (with a size of 0 bytes). Well, as you'll realize, this is a *pseudofile* (not a physical disk file) under `tracefs`; the majority of these pseudofiles have their size deliberately set to 0 as a hint that it's not a real file. It's a callback-based technology – reading it will cause the underlying kernel filesystem code to dynamically generate the data (or do whatever's appropriate). So let's copy the `trace` pseudofile's content to a regular file:

```
# cp trace /tmp/trc.txt
# ls -lh /tmp/trc.txt
-rw-r--r-- 1 root root 4.8M Jan 19 19:39 /tmp/trc.txt
```

Aha! It seems to have worked. The trace report file is quite large, yes? In this particular instance, in literally 1 second, we got 98,376 lines of tracing output (I checked with wc). Well, this is the kernel; *whatever code ran within the kernel in that 1 second is now in the trace report. This includes code that ran in that second on any and all CPU cores in kernel mode, including interrupt contexts.* This is great, but also illustrates one of the problems with ftrace (and tracing in general, especially within something as large and complex as the kernel!) – the output can be enormous. Learning to filter the functions traced is a key skill; worry not, we will!

Let's look up a few lines from the report; to make it a bit more interesting, I've shown you the actual trace data I got from line 24 onward... (you'll realize, of course, that the following output is from one sample run on my setup; it could, and very likely would, vary on your system):

```
# head -n40 /tmp/trc.txt
# tracer: function_graph
#
# CPU  DURATION                  FUNCTION CALLS
# |     |   |                      |   |   |   |
[...]
```

```
5)    1.156 us   |  tcp_update_skb_after_send();
5)    1.034 us   |  tcp_rate_skb_sent();
5)               |  tcp_event_new_data_sent() {
5)    1.107 us   |    tcp_rbtree_insert();
5)               |    tcp_rearm_rto() {
5)               |      sk_reset_timer() {
5)               |        mod_timer() {
5)               |          lock_timer_base() {
5)               |            _raw_spin_lock_irqsave() {
5)    0.855 us   |              preempt_count_add();
5)    2.754 us   |            }
5)    4.820 us   |          }
[...]
#
```

Also, do realize that, having saved the report under /tmp, it's volatile and won't survive a reboot. Remember to save your most important ftrace reports to a non-volatile location.

Interpreting the ftrace report output seen previously is easy; clearly, the first line tells us that the tracer employed is function_graph. The output is column-formatted and the header line clearly indicates the content of each of them. In this particular trace session (for at least the first portion), the kernel functions (seen on the extreme right) have executed on CPU core 5 (the first column; CPU core numbering begins at 0). The duration of each function's execution is shown in microseconds (yes, it's fast, even on a VM). Note the careful indentation of the function names; it allows us to literally understand the control flow, which is the whole idea behind using this tracer!

Trying it out – function_graph options plus latency-format (run two)

Our first trial run with ftrace was interesting but missing some key details. You can see the kernel code that ran (in that single second of time), which CPU core it ran on, and how long each function took, but think about this – who ran it?

> **Monolithic Design**
>
> There's no such concept as *the kernel runs the code within it*. As you will understand, one of the key aspects of a monolithic kernel design like Linux's is that kernel code executes in one of two contexts – **process or interrupt**. **Process context** is one where a process (or thread) issues a system call; now, the process itself switches into kernel mode and runs the code of the system call within the kernel (and possibly a driver). **Interrupt context** is the situation where a hardware interrupt causes a processor to immediately switch to a designated code path (the interrupt handler routine) – kernel/driver code that runs in the context of the interrupt (well, there's more to it – the so-called bottom-half mechanisms – tasklets and softirqs – run in an interrupt context as well). We covered system calls briefly in *Chapter 4, Debug via Instrumentation – Kprobes* in the *System calls and where they land in the kernel* section.

So, guess what? Ftrace can show you the context the kernel code ran in; you simply have to enable one of the many options ftrace provides. Several useful options to render the output can be found under the `options` directory under the `tracefs` mount point, as shown in the following screenshot:

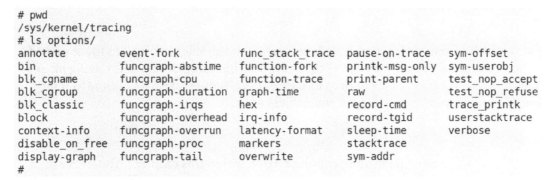

```
# pwd
/sys/kernel/tracing
# ls options/
annotate            event-fork          func_stack_trace    pause-on-trace      sym-offset
bin                 funcgraph-abstime   function-fork       printk-msg-only     sym-userobj
blk_cgname          funcgraph-cpu       function-trace      print-parent        test_nop_accept
blk_cgroup          funcgraph-duration  graph-time          print-parent        test_nop_refuse
blk_classic         funcgraph-irqs      hex                 record-cmd          trace_printk
block               funcgraph-overhead  irq-info            record-tgid         userstacktrace
context-info        funcgraph-overrun   latency-format      sleep-time          verbose
disable_on_free     funcgraph-proc      markers             stacktrace
display-graph       funcgraph-tail      overwrite           sym-addr
#
```

Figure 9.5 – A screenshot showing the content of the options directory under /sys/kernel/tracing

Again, we cover using some of these soon...

Resetting ftrace

If the ftrace system is in any kind of transient (or in-between) state, it's a good idea to reset everything to defaults, clear its internal trace (ring) buffers (thus freeing up memory as well), and so on. This can be done manually by writing a value (typically 0 or null) into several relevant tuning files within it. The `perf-tools` project has a script named `reset-ftrace` that resets ftrace to a known sane state: `https://github.com/brendangregg/perf-tools/blob/master/tools/reset-ftrace`.

We specified installing the `perf-tools[-unstable]` (and the `trace-cmd`) packages back in *Chapter 1, A General Introduction to Debugging Software*, in the *Installing required software packages* section.

So, let's make use of it now to help reset ftrace to reasonable defaults:

```
# reset-ftrace-perf
Reseting ftrace state...
current_tracer, before:
      1    nop
current_tracer, after:
      1    nop
[...]
```

It displays the before-and-after values of all the ftrace files it resets. There's a caveat – it doesn't reset *every* ftrace file (you'll see the ones it does at the bottom of the script). So, as an example, we'll manually reset one of them for now:

```
# echo 0 > options/funcgraph-proc
#
```

As well, it's also useful to leverage a really powerful (and simple) frontend to ftrace named `trace-cmd`, to help achieve a reset of the ftrace system. Internally, it turns tracing off and brings the system performance back to its native state. (It does take a bit of time to reset; there are option switches to reset internal buffers as well):

```
# trace-cmd reset
```

(We will cover `trace-cmd` in an upcoming section within this chapter.) Also, we've provided a pretty comprehensive `reset_ftrace()` function as part of this convenience script – `ch9/ftrace/ftrace_common.sh`.

> **Tip**
>
> Once reset, note that the ftrace system has the `tracing_on` file set to 1 (it's on) and `current_tracer` set to nop, thus effectively rendering tracing off. This implies that setting up a valid tracer plugin (such as `function_graph`) in the `current_tracer` file will cause tracing to begin immediately. Further, you can wipe out the current content of the trace buffer by doing `echo > trace` (as root).

Okay, back to the main point – now that ftrace is reset, how do we enable showing the context that runs kernel code? The relevant option file is this one – `options/funcgraph-proc` (**proc** is usually an abbreviation for **process**). Write 1 into it to enable ftrace to print the process context info:

```
# cat tracing_on
1
# echo 0 > tracing_on
# echo > trace
#
# echo function_graph > current_tracer
# echo 1 > options/funcgraph-proc
#
# echo 1 > tracing_on ; sleep 1; echo 0 > tracing_on
# cp trace /tmp/trc2.txt
#
# head /tmp/trc2.txt
# tracer: function_graph
#
# CPU  TASK/PID        DURATION                  FUNCTION CALLS
# |      |    |          |   |                   |   |    |    |
  2)   <idle>-0      |                   |  arch_cpu_idle_enter() {
  4)   bash-1153     |   3.225 us        |  mutex_unlock();
  2)   <idle>-0      |   0.980 us        |    tsc_verify_tsc_adjust();
  4)   bash-1153     |   0.621 us        |    __fsnotify_parent();
  2)   <idle>-0      |   0.549 us        |    local_touch_nmi();
  4)   bash-1153     |   0.581 us        |  preempt_count_add();
#
```

Figure 9.6 – A screenshot of our ftrace trial run with the funcgraph-proc option enabled

The figure makes it clear by highlighting the new option and the effect it has on the trace output. Here, in this instance, you can see that it happens to be the idle thread on CPU core 2 (with PID 0, seen as `<idle>-0`), and the `bash` process with PID `1153` on CPU core 4, which is executing the functions seen on the extreme right! (Yes, we can literally see the parallelism on two CPU cores here as a bonus!)

Several other function graph-related options exist; they're the files under the `options` directory that are prefixed with `funcgraph-`. Here, they are alongside their default values (a bit of `bash` foo is also helpful!):

```
# for f in options/funcgraph-* ; do echo -n "${f##*/}: "; cat
$f; done
funcgraph-abstime: 0
funcgraph-cpu: 1
funcgraph-duration: 1
funcgraph-irqs: 1
```

```
funcgraph-overhead: 1
funcgraph-overrun:  0
funcgraph-proc:  0
funcgraph-tail:  0
#
```

Do refer to the official kernel docs to get the details on what each of these means: *trace_ options*, under *Options for function tracer*: https://www.kernel.org/doc/html/ v5.10/trace/ftrace.html#trace-options.

Delving deeper into the latency trace info

This small example (in *Figure 9.6*) actually brings up a really interesting point! The CPU idle thread is scheduled to run when no other thread wants the processor; it might well be the case here. But don't take things for granted; it could also happen that the idle thread – or any other thread for that matter – gets interrupted by a hardware interrupt that then executes, but the context information (the column labeled TASK/PID) will still show the original context – the thread that got interrupted by the interrupt! So, how do we know for sure whether this code ran in a process or an interrupt context?

To know for sure, we must enable another really useful ftrace option – the latency- format one; the file is options/latency-format. Writing 1 into this file enables the latency format; what effect does this have? It causes an additional column (typically, after the CPU and TASK/PID ones) to appear that provides pretty deep insight into the state of the system when this kernel code executed (even helping you see why latencies occurred). Let's actually use it and then analyze the output. We first reset ftrace and then enable the function_graph tracer, along with both the funcgraph-proc and latency- format options:

```
# trace-cmd reset
# reset-ftrace-perf >/dev/null
# echo 0 > tracing_on
# echo > trace
# echo function_graph > current_tracer
# echo 1 > options/funcgraph-proc
# echo 1 > options/latency-format
#
# echo 1 > tracing_on ; sleep 1; echo 0 > tracing_on
# cp -f trace /tmp/trc3.txt
```

Figure 9.7 – A screenshot showing resetting ftrace and then focused on turning on the useful latency-format option for a sample 1-second trace

Here's the (truncated) ftrace report obtained (to make it interesting, I am showing a few different parts of the trace report):

```
 1 █ tracer: function_graph
 2 #
 3 # function_graph latency trace v1.1.5 on 5.10.60-prod01
 4 # --------------------------------------------------------------------------------
 5 # latency: 0 us, #166281/358344, CPU#0 | (M:preempt VP:0, KP:0, SP:0 HP:0 #P:6)
 6 #    -----------------
 7 #    | task:  0 (uid:0 nice:0 policy:0 rt_prio:0)
 8 #    -----------------
 9 #
10 #                       _-----=> irqs-off
11 #                      / _----=> need-resched
12 #                     | / _---=> hardirq/softirq
13 #                     || / _--=> preempt-depth
14 #                     ||| /
15 # CPU  TASK/PID        ||||      DURATION              FUNCTION CALLS
16 # |     |    |         ||||       |   |                |   |   |   |
17  1)  <idle>-0          d..1                           irq_enter_rcu() {
18  1)  <idle>-0          d..1      0.004 us               preempt_count_add(),
19  1)  <idle>-0          d..1                             tick_irq_enter() {
[...]
15105  0)  <idle>-0       d.h1    + 15.444 us           }
15106  0)  <idle>-0       d.h1                          __sysvec_apic_timer_interrupt() {
15107  0)  <idle>-0       d.h1                          hrtimer_interrupt() {
15108  0)  <idle>-0       d.h1                            _raw_spin_lock_irqsave() {
15109  0)  <idle>-0       d.h1      0.667 us               preempt_count_add();
15110  0)  <idle>-0       d.h2      1.809 us             }
15111  0)  <idle>-0       d.h2      0.957 us             ktime_get_update_offsets_now();
15112  0)  <idle>-0       d.h2                           __hrtimer_run_queues() {
15113  0)  <idle>-0       d.h2      0.544 us               __next_base();
15114  0)  <idle>-0       d.h2      1.168 us               __remove_hrtimer();
15115  0)  <idle>-0       d.h2                             _raw_spin_unlock_irqrestore() {
15116  0)  <idle>-0       d.h2      0.581 us               preempt_count_sub();
15117  0)  <idle>-0       d.h1      1.933 us             }
15118  0)  <idle>-0       d.h1                           tick_sched_timer() {
[...]
15192  0)  <idle>-0       d.h1    ! 100.461 us          }
15193  0)  <idle>-0       d.h1                          irq_exit_rcu() {
15194  0)  <idle>-0       d.h1      0.597 us             preempt_count_sub();
15195  0)  <idle>-0       d..1      0.812 us             ksoftirqd_running();
15196  0)  <idle>-0       d..1                           do_softirq_own_stack() {
15197  0)  <idle>-0       d..1                             __do_softirq() {
15198  0)  <idle>-0       d..1      0.557 us               preempt_count_add();
15199  0)  <idle>-0       ..s1                             run_rebalance_domains() {
15200  0)  <idle>-0       ..s1                               update_blocked_averages() {
15201  0)  <idle>-0       ..s1                                 _raw_spin_lock_irqsave() {
15202  0)  <idle>-0       d.s1      0.580 us                     preempt_count_add();
15203  0)  <idle>-0       d.s2      1.719 us                   }
15204  0)  <idle>-0       d.s2      0.633 us                   update_rq_clock();
15205  0)  <idle>-0       d.s2                               update_rt_rq_load_avg() {
15206  0)  <idle>-0       d.s2      0.625 us                     decay_load();
```

Figure 9.8 – Screenshots showing the effect of turning on the latency-format option; the latency format trace info column is highlighted

You may have noticed that the new output column (highlighted in the figure, the one between the TASK/PID and DURATION columns) looks familiar. We used pretty much precisely this format to encode information in our code base's PRINT_CTX() macro, defined within our convenient.h header. (Do note that here, as we use pr_debug() to emit output, it only shows up if you define the DEBUG symbol or employ the kernel's powerful dynamic debug framework to see the debug prints. We covered using dynamic debug and a lot more back in *Chapter 3*, *Debug via Instrumentation – printk and Friends*.)

To interpret this **latency trace info** (and the rest of the line), let's just consider a single line of output from *Figure 9.8* (the line – # 15,111 – is highlighted by a green horizontal bar in the middle screenshot in the figure); we reproduce it here:

```
0)    <idle>-0    |  d.h2  |   0.957 us    |              ktime_
get_update_offsets_now();
```

Let's interpret the preceding line column-wise, from left to right (it's one line, the function name on the extreme right might wrap around to the next line):

- First column – 0): The CPU core that the kernel function ran on (remember that CPU core numbering starts at 0).

- Second column – <idle-0>: The process context running the kernel function; do note that it could have run in the interrupt context (as we'll see in the next column). In cases like this, this second column shows the process context that was interrupted by the interrupt!

- Third column – d.h2: What we're really after here is the latency trace information (enabled via the options/latency-format option being set to 1). This information allows us to figure out in detail exactly what context the kernel function ran in – was it normal process context or an interrupt? Within the latter, was it a hardware interrupt (a hardirq), a softirq (bottom-half), a **Non Maskable Interrupt (NMI)**, or a hardirq preempting a softirq? Moreover, were interrupts enabled or disabled, was a reschedule pending at the time, and was the kernel in a non-preemptible state? It's really very detailed and precise! This column itself consists of four columns – here, their value happens to be d.h2, whose output is interpreted as follows:

  - First (latency trace info) column – IRQ (hardware interrupts) state – d:

    - d implies that IRQ's (hardware interrupts) are disabled.

    - . implies that IRQ's (hardware interrupts) are enabled.

- Second (latency trace info) column – the "need-resched" bit (*does the kernel need to reschedule?*):

  - N implies that both the TIF_NEED_RESCHED and PREEMPT_NEED_RESCHED bits are set (in effect, implying that a reschedule is needed on this preemptible kernel).

  - n implies that only the TIF_NEED_RESCHED bit is set.

  - p implies that only the PREEMPT_NEED_RESCHED bit is set.

  - . implies that both bits are cleared (the normal case, no reschedule is needed).

  > **FYI**
  >
  > The TIF_NEED_RESCHED bit when set implies that a schedule is pending (and will run soon – the N and n cases).
  >
  > The PREEMPT_NEED_RESCHED bit is explained in the x86 arch code, as follows. "We use the PREEMPT_NEED_RESCHED bit as an inverted NEED_RESCHED, such that a decrement hitting 0 means we can and should reschedule."

- Third (latency trace info) column – more on the running context (interestingly, the following list is *sorted by execution priority*):

  - Z implies that a NMI occurred inside – and thus preempted – a hard irq (hardware interrupt).

  - z implies that a NMI is running.

  - H implies that a hardirq occurred inside – and thus preempted – a softirq (a softirq is the internal kernel underlying implementation of an interrupt bottom-half mechanism)

  - h implies that a hardirq is running.

  - s implies that a softirq is running.

  - . implies a normal (process/thread) context is running.

- Fourth (latency trace info) column – the `preempt-depth`: This is also known as the `preempt_disabled` level. In effect, if positive, it implies that the kernel is in a non-preemptible state (when 0, the kernel is in a preemptible state). This matters when you're running a preemptible kernel (implying that even a kernel thread or kernel process context can be preempted!). As a simple case, consider a preemptible kernel and a thread context holding a spinlock. This is to be an atomic section; it shouldn't be preempted before it unlocks. To guarantee this, the preempt counter (in effect, what we're seeing here as `preempt-depth`) is incremented every time a spinlock is taken and decremented every time one of them is unlocked. Thus, while one or more spinlocks are held – in effect, while `preempt-depth` is positive – the context holding it shouldn't be preempted. You can see more on this here: `https://www.kernel.org/doc/Documentation/preempt-locking.txt`.

So, let's quickly interpret the latency trace info we're seeing here and now – `d.h2`:

- `d`: Hardware interrupts are disabled.

- `.`: Both the `TIF_NEED_RESCHED` and `PREEMPT_NEED_RESCHED` bits are cleared. This is the normal case, with no pending reschedule.

- `h`: The running context (here, the `<idle-0>` idle thread running on CPU 0) has been interrupted by a hardware interrupt, and its hardirq handler (or the so-called top-half, or **Interrupt Service Routine (ISR)**) is currently executing code on the processor core.

- `2`: The `preempt-depth` counter value is 2, implying that the kernel is currently marked as being in a non-preemptible state.

- Fourth column – `0.957 us`: This is the duration of the function call (in microseconds). When a function terminates (shown as a single close brace, }, in the last column), the duration is that of the entire function. Further, ftrace gives us a visual clue as to how long the function took by employing the notion of a delay marker (covered in detail in the upcoming *Viewing the context switch and delay markers* section). Also, the `options/graph-time` option (on by default) has, for the function graph tracer, the cumulative time spent in all nested functions displayed. Setting it to 0 will cause only individual function durations to be displayed.

- Fifth column – ktime_get_update_offsets_now();: This is the kernel function being executed. If only a single close brace, }, shows up here, it implies the termination of the overarching function. Setting this option switch – options/funcgraph-tail – to 1 will show the terminating function name as well (in the } /* function name */ format). You can also try using the % operator in vi [m] to jump to the function's start and thus see which function it represents (note, though, that this simple trick doesn't always work).

All this is very valuable information indeed!

The (upward and right-pointing) ASCII-art arrows at the header portion (see the top portion of *Figure 9.8*) briefly explain the latency trace info columns as well. In addition, the header portion of the ftrace report shows this line:

```
latency: 0 us, #166281/358344, CPU#0 | (M:preempt VP:0, KP:0,
SP:0 HP:0 #P:6)
```

The code that generates the preceding header line is here: https://elixir.bootlin.com/linux/v5.10.60/source/kernel/trace/trace.c#L3887. So, interpreting it via the code, we find the following.

If a latency tracer is being used to measure latency (delay) – for example, the irqsoff tracer – in those cases, the maximum latency seen during the trace session will be shown (here, it's 0, as we aren't running a latency tracer). It then shows the number of trace entries displayed in this report (166,281) and the total number of entries (358,344). Obviously, there were too many entries to hold. Increasing the size of the memory buffers is indicated (we will cover this in the *Other useful ftrace tuning knobs to be aware of* section). Also, the trace-cmd record frontend to ftrace allows us to use the -b <buffer-size-in-KB-per-cpu> option switch (we'll cover trace-cmd in detail in the *Using the trace-cmd, KernelShark, and perf-tools ftrace frontends* section). The next set of entries (in parentheses) begins with M for model – one of server, desktop, preempt, preemp_rt, or unknown. The next four entries – VP, KP, SP, and HP – are always 0 (they're reserved for later use). P specifies the number of online processor cores.

Figure 9.8 shows some portions of the ftrace report. These portions clearly indicate how kernel code ran with interrupts disabled, in a hardirq context, followed by switching into a softirq context (of course, it did many other things as well; we just can't show everything here).

Exercise
Let's say that the latency trace info field's value is dNs5. What does this mean?

So, this section also (more than) effectively answers our earlier puzzle – knowing whether code is executing in a process or interrupt context. Note that the latency trace format column also appears when you use any of the tracers related to latency measurement.

A quick aside – hardware interrupts aren't typically load-balanced by the kernel across all available CPU cores. Ftrace can help you understand on which core they execute, and thus you can load-balance them yourself (also look up /proc/irq/<irq#> and the irqbalance[-ui] utility). Note, though, that arbitrarily distributing an interrupt across CPU cores can cause CPU caching and thus performance issues! (You can find out more on hardware interrupt processing on Linux in my earlier *Linux Kernel Programming – Part 2* book in the chapter on hardware interrupt handling.)

Interpreting the delay markers

The first line of the third partial screenshot in *Figure 9.8* shows an ellipse surrounding the duration column. The line is reproduced here; check it out:

```
 0)    <idle>-0    |   d.h1 |  ! 100.461 us    |          }
```

The DURATION column (its default unit is microseconds) has an interesting annotation displayed at certain times to the left of the actual time duration. Here, for example, you can see the ! symbol to the immediate left of the time duration. This is one among the delay markers that can be shown, indicating that the function in question has taken a rather long time to execute! Here's how you interpret these delay markers (us, of course, means microseconds):

Delay marker	Duration of corresponding function execution (us)
$	Over 100,000,000 us (1 s)
@	Over 100,000 us (100 ms)
*	Over 10,000 us (10 ms)
#	Over 1,000 us (1 ms)
!	Over 100 us
+	Over 10 us
' '	<= 10 us

Table 9.2 – The ftrace delay markers

As indicated by the last row in the preceding table, the absence of a delay marker (or simply white space) to the immediate left of the duration indicates that the function's execution took less than 10 us (which is considered to be the normal case).

There's one last thing regarding the delay markers. When the tracer is function-graph (as was the case here), the function graph-specific option named options/funcgraph-overhead plays a role – the delay markers are only displayed if this is enabled (which is the default). To disable delay markers, write 0 into this file.

You might come across an arrow marker, =>, in the ftrace report. It's quite intuitive – it displays the fact that a *context switch* occurred.

As you'll have by now no doubt realized, in order to test ftrace, you can certainly put the commands we ran (such as those in *Figure 9.7*) into a script to avoid having to retype them and in effect have a custom tool at your service! Again, the Unix philosophy at work. (Actually, we've already done this for your convenience; you'll find, among others, very simple wrapper scripts here: ch9/ftrace; this particular one, tracing the kernel for 1 second with the function_graph tracer, is named ftrc_1s.sh)

Again, the really large number of ftrace options preclude our reproducing them individually here. You really must refer to the official kernel documentation on ftrace to view all options, their meaning, and possible values: https://www.kernel.org/doc/html/v5.10/trace/ftrace.html. Having said that, a few useful options to be aware of follows.

Other useful ftrace tuning knobs to be aware of

It is necessary to be aware of a few of the ftrace options. The following table is an attempt at a quick summarization:

Name of file under /sys/kernel/tracing/	What it's for
available_filter_functions	List of all functions that can possibly be traced
trace_options	Options that control the ftrace output and modify behavior – stack traces, and so on. (See the *Checking out trace_options* section.)
buffer_size_kb	The size of the per CPU trace buffer in kilobytes. Can modify as root. The buffer_total_size_kb pseudofile, on the other hand, is read-only and displays the total size of memory buffers for tracing. (See the ch9/ftrace/ping_ftrace.sh script for an example of how we calculate and set this option.)

max_graph_depth	When the tracer is function_graph, this specifies the maximum depth to which functions are traced. This is useful when you want to limit the depth and get an overview of what's running within the kernel. The default is 0, implying that it's unlimited.
saved_cmdlines	Enables saving the name (comm, for command) of the task along with the (kernel) PID. It's actually a cache of saved *PID comm* mappings kept by ftrace.
saved_cmdlines_size	The number of saved *PID comm* mappings kept by ftrace (as seen via the saved_cmdlines option). The default is 128. You can write an integer into this file to modify it.
saved_tgids	If record_tgids is set to 1 (it's 0 by default), the **Thread Group ID (TGID)** of the process – the PID value as seen by user space – is saved into a mapping cache (disabled by default).
trace_marker	Writing a string into this file causes it to appear in the trace ring buffer (and thus in the trace report). It's useful to synchronize user space with what's happening in the kernel. (Note that for this to work, you need to set the markers option in the trace_options file.)
tracing_thresh	This is meant for hardware latency tracers (such as hwlat). Traces get recorded only when this threshold is exceeded. When the hwlat tracer is used, it presets it to the value 10 (microseconds).

Table 9.3 – A summary of a few more ftrace (tracefs) files

The following notes flesh out some details.

Checking out trace_options

Reading the content of the `trace_options` `tracefs` pseudofile shows several ftrace options and their current values. Here are the default values on our 5.10.60 production kernel:

```
# cat trace_options
print-parent
nosym-offset
nosym-addr
noverbose
noraw
nohex
nobin
noblock
trace_printk
annotate
nouserstacktrace
nosym-userobj
noprintk-msg-only
context-info
nolatency-format
record-cmd
norecord-tgid
overwrite
nodisable_on_free
irq-info
markers
noevent-fork
nopause-on-trace
function-trace
nofunction-fork
nodisplay-graph
nostacktrace
notest_nop_accept
notest_nop_refuse
#
```

Figure 9.9 – A look at the trace_options pseudofile content under tracefs (with default values on x86_64 5.10.60)

The format works like this – the option name by itself implies that that option is enabled, and the option name prefixed with the word `no` implies that it's disabled. In effect, it's like this:

```
echo no<option-name> >  trace_options      : disable it
echo <option-name> >  trace_options        : enable it
```

For example, you can see in *Figure 9.9* the trace option called `nofunction-fork`. It specifies whether the children of a process will be traced as well. Clearly, the default is no. To trace all children, do this:

```
echo function-fork > trace_options
```

You'll recall from our detailed discussions on interpreting an Oops (*Chapter 7, Oops! Interpreting the Kernel Bug Diagnostic*, in the *Finding the code where the Oops occurred* section), one of the big clues as to where the issue is the format in which the instruction pointer content is displayed, as shown in the following example:

```
RIP: 0010:do_the_work+0x124/0x15e [oops_tryv2]
```

As we learned, the hex numbers following the + sign specify the offset from the beginning of the function and the length of the function (in bytes). In this example, it shows we're at an offset of `0x124` bytes from the beginning of the `do_the_work()` function, whose length is `0x15e` bytes. As *Chapter 7, Oops! Interpreting the Kernel Bug Diagnostic*, showed us, this format can be really useful. You can specify that all functions in the trace report are shown in this manner by doing the following:

```
echo sym-offset > trace_options
```

The default is `nosym-offset`, implying that it won't be shown this way.

> **A Caveat**
>
> The `sym-offset` and `sym-addr` options seem to apply to only the `function` tracer (and not `function_graph`). By the way, the `printk` format specifier, `%pF`, (and friends) also prints a function pointer in this manner! (See the useful `printk` format specifiers documented here: `https://www.kernel.org/doc/Documentation/printk-formats.txt`.)

Again, there are just too many `trace_options` options to cover individually here. I urge you to refer to the official kernel docs for details regarding `trace_options`: `https://www.kernel.org/doc/html/v5.10/trace/ftrace.html#trace-options`.

Useful ftrace filtering options

As you've surely noticed by now, the sheer volume of report data that ftrace can pump out can easily overwhelm you. Learning how to filter out stuff that's not required or secondary to your investigation is key!

Ftrace keeps all functions that it can trace in a file named `available_filter_functions`. On x86_64 and the 5.10 kernel, it's really pretty large. There are well over 48,000 functions within the kernel that can be traced!

```
# wc -l available_filter_functions
48660 available_filter_functions
```

Several powerful filters can be enabled via various `tracefs` files. The following table attempts to summarize some key ones; do study it carefully:

tracefs file to employ	How it filters	Comment/example
`set_ftrace_filter`	Functions to trace – write them into this file. Also, see the *More on the set_ftrace_filter file* section.	Dynamic ftrace! Only trace the function(s) specified (you can use globbing). This can be very slow when a large number of functions are specified (due to lengthy string processing). See the *Index-based function filtering* and *Module-based filtering* sections. Also, note that the surrounding function calls – the context in which a function was called, the parent functions, and the children it calls – are shown.
`set_ftrace_notrace`	Functions to never trace – write them into this file. Also, see the *More on the set_ftrace_filter file* section.	A complement of the preceding – don't trace the function(s) specified (you can use globbing). As with the preceding file, this can be very slow (due to lengthy string processing). See the *Index-based function filtering* and *Module-based filtering* sections.
`set_graph_function`	See the next column.	Same as `set_ftrace_filter` – trace only the functions written here. However, this is only relevant when the `function_graph` tracer is being employed.
`set_graph_notrace`	See the next column.	A complement of the preceding. Same as `set_ftrace_notrace` – never trace the functions written here. However, this is only relevant when the `function_graph` tracer is being employed.

`tracing_cpumask`	CPU cores (default – all online CPU cores are traced)	Write a bitmask specifying which CPU core(s) to trace to this file. Also, to see the trace on a given CPU, you can read from `per_cpu/cpun/trace[_pipe]` file, where n is the core number.
`set_ftrace_pid`	Process/thread PID(s) to trace.	Trace only kernel code paths executed by the PID(s) written to this file. Useful!
`set_ftrace_notrace_pid`	Process/thread PID(s) to not trace.	A complement of the preceding – do not trace code executed by these PID(s).
`set_event`	Filter only functions belonging to the set of events written here.	An alternate way to trace. These pertain to static tracepoints built into the kernel (for example, `net`, `sock`, `syscalls`, `tcp`, and so on). Please refer to the following for details: `https://www.kernel.org/doc/html/v5.10/trace/events.html#via-the-set-event-interface`.
`set_event_pid`	Process/thread PID(s) to trace when event tracing	With regard to event-based tracing, events will trace only kernel code paths executed by the PID(s) written to this file. Useful!
`set_event_notrace_pid`	Process/thread PID(s) to not trace when event tracing	A complement of the preceding – do not trace code executed by these PID(s) when performing event-based tracing.

Table 9.4 – A summary of some key ftrace filter options

The following sections help to flesh out this discussion. Please check them out, as they have some valuable details.

More on the set_ftrace_filter file

For function-based filtering, glob matching (commonly called **globbing**) can be very useful! You can specify a subset of functions (which must be present within available_ filter_functions, of course) using wildcards, like so:

- 'foo*': All function names that start with foo

- '*foo': All function names that end with foo

- '*foo*': All function names that contain the foo substring

- 'foo*bar': All function names that start with foo and end with bar

For example, to trace the ksys_write() and ksys_read() functions, we can do this:

```
echo "ksys_write" > set_ftrace_filter
echo "ksys_read" >> set_ftrace_filter
```

This also illustrates that you can employ the >> append notation with the usual semantics; here, both these functions (and their nested code by default) get traced.

There's actually a lot more control you can exert with the set_ftrace_filter file; instead of simply duplicating it here, you can view this (and a lot more!) by reading the content of the /sys/kernel/tracing/README file – a nice mini-HOWTO on ftrace! Here's the portion relevant to the set_ftrace_filter file:

```
cat /sys/kernel/tracing/README
tracing mini-HOWTO:
[...]
```

The following shows the relevant portion of the output:

```
available_filter_functions - list of functions that can be filtered on
set_ftrace_filter    - echo function name in here to only trace these
                       functions
         accepts: func_full_name or glob-matching-pattern
         modules: Can select a group via module
          Format: :mod:<module-name>
         example: echo :mod:ext3 > set_ftrace_filter
        triggers: a command to perform when function is hit
          Format: <function>:<trigger>[:count]
         trigger: traceon, traceoff
                  enable_event:<system>:<event>
                  disable_event:<system>:<event>
                  stacktrace
                  snapshot
                  dump
                  cpudump
         example: echo do_fault:traceoff > set_ftrace_filter
                  echo do_trap:traceoff:3 > set_ftrace_filter
         The first one will disable tracing every time do_fault is hit
         The second will disable tracing at most 3 times when do_trap is hit
           The first time do_trap is hit and it disables tracing, the
           counter will decrement to 2. If tracing is already disabled,
           the counter will not decrement. It only decrements when the
           trigger did work
         To remove trigger without count:
           echo '!<function>:<trigger> > set_ftrace_filter
         To remove trigger with a count:
           echo '!<function>:<trigger>:0 > set_ftrace_filter
set_ftrace_notrace    - echo function name in here to never trace.
         accepts: func_full_name, *func_end, func_begin*, *func_middle*
         modules: Can select a group via module command :mod:
         Does not accept triggers
```

Figure 9.10 – A partial screenshot of the content of the README mini-HOWTO file

So, here's another interesting thing – you can disable tracing when a certain function is hit (see the `traceoff` examples in *Figure 9.10*).

Index-based function filtering

String processing (as seen in the preceding section) can dramatically slow things down! So, an alternative is index-based filtering. The index is the numerical position of the function you want (or don't want, depending on the filter) in the `available_filter_functions` file. For example, let's use `sed` to look up the line numbers of all kernel functions that have the `tcp` string within them:

```
# grep -n tcp available_filter_functions |cut -f1 -d':'|tr '\n'
' '
```

```
3504 3505 30425 30426 30427 30428 30429 30430 30431 30432 30433
30434 30435 38537 38540 38541 38542 39133 39134 39198 [...]
43589 43590 43591 43593 #
```

It's a long list of line numbers (584 on 5.10.60):

```
# grep -n tcp available_filter_functions |cut -f1 -d':'|tr '\n'
' ' |wc -w
584
```

We use the `tr` utility to replace the newline with a space. Why? Because this is the format expected by the filter file. So, to set ftrace to trace only these functions (the ones with `tcp` in their name), we do the following:

```
grep -n tcp available_filter_functions   | cut -f1 -d':' | tr
'\n' ' ' >> set_ftrace_filter
```

We use this very technique for function filtering in one of our scripts (which we'll use in a later section) – `ch9/ftrace/ping_ftrace.sh`. Here's a bash function to do this generically (the parameter(s) are the substring or regular expression that the function name contains):

```
filterfunc_idx()
{
  [ $# -lt 1 ] && return
  local func
  for func in "$@"
  do
      echo $(grep -i -n ${func} available_filter_functions |cut
-f1 -d':'|tr '\n' ' ') >> set_ftrace_filter
  done
}
```

Here's a sample invocation:

```
filterfunc_idx read write net packet_ sock sk_ tcp udp \
  skb netdev netif_ napi icmp "^ip_" "xmit$" dev_ qdisc
```

Now, all kernel functions containing these substrings (or regular expressions) will be traced! Cool, huh?

How about the reverse? We'd perhaps like to filter out some functions, telling ftrace *to not trace them*. In the same script that we just referred to (ch9/ftrace/ping_ftrace. sh), we write this bash function to achieve this:

```
filterfunc_remove()
{
  [ $# -lt 1 ] && return
  local func
  for func in "$@"
  do
    echo "!${func}" >> set_ftrace_filter
    echo "${func}" >> set_graph_notrace
  done
}
```

Note the ! character prefixing the function string; this tells ftrace to not trace the matching functions. Similarly, writing the parameter into the set_graph_notrace file achieves the same (when the tracer is function_graph, as is the case here). Here's a sample invocation:

```
filterfunc_remove "*idle*" "tick_nohz_idle_stop_tick" "*_
rcu_*" "*down_write*" "*up_write*" [...]
```

These techniques not only achieve function filtering but are also fast; useful.

Module-based filtering

Write a string of the :mod:<module-name> form into set_ftrace_filter to allow functions from this module to be traced, as shown in the following example:

```
echo :mod:ext4 > set_ftrace_filter
```

This traces all functions within the ext4 kernel module. This example, and the usage of the mod keyword, is one case of using something called **filter commands**.

Filter commands

More powerfully, you can even employ so-called filter commands by writing in a certain format into the set_ftrace_filter file. The format is as follows:

```
echo '<function>:<command>:<parameter>' > set_ftrace_filter
```

Using the >> append form is also supported. A few commands are supported – mod, traceon/traceoff, snapshot, enable_event/disable_event, dump, cpudump, and stacktrace. Do refer to the official kernel docs for more details and examples of filter commands: https://www.kernel.org/doc/html/v5.10/trace/ftrace.html#filter-commands.

Note that filter commands being set do not affect ftrace filters! For example, setting a filter command to trace some module's functions or to toggle tracing on some condition does not affect the functions being traced via the set_ftrace_filter file.

Okay, you've seen a lot on how to set up and use ftrace – configuring the kernel, simple tracing, and plenty of the more powerful filtering maneuvers. Let's put this knowledge to practical use, leveraging ftrace to trace the kernel code paths taken when the following command – a single ping – is issued:

```
ping -c1 packtpub.com
```

Now, that would be interesting! Let's get this going.

Case 1 – tracing a single ping with raw ftrace

We'll employ the function_graph tracer throughout. The filtering of functions can be done in one of two broad ways here:

- Via the regular available_filter_functions file interface
- Via the set_event file interface

They're mutually exclusive; if we set our script's (ch9/ftrace/ping_ftrace.sh) FILTER_VIA_AVAIL_FUNCS variable to 1 (which is the default), then we filter via method 1, else via method 2. The first gives a much more detailed trace, showing all relevant functions (here, the networking-related ones), but it takes a bit more work – via index-based filtering – to set up and keep it quick. In our script, we keep filtering via the first method as the default:

```
// ch9/ftrace/ping_ftrace.sh
[...]
FILTER_VIA_AVAIL_FUNCS=1
echo "[+] Function filtering:"
if [ ${FILTER_VIA_AVAIL_FUNCS} -eq 1 ] ; then
    [...]
    # This is pretty FAST and yields good detail!
    filterfunc_idx read write net packet_ sock sk_ tcp udp skb
```

```
netdev \
    netif_ napi icmp "^ip_" "xmit$" dev_ qdisc [...]
    [...]
```

The usage and implementation of our `filterfunc_idx()` function was explained in the preceding *Index-based function filtering* section. Also, we use our `filterfunc_remove()` function to ensure that certain patterns of functions within the kernel aren't traced. Further, the script turns on tracing for any functions from the `e1000` module (the network driver, really):

```
KMOD=e1000
echo "[+] module filtering (for ${KMOD})"
if lsmod|grep ${KMOD} ; then
  echo "[+] setting filter command: :mod:${KMOD}"
  echo ":mod:${KMOD}" >> set_ftrace_filter
fi
```

A bit of complexity in the script occurs when implementing the code to run the `ping`. Why? Well, as far as is possible, we'd like to only trace the `ping` process on the CPU core on which it runs. To do so, we assign it to a specific CPU core by leveraging the `taskset` utility. We then tell ftrace the following:

- Which process to trace by setting `set_event_pid` to the PID of the `ping` process
- Which CPU core(s) to trace by setting the `tracing_cpumask` filter to a specified value, the one we set `taskset` to use.

Okay, now, let's think about this. To set the PID correctly, we first need the `ping` process running, but if it's running, it will execute (at least some) code before our script can set up and trace it. However, we can't obtain its PID until it's alive and running. A bit of a chicken-and-egg problem, isn't it?

So, we employ a wrapper script (named `runner`) to run the `ping` process. This script will synchronize with our main script (`ping_ftrace.sh`). It does so like this – the main script will run the `runner` script in the background and grab its PID, saving it in a variable (named `PID`). Then, the `runner` script will execute – via the shell's `exec` keyword – the `ping` process, thus ensuring that the PID of our `ping` process will be the same as it's PID (this is because, when an `exec` operation is performed, the successor process ends up with the same PID as that of the predecessor)!

Hang on, though. To properly synchronize, the `runner` script won't perform the `exec` operation until a so-called "trigger file" is created by the main script. It will then understand that only now tracing has become ready, and it will exec the `ping` process. The main script thus, once ready, creates the trigger file and then turns tracing on, tracing the target process.

When the process is done, it saves the report (I'll leave it to you to browse through the code of these scripts). If you (quite rightly) feel that this is a rather laborious way to trace a given task (process/thread), you'd be right. Our coverage on the `trace-cmd` frontend to ftrace will clearly reveal this!

Here's a sample run of our `ping_ftrace.sh` script:

```
$ sudo ./ping_ftrace.sh
[+] resetting ftrace
trace-cmd reset     (patience, pl...)
resetting set_ftrace_filter
resetting set_ftrace_notrace
resetting set_ftrace_notrace_pid
resetting set_ftrace_pid
resetting trace_options to defaults (as of 5.10.60)
resetting options/funcgraph-*
running '/usr/sbin/reset-ftrace-perf -q' now...
[+] tracer : function_graph
[+] setting options
[+] setting buffer size to 82 MB / cpu
[+] Function filtering:
 Regular filtering (via available_filter_functions):
 Setting filters for networking funcs only...
[+] filter: remove unwanted functions        (patience, pl...)
# of functions now being traced: 6649
[+] module filtering (for e1000)
e1000               143360  0
[+] setting filter command: :mod:e1000
[+] Setting up wrapper runner process now...
[+] Tracing PID 1556 on CPU 1 now ...
> runner:1556: triggered
PING packtpub.com (104.22.1.175) 56(84) bytes of data.
64 bytes from 104.22.1.175 (104.22.1.175): icmp_seq=1 ttl=63 time=15.2 ms

--- packtpub.com ping statistics ---
1 packets transmitted, 1 received, 0% packet loss, time 0ms
rtt min/avg/max/mdev = 15.167/15.167/15.167/0.000 ms
Ftrace report:
-rw-r--r-- 1 root root 272K Feb 25 13:23 /home/letsdebug/Linux-Kernel-Debugging/ch9/ftrace/ftrace_reports/pin
g_ftrace.sh_20220225.txt
$
```

Figure 9.11 – A screenshot showing our raw ftrace script tracing a single ping

You can see from the preceding screenshot that our scripts ran, that the `ping` process did its job and was traced, and the report was generated. The report obtained has a very decent size – just 272 KB here – implying that our filtering has indeed paid off.

FYI, this sample report is available here: ch9/ftrace/ping_ftrace_report.txt. (You will realize, though, that this report represents just one sample run on my setup; the results you see will likely vary, at least slightly.) It's still quite big; a few good places to start looking within this trace report are as follows:

- The call to __sys_socket() by the ping process.

- The transmit path (often called tx; more details follow). See the sock_sendmsg() API initiating the transmit path (you'll find it – among other instances – in our saved ping_ftrace_report.txt file at about 86% of the way down).

- The receive path (often called rx; more details follow).

The (very approximate) transmit path functionality (shown in a top-down fashion) includes these kernel functions – sock_sendmsg(), inet_sendmsg() (you can see these functions highlighted in the upper portion of *Figure 9.12*), udp_sendmsg(), udp_send_skb() (you can see the functions that follow them highlighted in the lower portion of *Figure 9.13*), ip_send_skb(), ip_output(), dev_queue_xmit(), dev_hard_start_xmit() (the last call leads to the e1000 network driver's transmit code routine; here, it's named e1000_xmit_frame()), and so on. The transmit and receive function lists here are certainly not exhaustive, just an indication of what you can expect to see.

Figure 9.12 shows a portion of the filtered report showing some of the transmit paths, where the ping process context is running this code within the kernel. (The leftmost column is merely the line number. Of course, this is what I got on one sample run. What you see may not precisely match this output):

```
2258  3024.237642 |   1)  ping-1869  | .... |            | sock_sendmsg() {
2259  3024.237643 |   1)  ping-1869  | .... |            | security_socket_sendmsg() {
2260  3024.237644 |   1)  ping-1869  | .... |            | apparmor_socket_sendmsg() {
2261  3024.237645 |   1)  ping-1869  | .... |   0.738 us |   aa_sk_perm();
2262  3024.237646 |   1)  ping-1869  | .... |   2.587 us | } /* apparmor_socket_sendmsg */
2263  3024.237647 |   1)  ping-1869  | .... |   4.816 us | } /* security_socket_sendmsg */
2264  3024.237649 |   1)  ping-1869  | .... |            | inet_sendmsg() {
2265  3024.237650 |   1)  ping-1869  | .... |            | inet_send_prepare() {
2266  3024.237652 |   1)  ping-1869  | .... |            | inet_autobind() {
2267  3024.237652 |   1)  ping-1869  | .... |   1.262 us | lock_sock_nested();
2268  3024.237655 |   1)  ping-1869  | .... |   1.135 us | _raw_write_lock_bh();
2269  3024.237658 |   1)  ping-1869  | ...1 |   1.537 us | sock_prot_inuse_add();
2270  3024.237661 |   1)  ping-1869  | ...1 |   1.115 us | _raw_write_unlock_bh();
2271  3024.237663 |   1)  ping-1869  | .... |   1.475 us | release_sock();
2272  3024.237665 |   1)  ping-1869  | .... | + 13.470 us | } /* inet_autobind */
2273  3024.237666 |   1)  ping-1869  | .... | + 15.571 us | } /* inet_send_prepare */
```

Figure 9.12 – A portion of the transmit portion of the ping trace via the raw ftrace regular filtering interface

Here is some more ftrace output from the transmit path:

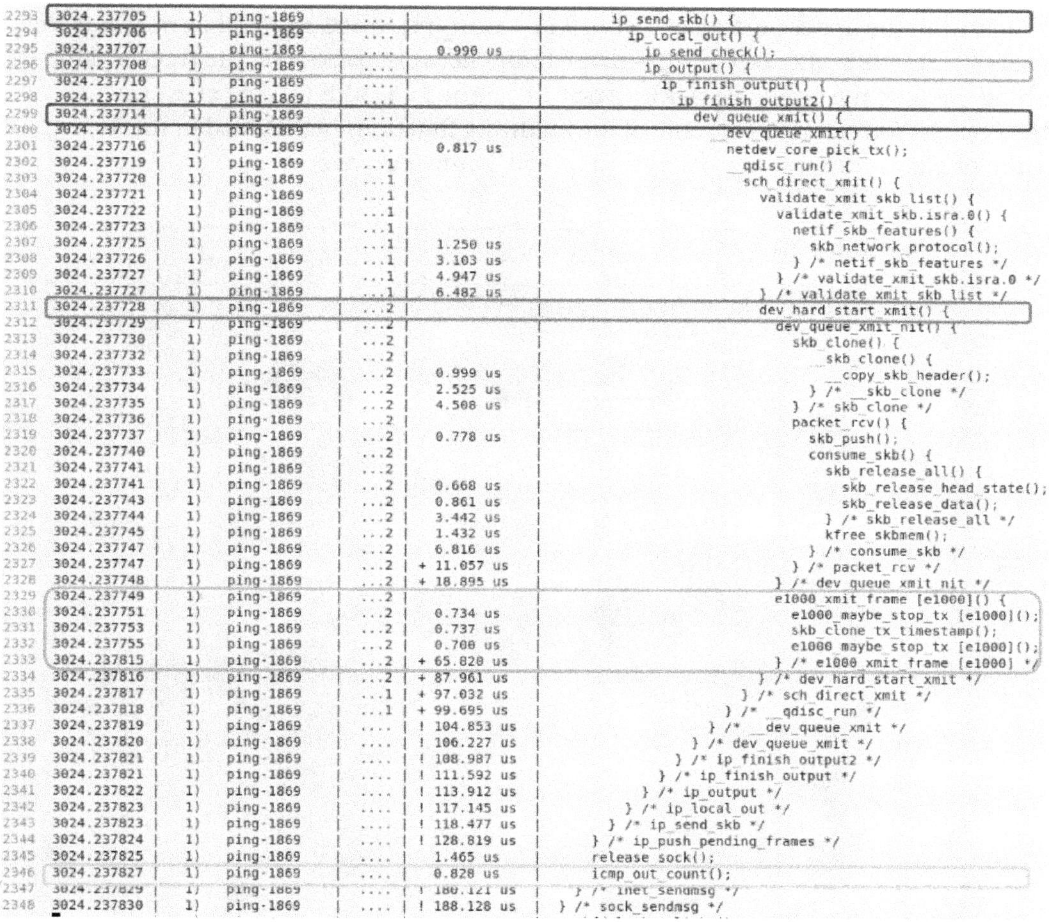

```
3024.237705 |  1)  ping-1869   | .... |              |  ip_send_skb() {
3024.237706 |  1)  ping-1869   | .... |              |    ip_local_out() {
3024.237707 |  1)  ping-1869   | .... |   0.990 us   |      ip_send_check();
3024.237708 |  1)  ping-1869   | .... |              |      ip_output() {
3024.237710 |  1)  ping-1869   | .... |              |        ip_finish_output() {
3024.237712 |  1)  ping-1869   | .... |              |          ip_finish_output2() {
3024.237714 |  1)  ping-1869   | .... |              |            dev_queue_xmit() {
3024.237715 |  1)  ping-1869   | .... |              |              __dev_queue_xmit() {
3024.237716 |  1)  ping-1869   | .... |   0.817 us   |                netdev_core_pick_tx();
3024.237719 |  1)  ping-1869   | ...1 |              |                __qdisc_run() {
3024.237720 |  1)  ping-1869   | ...1 |              |                  sch_direct_xmit() {
3024.237721 |  1)  ping-1869   | ...1 |              |                    validate_xmit_skb_list() {
3024.237722 |  1)  ping-1869   | ...1 |              |                      validate_xmit_skb.isra.0() {
3024.237723 |  1)  ping-1869   | ...1 |              |                        netif_skb_features() {
3024.237725 |  1)  ping-1869   | ...1 |   1.250 us   |                          skb_network_protocol();
3024.237726 |  1)  ping-1869   | ...1 |   3.103 us   |                        } /* netif_skb_features */
3024.237727 |  1)  ping-1869   | ...1 |   4.947 us   |                      } /* validate_xmit_skb.isra.0 */
3024.237727 |  1)  ping-1869   | ...1 |   6.482 us   |                    } /* validate_xmit_skb_list */
3024.237728 |  1)  ping-1869   | ...2 |              |                    dev_hard_start_xmit() {
3024.237729 |  1)  ping-1869   | ...2 |              |                      dev_queue_xmit_nit() {
3024.237730 |  1)  ping-1869   | ...2 |              |                        skb_clone() {
3024.237732 |  1)  ping-1869   | ...2 |              |                          __skb_clone() {
3024.237733 |  1)  ping-1869   | ...2 |   0.999 us   |                            __copy_skb_header();
3024.237734 |  1)  ping-1869   | ...2 |   2.525 us   |                          } /* __skb_clone */
3024.237735 |  1)  ping-1869   | ...2 |   4.508 us   |                        } /* skb_clone */
3024.237736 |  1)  ping-1869   | ...2 |              |                        packet_rcv() {
3024.237737 |  1)  ping-1869   | ...2 |   0.778 us   |                          skb_push();
3024.237740 |  1)  ping-1869   | ...2 |              |                          consume_skb() {
3024.237741 |  1)  ping-1869   | ...2 |              |                            skb_release_all() {
3024.237741 |  1)  ping-1869   | ...2 |   0.668 us   |                              skb_release_head_state();
3024.237743 |  1)  ping-1869   | ...2 |   0.861 us   |                              skb_release_data();
3024.237744 |  1)  ping-1869   | ...2 |   3.442 us   |                            } /* skb_release_all */
3024.237745 |  1)  ping-1869   | ...2 |   1.432 us   |                            kfree_skbmem();
3024.237747 |  1)  ping-1869   | ...2 |   6.816 us   |                          } /* consume_skb */
3024.237747 |  1)  ping-1869   | ...2 | + 11.057 us  |                        } /* packet_rcv */
3024.237748 |  1)  ping-1869   | ...2 | + 18.895 us  |                      } /* dev_queue_xmit_nit */
3024.237749 |  1)  ping-1869   | ...2 |              |                      e1000_xmit_frame [e1000]() {
3024.237751 |  1)  ping-1869   | ...2 |   0.734 us   |                        e1000_maybe_stop_tx [e1000]();
3024.237753 |  1)  ping-1869   | ...2 |   0.737 us   |                        skb_clone_tx_timestamp();
3024.237755 |  1)  ping-1869   | ...2 |   0.700 us   |                        e1000_maybe_stop_tx [e1000]();
3024.237815 |  1)  ping-1869   | ...2 | + 65.820 us  |                      } /* e1000_xmit_frame [e1000] */
3024.237816 |  1)  ping-1869   | ...2 | + 87.961 us  |                    } /* dev_hard_start_xmit */
3024.237817 |  1)  ping-1869   | ...1 | + 97.032 us  |                  } /* sch_direct_xmit */
3024.237818 |  1)  ping-1869   | ...1 | + 99.695 us  |                } /* __qdisc_run */
3024.237819 |  1)  ping-1869   | .... | ! 104.853 us |              } /* __dev_queue_xmit */
3024.237820 |  1)  ping-1869   | .... | ! 106.227 us |            } /* dev_queue_xmit */
3024.237821 |  1)  ping-1869   | .... | ! 108.987 us |          } /* ip_finish_output2 */
3024.237821 |  1)  ping-1869   | .... | ! 111.592 us |        } /* ip_finish_output */
3024.237822 |  1)  ping-1869   | .... | ! 113.912 us |      } /* ip_output */
3024.237823 |  1)  ping-1869   | .... | ! 117.145 us |    } /* ip_local_out */
3024.237823 |  1)  ping-1869   | .... | ! 118.477 us |  } /* ip_send_skb */
3024.237824 |  1)  ping-1869   | .... | ! 128.819 us |  } /* ip_push_pending_frames */
3024.237825 |  1)  ping-1869   | .... |   1.465 us   |  release_sock();
3024.237827 |  1)  ping-1869   | .... |   0.828 us   |  icmp_out_count();
3024.237829 |  1)  ping-1869   | .... | ! 180.121 us |  } /* inet_sendmsg */
3024.237830 |  1)  ping-1869   | .... | ! 188.128 us | } /* sock_sendmsg */
```

Figure 9.13 – Partial screenshots – a portion of the transmit portion of the ping trace via the raw ftrace regular filtering interface

Interesting, isn't it? This simple exercise shows us that ftrace is (also) literally a means to see the kernel in action (here, of course, we're seeing a portion of the kernel network stack in action), to be empirical, and to test theory.

In a similar fashion, on the network receive path functionality, we'll see some of the typical kernel routines that are invoked. You can see some of the following functions highlighted in the upper portion of *Figure 9.14* – `net_rx_action()` (this, in fact, is the kernel softirq – `NET_RX_SOFTIRQ` – that handles network packets on the receive path, pushing them up the protocol stack), `__netif_receive_skb()`, `ip_rcv()`, and `udp_rcv()`. You can also see some of the following functions highlighted in the lower portion of *Figure 9.15* – `sock_recvmsg()` and `inet_recvmsg()`:

```
2551  3824.253216 |   1)   ping-1869    | .....| + 27.963 us     |            } /* dev_queue_xmit */
2552  3824.253218 |   1)   ping-1869    | ..s1 |                 |          net_rx_action() {
2553  3824.253220 |   1)   ping-1869    | ..s1 |                 |            __netif_receive_skb() {
2554  3824.253221 |   1)   ping-1869    | ..s1 |                 |              __netif_receive_skb_one_core() {
2555  3824.253222 |   1)   ping-1869    | ..s1 | 1.311 us        |                __netif_receive_skb_core();
2556  3824.253224 |   1)   ping-1869    | ..s1 |                 |                ip_rcv() {
2557  3824.253225 |   1)   ping-1869    | ..s1 | 1.045 us        |                  ip_rcv_core();
2558  3824.253227 |   1)   ping-1869    | ..s1 |                 |                  ip_rcv_finish() {
2559  3824.253228 |   1)   ping-1869    | ..s1 | 0.868 us        |                    ip_rcv_finish_core.isra.0();
2560  3824.253229 |   1)   ping-1869    | ..s1 |                 |                    ip_local_deliver() {
2561  3824.253230 |   1)   ping-1869    | ..s1 |                 |                      ip_local_deliver_finish() {
2562  3824.253231 |   1)   ping-1869    | ..s1 |                 |                        ip_protocol_deliver_rcu() {
2563  3824.253232 |   1)   ping-1869    | ..s1 |                 |                          udp_rcv() {
2564  3824.253233 |   1)   ping-1869    | ..s1 |                 |                            __udp4_lib_rcv() {
2565  3824.253235 |   1)   ping-1869    | ..s1 |                 |                              udp4_lib_lookup() {
2566  3824.253236 |   1)   ping-1869    | ..s1 | 1.753 us        |                                udp4_lib_lookup2();
2567  3824.253238 |   1)   ping-1869    | ..s1 | 3.316 us        |                              } /* __udp4_lib_lookup */
2568  3824.253239 |   1)   ping-1869    | ..s1 |                 |                              udp_unicast_rcv_skb() {
2569  3824.253240 |   1)   ping-1869    | ..s1 |                 |                                udp_queue_rcv_skb() {
2570  3824.253240 |   1)   ping-1869    | ..s1 |                 |                                  udp_queue_rcv_one_skb() {
2571  3824.253242 |   1)   ping-1869    | ..s1 |                 |                                    sk_filter_trim_cap() {
2572  3824.253243 |   1)   ping-1869    | ..s1 |                 |                                      security_sock_rcv_skb() {
2573  3824.253244 |   1)   ping-1869    | ..s1 | 0.682 us        |                                        apparmor_socket_sock_rcv_skb();
2574  3824.253245 |   1)   ping-1869    | ..s1 | 2.214 us        |                                      } /* security_sock_rcv_skb */
2575  3824.253246 |   1)   ping-1869    | ..s1 | 4.543 us        |                                    } /* sk_filter_trim_cap */
2576  3824.253247 |   1)   ping-1869    | ..s1 | 0.947 us        |                                    skb_pull_rcsum();
2577  3824.253250 |   1)   ping-1869    | ..s1 |                 |                                    __udp_enqueue_schedule_skb() {
2578  3824.253251 |   1)   ping-1869    | ..s1 |                 |                                      sock_def_readable() {
2579  3824.253253 |   1)   ping-1869    | d.s2 | 1.869 us        |                                        _raw_read_lock_irqsave();
2580  3824.253257 |   1)   ping-1869    | d.s6 | 2.534 us        |                                        select_task_rq_fair();
2581  3824.253295 |   1)   ping-1869    | d.s3 | 1.471 us        |                                        _raw_read_unlock_irqrestore();
2582  3824.253297 |   1)   ping-1869    | ..s1 | + 46.191 us     |                                      } /* sock_def_readable */
2583  3824.253298 |   1)   ping-1869    | ..s1 | + 48.610 us     |                                    } /* __udp_enqueue_schedule_skb */
2584  3824.253298 |   1)   ping-1869    | ..s1 | + 58.163 us     |                                  } /* udp_queue_rcv_one_skb */
2585  3824.253299 |   1)   ping-1869    | ..s1 | + 59.576 us     |                                } /* udp_queue_rcv_skb */
2586  3824.253300 |   1)   ping-1869    | ..s1 | + 61.398 us     |                              } /* udp_unicast_rcv_skb */
2587  3824.253300 |   1)   ping-1869    | ..s1 | + 67.409 us     |                            } /* __udp4_lib_rcv */
2588  3824.253301 |   1)   ping-1869    | ..s1 | + 69.129 us     |                          } /* udp_rcv */
2589  3824.253302 |   1)   ping-1869    | ..s1 | + 71.250 us     |                        } /* ip_protocol_deliver_rcu */
2590  3824.253303 |   1)   ping-1869    | ..s1 | + 72.830 us     |                      } /* ip_local_deliver_finish */
2591  3824.253303 |   1)   ping-1869    | ..s1 | + 74.327 us     |                    } /* ip_local_deliver */
2592  3824.253304 |   1)   ping-1869    | ..s1 | + 77.403 us     |                  } /* ip_rcv_finish */
2593  3824.253305 |   1)   ping-1869    | ..s1 | + 80.581 us     |                } /* ip_rcv */
2594  3824.253305 |   1)   ping-1869    | ..s1 | + 84.347 us     |              } /* __netif_receive_skb_one_core */
2595  3824.253306 |   1)   ping-1869    | ..s1 | + 86.140 us     |            } /* __netif_receive_skb */
2596  3824.253308 |   1)   ping-1869    | ..s1 | 1.185 us        |            __kfree_skb_flush();
2597  3824.253309 |   1)   ping-1869    | ..s1 | + 91.088 us     |          } /* net_rx_action */
2598  3824.253311 |   1)   ping-1869    | .... | ! 125.340 us    |        } /* ip_finish_output2 */
2599  3824.253312 |   1)   ping-1869    | .... | ! 127.864 us    |      } /* ip_finish_output */
2600  3824.253312 |   1)   ping-1869    | .... | ! 130.798 us    |    } /* ip_output */
2601  3824.253313 |   1)   ping-1869    | .... | ! 134.884 us    |  } /* ip_local_out */
2602  3824.253314 |   1)   ping-1869    | .... | ! 136.662 us    | } /* ip_send_skb */
2603  3824.253314 |   1)   ping-1869    | .... | ! 149.710 us    | } /* udp_send_skb */
2604  3824.253315 |   1)   ping-1869    | .... | ! 167.463 us    | } /* udp_sendmsg */
2605  3824.253316 |   1)   ping-1869    | .... | ! 171.503 us    | } /* inet_sendmsg */
2606  3824.253317 |   1)   ping-1869    | .... | ! 178.015 us    | } /* sock_sendmsg */
```

Figure 9.14 – Partial screenshots – a portion of the receive portion of the ping trace via the raw ftrace regular filtering interface

Here is some more ftrace output from the receive path:

```
2632  3024.356151 |   1)   ping-1869   | ....|              sock_recvmsg() {
2633  3024.356152 |   1)   ping-1869   | ....|                security_socket_recvmsg() {
2634  3024.356154 |   1)   ping-1869   | ....|                  apparmor_socket_recvmsg() {
2635  3024.356155 |   1)   ping-1869   | ....|    2.265 us   |                    aa_sk_perm();
2636  3024.356157 |   1)   ping-1869   | ....|    3.719 us   |                  } /* apparmor_socket_recvmsg */
2637  3024.356158 |   1)   ping-1869   | ....|    5.898 us   |                } /* security_socket_recvmsg */
2638  3024.356159 |   1)   ping-1869   | ....|              inet_recvmsg() {
2639  3024.356160 |   1)   ping-1869   | ....|                udp_recvmsg() {
2640  3024.356161 |   1)   ping-1869   | ....|                  __skb_recv_udp() {
2641  3024.356162 |   1)   ping-1869   | ...1|    0.807 us   |                    __skb_try_recv_from_queue();
2642  3024.356164 |   1)   ping-1869   | ...1|    1.446 us   |                    udp_rmem_release();
2643  3024.356166 |   1)   ping-1869   | ....|    5.221 us   |                  } /* __skb_recv_udp */
2644  3024.356170 |   1)   ping-1869   | ....|    1.018 us   |                  lock_sock_nested();
2645  3024.356173 |   1)   ping-1869   | ....|    2.926 us   |                  release_sock();
2646  3024.356177 |   1)   ping-1869   | ....|                  skb_consume_udp() {
2647  3024.356178 |   1)   ping-1869   | ....|                    __consume_stateless_skb() {
2648  3024.356179 |   1)   ping-1869   | ....|                      skb_release_data() {
2649  3024.356180 |   1)   ping-1869   | ....|    2.715 us   |                        skb_free_head();
2650  3024.356183 |   1)   ping-1869   | ....|    4.676 us   |                      } /* skb_release_data */
2651  3024.356184 |   1)   ping-1869   | ....|    2.593 us   |                      kfree_skbmem();
2652  3024.356187 |   1)   ping-1869   | ....|    9.650 us   |                    } /* __consume_stateless_skb */
2653  3024.356187 |   1)   ping-1869   | ....| + 11.194 us   |                  } /* skb_consume_udp */
2654  3024.356188 |   1)   ping-1869   | ....| + 28.239 us   |                } /* udp_recvmsg */
2655  3024.356189 |   1)   ping-1869   | ....| + 30.201 us   |              } /* inet_recvmsg */
2656  3024.356189 |   1)   ping-1869   | ....| + 38.668 us   |            } /* sock_recvmsg */
2657  3024.356241 |   1)   ping-1869   | ....|              sock_close() {
```

Figure 9.15 – Partial screenshots – a portion of the receive portion of the ping trace via the raw ftrace regular filtering interface

I enjoy seeing the ftrace `function_graph` tracer report's function indentation dramatically move from left to right (as functions are invoked) and vice versa (as the functions return). This is great. At the very least, you can see how ftrace has allowed us to look deep into the kernel's network protocol stack!

Case 2 – tracing a single ping with raw ftrace via the set_event interface

Here, we change the way ftrace grabs (and even presents) information by employing an alternate means to specify what functions to trace – via the `set_event` interface. To use this method, you write the function(s) to trace into the `set_event` pseudofile. This won't be very different from what we just did in the previous section, using the `available_filter_functions` pseudofile, will it? The trick here is that we can specify *a whole class of functions to trace* by enabling a set of events; the events fall into classes such as `net`, `sock`, `skb`, and so on. How do we do this? Hang on a second...

Where do these events come from? Ah, they're the *kernel tracepoints*! You can see them all under the `events` directory within the `tracefs` mount point. The following screenshot makes this apparent:

```
# pwd
/sys/kernel/tracing
# ls events/
alarmtimer/     ftrace/         mce/            random/         task/
avc/            gpio/           mdio/           ras/            tcp/
block/          header_event    migrate/        raw_syscalls/   thermal/
bpf_test_run/   header_page     mmap/           rcu/            thermal_power_allocator/
bpf_trace/      huge_memory/    mmc/            regmap/         timer/
cgroup/         hwmon/          module/         regulator/      tlb/
clk/            i2c/            msr/            resctrl/        udp/
compaction/     initcall/       napi/           rpm/            vmscan/
cpuhp/          intel_iommu/    neigh/          rseq/           vsyscall/
devfreq/        interconnect/   net/            rtc/            wbt/
dma_fence/      iocost/         nmi/            sched/          workqueue/
drm/            iomap/          oom/            scsi/           writeback/
enable          iommu/          page_isolation/ signal/         x86_fpu/
exceptions/     io_uring/       pagemap/        skb/            xdp/
ext4/           irq/            page_pool/      smbus/          xen/
fib/            irq_matrix/     percpu/         sock/           xhci-hcd/
fib6/           irq_vectors/    power/          spi/
filelock/       jbd2/           printk/         swiotlb/
filemap/        kmem/           pwm/            sync_trace/
fs_dax/         libata/         qdisc/          syscalls/
#
```

Figure 9.16 – A screenshot showing the content of the events directory – all kernel tracepoints

We have, in fact, covered using kernel tracepoints in some detail with regard to dynamic kprobes in *Chapter 4, Debug via Instrumentation – Kprobes*, in the *Using the event tracing framework to trace built-in functions* section.

To illustrate the usage of a class of events, let's consider the one named `net`. (You can see it as a directory in *Figure 9.16*. Peeking within the `/sys/kernel/tracing/events/net` directory will reveal (again, as directories) all the kernel functions that can be traced via this class of tracepoints.) So, to tell ftrace that we want to trace all these network-related functions, we simply have to echo the `net:*` string into the `set_event` pseudofile!

The relevant code – where we set up to use the set_event interface to ftrace – from our script follows. Also, the only change required in the script is setting the FILTER_VIA_AVAIL_FUNCS variable to 0 (to try this case out, you'll have to manually make this edit in the script):

```
// ch9/ftrace/ping_ftrace.sh
[...]
FILTER_VIA_AVAIL_FUNCS=0
echo "[+] Function filtering:"
if [ ${FILTER_VIA_AVAIL_FUNCS} -eq 1 ] ; then
   [... already seen above ...]
else # filter via the set_event interface
  # This is FAST but doesn't yield as much detail!
  # We also seem to lose the function graph indentation (but do
gain seeing function parameters!)
  echo " Alternate event-based filtering (via set_event):"
  echo 'net:* sock:* skb:* tcp:* udp:* napi:* qdisc:* neigh:*
syscalls:*' >> set_event
 fi
```

Note how we try and trace only network-related kernel code (as well as all system calls, to lend context to the trace report). The remainder of the script's code is identical to what we saw in the previous section.

Here's a sampling of the output report (filtered to see only the `ping` process's work within the kernel) with this `set_event`-based approach:

```
403  9344.129025 |   1)    ping-1922   | ...1 |                | /* sys_socket(family: 2, type: 80802, protocol: 0) */
404  9344.129045 |   1)    ping-1922   | ...1 |                | /* sys_socket -> 0x5 */
405  9344.129049 |   1)    ping-1922   | ...1 |                | /* sys_setsockopt(fd: 5, level: 0, optname: b, optval: 7ffd9c843f04
, optlen: 4) */
406  9344.129055 |   1)    ping-1922   | ...1 |                | /* sys_setsockopt -> 0x0 */
407  9344.129058 |   1)    ping-1922   | ...1 |                | /* sys_connect(fd: 5, uservaddr: 782a59c864d4, addrlen: 10) */
408  9344.129080 |   1)    ping-1922   | ...1 |                | /* sys_connect -> 0x0 */
409  9344.129085 |   1)    ping-1922   | ...1 |                | /* sys_poll(ufds: 7ffd9c8440a8, nfds: 1, timeout_msecs: 0) */
410  9344.129089 |   1)    ping-1922   | ...1 |                | /* sys_poll -> 0x1 */
411  9344.129092 |   1)    ping-1922   | ...1 |                | /* sys_sendmmsg(fd: 5, mmsg: 7ffd8440d0, vlen: 2, flags: 4000) */
412  9344.129114 |   1)    ping-1922   | .... |                | /* net_dev_queue: dev=lo skbaddr=0000000024fe4664 len=83 */
413  9344.129117 |   1)    ping-1922   | .... |                | /* net_dev_start_xmit: dev=lo queue_mapping=0 skbaddr=0000000024fe4
664 vlan_tagged=0 vlan_proto=0x0000 vlan_tci=0x0000 protocol=0x0800 ip_summed=3 len=83 data_len=0 network_offset=14 transport_off
set_valid=1 transport_offset=34 tx_flags=0 gso_size=0 gso_segs=0 gso_type=0x0 */
414  9344.129120 |   1)    ping-1922   | .... |                | /* netif_rx_entry: dev=lo napi_id=0x2 queue_mapping=0 skbaddr=0000
00024fe4664 vlan_tagged=0 vlan_proto=0x0000 vlan_tci=0x0000 protocol=0x0800 ip_summed=3 hash=0x3d3df90e l4_hash=1 len=69 data_len
=0 truesize=768 mac_header_valid=1 mac_header=-14 nr_frags=0 gso_size=0 gso_type=0x0 */
415  9344.129122 |   1)    ping-1922   | .... |                | /* netif_rx: dev=lo skbaddr=0000000024fe4664 len=69 */
416  9344.129124 |   1)    ping-1922   | .... |                | /* netif_rx_exit: ret=0 */
417  9344.129125 |   1)    ping-1922   | .... |                | /* net_dev_xmit: dev=lo skbaddr=0000000024fe4664 len=83 rc=0 */
418  9344.129130 |   1)    ping-1922   | ..s1 |                | /* netif_receive_skb: dev=lo skbaddr=0000000024fe4664 len=69 */
419  9344.129180 |   1)    ping-1922   | ..s1 |                | /* napi_poll: napi poll on napi struct 00000000d4d867f1 for device
(no_device) work 1 budget 64 */
420  9344.129191 |   1)    ping-1922   | .... |                | /* net_dev_queue: dev=lo skbaddr=0000000021f442a5 len=83 */
421  9344.129192 |   1)    ping-1922   | .... |                | /* net_dev_start_xmit: dev=lo queue_mapping=0 skbaddr=0000000021f44
2a5 vlan_tagged=0 vlan_proto=0x0000 vlan_tci=0x0000 protocol=0x0800 ip_summed=3 len=83 data_len=0 network_offset=14 transport_off
set_valid=1 transport_offset=34 tx_flags=0 gso_size=0 gso_segs=0 gso_type=0x0 */
422  9344.129195 |   1)    ping-1922   | .... |                | /* netif_rx_entry: dev=lo napi_id=0x2 queue_mapping=0 skbaddr=00000
00021f442a5 vlan_tagged=0 vlan_proto=0x0000 vlan_tci=0x0000 protocol=0x0800 ip_summed=3 hash=0x3d3df90e l4_hash=1 len=69 data_len
=0 truesize=768 mac_header_valid=1 mac_header=-14 nr_frags=0 gso_size=0 gso_type=0x0 */
423  9344.129195 |   1)    ping-1922   | .... |                | /* netif_rx: dev=lo skbaddr=0000000021f442a5 len=69 */
424  9344.129196 |   1)    ping-1922   | .... |                | /* netif_rx_exit: ret=0 */
425  9344.129196 |   1)    ping-1922   | .... |                | /* net_dev_xmit: dev=lo skbaddr=0000000021f442a5 len=83 rc=0 */
426  9344.129198 |   1)    ping-1922   | ..s1 |                | /* netif_receive_skb: dev=lo skbaddr=0000000021f442a5 len=69 */
427  9344.129202 |   1)    ping-1922   | ..s1 |                | /* napi_poll: napi poll on napi struct 00000000d4d867f1 for device
(no_device) work 1 budget 64 */
```

Figure 9.17 – A partial screenshot of the (filtered) ftrace report for ping via the set_event interface

With this approach, it's interesting and useful to be able *to see each function's parameters along with their current value*! This is at the cost of not being able to see the call graph indentation, nor the level of detail regarding the trace (as compared with tracing the previous way, via the regular `available_filter_functions` interface). You can find this ftrace `set_event`-based report (a sample run) in the `ch9/ftrace/ping_ftrace_set_event_report.txt` file.

From a debugging perspective, being able to see function parameters can be really useful (as, quite often, incorrect parameters might be the underlying defect or contribute to it). Further, with the `trace-cmd` frontend as well (which we will cover in the next main section), not using the `function_graph` plugin auto-enables the printing of function parameters.

Using trace_printk() for debugging

The `trace_printk()` API is used to emit a string into the ftrace buffer. The syntax is identical to `printf()`. Here is an example:

```
trace_printk("myprj: at %s:%s():%d\n",__FILE__, __func__, __
LINE__);
```

Thus, it's typically used as a debugging aid, an instrumentation technique. But then why not simply employ `printk()` (or the `pr_foo()` wrapper or `dev_foo()` macros)? `trace_printk()` is *much faster*, writing only to RAM, and never to the console device. Thus, it's really useful for debugging fast code paths (interrupt code, for example), where `printk()` might be too slow (recall that we briefly talked about Heisenbugs in the introduction to *Chapter 8, Lock Debugging*). Also, the `printk()` buffer can be far too small at times; the ftrace buffers are (much) larger and tunable.

It's recommended that you use `trace_printk()` only for debug purposes. Now, if `trace_printk()` writes only to the ftrace buffer, how do you look up the content? Easy – simply read from the `trace` or `trace_pipe` files (*not* via `dmesg` or `journalctl`). The `trace_printk()` output is valid in all tracer plugins and works (like `printk()`) from any context – process or interrupt, and even NMIs. (By the way, it appears as a comment in the `function_graph` tracer report.)

Also, the kernel documentation mentions optimizations – for example, using `trace_puts()`. This only emits a literal string (which is often sufficient), as well as other optimization with `trace_printk()`: `https://www.kernel.org/doc/html/latest/driver-api/basics.html#c.trace_printk`.

`trace_printk()` can be disabled from writing into the trace buffer by writing `notrace_printk` into the `trace_options` file (it's enabled by default). Alternatively, it can be toggled by writing `0 / 1` into `options/trace_printk`.

Ftrace – miscellaneous remaining points via FAQs

Let's wrap up this content on kernel ftrace in a useful and familiar FAQ format:

- *Is there a documented quick way to get started with ftrace?*

  The ftrace subsystem includes a nice quick summary of using ftrace via a *tracing mini-HOWTO*; you can read it by doing the following:

  ```
  sudo cat /sys/kernel/tracing/README
  ```

- *I can't find some ftrace options or tracefs files on my system.*

  Remember that the `tracefs` pseudofiles and directories are an integral part of the kernel and thus there will be variances in what you see, based on the following:

  - The CPU architecture (typically x86_64 is the most rich and updated one)
  - The kernel version (here, it's based on the x86_64 arch and the 5.10.60 kernel)

- *How can I obtain trace data from the kernel as it's generated – that is, in a streaming fashion?*

  You can stream in ftrace data by reading from the `trace_pipe` pseudofile; you can simply read from it using `tail -f` or a custom script, or even filter it "live" by simply filtering the incoming trace data from `trace_pipe` via standard utilities such as `awk` and `grep`.

- *Within a tracing session with ftrace, can I toggle tracing on/off?*

  Toggling ftrace programmatically within a kernel (or a module for that matter) is easily done. Simply call these APIs (note that they're GPL-exported only):

  - `tracing_on()`: Turns tracing on
  - `tracing_off()`: Turns tracing off

  This is the programming equivalent of writing 1 or 0 into the `tracing_on` file, which you can use to toggle tracing via a script (running as root).

  Note that the entire ftrace system can be turned off by writing 0 to the `/proc/sys/kernel/ftrace_enabled` sysctl. This is obviously not to be done trivially. The kernel documentation (`https://www.kernel.org/doc/html/latest/trace/ftrace.html#ftrace-enabled`) has more details on this aspect.

- *Can ftrace help when a kernel Oops or panic occurs? How?*

  The powerful **kdump/kexec** infrastructure allows us to capture a snapshot of the entire kernel memory space when a crash – an Oops or panic – occurs. Subsequently, the **crash** tool allows you to perform post-mortem analysis of the kernel dump image (we will mention this technology briefly in this book's last chapter).

  However, even though this can be very helpful to debug a kernel crash, it doesn't actually provide any details on what occurred *before* the kernel crashed. This is where ftrace can, again, be very useful – we can set up ftrace to perform tracing prior to a known crash point. But once the system crashes, it could well be in an unusable state (completely frozen/hung); thus, you may not be able to even save the trace data to a file.

This is where the `ftrace_dump_on_oops` facility comes in. Enable it by writing 1 to the proc pseudofile,`/proc/sys/kernel/ftrace_dump_on_oops` (it's always 0 by default). This will have the kernel write the current content of the ftrace buffer(s) to the console device and the kernel log! Kdump will thus capture it along with the kernel dump image, and you'll now have not only the entire kernel state at the time of the crash but also the events leading up to the crash, as evidenced by the ftrace output. This can help in debugging the root cause of the crash.

The facility is also invokable at boot time via a kernel command-line parameter (which you can pass at boot time via the bootloader). The following screenshot from the kernel documentation on kernel parameters (`https://www.kernel.org/doc/html/latest/admin-guide/kernel-parameters.html`) makes its purpose amply clear:

```
ftrace_dump_on_oops[=orig_cpu]
                [FTRACE] will dump the trace buffers on oops.
                If no parameter is passed, ftrace will dump
                buffers of all CPUs, but if you pass orig_cpu, it will
                dump only the buffer of the CPU that triggered the
                oops.
```

Figure 9.18 – A partial screenshot of kernel parameters showing the ftrace_dump_on_oops one

This is interesting. Using `ftrace_dump_on_oops=orig_cpu` can often be very useful. Only the relevant ftrace buffer – the one for the CPU where the Oops got triggered – will get dumped to the kernel log (and console).

> **Tip – Kernel Parameters Relevant to Ftrace**
>
> Ftrace can be programmed to start collecting trace data as early as possible after boot by passing the `ftrace=[tracer]` kernel parameter (where `[tracer]` is the name of the tracer plugin to employ) to help you debug early boot issues. Similarly, several other ftrace-related kernel parameters are available. To see them, navigate to the official kernel docs on kernel command-line parameters (`https://www.kernel.org/doc/html/latest/admin-guide/kernel-parameters.html`) and search for the `[FTRACE]` string.

- *What's the irqsoff and the other latency-measurement tracers for?*

  It's best to not disable hardware interrupts; sometimes though, it becomes necessary to. For example, the critical section of a spinlock – the code between the spinlock being taken and released – has interrupts disabled (in order to guarantee its correct functioning). Keeping interrupts disabled for long-ish periods of time though – anything more than, say, 100 microseconds – can certainly contribute to system latencies and lags. The `irqsoff` tracer can measure the longest time for which hardware interrupts are turned off; even better, it allows you to see where exactly this occurred as well.

  Usage details regarding the `irqsoff` tracer have already been covered in my earlier (free) ebook *Linux Kernel Programming – Part 2*, in *Chapter 4, Handling Hardware Interrupts*, in the *Using Ftrace to get a handle on system latencies* section. Do check it out.

  The official kernel documentation on ftrace (`https://www.kernel.org/doc/html/latest/trace/ftrace.html#irqsoff`) does indeed cover the meaning and specifics of measuring latencies via these latency-measurement-related tracers. Please do check it out. Here are the latency measurement-related ftrace tracers:

  - `irqsoff`: Measures and reports the maximum duration for which hardware interrupts (IRQs) are turned off (disabled).

  - `preemptoff`: Measures and reports the maximum duration for which kernel preemption is turned off.

  - `preemptirqsoff`: Measures and reports the maximum duration for which hardware IRQs and/or kernel preemption are turned off. In effect, that is the maximum of both of the preceding tracers, and, in actuality, the total time during which the kernel cannot schedule anything!

  - `wakeup`: Measures and reports the schedule latency – the time that elapses between a task being awoken and the time to which it actually runs. It's measured for the highest priority non-real-time task only.

  - `wakeup_rt`: Same as the previous, except that it measures and reports the schedule latency for the highest priority real-time task currently on the system. This is an important metric for real time.

  As mentioned (in the following section), the first three tracers listed previously are often used to check whether drivers have left hardware interrupts or kernel preemption on for too long.

> **IRQs Off/Kernel Preemption Off – How Long Is Too Long?**
>
> In general, anything in the range of *tens of milliseconds* – in effect, anything over 10 milliseconds – is considered too long for hardware interrupts and/or kernel preemption to be turned off.

Here's a quick pro tip – monitor your project's `irqqsoff` and `preemptoff` worst-case times with the above mentioned latency measurement tracers.

- *Can I perform more than one ftrace recording/reporting session simultaneously on the same kernel?*

  Ftrace has an *instances* model. It allows for more than one trace to be done at a time! Simply create a directory under the `/sys/kernel/tracing/instances/` directory (with `mkdir`) and proceed to use it, just as you would with normal ftrace. Each instance has its own set of buffers, tracers, filters, and so on, allowing multiple simultaneous tracing when required. For more information, this presentation by Steven Rostedt covers using ftrace instances: *Tracing with ftrace – Critical tooling for Linux Development*, June 2021: `https://linuxfoundation.org/wp-content/uploads/ftrace-mentorship-2021.pdf`.

Ftrace use cases

Here, we will mention a few of the many ways that ftrace has been (or can be) leveraged, with a focus on debugging.

Checking kernel stack utilization and possible overflow with ftrace

As you'll know, every (user-mode) thread alive has two stacks – a user-mode stack and a kernel-mode stack. The user-mode stack is dynamic and large. (The maximum size it can grow to is a resource limit, `RLIMIT_STACK`, typically 8 MB on vanilla Linux. Also, kernel threads only have a kernel-mode stack, of course.) The kernel-mode stack, however, *is fixed in size and is small* – typically, just 8 KB on 32-bit systems and 16 KB on 64-bit systems. Overflowing the kernel-mode stack is, of course, a memory-related bug and will usually cause a system to abruptly lock up or even panic. It's a dangerous thing.

> **Tip**
>
> Enabling `CONFIG_VMAP_STACK` (essentially, using the kernel vmalloc region for kernel stacks) can be useful. It enables the kernel to set up a guard page to catch any overflow gracefully and report it via an Oops; the offending process context is killed as well. Also, enabling `CONFIG_THREAD_INFO_IN_TASK` helps mitigate the problems that a stack overflow bug can cause. See the *Further reading* section for more info on these kernel configs.

So, monitoring/instrumenting a kernel-mode stack size at runtime can be a useful task to carry out, to flag any outliers! Ftrace has a way to do so – the so-called **stack tracing** (or stack tracer) functionality. Enable it by setting CONFIG_STACK_TRACER=y in the kernel config (it's typically set by default). The tracer is controlled via the proc pseudofile /proc/sys/kernel/stack_tracer_enabled, and is turned off by default.

Here's a quick sample run where we'll turn on ftrace's stack tracer, do a sample tracing session, and see which kernel functions had the highest kernel-mode stack utilization (note that we're running as root):

1. Turn the ftrace stack tracer on:

    ```
    echo 1 > /proc/sys/kernel/stack_tracer_enabled
    ```

2. Perform a tracing session. We make use of our very simple script, ch9/ftrace/ftrc_1s.sh, which traces whatever executes within the kernel for 1 second:

    ```
    cd /sys/kernel/tracing
    <...>/ch9/ftrace/ftrc_1s.sh
    [...]
    ```

3. Look up the maximum kernel stack utilization and the details:

    ```
    cat stack_max_size
    cat stack_trace
    ```

The following screenshot displays a sample run. Here, the maximum kernel stack size utilization turned out to be over 4,000 bytes:

```
# ~/lkdsrc/ch9/ftrace/ftrc_1s.sh
trace-cmd reset
resetting set_ftrace_filter
resetting set_ftrace_notrace
resetting set_ftrace_notrace_pid
resetting set_ftrace_pid
resetting trace_options to defaults (as of 5.10.60)
resetting options/funcgraph-*
running '/usr/sbin/reset-ftrace-perf -q' now...
Tracing with function_graph for 1s ...
-rw-r--r-- 1 root root 371K Jan 28 12:40 /root/ftrace_reports/ftrc_1s.sh_20220128_124002.txt
#
# cat stack_max_size
4224
#
# cat stack_trace
        Depth    Size   Location    (35 entries)
        -----    ----   --------
  0)    4280       64   decay_load+0x5/0xa0
  1)    4216       96   __update_load_avg_se+0x22b/0x2c0
  2)    4120       88   update_load_avg+0x2c9/0x6f0
  3)    4032      136   update_blocked_averages+0x4c5/0x6a0
  4)    3896       24   update_nohz_stats+0x44/0x60
  5)    3872      296   update_sd_lb_stats.constprop.0+0x433/0xff0
  6)    3576      256   find_busiest_group+0x4d/0x370
  7)    3320      336   load_balance+0x168/0x1630
  8)    2984       96   newidle_balance+0x31a/0x470
  9)    2888       72   pick_next_task_fair+0x41/0x470
 10)    2816      128   __schedule+0x32e/0xc90
 11)    2688       32   schedule+0x4e/0xf0
 12)    2656       24   io_schedule+0x16/0x40
```

Figure 9.19 – A partial screenshot showing a sample kernel stack utilization via ftrace's stack tracer

The official kernel documentation has information on the stack tracer: https://www.kernel.org/doc/html/v5.10/trace/ftrace.html#stack-trace.

How the AOSP uses ftrace

The **Android Open Source Project** (**AOSP**) indeed uses ftrace to help debug kernel/driver issues. (Internally, it uses what is essentially wrapper tooling – *atrace*, *systrace*, and *Catapult* – over ftrace, though ftrace can be used directly as well.)

The AOSP describes using dynamic ftrace (just as we have been doing) to debug and find the root cause of difficult-to-figure performance-related defects. A brief quote – from `https://source.android.com/devices/tech/debug/ftrace` – is in order:

> *"However, every single difficult performance bug in 2015 and 2016 was ultimately root-caused using dynamic ftrace. It is especially powerful for debugging uninterruptible sleeps because you can get a stack trace in the kernel every time you hit the function triggering uninterruptible sleep. You can also debug sections with interrupts and preemptions disabled, which can be very useful for proving issues. [...] irqsoff and preemptoff are primarily useful for confirming that drivers may be leaving interrupts or preemption turned off for too long."*

We, in fact, just talked about using the `irqsoff`, `preemptoff`, and `preemptirqsoff` tracers in the previous section.

An actual use case – an Android smartphone, a Pixel XL, after taking a **High Dynamic Range** (**HDR**) photo and immediately rotating the viewfinder, resulted in *jank* – was root-caused using ftrace: `https://source.android.com/devices/tech/debug/ftrace#expandable-1`.

Similarly, the AOSP documentation also refers to actual cases where the following occurs:

- Drivers can leave hardware IRQs and/or preemption disabled for too long, causing performance issues (`https://source.android.com/devices/tech/debug/jank_jitter#drivers`).

- Drivers can have long softirqs, again causing performance issues (why? As softirqs disable kernel preemption (`https://source.android.com/devices/tech/debug/jank_jitter#long-softirqs`).

This is interesting stuff.

We mention one more interesting use case – that of using the powerful and user-friendly `perf-tools` scripts (another frontend to ftrace) – to help debug performance issues on Netflix Linux (Ubuntu) cloud instances. We will discuss this in a later section – *Investigating a database disk I/O issue on Netflix cloud instances with perf-tools*.

On an unrelated note, perhaps you'll find by studying an ftrace report several – perhaps too many – calls to security-related interfaces, typically enforced via **Linux Security Modules (LSMs)**, such as SELinux, AppArmor, Smack, TOMOYO, and so on. In a highly performance-sensitive app (or project – for example, a near real-time system), this might indicate the need to disable these security interfaces (via kernel configuration – if possible, at least during the time-critical code paths). This can be especially true when several LSMs are enabled.

Using the trace-cmd, KernelShark, and perf-tools ftrace frontends

There's no doubt that the Linux kernel ftrace infrastructure is immensely powerful, enabling you to look deep inside the kernel, throwing light into the dark corners of the system, as it were. This power does come at the cost of a somewhat steep learning curve – lots of sysfs-based tuning and options knobs that you need to be intimately aware of, plus the burden of filtering a possibly huge amount of noise in the resulting traces (as you'd have already learned from the previous sections of this chapter!). Steven Rostedt thus built a powerful and elegant command-line-based frontend to ftrace, `trace-cmd`. What's more, there's a true GUI frontend to `trace-cmd` itself, the **KernelShark** program. It parses the trace data recorded (`trace.dat` by default) by `trace-cmd` and displays it in a more human-digestible GUI. In a similar manner, Brendan Gregg has built the `perf-tools` script-based frontend project to ftrace as well.

An introduction to using trace-cmd

The `trace-cmd` utility – in the style of modern Linux console software such as `git` – has several sub-commands. They allow you to easily record a tracing session, for the entire system or a particular process only (optionally, as well as its descendants), and generate a report. It can do a lot more – control ftrace config parameters, clear and reset ftrace, see the current status, and list all available events, plugins, and options. It can even perform profiling, show a histogram of the trace, take a snapshot, listen on a network socket for clients, and more. As `trace-cmd` works upon the underlying ftrace kernel subsystem, you typically need root access when running its sub-commands. Here, we're working with the `trace-cmd` version, available as an Ubuntu 20.04 LTS package at the time of writing – version 2.8.3.

Getting help

The trace-cmd utility is well documented. A few ways to get help include the following:

- Every trace-cmd sub-command has its very own man page – for example, to read the man page on the record sub-command, type man trace-cmd-record. Of course, man trace-cmd gives an overview of the tool and every sub-command.

- To get a quick help screen, type the command followed by -h – for example, trace-cmd record -h.

- Several excellent tutorials are available. Refer to the *Further reading* section of this chapter for some.

Running trace-cmd and checking for available man pages related to it (by using bash auto-complete) are shown in the following screenshot:

```
$ trace-cmd

trace-cmd version 2.8.3

usage:
  trace-cmd [COMMAND] ...

  commands:
     record - record a trace into a trace.dat file
     start - start tracing without recording into a file
     extract - extract a trace from the kernel
     stop - stop the kernel from recording trace data
     restart - restart the kernel trace data recording
     show - show the contents of the kernel tracing buffer
     reset - disable all kernel tracing and clear the trace buffers
     clear - clear the trace buffers
     report - read out the trace stored in a trace.dat file
     stream - Start tracing and read the output directly
     profile - Start profiling and read the output directly
     hist - show a histogram of the trace.dat information
     stat - show the status of the running tracing (ftrace) system
     split - parse a trace.dat file into smaller file(s)
     options - list the plugin options available for trace-cmd report
     listen - listen on a network socket for trace clients
     list - list the available events, plugins or options
     restore - restore a crashed record
     snapshot - take snapshot of running trace
     stack - output, enable or disable kernel stack tracing
     check-events - parse trace event formats

$ man trace-cmd-
trace-cmd-check-events  trace-cmd-profile    trace-cmd-split
trace-cmd-extract       trace-cmd-record     trace-cmd-stack
trace-cmd-hist          trace-cmd-report     trace-cmd-start
trace-cmd-list          trace-cmd-reset      trace-cmd-stat
trace-cmd-listen        trace-cmd-restore    trace-cmd-stop
trace-cmd-mem           trace-cmd-show       trace-cmd-stream
trace-cmd-options       trace-cmd-snapshot
```

Figure 9.20 – A screenshot showing trace-cmd's brief help screen and available man pages

A simple first tracing session with trace-cmd

Here, we present steps to carry out a very simple tracing session via `trace-cmd`. As space is limited, we won't repeat what the man pages (and other docs) already explain in depth. We will leave it to you to read through the deeper details. Let's jump right into it:

1. Reset the ftrace tracing subsystem (optional):

   ```
   sudo trace-cmd reset
   ```

2. Record a trace. Let's record everything that goes on in the kernel for 1 second, using the powerful `function_graph` plugin (or tracer), specified via the –p option switch. The `-F <command>` option switch has `trace-cmd` trace *only* that command (adding `-c` before the `-F` switch will also trace its descendants, if there are any):

   ```
   sudo trace-cmd record -p function_graph -F sleep 1
   ```

3. Save the trace report (in ASCII text format). The `-l` option adds a column showing the really useful latency output format (we covered this in the *Delving deeper into the latency trace info* section). In addition to the four latency info columns we already saw, `trace-cmd` prefixes an additional column – the CPU core that the function ran upon:

   ```
   sudo trace-cmd report -l > sleep1.txt
   ```

Alternatively, `trace-cmd show` will show you the current content of the ftrace buffers. Also, note that in *step 2*, instead of specifying a command to trace (via the `-F` switch), you can specify a process to trace via the `-P <PID>` option.

It's also worth noting that the second step produces a binary trace file named, by default, `trace.dat` (you'll find we use it with the KernelShark GUI frontend). Do try out these simple steps and trace the kernel easily! You'll quickly realize how much easier this is compared with directly working with raw ftrace. Of course, on constrained embedded systems, setting up frontends such as `trace-cmd` may not be viable at all (it really does depend on your project/product); thus, knowing how to leverage raw ftrace is indeed still important!

> **Tip**
> It's advisable to not run `trace-cmd` from within the `tracefs` (`/sys/kernel/[debug]/tracing`) directory. It can fail as it attempts to write the trace data (you'll have to override this with the `-o` option switch, and so on).

Viewing and leveraging all available events

It gets a lot more powerful. `trace-cmd list` shows you all available events (as well as plugins and other options) that can be leveraged while recording a trace. While doing this reveals all possible events that can be traced, the list is huge – over 1,400 of them at the time of writing (with the 5.10 kernel series – try it out and see for yourself). Here's a truncated view:

```
$ sudo trace-cmd list
events:
drm:drm_vblank_event
drm:drm_vblank_event_queued
drm:drm_vblank_event_delivered
initcall:initcall_finish
initcall:initcall_start
initcall:initcall_level
vsyscall:emulate_vsyscall
xen:xen_cpu_set_ldt
[...]
tracers:
hwlat blk mmiotrace function_graph wakeup_dl wakeup_rt wakeup
function nop
options:
print-parent
nosym-offset
[...]
```

To see a sorted list of just the event labels – which are similar to event classes – and not each and every function associated with each event, in an abbreviated format, we firstly employ the `-e` option to `trace-cmd list` (show only events) and perform some quick bash magic:

```
$ sudo trace-cmd list -e | awk -F':' 'NF==2 {print $1}' | sort
| uniq | tr '\n' ' '
alarmtimer asoc avc block bpf_test_run bpf_trace bridge
cfg80211 cgroup clk compaction cpuhp cros_ec devfreq devlink
dma_fence drm error_report exceptions ext4 fib fib6 filelock
filemap fs_dax gpio gvt hda hda_controller hda_intel huge_
memory hwmon hyperv i2c i915 initcall intel_iommu intel_ish
interconnect iocost iomap iommu io_uring irq irq_matrix irq_
vectors iwlwifi iwlwifi_data iwlwifi_io iwlwifi_msg iwlwifi_
```

```
ucode jbd2 kmem kvm kvmmmu libata mac80211 mac80211_msg mce
mdio mei migrate mmap mmap_lock mmc module mptcp msr napi
neigh net netlink nmi nvme oom page_isolation pagemap page_
pool percpu power printk pwm qdisc random ras raw_syscalls rcu
regmap regulator resctrl rpm rseq rtc sched scsi signal skb
smbus sock spi swiotlb sync_trace syscalls task tcp thermal
thermal_power_allocator timer tlb ucsi udp v4l2 vb2 vmscan
vsyscall wbt workqueue writeback x86_fpu xdp xen xhci-hcd $
```

The precise event classes seen depend on the architecture, the kernel version, and the kernel config. Now, the wonderful thing is that you can pick one or more of these event classes and have `trace-cmd record` *only trace and report the functionality corresponding to them by using the* `-e` *option switch.* Here's an example:

```
trace-cmd record <...> -e net -e sock -e syscalls
```

As you'll guess, this has `trace-cmd` record all network, socket, and system call-related tracing events (functions) only, that occur within the kernel during its recording run.

The `trace-cmd list` sub-command can show interesting stuff – for example, `trace-cmd list -t` shows all available tracers (completely equivalent to `cat /sys/kernel/tracing/available_tracers`). To see all that it can show, display the `list` sub-command help screen as follows:

```
# trace-cmd list -h
trace-cmd version 2.8.3
usage:
trace-cmd list [-e [regex]] [-t] [-o] [-f [regex]]
          -e list available events
            -F show event format
            -R show event triggers
            -l show event filters
          -t list available tracers
          -o list available options
          -f [regex] list available functions to filter on
          -P list loaded plugin files (by path)
          -O list plugin options
          -B list defined buffer instances
          -C list the defined clocks (and active one)
#
```

Do look up man `trace-cmd-list` to understand the details. Also, if you want to trace particular functions (and not all ones that correspond to an encompassing event, which we did with the `-e <event1> -e <event2>` option to `trace-cmd record`), then execute `trace-cmd record [...] -l <func1> -l <func2> [...]`.

Case 3.1 – tracing a single ping with trace-cmd

Performing a trace of a single ping, in a manner very similar to what we achieved with ftrace-ing via the `set_event` interface (we covered this in the section *Case 2 – tracing a single ping with raw ftrace via the set_event interface*), with function parameters being revealed, can be easily done in simply two steps with `trace-cmd`:

1. The recording of data:

    ```
    sudo trace-cmd record -q -e net -e sock -e skb -e tcp -e
    udp -F ping -c1 packtpub.com
    ```

2. The reporting of the trace:

    ```
    sudo trace-cmd report -l -q > reportfile.txt
    ```

If, in the recording step, you add the `-p function_graph` parameter, you'll get the report with function call graph indentation but without any function parameters (as you will now realize, both ways are useful).

This single ping trace via `trace-cmd` has been encapsulated via a simple bash script – `ch9/tracecmd/trccmd_1ping.sh`. When running, the script requires, via an option switch, a decision on whether to trace so that a function graph-style report or function parameters (and their current values) are displayed in the trace report. Do try it out!

Kernel modules and trace-cmd

Ftrace has the ability to automatically recognize any and all functions within kernel modules! This is excellent; thus, `trace-cmd` – being a frontend to ftrace – also automatically recognizes them. To test this, I simply loaded up a module we used earlier (`ch5/kmembugs_test/test_kmembugs.ko`). Then, we use `trace-cmd list -f`, grepping for the presence of this module's functions. They do indeed show up:

```
# trace-cmd list -f |grep "test_kmembugs]$" |head
irq_work_leaky [test_kmembugs]
delay_sec [test_kmembugs]
umr [test_kmembugs]
umr_slub [test_kmembugs]
uar [test_kmembugs]
leak_simple1 [test_kmembugs]
leak_simple2 [test_kmembugs]
leak_simple3 [test_kmembugs]
global_mem_oob_right [test_kmembugs]
global_mem_oob_left [test_kmembugs]
#
```

Figure 9.21 – A screenshot showing how trace-cmd automatically recognizes module functions available to it for tracing

Now, to trace a particular module's functions, do the following:

```
trace-cmd record [...] --module <module-name> [...]
```

Also, as an aside, I'm working upon a wrapper script over the `trace-cmd` utility called `trccmd`. Please see the GitHub repository of this small project here: `https://github.com/kaiwan/trccmd`. As an example, here's this utility being used to trace the flow of a single `ping` packet:

```
./trccmd -F 'ping -c1 packtpub.com' -e 'net sock skb tcp udp'
```

Right, let's move on to graphically visualizing our hard work!

Using the KernelShark GUI

KernelShark is an excellent GUI frontend to the output produced by `trace-cmd`. More specifically, it parses the binary `trace.dat` file produced by either the `trace-cmd record` or `trace-cmd extract` sub-commands.

Getting a trace.dat type output from raw ftrace

This might leave you wondering – what if you're using raw ftrace to trace (not `trace-cmd`) and still want to visualize the trace with KernelShark? It's easy – you simply have to use `trace-cmd extract` to extract the raw trace buffer content to a file – it will be in the expected binary format! Follow along with this example (as root):

```
cd /sys/kernel/tracing
trace-cmd reset ; echo > trace
echo function_graph > current_tracer
echo 1 > tracing_on ; sleep .5 ; echo 0 > tracing_on
trace-cmd extract -o </path/to/>trc.dat
```

Now, the `trc.dat` file can be provided as input to KernelShark.

Moving along, the latest version of KernelShark (at the time of writing, in March 2022) is 2.1.0. It's moved on from GTK+ 2.0 to Qt 5. (The 1.0 release got an LWN article for itself: *KernelShark releases version 1.0*, Jake Edge, July 2019: `https://lwn.net/Articles/794846/`.)

Being pretty new, the combination of the latest `trace-cmd` version (3.0.-dev, at the time of writing) and KernelShark 2.1.0 gave me some trouble; hence, I am just going with the older distro-package (Ubuntu 20.04 LTS) releases here – `trace-cmd` version 2.8.3 and KernelShark version 0.9.8.

Very useful and detailed documentation for KernelShark is available here: `https://kernelshark.org/Documentation.html`.

Case 3.2 – viewing the single ping with KernelShark

We come back to our favorite trace test – that of tracing a single ping! Of course, the whole idea is that this time, we'll visualize the trace report via KernelShark. To do so, we first execute our simple bash script (`ch9/tracecmd/trccmd_1ping.sh`) that will capture the trace data and write it to the `trace.dat` file:

```
cd <booksrc>/ch9/tracecmd
./trccmd_1ping.sh -f
[...]
```

(We covered the basics on this in the earlier *Case 3.1 – tracing a single ping with trace-cmd* section; the -f option provided here has the recording done via the function_graph tracer plugin). The ASCII text report file we generate (ping_trccmd.txt, here) is of no use to KernelShark. It instead uses the binary trace.dat report file (which also gets generated by trace-cmd).

KernelShark is essentially a trace reader. It parses and displays the content of the trace.dat file in a useful GUI. The picking up of the trace.dat file is automatic when you run KernelShark from the directory where a trace.dat file is present. Alternatively, you can always override this and pass the relevant binary trace file via the -i parameter, or even open it from the GUI's **File | Open** menu. Here's a screenshot of the KernelShark GUI, visualizing our single ping trace:

Figure 9.22 – A screenshot of the KernelShark GUI visualizing the single ping; the Events filter dialog is seen as well

Note how usefully we can filter the output. Here, I've applied a few filters:

- **CPU**: Set to CPU 1 only (or whichever are appropriate). Access it via **Plots | CPUs**.

- **Tasks**: Set to the `ping-[PID]` task only. Access it via **Plots | Tasks**.

- **Events**: This is very useful. It's set to filter events of interest – we eliminate all ftrace events except `funcgraph_entry`; this one allows us to see the names of kernel functions as they're entered in the list view. Access it via **Filter | Show events**. (*Here's a quick tip* – as you probably know, all kernel events can be seen under `/sys/kernel/tracing/events/`.)

Very powerful indeed!

There are two major tiled widgets – the *graph* and *list* views. The former – the upper portion of the GUI – shows the kernel flow graphically, with vertical tick marks indicating events. The list widget (the lower pane) is literally a list of the events – essentially, it's the raw ftrace/`trace-cmd` output. Above the graph region is the "Pointer", navigation/zoom, and two **Marker** widgets. Between the graph and list regions is a widget, allowing you to search and filter on any of the available columns. Again, it's very intuitive, so do try it out. The KernelShark doc clearly explains the GUI layout. Here's a few elements of the GUI (note that this screenshot is a different session from the previous one):

Figure 9.23 – A partial screenshot showing the upper portion of the KernelShark GUI

Here's a quick run-through of some key elements of *Figure 9.23*:

- The so-called "Pointer" – this shows the current location in the timeline. As you move the mouse over events, the information pertaining to that event – in effect, the last column in the list view (labeled `Info`) – is seen to the right of the Pointer. (You can see the mouse pointer on the graph and the corresponding event info to the right of the `Pointer:` widget – it shows that the mouse pointer is currently on the `ping` process, on entry to the `write()` system call.)

- The buttons to zoom and move (<, +, -, and >):

  - The < button moves the graph left.

  - The + button zooms in and the − button zooms out (as does scrolling the mouse).

  - The > button moves the graph right.

- The ++ button zooms into the graph to the maximum extent, and the - - button zooms out to the full timeline width.

- The two `Marker` widgets can be very helpful, allowing you to focus on a particular section of the code path and to see the time delta between the two. Using them is easy – for example, to set `Marker B`, click on it first, and then double-click anywhere on the graph or list. This sets it (do the same for `Marker A`), and when both are set, the time delta also shows up!

A few interesting gems turn up while reading the KernelShark HTML doc; Here's one:

> *"The hollow green bar that is shown in front of some events in the task plot represents when the task was woken up from a sleeping state to when it actually ran. The hollow red bar between some events shows that the task was preempted by another task even though that task was still runnable.*
>
> *Since the hollow green bar shows the wake up latency of the task, the A,B markers can be used to measure that time."*

Also, detailed custom filtering can be done via the **Filters | TEP Advance Filtering** (or **Advanced Filtering,** in older versions) menu; documentation on this can be found in the KernelShark HTML doc in the *Advanced Event Filter* section.

As we saw with ftrace, KernelShark too is used professionally to debug both performance issues as well as help root-cause defects. Here's an article by (who else?) Steven Rostedt: *Using KernelShark to analyze the real-time scheduler*, Feb 2011: `https://lwn.net/Articles/425583/`. As with ftrace's (and perf's) other frontends, KernelShark is moving away from being "the one GUI" solution to being merely one of any number of frontends that can take advantage of a framework where libraries will provide interfaces to access raw trace data (as *Figure 9.2* hints at – although it doesn't explicitly include KernelShark).

An introduction to using perf-tools

The `perf-tools` project is a collection of (mostly `bash`) scripts that are essentially wrappers over the kernel's ftrace and the `perf_events` (perf) infrastructure. They help automate much of the work when performing performance analysis/observability/debugging at the level of the kernel (and userspace, to an extent). The primary author is Brendan Gregg. This is the GitHub repository of the project: `https://github.com/brendangregg/perf-tools`.

It's not new to us – we covered, pretty in depth, the usage of the `kprobe[-perf]` tool within the `perf-tools` collection in *Chapter 4*, *Debug via Instrumentation – Kprobes*, in the *The easier way – dynamic kprobes or kprobe-based event tracing* section.

Once you've installed the `perf-tools[-unstable]` package, the scripts are typically installed in `/usr/sbin`. Let's check it out:

```
$ (cd /usr/sbin; ls *-perf)
bitesize-perf     execsnoop-perf     funcgraph-perf
functrace-perf    iosnoop-perf       kprobe-perf
perf-stat-hist-perf
syscount-perf     tpoint-perf        cachestat-perf
funccount-perf    funcslower-perf    iolatency-perf
killsnoop-perf    opensnoop-perf     reset-ftrace-perf
tcpretrans-perf uprobe-perf
$
```

These tools tend to help with performance observability (and debug) at various portions of the Linux stack. A picture is, of course, worth a thousand words; thus, I reproduce this useful diagram from the `perf-tools` GitHub repository here: `https://github.com/brendangregg/perf-tools/raw/master/images/perf-tools_2016.png`:

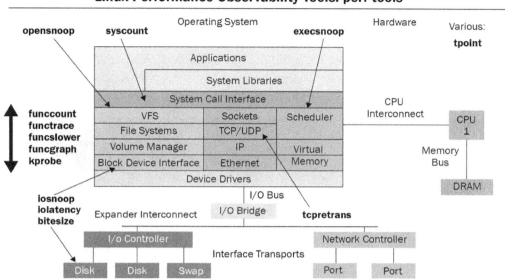

Figure 9.24 – The perf-tools collection of scripts (pic credit – the perf-tools GitHub repository)

Glancing at the diagram, you can see how these tools can be employed at various portions of the stack!

A big plus is that these tools are documented really well. Each has its own man page. Further, when any of these tools is run on the command line with the -h option switch, they shows a brief summary, often with very helpful one-line example usage (see the upper portion of *Figure 9.25* as an example). Due to limited space, we'll check out just a couple of examples (again, we covered using the kprobe [-perf] script in an earlier chapter).

Tracing all open()s via perf-tool's opensnoop

You'll recall how, back in *Chapter 4, Debug via Instrumentation – Kprobes*, via a multitude of ways, we worked hard to figure out which files were being opened (via the open() system call, which becomes the do_sys_open() function within the kernel). Let's revisit this, this time using ftrace! We could use raw ftrace or trace-cmd to quite easily do this, but let's use the (even easier to use) perf-tools wrapper script, opensnoop[-perf]! It does the job handily. Needless to say, run it as root:

```
# opensnoop-perf -h
USAGE: opensnoop [-htx] [-d secs] [-p PID] [-L TID] [-n name] [filename]
                   -d seconds     # trace duration, and use buffers
                   -n name        # process name to match on open
                   -p PID         # PID to match on open
                   -L TID         # PID to match on open
                   -t             # include time (seconds)
                   -x             # only show failed opens
                   -h             # this usage message
                   filename       # match filename (partials, REs, ok)
     eg,
        opensnoop                 # watch open()s live (unbuffered)
        opensnoop -d 1            # trace 1 sec (buffered)
        opensnoop -p 181          # trace I/O issued by PID 181 only
        opensnoop conf            # trace filenames containing "conf"
        opensnoop 'log$'          # filenames ending in "log"

See the man page and example file for more info.
#
#
# opensnoop-perf 'conf$' 2>/dev/null
Tracing open()s for filenames containing "conf$". Ctrl-C to end.
COMM              PID      FD FILE
tlp               readconfs  0x3 /usr/share/tlp/defaults.conf
tlp               readconfs  0x3 /etc/tlp.d/00-template.conf
tlp               readconfs  0x3 /etc/tlp.conf
tlp               readconfs  0x3 /usr/share/tlp/defaults.conf
tlp               readconfs  0x3 /etc/tlp.d/00-template.conf
tlp               readconfs  0x3 /etc/tlp.conf
tlp               readconfs  0x3 /usr/share/tlp/defaults.conf
tlp               readconfs  0x3 /etc/tlp.d/00-template.conf
tlp               readconfs  0x3 /etc/tlp.conf
^C
Ending tracing...
#
```

Figure 9.25 – A screenshot showing the help screen – with examples! – of opensnoop[-perf], along with a quick example, tracing the open system call of all files ending in conf system-wide

> **Tip – Digging In**
>
> I suggest you look at the code of some of these `perf-tools` scripts. `funcgraph[-perf]` is a good one: `https://github.com/brendangregg/perf-tools/blob/master/kernel/funcgraph`. It's a useful `bash` script wrapper over precisely what we learned to do earlier in this chapter – use raw ftrace via the `function_graph` tracer.

Also, recall how in *Chapter 4*, *Debug via Instrumentation – Kprobes*, in the *Observability with eBPF tools – an introduction* section, we used one of the powerful **BPF Compiler Collection** (**BCC**) frontends, `opensnoop-bpfcc`, to figure out which files are being opened by which process/thread.

Tracing functions that are latency outliers via perf-tool's funcslower

Here's one more quick `perf-tools` example – finding functions that are *latency outliers* with the `funcslower[-perf]` tool! To try this, I check for the `mutex_lock()` kernel function, taking longer than 50 microseconds to complete (I ran this on my native x86_64 laptop, running Ubuntu 20.04 LTS):

```
Linux-Kernel-Debugging $ sudo funcslower-perf -a mutex_lock 50
Tracing "mutex_lock" slower than 50 us... Ctrl-C to end.
# tracer: function_graph
#
#      TIME      CPU  TASK/PID        DURATION              FUNCTION CALLS
#       |        |     |   |           |   |                 |   |   |   |
284741.775198 |   10) Qt bear-11719  | + 54.044 us   | } /* mutex_lock */
284741.775400 |   10) Qt bear-11719  | + 61.039 us   | } /* mutex_lock */
284741.775507 |    0) Qt bear-2678454 | ! 106.166 us  | } /* mutex_lock */
   10) Qt bear-11719  => chrome-3780976
284761.091939 |   10) chrome-3780976 | + 52.208 us   | } /* mutex_lock */
284794.433903 |   11) VizComp-13360  | ! 302.772 us  | } /* mutex_lock */
284811.775269 |    0) Qt bear-2678454 | + 84.911 us   | } /* mutex_lock */
284811.775321 |    6) Qt bear-11719  | + 51.145 us   | } /* mutex_lock */
284811.775503 |    6) Qt bear-11719  | + 60.297 us   | } /* mutex_lock */
284811.775570 |    0) Qt bear-2678454 | + 65.780 us   | } /* mutex_lock */
284821.775447 |    0) Qt bear-2678454 | ! 101.478 us  | } /* mutex_lock */
284821.775560 |    6) Qt bear-11719  | ! 112.713 us  | } /* mutex_lock */
284825.251178 |   10) kworker-3759943 | * 32702.53 us | } /* mutex_lock */
284831.775498 |    6) Qt bear-11719  | + 53.848 us   | } /* mutex_lock */
284837.937573 |    1) gnome-s-11328  | ! 144.973 us  | } /* mutex_lock */
284851.775317 |    6) Qt bear-11719  | + 50.153 us   | } /* mutex_lock */
284851.775515 |    6) Qt bear-11719  | + 60.809 us   | } /* mutex_lock */

^C
Ending tracing...
```

Figure 9.26 – A partial screenshot showing how the funcslower[-perf] tool catches function outliers

Note the big fat outlier here – a `kworker` thread taking over 32 milliseconds, a likely corner case! Again, this shows that vanilla Linux is by no means a **Real-Time Operating System** (**RTOS**). This has me unable to resist pointing out that Linux *can* indeed run as an RTOS – look up the **Real-Time Linux** (**RTL**) wiki site and patches.

The `perf-tools` GitHub site has example content on most, if not all, the perf tools – you can check out interesting examples using `funcslower[-perf]` here: `https://github.com/brendangregg/perf-tools/blob/master/examples/funcslower_example.txt`. Several more interesting screenshots and links with respect to other tools can also be found here, so do check it out.

Don't forget eBPF and its frontends

Note that many of these `perf-tools` wrapper scripts are now superseded by the more recent and powerful *eBPF* technology. Brendan Gregg's answer to this is his newer frontend to eBPF – the `*-bpfcc` toolset! (You can read more on it here: `https://www.brendangregg.com/ebpf.html`.) Recall how, in *Chapter 4*, *Debug via Instrumentation – Kprobes*, in the *Observability with eBPF tools – an introduction* section, when we tried to figure out who was issuing the `execve()` system call to execute a process, the `perf-tools execsnoop-perf` wrapper script didn't quite cut it. The `execsnoop-bpfcc` BCC frontend wrapper script worked well instead.

Investigating a disk I/O issue on Netflix cloud instances with perf-tools

Brendan Gregg describes using ftrace via his powerful and user-friendly `perf-tools` scripts to help debug performance issues on Netflix Linux (Ubuntu) cloud instances. An article by him on this topic (you'll find the link to it shortly), although pretty old (August 2014), clearly illustrates how powerful ftrace – and the `perf-tools` frontend – can be at digging deep into and figuring out performance issues.

In this article, he shows how exactly he figured out an issue with a Cassandra database that was experiencing abnormally heavy disk I/O. It was caused by an initially incorrect disk readahead setting. At first, it seemed that even after tuning the readahead values to saner ones, it had no effect on the disk I/O. Digging deeper with the `perf-tools` scripts (he uses many interesting ones – `iosnoop[-perf]`, `tpoint[-perf]`, `funccount[-perf]`, `funcslower[-perf]`, `kprobe[-perf]`, and `funcgraph[-perf]`), he found that the tuning had no effect, as the initialization of the disk readahead setting took place within the context of the `open()` system call (but Cassandra was still running). Restarting Cassandra on the instance had the readahead value initialized to the correct value, the disk I/O dropped, and all was good again.

The article can be found here: *ftrace: The Hidden Light Switch*, Brendan Gregg, August 2014: `https://lwn.net/Articles/608497/`.

An introduction to kernel tracing with LTTng and Trace Compass

The **Linux Trace Toolkit – next generation** (**LTTng**) is a powerful and popular tracing system for the Linux kernel as well as userspace apps and libraries; it's open source, released under the Lesser GPL (modules and libraries), the GPL (tooling), and some components under the MIT license. Its original version (LTT) dates back to 2005, and LTTng is actively maintained. It has made a name for itself in helping track down performance and debug issues on multicore parallel and real-time systems. (Here, we're using the latest stable version at the time of writing – v2.13.)

The LTTng website (`https://lttng.org/`) does an excellent job documenting all aspects (learn what exactly tracing is at `https://lttng.org/docs/v2.13/#doc-what-is-tracing`). Due to space constraints, we shall simply refer you to the appropriate links. To install LTTng, please see this link: `https://lttng.org/docs/v2.13/#doc-installing-lttng`.

Tip – LTTng Package Installation for Ubuntu 20.04

Though you won't get the latest version, it's easy to simply install these LTTng packages like this – `sudo apt install lttng-tools lttng-modules-dkms -y`. (Using this technique, at the time of writing, I got LTTng version 2.11 and 2.12 for the modules.)

A quick introduction to recording a kernel tracing session with LTTng

Once installed, please do read the *Quick start* guide on the LTTng website: `https://lttng.org/docs/v2.13/#doc-getting-started`. As this is a book on kernel debugging, we will only make use of LTTng for kernel tracing (it has the capability to perform user mode tracing as well). Thus, I suggest you read these sections – at least to begin with:

- Record Linux kernel events: `https://lttng.org/docs/v2.13/#doc-tracing-the-linux-kernel`

- View and analyze the recorded events: `https://lttng.org/docs/v2.13/#doc-viewing-and-analyzing-your-traces`

To very briefly summarize recording a kernel tracing session with LTTng, perform the following steps (all performed as root):

1. Create a session:

    ```
    lttng create <session-name> --output=~/my_lttng_traces/
    ```

    If the --output parameter isn't provided, it defaults to saving it in ~/lttng_traces/.

2. Set up kernel events to trace. Here, we'll be simplistic and simply trace all kernel events (which could result in large raw data files being saved though):

    ```
    lttng enable-event --kernel --all
    ```

3. Perform the recording:

    ```
    lttng start
    ```

    Do whatever's necessary on the system to reproduce your issue (or simply do something for now).

4. Stop recording (optional):

    ```
    lttng stop
    ```

5. Destroy the recording session. Relax – this doesn't delete the raw trace data (also, this step implicitly stops the recording session):

    ```
    lttng destroy
    ```

6. Make the raw trace data accessible to other users (optional):

    ```
    sudo chown -R $(whoami):$(whoami) ~/my_lttng_traces
    ```

I've made a small (lightly tested, so no promises!) attempt at these steps via a wrapper bash script (ch9/lttng/lttng_trc.sh). It's just to get you started quickly. It expects a session name followed by either 0, implying the entire kernel gets traced, or the name of a program to execute, and it traces all kernel events as it executes (of course, it's simplistic – the trace isn't exclusive to the process):

```
$ cd <lkd_src>/ch9/lttng ; sudo ./lttng_trc.sh
Usage: lttng_trc.sh session-name program-to-trace-with-LTTng|0
  1st parameter: name of the session
  2nd parameter, ...:
```

```
    If '0' is passed, we just do a trace of the entire system
(all kevents),
    else we do a trace of the particular process (all kevents).
Eg. sudo ./lttng_trc.sh ps1 ps -LA
[NOTE: other stuff running _also_ gets traced (this is
non-exclusive)].
$
```

As a quick usage example, let's trace a single `ping` packet using LTTng (what a surprise)! The following screenshot shows its execution:

```
lttng $ sudo ./lttng_trc.sh ping1 ping -c1 packtpub.com
Session name :: "ping1"
[+] (Minimal) Checking for LTTng support ... [OK]
[+] lttng create lttng_ping1_08Mar22_1104 --output=/tmp/lttng_ping1_08Mar22_1104
Session lttng_ping1_08Mar22_1104 created.
Traces will be output to /tmp/lttng_ping1_08Mar22_1104
[+] lttng enable events ...
All kernel events are enabled in channel channel0
ust event lttng_ust_tracef:* created in channel channel0
@@@ lttng_trc.sh: Tracing "ping -c1 packtpub.com" now ... @@@
Tuesday 08 March 2022 11:04:18 AM IST
1646717658.985523388
Tracing started for session lttng_ping1_08Mar22_1104
PING packtpub.com (104.22.0.175) 56(84) bytes of data.
64 bytes from 104.22.0.175 (104.22.0.175): icmp_seq=1 ttl=58 time=14.6 ms

--- packtpub.com ping statistics ---
1 packets transmitted, 1 received, 0% packet loss, time 0ms
rtt min/avg/max/mdev = 14.563/14.563/14.563/0.000 ms
Waiting for data availability.
Tracing stopped for session lttng_ping1_08Mar22_1104
Tuesday 08 March 2022 11:04:19 AM IST
1646717659.517192093
Tuesday 08 March 2022 11:04:19 AM IST
1646717659.521628654
[+] cleaning up...
lttng_trc.sh: done. Trace files in /tmp/lttng_ping1_08Mar22_1104 ; size:
5       /tmp/lttng_ping1_08Mar22_1104
Destroying session lttng_ping1_08Mar22_1104..
Session lttng_ping1_08Mar22_1104 destroyed
 [+] ...generating compressed tar file of trace now, pl wait ...
tar: Removing leading `/' from member names
-rw-r--r-- 1 root root 755K Mar  8 11:04 lttng_ping1_08Mar22_1104.tar.gz
lttng $
```

Figure 9.27 – A screenshot showing the execution of our simple LTTng kernel trace wrapper script

The script makes a few validity checks and then performs kernel-level tracing – while the user app (here, `ping`) executes – with all the kernel events enabled. It sets up so that the actual trace data is saved under `/tmp/lttng_<sessionname>_<timestamp>`. Further, once done, it archives and compresses this data – you can see the file at the end of the screenshot (here, it's named `lttng_ping1_08Mar22_1104.tar.gz`). This enables you to transfer and analyze the trace on a different system.

Analyzing LTTng traces on the command line – a mention

LTTng includes a set of libraries and tools to analyze its raw trace data. The primary tool is called Babeltrace 2, which is a command-line-based utility. I refer you to this link on the LTTng website to delve into how exactly to use it: *Use the babeltrace2 command-line tool*: `https://lttng.org/docs/v2.13/#doc-viewing-and-analyzing-your-traces-bt`. Being console-based, the output can be overwhelming. (The `babeltrace` output from the single `ping` trace just performed yielded over 123,000 lines of information!)

LTTng has another set of powerful tools to interpret and analyze its raw trace data called the **LTTng analyses** project. Though command-line-based, it provides intuitive Python-based interfaces to help visualize the trace session. You can learn more on this here: `https://github.com/lttng/lttng-analyses`.

Using the Trace Compass GUI to visualize the single ping LTTng trace

A very visually appealing and popular GUI interface for interpreting and analyzing LTTng traces is via the superb **Trace Compass GUI**. Trace Compass is an Eclipse-based project. Do look up its excellent site, for installation, documentation, and even screenshots: `https://www.eclipse.org/tracecompass/`. Here, we only introduce the usage of the Trace Compass GUI.

Once installed, simply run Trace Compass, go to the **File | Open Trace...** menu, and select the directory where your LTTng tracing session was saved. Trace Compass parses and displays it – here's a portion of the GUI (I also popped up the **Legend** dialog so that you can understand the color-coding applied in the upper pane, the graph area):

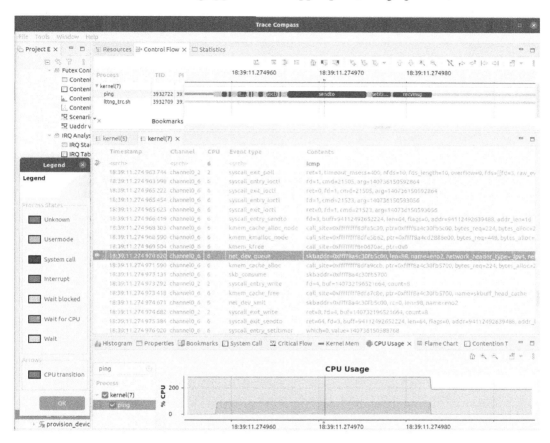

Figure 9.28 – A truncated screenshot of the superb Trace Compass GUI

There's a lot of customization possible with Trace Compass's views and perspectives – try and spend some time with it. In this particular case, to help zoom into the relevant region of the trace, I filtered in the following manner. Type 6 into the **CPU** column search widget (as that's the CPU core that `ping` happened to run upon in this particular section). Also, type the `icmp` string into the **Contents** column search widget – now lines containing it get matched. Here, there's only one – the `net_dev_queue` kernel event seen in the screenshot. This may not always work, though; if not, try searching the **Contents** column for known events, such as `net_dev_xmit`. (*Here's a quick tip* – as you know, all kernel events can be seen under `/sys/kernel/tracing/events/`).

Right-click on the line item or event of choice. Here, I clicked on the item seen highlighted in the screenshot – `net_dev_queue`. Select the **Copy to Clipboard** menu item. Pasting it, this is what I see (the text has wrapped):

```
Timestamp      Channel       CPU  Event type       Contents     TID
Prio      PID       Source
18:39:11.274 970 channel0_6 6    net_dev_queue
skbaddr=0xffff8a4c30fb5c00, len=98, name=eno2, network_header_
type=_ipv4, network_header=ipv4=[version=4, ihl=5, tos=0,
tot_len=84, id=0x2950, frag_off=16384, ttl=64, protocol=_
icmp, checksum=0xe6db, saddr_padding=[], saddr=[192, 168, 1,
16], daddr_padding=[], daddr=[104, 22, 0, 175], transport_
header_type=_icmp, transport_header=icmp=[type=8, code=0,
checksum=393, gateway=720897]]    3932722  20  3932722
```

As mentioned before, knowing the value of parameters at runtime can be crucial to debugging a given scenario. In addition, other information is provided (this is due to how the kernel events subsystem works).

To help see (a part of) the `ping` process's execution timeline more clearly, here's a zoomed (and truncated) screenshot of the (interesting!) graph area:

Figure 9.29 – A screenshot showing a part of the ping process's timeline

You can see we've filtered down to the `ping` process (on the left side). Its timeline and the functions in its execution code path are seen to its right. Note the very useful *color coding* – blue represents a system call, green, userspace code (the **Legend** dialog seen in *Figure 9.28* displays the current color-coding settings). Placing the mouse over any part of the graph shows more information on it (including its duration).

With LTTng and Trace Compass, we find that it makes the *state* of the concerned thread (as well as all others) very clear, by appropriately color-coding it. This helps you understand the overall context visually. You can literally see what is happening – is it blocking on I/O, is it running user/kernel code paths, or is a softirq or hardirq running? All is clear. For example, the left extreme of `ping` in the preceding figure has a brown color bar representing it, waiting for the CPU (in a non-blocking manner; clicking on it, I can see that it's in the `sched_switch` event for 48.6 us). The lemon yellow color bar on the extreme right shows that it's blocked on I/O here. Again, clicking on it, I can see that it happens to be blocked on the `power_cpu_idle` event for 14.6 us). On the other hand, KernelShark (and thus ftrace and `trace-cmd`) clearly detail the context – via the super useful *latency format* info columns – which is missing in Trace Compass's (and LTTng's) information.

Exercises

Trace the kernel flow of the following:

- The classic K&R "Hello, world" app

- A simple "Hello, world" kernel module

Do this using (raw) ftrace, the `trace-cmd` frontend, and (possibly) my `trccmd` frontend. (I could provide solutions but it's pointless. The end result here isn't the thing that really matters; it's the process.). Visualize your traces with KernelShark. You can also try exporting the trace into the **Common Trace Format (CTF)** employed by Trace Compass and visualize it there as well.

Summary

Well done on completing this long and useful topic on tracing within the Linux kernel. We began this chapter with an overview of the many available tracing mechanisms on the Linux kernel – the first couple of figures nicely summarized this. A large portion of this chapter dealt with how you can leverage the powerful ftrace kernel infrastructure. It is high-performance and minimally invasive, with pretty much no dependencies, making it ideal for even constrained embedded systems!

To make it easier though, several useful frontends for ftrace exist. We covered using `trace-cmd`, the KernelShark GUI, and the `perf-tools` project. We finished the chapter with an introduction to using LTTng for kernel tracing and the Trace Compass GUI for visualizing the trace.

You'll tend to find that one tracing/visualization tool may be superior to another in some respects but is inferior in others. This, of course, is very typical (of pretty much everything – trade-offs, right). Remember, as Fred Brooks has told us back in 1975 (in his timeless book *The Mythical Man Month*) **there's no silver bullet**! Learning to use several powerful tools will stand you in good stead.

Do try working with all this technology on your own (and try the few specified exercises)! Then, the next chapter awaits – kernel panic and more. Don't panic! We'll get through it.

Further reading

- *Unified Tracing Platform – Bringing tracing together*, Steven Rostedt, VMware, 2019: `https://static.sched.com/hosted_files/osseu19/5f/unified-tracing-platform-oss-eu-2019.pdf`

- *Unifying kernel tracing*, Jack Edge, Oct 2019: `https://lwn.net/Articles/803347/`

- *Linux tracing systems & how they fit together*, Julia Evans (@b0rk), July 2017: `https://jvns.ca/blog/2017/07/05/linux-tracing-systems/`

- *Using the Linux Tracing Infrastructure*, Jan Altenberg, Linutronix GmbH, Nov 2017: `https://events.static.linuxfound.org/sites/events/files/slides/praesentation_0.pdf`

- The comprehensive kernel index – all articles on *tracing* on LWN: `https://lwn.net/Kernel/Index/#Tracing`

- Ftrace:

  - Official kernel documentation – very detailed and comprehensive: *ftrace - Function Tracer*: `https://www.kernel.org/doc/html/v5.10/trace/ftrace.html#ftrace-function-tracer`

  - The LWN kernel index and ftrace: `https://lwn.net/Kernel/Index/#Ftrace`

  - *Ftrace: The hidden light switch*, Brendan Gregg, August 2014: `https://lwn.net/Articles/608497/`

  - Ftrace internals: *Two kernel mysteries and the most technical talk I've ever seen*, Brendan Gregg, Oct 2019: `https://www.brendangregg.com/blog/2019-10-15/kernelrecipes-kernel-ftrace-internals.html`

- Older but definitely useful:

  - *Debugging the kernel using Ftrace - part 1*, Steven Rostedt, December 2009: `https://lwn.net/Articles/365835/`

  - *Debugging the kernel using Ftrace - part 2*, Steven Rostedt, December 2009: `https://lwn.net/Articles/366796/`

  - *Secrets of the Ftrace function tracer*, Steven Rostedt, January 2010: `https://lwn.net/Articles/370423/`

- *Welcome to ftrace & the Start of Your Journey to Understanding the Linux Kernel!*, Steven Rostedt, November 2019: `https://blogs.vmware.com/opensource/2019/11/12/ftrace-linux-kernel/`

- *Debugging the kernel using Ftrace*, Programmer Group, November 2021: `https://programmer.group/debugging-the-kernel-using-ftrace.html`

- *ftrace: trace your kernel functions!*, Julia Evans, March 2017: `https://jvns.ca/blog/2017/03/19/getting-started-with-ftrace/`

- Ftrace cheat sheets:

  - *Ftrace Favorites Cheat Sheet - Fun Commands to Try with Ftrace*: `http://linux-tipps.blogspot.com/2011/05/ftrace-favorites-cheat-sheet-fun.html`

  - *Kernel Tracing Cheat Sheet*: `https://lzone.de/cheat-sheet/Kernel%20Tracing`

  - *Linux Tracing Workshops Materials*: `https://github.com/goldshtn/linux-tracing-workshop`

- *Virtually mapped kernel stacks*, Jon Corbet, June 2016, LWN: `https://lwn.net/Articles/692208/`

- *Virtually mapped stacks 2: thread_info strikes back*, Jon Corbet, June 2016, LWN: `https://lwn.net/Articles/692953/`

- The trace-cmd frontend:

  - *trace-cmd: A front-end for Ftrace*, Steven Rostedt, LWN, October 2010: `https://lwn.net/Articles/410200/`

  - *Kernel tracing with trace-cmd*, G Kamathe, RedHat, July 2021: `https://opensource.com/article/21/7/linux-kernel-trace-cmd`

- The KernelShark GUI:

  - Useful – the "official" KernelShark HTML documentation page: `https://www.kernelshark.org/Documentation.html`

  - A colorful presentation – *Swimming with the New KernelShark*, Yordan Karadzhov, VMware, 2018: `https://events19.linuxfoundation.org/wp-content/uploads/2017/12/Swimming-with-the-New-KernelShark-Yordan-Karadzhov-VMware.pdf`

  - *KernelShark (quick tutorial)*, Steven Rostedt, ELC 2011: `https://elinux.org/images/6/64/Elc2011_rostedt.pdf`

- The `perf-tools` wrapper scripts over ftrace and `perf[_events]`:

  - GitHub repository: `https://github.com/brendangregg/perf-tools/`

  - Examples of most `perf-tools` scripts: `https://github.com/brendangregg/perf-tools/tree/master/examples`

- *Linux Performance Analysis: New Tools and Old Secrets*, November 2014, Brendan Gregg, at USENIX LISA14:

  - Video presentation: `https://www.usenix.org/conference/lisa14/conference-program/presentation/gregg`

  - Slides: `https://www.slideshare.net/brendangregg/linux-performance-analysis-new-tools-and-old-secrets`

- You can find useful links to *eBPF* and its frontends in *Chapter 4, Debug via Instrumentation – Kprobes*, in the *Further reading* section

- LTTng:

  - The LTTng main website: `https://lttng.org/`

  - The LTTng site's *Quick start* guide: `https://lttng.org/docs/v2.13/#doc-getting-started`

  - *Babeltrace 2: The command-line interface (CLI)*: `https://lttng.org/blog/2020/06/01/bt2-cli/`

  - The LTTng analyses project: `https://github.com/lttng/lttng-analyses#lttng-analyses`

  - *Finding the Root Cause of a Web Request Latency*, Julien Desfossez, February 2015: `https://lttng.org/blog/2015/02/04/web-request-latency-root-cause/`

- *Tutorial: Remotely tracing an embedded Linux system*, C Babeux, March 2016: `https://lttng.org/blog/2016/03/07/tutorial-remote-tracing/`
- *LTTng: The Linux Trace Toolkit Next Generation – A Comprehensive User's Guide (version 2.3 edition)*, Daniel U. Thibault, DRDC Valcartier Research Centre: `https://cradpdf.drdc-rddc.gc.ca/PDFS/unc246/p804561_A1b.pdf`

- Trace Compass GUI:

  - Trace Compass website: `https://www.eclipse.org/tracecompass/#home`
  - Alternate tracing tools: `https://lttng.org/docs/v2.13/#doc-lttng-alternatives`

- Miscellaneous:

  - Boot-time tracing via ftrace: `https://www.kernel.org/doc/html/latest/trace/boottime-trace.html`
  - *Trace Linux System Calls with Least Impact on Performance in Production*, December 2020: `https://en.pingcap.com/blog/how-to-trace-linux-system-calls-in-production-with-minimal-impact-on-performance/`

10
Kernel Panic, Lockups, and Hangs

It's unpleasant – that queasy feeling deep in the pit of your stomach, the cold sweat forming on your brow – when you get that dreaded kernel panic message on the console, and those absolute, unforgiving pixels, with the hard cold eye of a god, tell you that the system is effectively dead:

```
Kernel panic - not syncing: [...]
```

Why, oh why? – your lamentations are futile. Unless, unless... you don't panic (pun intended), read this chapter, figure out what's going on (by writing your own custom panic handler to help with that), and get on with your life, dude!

In addition to understanding and dealing with kernel panics, we also delve into the causes of kernel lockups, hung tasks, stalls, and how to configure a kernel to detect them. In this chapter, we're going to cover the following main topics:

- Panic! – what happens when a kernel panics
- Writing a custom kernel panic handler routine
- Detecting lockups and CPU stalls in the kernel
- Employing the kernel's hung task and workqueue stall detectors

Technical requirements

The technical requirements and workspace remain identical to what's described in *Chapter 1, A General Introduction to Debugging Software*. The code examples can be found within the book's GitHub repository here: `https://github.com/PacktPublishing/Linux-Kernel-Debugging`.

Panic! – what happens when a kernel panics

To conquer the beast, you must first understand it. In that spirit, let's panic!

The primary panic handling code in the kernel lies here: `kernel/panic.c:panic()`. The `panic()` function – the heart of it – receives, as parameters, a variable argument list – a `printf`-style format specifier and associated variables (whose values will be printed):

```
// kernel/panic.c
/**
*   panic - halt the system
*   @fmt: The text string to print
*
*   Display a message, then perform cleanups.
*   This function never returns.
*/
void panic(const char *fmt, ...)
{ [...]
```

This function should (quite obviously) never be lightly invoked; calling it implies that the kernel is in an unusable, unusable state; once called, the system effectively comes to a grinding halt.

Let's panic

Here, with a view to being empirical and experimenting (on our test VM, of course), let's just call `panic()` and see what happens. Our enviously simple module does so; here's its code (besides the boilerplate `#include` and module macros):

```
// ch10/letspanic/letspanic.c
static int myglobalstate = 0xeee;
static int __init letspanic_init(void)
{
    pr_warn("Hello, panic world\n");
```

```
        panic("whoa, a kernel panic! myglobalstate = 0x%x",
        myglobalstate);
        return 0;          /* success */
}
module_init(letspanic_init);
```

There's no need for a cleanup handler in this module, so we don't register one. Great, I build and insmod it on my trusty x86_64 Ubuntu 20.04 LTS (which I've logged into over ssh) running our custom production 5.10.60-prod01 kernel:

```
$ sudo insmod ./letspanic.ko
[... <panicked, and hung> ... ]
```

Immediately upon insmod, the system simply hung; no printks appeared on the console (on which I had SSHed in), nor on the graphical VirtualBox interface! It has obviously panicked, yes, but in order to debug why, we need to – at the very least – be able to see the details emitted by the kernel code within the panic code path. You'll very soon see what details it prints (hint – it's much like the details seen in a kernel Oops diagnostic, which we covered in depth in *Chapter 7, Oops! Interpreting the Kernel Bug Diagnostic*). So, what do we do now? The short answer – we use *netconsole*! Before that though, let's quickly mention the kernel's SysRq feature.

Creating a panic via the command line

There's an alternate easy non-programming way to generate a kernel panic – we take advantage of the kernel's **Magic SysRq** facility and the kernel.panic_on_oops sysctl, like so (as root):

```
echo 1 > /proc/sys/kernel/panic_on_oops
echo 1 > /proc/sys/kernel/sysrq
echo c > /proc/sysrq-trigger
```

It's simple – the first command sets up the kernel to panic on an Oops. The second command is to play safe and enable the kernel's Magic SysRq features (in case it isn't already enabled – for security, perhaps). The third command has the kernel's Magic SysRq feature trigger a crash!

What's this Magic SysRq thingy anyway?

In a nutshell, the kernel Magic SysRq facility is a keyboard-enabled hotkey interface allowing users (typically the sysad or developer) to force the kernel to take certain code paths. These effectively become similar to *backdoors* into the kernel, useful for debugging system hangs and such.

It must first be enabled (CONFIG_MAGIC_SYSRQ=y). For security, it can be turned off or tuned to allow only certain functionality; to do so, as root, do the following:

- To turn it off, write 0 into the /proc/sys/kernel/sysrq pseudofile.

- To enable all features, write 1 into it.

- A combination via a bitmask can be written as well.

The default value of the bitmask is the value of the kernel config CONFIG_MAGIC_SYSRQ_DEFAULT_ENABLE (typically, 1).

It allows you to do pretty radical things – force a crash (c), a cold reboot (b), a power off (o), forcibly invoke the **Out Of Memory (OOM)**-killer (f), force an emergency sync (s), unmount all filesystems (u), and so on. It can really help with debugging – it allows you to see all active task backtraces on all CPU cores (l), show CPU registers (p), show kernel timers (q), show blocked tasks (w), dump all ftrace buffers (z), and so on. The letter in parentheses is the one to use for that functionality. It can be used in two ways:

- Interactively, by pressing the particular key combination for your system (on x86, it's Alt + SysRq + <letter>; note that on some keyboards, the SysRq key is the same as the *Prt Sc* key).

- Non-interactively, by echoing the letter to the /proc/sysrq-trigger pseudofile (see the simple demo in the following figure). Echoing the ? letter results in a help screen of sorts being written via the kernel printk:

```
# echo 1 > /proc/sys/kernel/sysrq
# echo ? > /proc/sysrq-trigger ; dmesg |tail -n1
[157150.167020] sysrq: HELP : loglevel(0-9) reboot(b) crash(c) terminate-all-tasks(e) m
emory-full-oom-kill(f) kill-all-tasks(i) thaw-filesystems(j) sak(k) show-backtrace-all-
active-cpus(l) show-memory-usage(m) nice-all-RT-tasks(n) poweroff(o) show-registers(p)
show-all-timers(q) unraw(r) sync(s) show-task-states(t) unmount(u) force-fb(v) show-blo
cked-tasks(w) dump-ftrace-buffer(z)
#
```

Figure 10.1 – A screenshot showing how the kernel Magic SysRq feature can be enabled and queried

The official kernel documentation can be found here: *Linux Magic System Request Key Hacks*: `https://www.kernel.org/doc/html/latest/admin-guide/sysrq.html#linux-magic-system-request-key-hacks`. It's comprehensive, so do check it out!

Magic SysRq and the kernel `panic_on_oops` sysctl does indeed let us panic the kernel, but we'd like to do so in code, via a module. Hence, we did so previously (a demo of doing it this way, via kernel `panic_on_oops` and Magic SysRq, is shown in the *Trying it out – our custom panic handler module* section).

To the rescue with netconsole

Hopefully you recall that the kernel netconsole code (often deployed as a module) transmits all kernel printks over the network to a receiver system (both source and destination being specified via the usual `IP:port#` style addresses). We won't repeat the how-to part of it, as we already covered how to use netconsole in *Chapter 7, Oops! Interpreting the Kernel Bug Diagnostic* in the *An oops on an ARM Linux system and using netconsole* section.

So, I set up my VM (where I am running our `letspanic` module) as the sender (of course, I need to have netconsole configured here, and it is) and my host system (a native x86_64 Ubuntu system) as the receiver. For your convenience, when loading the netconsole driver as a module, this is the format of the key parameter, named `netconsole`:

```
netconsole=[+] [src-port]@[src-ip]/[<dev>],[tgt-port]@<tgt-ip>/
[tgt-macaddr]
```

Here's the very brief setup detail (we leave the source and destination ports to their default values):

- Sender: an x86_64 Ubuntu 20.04 LTS VM running our custom `5.10.60-prod01` kernel (type this on one line):

    ```
    sudo modprobe netconsole netconsole=@192.168.1.20/
    enp0s8,@192.168.1.101/
    ```

- Receiver: a native x86_64 Ubuntu 20.04 LTS system running the standard Ubuntu kernel:

    ```
    netcat -d -u -l 6666 | tee -a klog_from_vm.txt
    ```

Of course, these are IP addresses and network interface names I encountered. Please replace them appropriately on your system. The `netcat` process blocks on receiving the incoming packets from the sender system and displays them (also writing them into the log file over `tee`). Here's a screenshot that clearly shows their interplay:

```
~ $ netcat -d -u -l 6666 | tee -a klog_from_vm.txt
[  919.864326] Kernel panic - not syncing: whoa, a kernel panic! myglobalstate = 0xeee
[  919.864395] CPU: 5 PID: 1751 Comm: insmod Tainted: G          OE     5.10.60-prod01 #6
[  919.864439] Hardware name: innotek GmbH VirtualBox/VirtualBox, BIOS VirtualBox 12/01/2006
[  919.864487] Call Trace:
[  919.864511]  dump_stack+0x76/0x94
[  919.864574]  panic+0x1ac/0x382
[  919.864602]  ? printk+0x58/0x6f
[  919.864637]  ? 0xffffffffc06f4000
[  919.864652]  letspanic_init+0x39/0x1000 [letspanic]
[  919.864694]  do_one_initcall+0x48/0x210
[  919.864731]  ? kmem_cache_alloc_trace+0x3ae/0x450
[  919.864786]  do_init_module+0x62/0x240
[  919.864801]  load_module+0x2a04/0x3080
[  919.864826]  ? security_kernel_post_read_file+0x5c/0x70
[  919.864876]  __do_sys_finit_module+0xc2/0x120
[  919.864915]  ? __do_sys_finit_module+0xc2/0x120
[  919.864934]  __x64_sys_finit_module+0x1a/0x20
[  919.864973]  do_syscall_64+0x38/0x90
[  919.864989]  entry_SYSCALL_64_after_hwframe+0x44/0xa9
[  919.865008] RIP: 0033:0x7c74cba5a76d
[  919.865025] Code: 00 c3 66 2e 0f 1f 84 00 00 00 00 00 90 f3 0f 1e fa 48 89 f8 48 89 f7 48 89 d6 48 89 ca 4d 89
c2 4d 89 c8 4c 8b 4c 24 08 0f 05 <48> 3d 01 f0 ff ff 73 01 c3 48 8b 0d f3 36 0d 00 f7 d8 64 89 01 48
[  919.865085] RSP: 002b:00007ffe183750d8 EFLAGS: 00000246 ORIG_RAX: 0000000000000139
[  919.865118] RAX: ffffffffffffffda RBX: 00005bce652397d0 RCX: 00007c74cba5a76d
[  919.865146] RDX: 0000000000000000 RSI: 00005bce632ec358 RDI: 0000000000000003
[  919.865172] RBP: 0000000000000000 R08: 0000000000000000 R09: 00007c74cbb31580
[  919.865197] R10: 0000000000000003 R11: 0000000000000246 R12: 00005bce632ec358
[  919.865222] R13: 0000000000000000 R14: 00005bce65239780 R15: 0000000000000000
[  919.865300] Kernel Offset: 0x1200000 from 0xffffffff81000000 (relocation range: 0xffffffff80000000-0xffffffffbf
ffffff)
[  919.865337] ---[ end Kernel panic - not syncing: whoa, a kernel panic! myglobalstate = 0xeee ]---

$ sudo insmod ./letspanic.ko
```

Figure 10.2 – A screenshot showing the receiver window on top (netcat receiving the kernel printks) from the sender system (below), which has insmoded our letspanic module

This is fantastic! Now, we can clearly see the kernel printks emitted by the panic handler routine deep within the kernel.

Interpreting the panic output

As already mentioned, interpreting it is easy, as it pretty much follows the Oops diagnostic output format. Following along with *Figure 10.2*, you can see the dreaded `Kernel panic - not syncing` message at the very top, followed by our message – the parameter to `panic()` – which is sent out at the `KERN_EMERG` printk log level, the highest possible (remember that this will cause the panic message to be immediately broadcast on all console devices). Thus, the line looks like this:

```
Kernel panic - not syncing: whoa, a kernel panic! myglobalstate
= 0xeee
```

This is followed by the *usual* stuff:

- The process context (here, it's `insmod`, of course), the tainted flags, and the kernel version.

- A hardware detail line.

- If enabled (if `CONFIG_DEBUG_BUGVERBOSE` is on, it typically is), display the call stack (via the kernel's `dump_stack()` routine). This, of course, becomes one of the big clues as to how we got here, how we panicked – reading the kernel-mode stack in a bottom-up fashion shows us how we got here (as usual, ignore call frames prefixed with a ? symbol).

- This is followed by the instruction pointer (`RIP`) value, and both the machine code on the processor and the CPU register values at this point in time.

- In kernels from 3.14 onward that use the **Kernel Address Space Layout Randomization (KASLR)** feature as a security measure, the kernel offset is displayed (via an arch-specific function, `dump_kernel_offset()`, which, very interestingly, is invoked via something called the chain notifier mechanism, which we'll delve into in the following main section).

- The panic is capped with an end message, pretty much the same as the start one:

  ```
  ---[ end Kernel panic - not syncing: whoa, a kernel
  panic! myglobalstate = 0xeee ]---
  ```

Remember that we covered the detailed interpretation of the first four preceding bullet points in *Chapter 7, Oops! Interpreting the Kernel Bug Diagnostic*, in the *Devil in the details – decoding the oops* section.

So, where exactly within the kernel code does the sinister `Kernel panic - not syncing:...` message come from? Here's the beginning portion of the `panic()` code on the 5.10.60 kernel (https://elixir.bootlin.com/linux/v5.10.60/source/kernel/panic.c#L177):

```
void panic(const char *fmt, ...)
{
    static char buf[1024];
    va_list args;
    [...]
    pr_emerg("Kernel panic - not syncing: %s\n", buf);
    [...]
}
```

It's an exported function, so it's callable by modules. Also, quite clearly, after performing some tasks (not shown here) early on, it emits an emergency log-level (KERN_EMERG) printk message to the kernel logs and console device(s) (if configured), proclaiming the fact that the kernel has panicked! It appends any message passed to it, then performs what cleanup it can, and dumps useful system state information to all registered console devices (as we just saw).

> **Why Is the Phrase "not syncing" in the Kernel Panic Message?**
>
> The not syncing phrase means precisely that – buffers containing device data are deliberately not flushed – or synchronized (synced) – to disk (or flash, or whatever). This action, if performed, can actually make a bad situation worse, even corrupting data; thus, it's avoided.

You've realized that the system is now in an undefined unstable state; thus, the panic code does what it can with the ever-present possibility of an inadvertent complete lockup or failure. Again, this is why we can only do the bare minimum. Pretty much the entire kernel panic code path is run on a single CPU core. Again, this is to avoid complexity and possible deadlocks; in a similar fashion, local interrupts are disabled and kernel preemption is turned off (the code has pretty detailed comments – take a look).

When possible – and especially when the CONFIG_DEBUG_BUGVERBOSE config is on – the panic function tries to emit as much relevant system information as possible (as we saw – refer to *Figure 10.2* and the related notes). To this end, a function named panic_ print_sys_info() is invoked; it uses a bitmask (which you can set via the panic_ print kernel parameter) to determine and show more system information – things such as all tasks info, memory, timer, lock, ftrace info, and all kernel printks. However, the default value of the bitmask is 0, implying that it doesn't show any of these. This extra information could indeed prove very useful; in the following section, we show how to set this bitmask.

Within `panic()`, once this critical info dump is done, the last thing the function does is loop infinitely on the single enabled processor core; within this loop, it resets the **Non-Maskable Interrupt (NMI)** watchdog (as interrupts are now disabled) and then periodically invokes an arch-dependent function called `panic_blink()`. On the x86, if enabled, this hooks into the keyboard/mouse driver here – `drivers/input/serio/i8042.c:i8042_panic_blink()`. This code causes the keyboard LEDs to blink, alerting a user running a GUI (such as *X*) to realize that the system isn't just *soft hung* but panicked. Here's the last code paragraph in the kernel `panic()` function, just after the end message:

```
        pr_emerg("---[ end Kernel panic - not syncing: %s ]---\n",
buf);

        /* Do not scroll important messages printed above */
        suppress_printk = 1;
        local_irq_enable();
        for (i = 0; ; i += PANIC_TIMER_STEP) {
            touch_softlockup_watchdog();
            if (i >= i_next) {
                i += panic_blink(state ^= 1);
                i_next = i + 3600 / PANIC_BLINK_SPD;
            }
            mdelay(PANIC_TIMER_STEP);

        }
```

This, again, is completely deliberate – we want to ensure that the critical and precious debug info printed on the console(s) doesn't simply scroll away and disappear from view (as, of course, you can't scroll up, down, or do anything; the system is effectively dead now).

More Ways to Collect Panic Messages

Many Android devices make use of the Linux kernel's upstream `pstore` and `ramoops` support to enable you to collect the kernel log on a kernel panic. Of course, this implies a system containing persistent RAM and/or block devices that the `pstore` abstraction layer can use. Thus, `pstore` and `ramoops` can be viewed as being somewhat analogous to `kexec/kdump` in terms of being able to collect system information upon a kernel crash or panic and store it for later retrieval and analysis.

Also, the **Intelligent Platform Management Interface (IPMI)** is a standardized way to monitor and control a system's sensors. It includes panic and watchdog tuning. See the *Further reading* section for more links on all these.

Hang on, though – the panic code *can* take other code paths from what we saw just previously:

- When the kernel `kexec/kdump` feature is enabled, and the kernel has panicked or Oops'ed, to warm-boot into a secondary so-called **dump-capture kernel** (thus allowing the content of the kernel RAM to be saved to a snapshot and examined later!). In other words, the `panic()` function is a trigger point for invoking this functionality, which will (ultimately) invoke the `kexec` facility within the kernel to warm-boot the system into the dump-capture kernel (we'll briefly talk about this in the book's last chapter).

- When a custom panic handler is installed via the panic notifier chain mechanism, it gets called in addition to the regular panic handling code. Interesting! We will cover how you can do just this in the section that follows.

When the `panic=n` kernel parameter is set, it implies a panic timeout and reboot (more on this follows). Right, now that we can interpret the kernel's panic diagnostic, let's move along.

Kernel parameters, tunables, and configs that affect kernel panic

Here, we present in a convenient summary format a few kernel parameters (passed via the bootloader) as well as some possible sysctl tunables and kernel config macros that affect the kernel panic code path (a few more regarding lockups and hung tasks will be covered in an upcoming section):

Kernel boot-time parameter	Equivalent kernel sysctl knob	Equivalent kernel config macro (CONFIG_<FOO>)	Purpose
oops=panic	kernel.panic_on_oops	PANIC_ON_OOPS	Set this to 1 (it's 0 by default) to have the kernel panic when an Oops occurs; this can be valuable on some types of production systems, alerting all stakeholders that we have a show-stopper level kernel/driver bug.
panic=n	kernel.panic	PANIC_TIMEOUT	When n is positive, the kernel attempts to reboot the system n seconds after panic (which is especially useful for deeply embedded systems); if 0 is passed, it implies doing nothing, just waiting forever (this is the default), while a negative value implies an immediate reboot. Note that this option might require arch-specific support.
panic_print=<bitmask>	kernel.panic_print	<none>	Bitmask – the panic_print_sys_info() function (invoked as part of the panic() code path) interprets the bits and accordingly prints system-level details. The bits and how they're interpreted are shown in Table 10.2.
panic_on_taint=<bitmask>, [,nousertaint]	<none>	<none>	Bitmask – if passed, have the kernel panic when any of the set of tainted flags here match the tainted bitmask. More details can be found here: https://www.kernel.org/doc/html/latest/admin-guide/sysctl/kernel.html?highlight=hung_task#tainted.
panic_on_warn=	kernel.panic_on_warn	<none>	Set this to 1 (it's 0 by default) to have the kernel panic when any kernel warning is issued (via the WARN*() macros). This can be useful to have a kernel dump image generated on any warning.

Kernel boot-time parameter	Equivalent kernel sysctl knob	Equivalent kernel config macro (CONFIG_<FOO>)	Purpose
<none>	`kernel.panic_on_stackoverflow`	DEBUG_ STACKOVERFLOW	Set this to 1 (it's 0 by default) to have the kernel panic when a kernel, exception, or IRQ stack overflow is detected. This requires CONFIG_ DEBUG_STACKOVERFLOW =y.
`unknown_nmi_ panic`	<none>	<none>	X86 only. This has the kernel panic when an unknown NMI fires.
`crash_kexec_ post_notifiers`	<none>	<none>	Boolean (False by default) – when set, the panic code path first runs all registered panic notifiers, dumps the kernel log, and then runs kdump. This isn't actually necessary – a kdump-configured kernel will have kdump/kexec run after a panic (doing this, in fact, increases the risk of instability).
<none>	`vm.memory_ failure_recovery`	<none>	If set to 1 (the default is 0), attempt memory failure recovery; otherwise, panic if memory fails (depends on CONFIG_MEMORY_FAILURE=y).
<none>	`vm.panic_on_oom`	<none>	If set to 1 (the default is 0), the kernel panics when the OOM situation occurs; otherwise, the OOM-killer facility typically kills the rogue process(es) and the system survives. If set to 2, the kernel always panics on OOM (see man 5 proc for more detail).
<none>	<none>	SCHED_STACK_ END_CHECK	If enabled, and, on calls to schedule(), a stack overrun is detected, have the kernel panic immediately (the default is n). This requires DEBUG_KERNEL=y.

Table 10.1 – A summary table of the kernel parameters, sysctl tuning knobs, and config macros related to kernel panic handling

For sysctl knobs, the `kernel.foo` syntax implies that you'll find the tuning pseudofile, `foo`, in the `/proc/sys/kernel` directory.

The interpretation of the bits within the `panic_print` kernel parameter's bitmask (mentioned previously) is as follows:

Bit #	Bit name	Meaning
0	`PANIC_PRINT_TASK_INFO`	If set, show info pertaining to all alive tasks. This includes the task name, state, PID, PPID, flags, and a call trace of its kernel-mode stack.
1	`PANIC_PRINT_MEM_INFO`	If set, show info on system memory.
2	`PANIC_PRINT_TIMER_INFO`	If set, show info on kernel timers.
3	`PANIC_PRINT_LOCK_INFO`	If set, and `CONFIG_LOCKDEP` is enabled, show kernel locks info.
4	`PANIC_PRINT_FTRACE_INFO`	If set, dump the ftrace buffer content.
5	`PANIC_PRINT_ALL_PRINTK_MSG`	If set, dump all the kernel printks.

Table 10.2 – Use the panic_print kernel parameter to set bits to get additional system info on kernel panic

The default value of the `panic_print` bitmask is 0, implying that no additional system info is printed during panic. Set the bits appropriately to display whatever details you'd like. So, for example, to show *all* the preceding details, we append `panic_print=0x3f` to the kernel parameter list when booting. Depending on your project, *these additional details can prove very useful when debugging a kernel panic!*

FYI, the official kernel documentation very clearly documents all kernel sysctl knobs (tunables) here: `https://www.kernel.org/doc/html/latest/admin-guide/sysctl/kernel.html`.

> **Exercise**
>
> Pass the `panic_print=n` kernel parameter setting n to an appropriate value (see *Table 10.2*). Then, run the `letsdebug` kernel module. The kernel will panic. Verify that you get (perhaps via netconsole) the additional system information details you requested via the `panic_print` bitmask.

All right, now that you understand what happens within the kernel on panic, let's move on to doing our own thing if and when it occurs.

Writing a custom kernel panic handler routine

The Linux kernel has a powerful feature named **notifier chains** (the word *chains* implying the usage of linked lists). It's essentially based upon a *publish-and-subscribe model*. The subscriber is the component that wants to know when a given asynchronous event occurs; the publisher is the one that pushes the notification that the event did occur. Quite obviously, the subscriber(s) register interest in a given event and supply a callback function. When the event occurs, the notification mechanism invokes the callback. When someone registers itself with a notifier chain, they have subscribed to it and specified a callback function. When the relevant event occurs, all subscribers' callback functions for that notifier chain are invoked (there's even a way to specify your priority and pass some data along, which we'll get to soon enough). We shall make use of one of the kernel's predefined notifier chains – the *panic notifier chain* – to register our custom panic handler.

Linux kernel panic notifier chains – the basics

First, though, we should understand some basics regarding the notifier chains. The Linux kernel supports four different types. The classification is based on the context in which the callback function executes (process or interrupt) and, thus, whether it can be blocking in nature or not (atomic). These are the four types of notifier chains:

- **Atomic**: Chain callbacks run in atomic context and cannot block (internally uses a spinlock for critical region protection).

- **Blocking**: Chain callbacks run in process context and can block (internally uses a read-write semaphore locking primitive to implement blocking behavior).

- **Sleepable RCU (SRCU)**: Chain callbacks run in process context and can block (internally uses the more sophisticated **Read-Copy-Update** (**RCU**) mechanism to implement lock-free semantics; here, the read-side critical section can block/sleep). This type is good for cases where the callbacks occur often and the removal of the notifier block is rare.

- **Raw**: Chain callbacks can run in any context, and may or may not block. No restrictions are enforced. It's all left to the caller, who must provide locking/protection as required.

The `include/linux/notifier.h` header contains very useful comments regarding notifier chain types and more. Do check it out. For example, it mentions current (and potential future) users of this powerful mechanism. I can't resist showing you via a screenshot taken from here: `https://elixir.bootlin.com/linux/v5.10.60/source/include/linux/notifier.h`:

```
204    /*
205     *      Declared notifiers so far. I can imagine quite a few more chains
206     *      over time (eg laptop power reset chains, reboot chain (to clean
207     *      device units up), device [un]mount chain, module load/unload chain,
208     *      low memory chain, screenblank chain (for plug in modular screenblankers)
209     *      VC switch chains (for loadable kernel svgalib VC switch helpers) etc...
210     */
211
212    /* CPU notfiers are defined in include/linux/cpu.h. */
213
214    /* netdevice notifiers are defined in include/linux/netdevice.h */
215
216    /* reboot notifiers are defined in include/linux/reboot.h. */
217
218    /* Hibernation and suspend events are defined in include/linux/suspend.h. */
219
220    /* Virtual Terminal events are defined in include/linux/vt.h. */
```

Figure 10.3 – A partial screenshot of the notifier.h header, showing the current and potential users of the notifier chains mechanism

Users (subscribers) are expected to use the registration APIs provided to register and, when done, unregister (although there are restrictions on when you can unregister in some cases – see the `notifier.h` header). For example, a network driver can elect to subscribe to the `netdevice` notifier chain in order to be notified whenever interesting events on a network device occur (stuff such as the network device coming up or going down, changing its name, and so on). You can see available netdevice events within the enum here: `include/linux/netdevice.h:netdev_cmd`. For example, the netconsole driver uses this facility to be notified of netdevice events. This is its netdevice chain callback function: `drivers/net/netconsole.c:netconsole_netdev_event()`. Another interesting use case of notifier chains is the reboot notifier chain (set up via the `register_reboot_notifier()` function). It can be used, for example, to properly shut down a **Direct Memory Access** (**DMA**) operation when an unexpected reboot occurs.

We don't propose covering more on the internals or other uses of notifier chains, instead choosing to focus on what matters here – setting up our own panic handler via this mechanism. Do refer to links within the *Further reading* section for more on notifier chains in general. Now, let's get on to the key part of this section.

Setting up our custom panic handler within a module

We will get hands-on here, first understanding the relevant data structures and APIs, and then writing and running module code to set up our very own custom kernel panic handler!

Understanding the atomic notifier chain APIs and the notifier_block structure

We will develop a kernel module that will employ the kernel's predefined panic notifier chain, named `panic_notifier_list`, in order to hook into the kernel panic. It's declared here:

```
// kernel/panic.c
ATOMIC_NOTIFIER_HEAD(panic_notifier_list);
EXPORT_SYMBOL(panic_notifier_list);
```

Clearly, it belongs to the atomic variety of notifier chains, implying that our callback cannot block in any manner.

Registering with an atomic notifier chain

To hook into it, we must register with it. This is achieved via the API:

```
int atomic_notifier_chain_register(struct atomic_notifier_head
*, struct notifier_block *);
```

It's actually a simple wrapper over the generic `notifier_chain_register()` API, which it invokes within a `spin_lock_irqsave()`/`spin_unlock_irqrestore()` pair of locking primitives. The first parameter to `atomic_notifier_chain_register()` specifies the notifier chain you want to register with – in our case, we'll specify it as `panic_notifier_list` – and the second parameter is a pointer to a `notifier_block` structure.

Understanding the notifier_block data structure and the callback handler

The `notifier_block` structure is the centerpiece structure of the notifier chain framework. It's defined as follows:

```
// include/linux/notifier.h
struct notifier_block {
    notifier_fn_t notifier_call;
    struct notifier_block __rcu *next;
    int priority;
};
```

The first member is the really key one, a function pointer. *It's the callback function*, the function that will be invoked via the framework when the asynchronous event occurs! Here's its signature:

```
typedef int (*notifier_fn_t)(struct notifier_block *nb,
unsigned long action, void *data);
```

Thus, the parameters you'll receive in the callback handler are as follows:

- `struct notifier_block *nb`: The pointer to the same `notifier_block` data structure used to set up the notifier.

- `unsigned long action`: This is actually a value specifying how or why we got here, a clue as to what caused the kernel panic. It's an enum named `die_val` and is arch-specific:

```
// arch/x86/include/asm/kdebug.h
enum die_val {
    DIE_OOPS = 1,
    DIE_INT3, DIE_DEBUG,       DIE_PANIC, DIE_NMI,
    DIE_DIE,  DIE_KERNELDEBUG, DIE_TRAP,  DIE_GPF,
    DIE_CALL, DIE_PAGE_FAULT,  DIE_NMIUNKNOWN,
};
```

Note that most driver authors, in their callback handler, seem to name this parameter either `val` or `event`. (Also note that the `INT 3` software interrupt is the classic breakpoint instruction on the x86.)

- `void *data`: This is actually an interesting structure, `struct die_args`, passed via a pointer here. Here's its definition:

```
// include/linux/kdebug.h
struct die_args {
    struct pt_regs *regs;
    const char *str;
    long err;
    int trapnr;
    int signr;
};
```

Among its members is the string passed to the `panic()` function, which is usually what the `data` parameter evaluates to. You can look up its definition in `include/linux/kdebug.h` and its setup via the notifier framework here: `kernel/notifier.c:notify_die()`. Example usage of this structure within a panic callback is within this Hyper-V driver from Microsoft: `drivers/hv/vmbus_drv.c:hyperv_die_event()`. It retrieves the CPU registers (via the familiar `struct pt_regs *`) from here, and employs the previous `action` parameter (it names it `val`) to verify that it's in the panic handler due to an Oops.

Let's get back to the `notifier_block` data structure. The second member is the usual `next` pointer to the next node in the notifier chain. Leaving it as `NULL` has the kernel notifier framework handle it appropriately.

The third and final member, `priority`, is clearly a prioritization. Setting it to `INT_MAX` informs the framework to invoke your callback as early as possible. We usually leave it undefined, though. Note that the kernel uprobes framework sets the priority of its exception notifier callback to `INT_MAX-1`:

```
// kernel/events/uprobes.c
static struct notifier_block uprobe_exception_nb = {
    .notifier_call = arch_uprobe_exception_notify,
    .priority = INT_MAX-1, /* notified after kprobes, kgdb */
};
```

Note though that uprobes registers to the kernel *die chain* (via the `register_die_notifier()` API) – an interesting notifier chain, where callbacks are invoked when a CPU exception occurs in kernel mode. This can be another useful way to get relevant details when your system receives unexpected CPU exceptions in kernel mode!

Finally, after your callback function – the subscriber – performs its work, it returns a specific value, indicating whether all is well or not. These are the possible return values (you must use one of them):

- `NOTIFY_OK`: The handler is done – notification correctly handled. This is the typical one to return when all goes well.

- `NOTIFY_DONE`: The handler is done – don't want any further notifications.

- `NOTIFY_STOP`: The handler is done – stop any further callbacks.

- `NOTIFY_BAD`: The handler signals that something went wrong – don't want any further notifications (the kernel mentions this as being considered a *bad/veto action*).

Of course, you have to pair the notifier chain registration with a corresponding unregister. This is the API to use:

```
int atomic_notifier_chain_unregister(struct atomic_notifier_
head *nh, struct notifier_block *n);
```

For a kernel panic, it won't be invoked. As good coding practice, we do this for our panic handler in the module cleanup method.

Our custom panic handler module – viewing the code

So, here it is – the relevant code from our custom kernel panic handler module! (Do browse through the complete code from the book's GitHub repo.) Let's begin by seeing the registration of the custom handler with the kernel panic notifier list in the module's *init* method:

```
// ch10/panic_notifier/panic_notifier_lkm.c
/* The atomic_notifier_chain_[un]register() api's are GPL-
exported! */
MODULE_LICENSE("Dual MIT/GPL"); [...]
static struct notifier_block mypanic_nb = {
    .notifier_call = mypanic_handler,
/*   .priority = INT_MAX   */
};
static int __init panic_notifier_lkm_init(void)
{
    atomic_notifier_chain_register(&panic_notifier_list,
&mypanic_nb);
```

Next, we have the actual panic handler routine(s):

```
/* Do what's required here for the product/project,
 * but keep it simple. Left essentially empty here.. */
static void dev_ring_alarm(void)
{
    pr_emerg("!!! ALARM !!!\n");
}
static int mypanic_handler(struct notifier_block *nb, unsigned
long val, void *data)
{
```

```
    pr_emerg("\n*********** Panic : SOUNDING ALARM
***********\n\
val = %lu\n\
data(str) = \"%s\"\n", val, (char *)data);
    dev_ring_alarm();
    return NOTIFY_OK;
}
```

Do note the following:

- Our custom panic handler emits a printk at KERN_EMERG, ensuring as much as possible that it's seen.

- The data parameter evaluates to the message passed to the panic() function. In this case, as we trigger the panic via it, it's the SysRq crash triggering the code's message (sysrq triggered crash).

- We call a dev_ring_alarm() function. Note that it's simply a dummy placeholder – in your actual project or product, do what's (minimally) required here. For example, an embedded device controlling a laser on a factory floor might want to switch off the laser end and sound a physical alarm of some sort to indicate that the system is unusable, or whatever makes sense, *constrained* by the key fact that the system *is* in an unstable precarious state!

- We return NOTIFY_OK, signaling that all is well (as can be).

Okay, let's just do it!

Trying it out – our custom panic handler module

We have a simple script to trigger an oops – via the kernel Magic SysRq c option – and set the kernel.oops_on_panic to 1 to convert this Oops into a kernel panic! Here's the script:

```
$ cat ../cause_oops_panic.sh
sudo sh -c "echo 1 > /proc/sys/kernel/panic_on_oops"
sudo sh -c "echo 1 > /proc/sys/kernel/sysrq"
sync; sleep .5
sudo sh -c "echo c > /proc/sysrq-trigger"
$
```

Careful, though – don't run it until you've set up netconsole (to capture kernel printks from this system onto a receiver system). For this too, we employ a simple wrapper script, ch10/netcon (I'll leave you to browse through it). We run it first, passing the receiver system's IP address as a parameter. It sets up netconsole accordingly:

```
$ ../netcon 192.168.1.8
[...]
```

You can see the dmesg output regarding netconsole in the following screenshot:

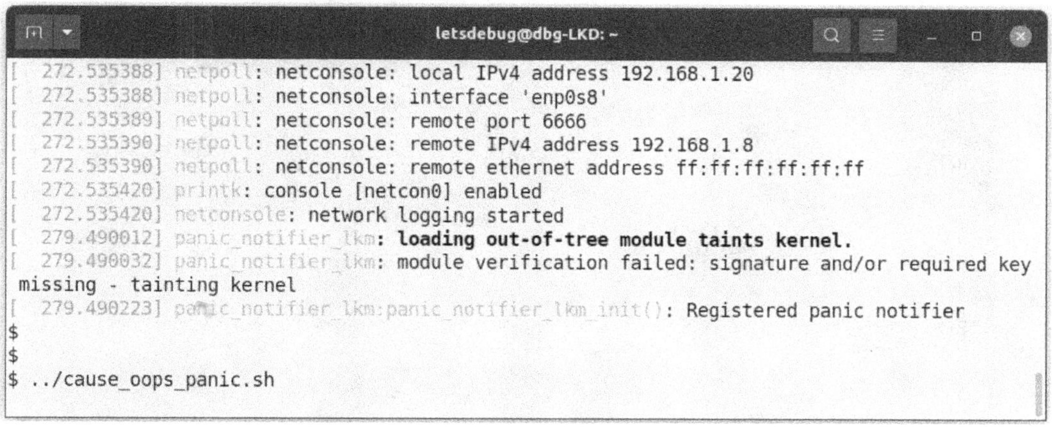

```
                              letsdebug@dbg-LKD: ~
[  272.535388] netpoll: netconsole: local IPv4 address 192.168.1.20
[  272.535388] netpoll: netconsole: interface 'enp0s8'
[  272.535389] netpoll: netconsole: remote port 6666
[  272.535390] netpoll: netconsole: remote IPv4 address 192.168.1.8
[  272.535390] netpoll: netconsole: remote ethernet address ff:ff:ff:ff:ff:ff
[  272.535420] printk: console [netcon0] enabled
[  272.535420] netconsole: network logging started
[  279.490012] panic_notifier_lkm: loading out-of-tree module taints kernel.
[  279.490032] panic_notifier_lkm: module verification failed: signature and/or required key
missing - tainting kernel
[  279.490223] panic_notifier_lkm:panic_notifier_lkm_init(): Registered panic notifier
$
$
$ ../cause_oops_panic.sh
```

Figure 10.4 – A screenshot showing the guest VM where netconsole is set up

Also, do ensure that you're running netcat on the receiver system (in the usual manner; I use netcat -d -u -l 6666).

Once triggered via our `../cause_panic_oops.sh` script, the kernel panic causes our custom panic handler – registered to the panic notifier list – to be invoked. The netcat utility spews out the remote kernel printks on panic:

```
~ $ netcat -d -u -l 6666
[ 293.076610] sysrq: Trigger a crash
[ 293.076644] Kernel panic - not syncing: sysrq triggered crash
[ 293.076663] CPU: 5 PID: 2467 Comm: sh Tainted: G           OE     5.10.60-prod01 #6
[ 293.076684] Hardware name: innotek GmbH VirtualBox/VirtualBox, BIOS VirtualBox 12/01/2006
[ 293.076718] Call Trace:
[ 293.076739]  dump_stack+0x76/0x94
[ 293.076753]  panic+0x1ac/0x382
[ 293.076821]  sysrq_handle_crash+0x1a/0x20
[ 293.076839]  __handle_sysrq+0xf8/0x170
[ 293.076895]  ? common_file_perm+0x78/0x1a0
[ 293.076990]  write_sysrq_trigger+0x28/0x40
[ 293.077030]  proc_reg_write+0x66/0x90
[ 293.077072]  vfs_write+0xca/0x2c0
[ 293.077104]  ksys_write+0x67/0xe0
[ 293.077117]  __x64_sys_write+0x1a/0x20
[ 293.077179]  do_syscall_64+0x38/0x90
[ 293.077226]  entry_SYSCALL_64_after_hwframe+0x44/0xa9
[ 293.077278] RIP: 0033:0x779eec7000a7
[ 293.077331] Code: 64 89 02 48 c7 c0 ff ff ff ff eb bb 0f 1f 80 00 00 00 00 f3 0f 1e fa 64
8b 04 25 18 00 00 00 85 c0 75 10 b8 01 00 00 00 0f 05 <48> 3d 00 f0 ff ff 77 51 c3 48 83 ec
28 48 89 54 24 18 48 89 74 24
[ 293.077425] RSP: 002b:00007ffe732d9078 EFLAGS: 00000246 ORIG_RAX: 0000000000000001
[ 293.077481] RAX: ffffffffffffffda RBX: 00006081643436f0 RCX: 0000779eec7000a7
[ 293.077529] RDX: 0000000000000002 RSI: 00006081643436f0 RDI: 0000000000000001
[ 293.077597] RBP: 0000000000000002 R08: 00006081643436f0 R09: 000000000000007c
[ 293.077620] R10: 00000000000001b6 R11: 0000000000000246 R12: 0000000000000001
[ 293.077642] R13: 0000000000000002 R14: 7fffffffffffffff R15: 00007ffe732d9240
[ 293.077747] Kernel Offset: 0x31200000 from 0xffffffff81000000 (relocation range: 0xffffffff
ff80000000-0xffffffffbfffffff)
[ 293.077804] panic_notifier_lkm:mypanic_handler():
[ 293.077804] ************ Panic : SOUNDING ALARM ************
[ 293.077804] val = 0
[ 293.077804] data(str) = "sysrq triggered crash"
[ 293.077849] panic_notifier_lkm:dev_ring_alarm(): !!! ALARM !!!
[ 293.078217] ---[ end Kernel panic - not syncing: sysrq triggered crash ]---
```

Figure 10.5 – A partial screenshot – the host, where netcat receives and prints to stdout the kernel printks from the guest VM's kernel panic; note the output from our custom panic handler!

Clearly, you can see (*Figure 10.5*) that this time, as we used the Magic SysRq crash triggering feature, this is what is reflected in the kernel panic message and the kernel stack backtrace. The interesting thing, though, highlighted toward the bottom of *Figure 10.5*, is the output from our custom panic handler – it's clearly visible! – followed by the capped end message (`---[ end Kernel panic - ... ]---`) from the kernel.

Again, a reminder – be careful what you do within your panic handler routine. *Keep it to a bare minimum and test it.* This kernel comment emphasizes the point:

```
// kernel/panic.c:panic()
 * Note: since some panic_notifiers can make crashed kernel
 * more unstable, it can increase risks of the kdump failure
too.
```

The kernel tree has several instances of the panic notifier chain being employed (mostly by various drivers and watchdogs). As a quick experiment, I used `cscope` to search for the `atomic_notifier_chain_register(&panic_notifier_list` string, within the 5.10.60 kernel source tree. This partial screenshot shows that it obtained 29 matches (with the left column revealing the source filename):

```
Text string: atomic_notifier_chain_register(&panic_notifier_list

   File                    Line
0  setup.c                 1259 atomic_notifier_chain_register(&panic_notifier_list,
1  enlighten.c              314 atomic_notifier_chain_register(&panic_notifier_list, &xen_panic_block);
2  brcmstb_gisb.c           492 atomic_notifier_chain_register(&panic_notifier_list,
3  ipmi_msghandler.c       5163 atomic_notifier_chain_register(&panic_notifier_list, &panic_block);
4  altera_edac.c           2117 atomic_notifier_chain_register(&panic_notifier_list,
5  gsmi.c                  1021 atomic_notifier_chain_register(&panic_notifier_list,
6  vmbus_drv.c             1501 atomic_notifier_chain_register(&panic_notifier_list,
7  coresight-cpu-debug.c    536 ret = atomic_notifier_chain_register(&panic_notifier_list,
8  ledtrig-activity.c       249 atomic_notifier_chain_register(&panic_notifier_list,
9  ledtrig-heartbeat.c      192 atomic_notifier_chain_register(&panic_notifier_list,
a  ledtrig-panic.c           66 atomic_notifier_chain_register(&panic_notifier_list,
b  heartbeat.c               41 atomic_notifier_chain_register(&panic_notifier_list, &panic_notifier);
c  pvpanic.c                110 atomic_notifier_chain_register(&panic_notifier_list,
d  pvpanic.c                150 atomic_notifier_chain_register(&panic_notifier_list,
e  ipa_smp2p.c              138 return atomic_notifier_chain_register(&panic_notifier_list,
f  power.c                  232 atomic_notifier_chain_register(&panic_notifier_list,
g  ltc2952-poweroff.c       275 atomic_notifier_chain_register(&panic_notifier_list,
h  remoteproc_core.c       2450 atomic_notifier_chain_register(&panic_notifier_list, &rproc_panic_nb);
i  con3215.c                952 atomic_notifier_chain_register(&panic_notifier_list, &on_panic_nb);
j  con3270.c                643 atomic_notifier_chain_register(&panic_notifier_list, &on_panic_nb);
k  sclp.c                  1249 rc = atomic_notifier_chain_register(&panic_notifier_list,
l  sclp_con.c               348 atomic_notifier_chain_register(&panic_notifier_list, &on_panic_nb);
m  sclp_vt220.c             890 atomic_notifier_chain_register(&panic_notifier_list, &on_panic_nb);
n  pm-arm.c                 802 atomic_notifier_chain_register(&panic_notifier_list,
o  olpc_dcon.c              655 atomic_notifier_chain_register(&panic_notifier_list, &dcon_panic_nb);
p  hyperv_fb.c             1257 atomic_notifier_chain_register(&panic_notifier_list,
q  hung_task.c              306 atomic_notifier_chain_register(&panic_notifier_list, &panic_block);

* Lines 1-28 of 29, 2 more - press the space bar to display more *
```

Figure 10.6 – A partial screenshot showing various users of the panic notifier chain within the kernel

Now, armed with your custom kernel panic handler, let's tackle how we detect lockups within the kernel!

Detecting lockups and CPU stalls in the kernel

The meaning of *lockup* is obvious. The system, and one or more CPU cores, remain in an unresponsive state for a significant period of time. In this section, we'll first briefly learn about watchdogs and move on to learn how to leverage the kernel to detect both hard and soft lockups.

A short note on watchdogs

A **watchdog** or **watchdog timer** (**WDT**) is essentially a program that monitors a system's health and, on finding it lacking in some way, has the ability to reboot the system. Hardware watchdogs latch into the board circuitry and thus have the ability to reset the system when required. Their drivers tend to be very board-specific.

The Linux kernel provides a generic watchdog driver framework, allowing driver authors to fairly easily implement watchdog drivers for specific hardware watchdog chipsets. You can find the framework explained in some detail in the official kernel documentation here: *The Linux WatchDog Timer Driver Core kernel API*: `https://www.kernel.org/doc/html/latest/watchdog/watchdog-kernel-api.html#the-linux-watchdog-timer-driver-core-kernel-api`. As this isn't a book on writing Linux device drivers, we won't go into more detail.

There is a facility to employ a userspace watchdog daemon process as well. (On Ubuntu at least, the package and utility are simply named `watchdog`. You'll have to configure and run it.) Its job is to monitor various system parameters, perform the heartbeat ping functionality (typically, by writing something at least once a minute into the device file for the kernel watchdog driver, `/dev/watchdog`), and communicate with it using various predefined ioctls. The kernel documentation on this is here: *The Linux Watchdog driver API*: `https://www.kernel.org/doc/html/latest/watchdog/watchdog-api.html#the-linux-watchdog-driver-api`.

You can configure and tune several system parameters to values appropriate to your system. Details on the user mode watchdog daemon can be found in these man pages: `watchdog(8)` and `watchdog.conf(5)`. These help configure the watchdog to trigger – and thus reboot the system – based on various system parameters (for example, setting a bare minimum number of free RAM pages, a maximum heartbeat interval between two writes to the device file, a process – specified by a PID file – that must always be alive, the maximum load allowed on the system, and system temperature thresholds). It's very interesting to peruse the man page; all parameters that can possibly be monitored are shown. Watchdog-based monitoring can indeed prove very useful for many types of products, especially ones that aren't human-interactive (remote servers, deeply embedded systems, many kinds of IoT edge devices, and so on).

We enable the software watchdog in our custom production kernel (CONFIG_SOFT_
WATCHDOG=m; look for it and many available hardware watchdogs under Device
Drivers | Watchdog Timer Support within the make menuconfig kernel UI).
As we selected it as a module, it gets built, and the module is named, quite appropriately,
softdog. Do note though that being a pure software watchdog, it may not reboot the
system in some situations. (If interested, look up the official kernel documentation on the
various module control parameters you can specify for the softdog software watchdog,
as well as known hardware ones, in the kernel here: *WatchDog Module Parameters*:
https://www.kernel.org/doc/html/latest/watchdog/watchdog-
parameters.html#watchdog-module-parameters.)

Running the softdog watchdog and the user watchdog daemon

As an experiment, we will load the softdog software watchdog driver (all defaults)
on my x86_64 Ubuntu VM and then (manually) run the watchdog service daemon in
verbose mode (I did tweak a few parameters in its config file, /etc/watchdog.conf):

```
$ sudo modprobe softdog
$ sudo watchdog --verbose &
[...]
watchdog: String 'watchdog-device' found as '/dev/watchdog'
watchdog: Variable 'realtime' found as 'yes' = 1
watchdog: Integer 'priority' found = 1
[1]+  Done                      watchdog --verbose
```

Right, let's verify that it's running:

```
# ps -e | grep watch
    111 ?          00:00:00 watchdogd
  10106 ?          00:00:00 watchdog
```

The first line in the `ps` output is actually the `watchdogd` kernel thread. The second one is the software userspace watchdog daemon process we just ran. Here's some of the initial output from the user-mode watchdog daemon process:

```
Integer 'retry-timeout' found = 60
Integer 'repair-maximum' found = 2
String 'watchdog-device' found as '/dev/watchdog'
Variable 'realtime' found as 'yes' = 1
Integer 'priority' found = 1
starting daemon (5.15)
int=1s realtime=yes sync=no load=0,0,0 soft=no
memory not checked
ping
file
pidfile
interface
temperature
no test binary files
no repair binary files
error retry time-out = 60 seconds
repair attempts = 2
alive=/dev/watchdog heartbeat=[none] to=root no_act=no force=no
watchdog now set to 60 seconds
hardware watchdog identity
still alive after 1 interval(s)
still alive after 2 interval(s)
still alive after 3 interval(s)
```

Figure 10.7 – A partial screenshot showing the initial output from the watchdog daemon process when run in verbose mode

In a similar fashion, systemd-based systems can also perform watchdog monitoring (see the watchdog-related entries within /etc/systemd/system.conf). Also, it's important to note that while watchdogs are useful in production, they might need to be turned off during debug (for example, when running an interactive kernel debugger); otherwise, they might trigger and cause a system reboot. Okay, I'll leave it to you to explore further with this. Let's move on to learning about an interesting application of watchdogs – the kernel lockup detectors!

Employing the kernel's hard and soft lockup detector

Software (hardware too) isn't perfect. I am betting you've experienced a system that mysteriously hangs. The system probably isn't completely dead or panicked; it's simply hung and become unresponsive. This, in general, is termed a *lockup*. The Linux kernel has the ability to detect lockups, and we aim to examine this.

The reason I mentioned the watchdog (in the prior section) is that the Linux kernel leverages the NMI watchdog facility (as well as the perf subsystem) to detect both hard and soft lockups (we'll cover what these mean very soon). The kernel can be configured to detect both hard and soft lockups. The relevant menu (via the usual `make menuconfig` UI) is here: **Kernel hacking | Debug Oops, Lockups and Hangs**. Here's a screenshot of the same (on our custom 5.10.60 production kernel):

```
.config - Linux/x86 5.10.60 Kernel Configuration
 > Kernel hacking > Debug Oops, Lockups and Hangs
                       Debug Oops, Lockups and Hangs
   Arrow keys navigate the menu.  <Enter> selects submenus ---> (or empty submenus ----).
   Highlighted letters are hotkeys.  Pressing <Y> includes, <N> excludes, <M> modularizes
   features.  Press <Esc><Esc> to exit, <?> for Help, </> for Search.  Legend: [*] built-in  [ ]
   excluded  <M> module  < > module capable

            [ ] Panic on Oops
            (0) panic timeout
            -*- Detect Soft Lockups
            [ ]     Panic (Reboot) On Soft Lockups
            [*] Detect Hard Lockups
            [ ]     Panic (Reboot) On Hard Lockups
            [*] Detect Hung Tasks
            (120) Default timeout for hung task detection (in seconds)
            [ ]     Panic (Reboot) On Hung Tasks
            [*] Detect Workqueue Stalls
            < > Test module to generate lockups
```

Figure 10.8 – A partial screenshot showing the kernel config UI for debugging Oops, lockups, and hangs

Glancing at the kernel config in *Figure 10.8*, you may wonder why, this being a so-called production kernel, we haven't enabled things such as panic on oops and panic on soft and hard lockup? Good question! I leave the *panic* options off as, although we claim that this is a production kernel, it isn't really in the sense that we use it throughout the book to demonstrate things. On an actual project or product, enabling them is definitely something to consider. Does the system need to be auto-rebooted in case it locks up, hangs, Oops'es, or panics? If yes, then enable the panic-on configs and pass the `panic=n` kernel parameter to have the system reboot n seconds after panicking.

The kernel config to detect both hard and soft lockups as well as hung tasks and workqueue stalls is indeed enabled. The relevant kernel config options, boot parameters, and kernel sysctl knobs are summarized in *Table 10.3* (it may be useful to refer back to the table as you cover more of the material):

Kernel boot-time parameter	Equivalent kernel sysctl knob	Purpose
`nmi_watchdog`	`kernel.nmi_watchdog`	[X86, SMP]: • Pass/set to `0` – turn off the NMI watchdog and hard lockup detection (default). • Pass/set to `1` – turn on the NMI watchdog and hard lockup detection. Note that NMI interrupts can occur often and cause fairly significant overhead.
`<none>`	`kernel.soft_watchdog`	• Set to `0` – turn off soft lockup detection. • Set to `1` – turn on soft lockup detection (typically on as `CONFIG_SOFTLOCKUP_DETECTOR=y` by default).
`<none>`	`kernel.watchdog`	Enable/disable both hard/soft lockup detection: • Set to `0` to disable both. • Set to `1` to enable both. Note that the value when read is the bitwise `OR` of `nmi_watchdog` and `soft_watchdog` (so if either is enabled, it shows as `1`).
`<none>`	`kernel.watchdog_cpumask`	The CPU cores that the watchdog runs on (the default is all active cores). They are affected by the `nohz_full=` kernel parameter (see `https://www.kernel.org/doc/html/latest/admin-guide/sysctl/kernel.html#watchdog-cpumask` for details).
`watchdog_thresh`	`kernel.watchdog_thresh`	Passing/writing an integer n here sets the watchdog threshold – in effect, the hard lockup timeout – to that many seconds (the default is 10). Soft lockup is 2*n seconds. Passing/writing 0 disables both hard/soft lockup detection.
`nowatchdog / nosoftlockup`	Equivalent to writing 0 to `kernel.watchdog` and `kernel.soft_watchdog`	Sometimes, a watchdog can work against you (especially during dev/debug). Passing this parameter disables both hard lockup (NMI watchdog) and soft lockup detection. Also, passing the `nosoftlockup` boot parameter disables soft lockup detection.

Table 10.3 – A summary of the watchdog settings affecting kernel hard/soft lockup detection

The settings on your box regarding the watchdog can be verified using the `sysctl` utility (note that `nmi_watchdog` refers to hard lockup and `soft_watchdog` to soft lockup detection, not to the `softdog` module):

```
$ sudo sysctl -a | grep watchdog
kernel.nmi_watchdog = 0
kernel.soft_watchdog = 1
kernel.watchdog = 1
kernel.watchdog_cpumask = 0-5
kernel.watchdog_thresh = 10
$
```

The `nmi_watchdog` value shows as 0 as there's no hardware watchdog chip available. `soft_watchdog` is always available (as is the kernel's built-in watchdog support). Let's get to what exactly all of this means!

What's a soft lockup bug?

A **soft lockup** is a bug wherein a task running in kernel mode remains in a tight loop or is somehow stuck to the processor, for a long time, not allowing other tasks to get scheduled on that core. The default timeout for a *hard lockup* is the value of the `kernel.watchdog_thresh` sysctl – it's 10 seconds by default – and that of a soft lockup is twice this (that is, 20 seconds). This, of course, is tunable (as root). Let's look up the value on my Ubuntu 20.04 LTS VM:

```
$ cat /proc/sys/kernel/watchdog_thresh
10
```

Thus, the actual soft lockup timeout value is 2*10 – 20 seconds. Writing an integer into the `watchdog_thresh` kernel sysctl modifies the threshold to that value (in seconds). Writing 0 disables checking.

When a soft lockup's detected, what happens?

Panic – if the `softlockup_panic` kernel (boot-time) parameter is set to 1, the `kernel.softlockup_panic` sysctl is 1, or if the kernel config is BOOTPARAM_SOFTLOCKUP_PANIC=1, then the kernel panics! The typical default is 0.

If the preceding isn't true – that is, the kernel doesn't panic on soft lockup – it emits a warning message to the kernel log, showing details of the hung task. The kernel stack trace is dumped as well (allowing us to see how it got to this point!)

It's important to note that in the latter case (where it doesn't panic), the buggy task continues to hang the affected CPU core.

Triggering a soft lockup on the x86_64

Can we trigger a soft lockup? Of course. Simply do something to wreak havoc on a poor CPU core, causing it to spin for a long while in kernel mode! As an example, I added a few lines of code to do just this to a kernel thread demo module, created for my earlier *Linux Kernel Programming Part 2* book (the original code is here: `https://github.com/PacktPublishing/Linux-Kernel-Programming-Part-2/tree/main/ch5/kthread_simple`).

To save space, we just run the module with the (slightly) modified code. You can see the added buggy code (the full code is here: `ch10/kthread_stuck`) and the visible effect it has. After over 20 seconds, the kernel watchdog detects the soft lockup and jumps in, emitting a BUG() message! Running it with no module parameter specified has it use the default – the test for the *soft lockup* (we'll test the hard lockup shortly):

```
while(!kthread_should_stop()) {
//- - - - - - - - - - - - - - - - - - - - - - - - - - - -
    pr_info("DELIBERATELY spinning on CPU core now...\n");

    if (likely(lockup_type == DO_SOFT_LOCKUP))
        spin_lock(&spinlock);
    else
        spin_lock_irq(&spinlock);

    while (i < 10000000000) { // adjust these arbit #s for your system if reqd..
        i ++;
        if (!(i%50000000))
            PRINT_CTX();
    }

    if (likely(lockup_type == DO_SOFT_LOCKUP))
        spin_unlock(&spinlock);
    else
        spin_unlock_irq(&spinlock);
//- - - - - - - - - - - - - - - - - - - - - - - - - - - -

    pr_info("FYI, I, kernel thread PID %d, am going to sleep now...\n",
        current->pid);
    set_current_state(TASK_INTERRUPTIBLE);
    schedule(); // yield the processor, go to sleep...
```

```
Message from syslogd@dbg-LKD at Mar 25 19:56:09 ...y due to either the
kernel:[ 1528.659809] watchdog: BUG: soft lockup - CPU#2 stuck for 22s! [lkd/kt_stuck:3530]
        */
```

Figure 10.9 – A partial screenshot showing deliberately buggy CPU-intensive code to cause a soft lockup bug; see – at the bottom – the BUG() message at KERN_EMERG overwriting the console

In addition to the kernel watchdog's `BUG: soft lockup ...` message at `KERN_EMERG`, the watchdog also emits the usual diagnostics by invoking `dump_stack()` and related routines. (You'll see the modules in the kernel memory, the context info, the kernel status information, the hardware info, the CPU register dump, the machine code running on core, and – a key portion – in the kernel mode stack backtrace, the call stack.) Our useful `convenient.h:PRINT_CTX()` macro helps reveal the system state. Here's an example of output from it while the soft lockup was taking place:

```
002) [lkd/kt_stuck]:3530   |  .N.1   /* simple_kthread() */
```

The `.N.1` ftrace latency format-like string reveals that, as the first of these four columns is a period, it's running *with hardware interrupts enabled*. This is because, when we test the soft lockup, we invoke `spin_lock()` and not the IRQ-disabling `spin_lock_irq()` routine (along with their unlock counterpart, of course). Great!

Don't Forget the Spinlock!

A key point to remember is that we perform the CPU-intensive code path in a loop, around which we take a spinlock. Why? Remember that the spinlock, particularly the `spin_lock_irq[save]()` variants, besides having a loser context `spin`, while the lock owner runs the code of the critical section (here, the body of the `while` loop), *also disables hardware interrupts. Disabling interrupts has the nice side effect of disabling kernel preemption*; thus, the code runs pretty much guaranteed without preemption of any sort, not even by hardware interrupts! In other words, automatically.

For emulating a hard lockup, this is exactly what we want. But, then, think about this – how will the kernel watchdog detect it? Ah, that's because it traps into the NMI and checks for lockup in the NMI handler! And, of course, the *NMI does preempt and interrupt* the code, as it's by definition a non-maskable interrupt.

(Again, what we're doing here as an experiment is the exact opposite of what's recommended – keep the critical section within the spinlock as short as possible, as we discussed in *Chapter 8, Lock Debugging*, in the *Identifying some locking defects from various blog articles and the like* section. We do so here to deliberately cause a soft or hard lockup, as a learning exercise.)

If interested, you'll find the kernel code implementation of the soft lockup detection here: `kernel/watchdog.c:watchdog_timer_fn()`.

Also, interestingly, attempting `rmmod` (without first sending the `SIGINT` or `SIGQUIT` signals to our kthread to have it die, as we've programmed it that way) in our buggy module can, in about 2 minutes, have the `rmmod` process detected as being a *hung task!* We will discuss hung task detection in the next major section of this chapter. Now, let's move on to the next type of lockup...

What's a hard lockup bug?

A **hard lockup** is a bug wherein a CPU core running in kernel mode remains in a tight loop, or somehow stuck, for a long time, not allowing other hardware interrupts to run on that core. As already mentioned, the default timeout for a hard lockup is the value of the `kernel.watchdog_thresh` kernel sysctl, set to 10 seconds by default. This, of course, is tunable (as root).

When a hard lockup's detected, what happens?

Panic – if the `nmi_watchdog=1` kernel boot parameter (and if the system supports a hardware watchdog), the `kernel.hardlockup_panic` sysctl is set to 1, or the kernel config is `BOOTPARAM_HARDLOCKUP_PANIC=y`, then the kernel panics! The typical default is that panic is off.

If the preceding isn't true – that is, the kernel doesn't panic on hard lockup (default) – it emits a warning message to the kernel log, showing details of the system state. The kernel stack trace is dumped as well (allowing us to see how it got to this point). If the `hardlockup_all_cpu_backtrace=1` kernel boot parameter is passed, the kernel generates a kernel stack backtrace on all CPUs.

It's important to note that in the latter case (where it doesn't panic), the buggy code continues to (hard) hang the affected CPU core.

There's more to this – the kernel RCU lock-free feature can result in CPU stalls as well.

RCU and RCU CPU stalls

The Linux kernel's **Read-Copy-Update** (**RCU**) infrastructure is a powerful way to perform lock-free work within the kernel. It's important to realize that, similar to hard lockup, warnings can occur due to RCU CPU stalls as well. The `RCU_CPU_STALL_TIMEOUT` kernel config determines the RCU grace period. On 5.10, it's 60 seconds by default, with a range of 3 to 300. If the RCU grace period exceeds the number of seconds specified by this config, a CPU RCU stall warning is emitted, with the possibility of more occurring when the problem persists. A very brief conceptual introduction to RCU follows, so do check it out.

Conceptually understanding RCU in a nutshell

The RCU implementation works by essentially having readers work upon shared data simultaneously *without* using locking, atomic operators, increments to a variable, or even (with the exception of the Alpha processor) memory barriers! Thus, in mostly read situations, performance remains high – the main benefit of using RCU. How does it work?

Imagine several readers (say, threads R1, R2, and R3) enter a section of code where they work upon shared data in parallel – an RCU read-side critical section. When a writer thread comes along, realizing it's an RCU critical section, the writer makes a copy of the data item being referenced and modifies it. The existing readers continue to work upon the original item. Then, the writer atomically updates the original pointer to refer to the new (just modified) data item (while R1, R2, and R3 continue to work upon the original one). The writer must then free (destroy) the original data item. This, of course, can't be done until all readers currently accessing it finish.

How will it know? The RCU implementation has the writer wait for all current readers to cycle off the CPU by checking when they yield the processor – that is, invoke the scheduler and thus move off the CPU core! Now, the writer allows a *grace period* (as long as a minute!) to elapse – allowing any sluggish readers to complete – and then destroys (frees) the original data item, and all is well. (Note that in an uncommon case, that of parallel RCU writers, they can avoid stepping on each others toes by using some sort of locking primitive, typically spinlocks.)

The official kernel documentation, *Using RCU's CPU Stall Detector* (`https://www.kernel.org/doc/html/latest/RCU/stallwarn.html#using-rcu-s-cpu-stall-detector`), mentions the several causes that can result in an RCU CPU stall warning. Among them is looping on a CPU for a long while with interrupts, preemption, or bottom halves disabled (there are many more reasons; do look up the kernel documentation). That's why we got into RCU CPU stalls here. Among the conditions that make them occur is the one we're dealing with – disabling interrupts for a long while!

Triggering a hard lockup/RCU CPU stalls on a native x86_64

Can we trigger a hard lockup and/or an RCU CPU stall bug? Indeed we can, but even on an x86_64, there's at least a few pre-conditions:

- The hard lockup can only be detected by the NMI on a native x86_64, so you should be running Linux on one (a guest VM won't do).
- The NMI and the NMI watchdog must be enabled by adding the `nmi_watchdog=1` string to your kernel boot parameter list.

- Explicitly enable the NMI watchdog in the kernel by writing 1 to the `kernel.nmi_watchdog` sysctl.

- The `CONFIG_RCU_CPU_STALL_TIMEOUT` kernel config should have a value in the range of 3 to 300 – the number of seconds after which an RCP CPU stall is deemed to have occurred.

Once these are satisfied, `sysctl` should reflect it:

```
# sysctl -a | grep watchdog
kernel.nmi_watchdog = 1
kernel.soft_watchdog = 1
kernel.watchdog = 1
kernel.watchdog_cpumask = 0-11
kernel.watchdog_thresh = 10
```

Additionally, on my system, I have CONFIG_RCU_STALL_COMMON=y and CONFIG_RCU_CPU_STALL_TIMEOUT=60.

To test for hard lockup/RCU CPU stalls, fire up our demo module (`ch10/kthread_stuck`), this time passing the `lockup_type=2` module parameter. This parameter value has our kthread spin on the CPU in a tight loop while holding a spinlock with IRQs and preemption disabled (the `spin_lock_irq()` variant). After some time elapses, the kernel log should reveal the NMI interrupt (and indeed the NMI backtrace) having fired due to the hard lockup or RCU CPU stall bug our module causes.

It's entirely possible that the actual warning is due to the fact that RCU CPU stalls are detected (which happened when I tested it)! This is because the kernel's RCU stall detection code deems that an RCU CPU stall has occurred (among several other reasons) when code spins on a CPU core for a long while *with interrupts, preemption, or bottom halves disabled.* Our code does indeed spin for a long while with interrupts and preemption disabled (as we employ the IRQ/preempt-disabling version of the spinlock). The kernel log reveals the RCU stall being detected:

```
rcu:   INFO: rcu_sched detected stalls on CPUs/tasks:
rcu:        3-...0: (1 GPs behind)
idle=462/1/0x4000000000000000 softirq=60126/60127 fqs=6463
(detected by 2, t=15003 jiffies, g=127897, q=1345272)
Sending NMI from CPU 2 to CPUs 3:
NMI backtrace for cpu 3
```

```
CPU: 3 PID: 16351 Comm: lkd/kt_stuck Tainted: P         W   OEL
5.13.0-37-generic #42~20.04.1-Ubuntu
[...]
```

More information on interpreting the kernel's RCU stall warnings is documented in the official kernel documentation article on RCU CPU stall detection mentioned earlier. Also, the `kernel.panic_on_rcu_stall` kernel sysctl can be set to 1 to enable panic on the RCU stall. It's off by default. The `kernel.panic_on_rcu_stall` sysctl (available from 5.11 only) allows configuring the number of times the RCU stall must occur before the kernel panics.

To round off this topic, note the following:

- The kernel provides a much more sophisticated module to help test watchdogs, lockups, hangs, RCU CPU stalls, and more. Enable it by setting `CONFIG_TEST_LOCKUP`=m (or y). The module will be named `test_lockup` (its code is here: `lib/test_lockup.c`).

- Specifically for deep RCU testing, the kernel also has an *RCU torture* facility. The official kernel doc regarding it is here: `https://www.kernel.org/doc/html/latest/RCU/torture.html#rcu-torture-test-operation`.

- The official kernel documentation goes into details regarding the implementation of the hard/soft lockup detection: *Softlockup detector and hardlockup detector (aka nmi_watchdog)*: `https://www.kernel.org/doc/html/v5.10/admin-guide/lockup-watchdogs.html#implementation`. Also, the kernel code implementing it is here: `kernel/watchdog.c`.

Check out this table summarizing the various boot parameters, kernel sysctl knobs, and kernel configs relevant to hard/soft lockup:

Kernel boot-time parameter	Equivalent kernel sysctl knob	Equivalent kernel config macro (CONFIG_<FOO>)	Purpose
`softlockup_panic=[0|1]`	`kernel.softlockup_panic`	`BOOTPARAM_SOFTLOCKUP_PANIC`	If set to 1 (the default is 0), this enables kernel panic when buggy code loops in kernel mode for over 20 seconds (configurable via `/proc/sys/kernel/watchdog_thresh`), not allowing other tasks to run [1].
`softlockup_all_cpu_backtrace=[0|1]`	`kernel.softlockup_all_cpu_backtrace`	`<none>`	If passed as or set to 1 (the sysctl default is 0), the kernel stack back-traces of all active CPU cores will be sent to the kernel log when a soft lockup is detected.
`nosoftlockup`	`kernel.soft_watchdog`	`SOFTLOCKUP_DETECTOR`	Passing this parameter or setting the sysctl to 0 disables the kernel soft-lockup detector.
`nmi_watchdog=[0|1]`	`kernel.hardlockup_panic`	`BOOTPARAM_HARDLOCKUP_PANIC=[y|n]`	If set to 1 (the default is 0), this enables kernel panic when buggy code loops in kernel mode for over 10 seconds (configurable via `/proc/sys/kernel/watchdog_thresh`), not allowing other hardware interrupts to run [1].
`hardlockup_all_cpu_backtrace=[0|1]`	`kernel.hardlockup_all_cpu_backtrace`	`<none>`	If set to 1, this has the kernel generate a kernel stack backtrace on all CPUs on a hard lockup.
`hung_task_panic=[0|1]`	`kernel.hung_task_panic`	`BOOTPARAM_HUNG_TASK_PANIC=[y|n]`	If passed as or set to 1 (the default is 0), this enables kernel panic when a `task_struct`'s `state` member is left as `TASK_UNINTERUPTIBLE` (D, as seen in `ps -le` [L], the second column) [2].

Table 10.4 – A summary of the boot parameters, kernel sysctl knobs, and kernel configs relevant to hard/soft lockup

Note the following citations in the table:

- [1] – For both the soft lockup and hung task cases, the kernel's config specifies the following:

```
// lib/Kconfig.debug
[...]
The panic can be used in combination with panic_timeout
to cause the system to reboot automatically after a hung
task has been detected. This feature is useful for high-
availability systems that have uptime guarantees and
where hung tasks must be resolved ASAP.
```

- [2] – A simple bit of bash magic can help us see all threads whose state is D – uninterruptible sleep (TASK_UNINTERRUPTIBLE):

```
ps -leL | awk '{printf("%s %s\n", $2, $14)}' | grep "^D"
```

A side effect of watchdogs being able to reboot a system when necessary is, of course, the fact that, when debugging, you don't get a chance to capture key information before the reboot occurs. To this end, disabling watchdog(s) (and even RCU CPU stall detection) during debug might be a good idea. Some (x86/ARM-based) systems even provide pre-timeout notifications, enabling you to save key state information before reboot!

Practically speaking, there are a few things you can check. The nowatchdog kernel parameter turns off both the hardware NMI watchdog as well as soft-lockup functionality at boot. The kernel documentation on watchdog parameters gives insight into various options you can play with for real-world hardware watchdog drivers: https://www.kernel.org/doc/Documentation/watchdog/watchdog-parameters.rst.

Another possibly useful thing to keep in mind when working on a device driver is that the kernel typically provides callback mechanisms for power management events such as suspend and shutdown. Take advantage of them to perhaps save state info when a certain abnormal condition is detected, perform necessary tasks such as aborting a DMA transfer, and so on.

Great! Now, let's complete this chapter by learning how to leverage a kernel's hung task and workqueue stall detectors.

Employing the kernel's hung task and workqueue stall detectors

A hung task is one that's become unresponsive. Similarly, the kernel can also, on occasion, suffer from some types of stalls (workqueue and RCU). In this section, we will examine how we can leverage these features, allowing us to detect them so that an action – such as triggering a panic or emitting a warning with stack backtraces – can be taken. Obviously, the warnings logged can then help you, the developer, understand what occurred and work to fix it.

Leveraging the kernel hung task detector

Configuring the kernel via the usual make menuconfig UI, under the **Kernel hacking | Debug Oops, Lockups and Hangs** menu (refer to *Figure 10.8*), you'll find entries labeled as follows:

```
[*] Detect Hung Tasks
(120) Default timeout for hung task detection (in seconds)
[ ]    Panic (Reboot) On Hung Tasks
```

These are what we discuss here. The whole idea, when enabled, is to allow the kernel to be able to detect tasks (processes and/or threads) that have become non-responsive and stuck in an uninterruptible sleep for a long while. The state of the task (within its task structure's state member) is named TASK_UNINTERRUPTIBLE, which implies that it can't be disturbed by any signal from userspace. As seen, the default timeout to consider it as hung is 120 seconds. This is tunable, of course, by either of the following:

- Changing the value of CONFIG_DEFAULT_HUNG_TASK_TIMEOUT (corresponding to the menu's second line shown previously).

- Modifying the value in the kernel.hung_task_timeout_secs sysctl. Setting it to 0 disables the check.

The CONFIG_DETECT_HUNG_TASK kernel config option is turned on for a debug kernel by default. It can be very useful in detecting hung non-responsive tasks even on production systems, as the overhead is considered minimal.

The third line in the menu seen previously corresponds to whether the kernel should panic when a hung task is detected. It's the CONFIG_BOOTPARAM_HUNG_TASK_PANIC config and is set to off by default. This behavior is also settable via the kernel.hung_task_panic sysctl.

Note that the detection of hung tasks is implemented by having a kernel thread, `khungtaskd`, continually scan for them.

We will round off this topic by showing the meaning of various kernel sysctl tunables relevant to hung task detection (all of which depend on `CONFIG_DETECT_HUNG_TASK` being enabled):

- `hung_task_all_cpu_backtrace`: If set to 1 (the default is 0), the kernel sends the NMI interrupt to all cores, triggering a stack backtrace when a hung task is detected. This requires `CONFIG_DETECT_HUNG_TASK` and `CONFIG_SMP` to be enabled.

- `hung_task_check_count`: The upper bound on the number of tasks checked. It can be useful to dampen this value on a resource-constrained (embedded) system. Interestingly, the value does tend to be arch-specific (on an ARM-32 compiled Raspberry Pi, for example, the value is 32,768, while on an x86_64, it's 4,194,304).

- `hung_task_check_interval_secs`: Typically, 0, implying that the timeout value for hung tasks is the `kernel.hung_task_timeout_secs` sysctl. If positive, then it overrides it and checks for hung tasks at this interval (seconds). The legal range is {`0:LONG_MAX/HZ`}.

- `hung_task_timeout_secs`: The essential hung task facility – when a task remains in an uninterruptible sleep (the D state, as seen by `ps -l`) for more than this amount of time (in seconds), it triggers a kernel warning (and possibly panic – see the following). The legal range is {`0:LONG_MAX/HZ`}.

- `hung_task_panic`: If set to 1 (the default is 0), the kernel panics when a hung task is detected. If 0, the task remains in the hung (D) state.

- `hung_task_warnings`: The maximum number of warnings to report (defaults to 10). Once a hung task is detected, it's decremented by 1. A value of -1 implies that infinite warnings can occur.

Here is an example of the same, looking up these (default) values on my x86_64 guest VM:

```
$ sudo sysctl -a|grep hung_task
kernel.hung_task_all_cpu_backtrace = 0
kernel.hung_task_check_count = 4194304
kernel.hung_task_check_interval_secs = 0
kernel.hung_task_panic = 0
kernel.hung_task_timeout_secs = 120
kernel.hung_task_warnings = 10
$
```

All right, let's now move on to the final portion of this topic!

Detecting workqueue stalls

The kernel workqueue infrastructure can be of immense help to a driver (and other) authors, allowing them to have work consumed in a process (blocking) context very easily (it internally manages pools of kernel worker threads to achieve this). One of the issues with using them, however, is the fact that work can get stalled (delayed) on occasion to unacceptable levels, and thus significantly affect performance. Thus, the kernel provides a means of detecting *workqueue stalls*.

This feature is enabled by selecting the CONFIG_WQ_WATCHDOG=y kernel config (you'll find it within the make menuconfig UI under **Kernel hacking | Debug Oops, Lockups and Hangs | Detect Workqueue Stalls** (refer to *Figure 10.8*)). Once set to y, if a workqueue's worker pool fails to progress on a work item, a warning message (at KERN_ WARN) is emitted to the kernel log, along with the workqueue internal state information.

The time beyond which the workqueue stall detection occurs is governed by the workqueue.watchdog_thresh kernel boot parameter as well as the corresponding sysfs file. It's 30 seconds by default. Writing (or setting) 0 here disables workqueue stall checks.

Triggering a workqueue stall

A simple experiment to test workqueue lockup is by inserting a couple of lines of CPU-intensive code in our kernel-default workqueue's work function (the original code is from my earlier *Linux Kernel Programming – Part 2* book and the relevant code (copied here) is from here: https://github.com/PacktPublishing/Linux-Kernel-Programming-Part-2/tree/main/ch5/workq_simple. Do remember to test stuff like this on a multicore system!). The added buggy code is clear in *Figure 10.10* – it's the few lines from line number 96 to line number 101 (the source is in ch10/workq_ stall, as it's possible that exact line numbers can vary):

```
78 {
79     struct st_ctx *priv = container_of(work, struct st_ctx, work);
80     u64 i = 0;
81
82     t2 = ktime_get_real_ns();
83     pr_info("In our workq function: data=%d\n", priv->data);
84     PRINT_CTX();
85     SHOW_DELTA(t2, t1);
86
87     /* Deliberately spin for a loooong while... causing the kernel softlockup
88      * detector to swing into action!
89      */
90     pr_info("Deliberately locking up the cpu now!\n");
91     //mdelay(1000*30);
92     while (1)
93         i += 3;
94 }
95
Message from syslogd@dbg-LKD at Mar 24 18:49:39 ...
kernel:[29612.080043] BUG: workqueue lockup - pool cpus=2 node=0 flags=0x0 nice=0 stuck for
166s!  ctx.data = INITIAL_VALUE;
99
Message from syslogd@dbg-LKD at Mar 24 18:50:10 ...
kernel:[29642.797043] BUG: workqueue lockup - pool cpus=2 node=0 flags=0x0 nice=0 stuck for
197s!
103     /* Initialize our kernel timer */
Message from syslogd@dbg-LKD at Mar 24 18:50:40 ...s(exp_ms);
kernel:[29673.522018] BUG: workqueue lockup - pool cpus=2 node=0 flags=0x0 nice=0 stuck for
228s!                                                          95,0-1          73%


Message from syslogd@dbg-LKD at Mar 24 18:51:11 ...
kernel:[29704.249864] BUG: workqueue lockup - pool cpus=2 node=0 flags=0x0 nice=0 stuck for
258s!
```

Figure 10.10 – A partial screenshot showing how our buggy code – deliberately locking up the CPU – causes the BUG: workqueue lockup, overwriting the console display

Clearly, the kernel's workqueue stall detection code senses the issue and emits emergency level printks (the code detecting this is here: `https://elixir.bootlin.com/linux/v5.10.60/source/kernel/workqueue.c#L5806`). Additionally, while this is going on, running a utility such as `top -i` will reveal how pretty much 100% of the CPU is being consumed by a kernel worker thread (typically, belonging to the kernel-default worker pool).

Summary

Congratulations on completing this chapter! By now, you should have your own custom panic handler reading and raring to go!

To quickly summarize, in this chapter, we covered what a kernel panic is, interpreted its log output, and importantly, learned how to leverage the kernel's powerful notifier chain infrastructure to develop our own custom kernel panic handler.

We then moved on to what kernel lockup – hard, soft, and RCU CPU stalls – means and how to configure the kernel to detect it (with small examples to show what it looks like when it locks up!). The final section covered how to detect hung tasks (unresponsive tasks that remain in the D state for a long while) and workqueue stalls.

Once issues like this are detected, examining the kernel log (where, typically, you'll have the kernel warning and CPU backtraces) can provide you with valuable clues as to where an issue lies, thus helping you fix it.

I'll see you in the next chapter, where we will learn to leverage kernel GDB tooling to interactively debug kernel code.

Further reading

- Official kernel documentation on collecting kernel logs via the kernel ramoops and pstore facilities:

  - Ramoops oops/panic logger: `https://www.kernel.org/doc/html/latest/admin-guide/ramoops.html#ramoops-oops-panic-logger`

  - pstore block oops/panic logger: `https://www.kernel.org/doc/html/latest/admin-guide/pstore-blk.html?highlight=pstore#pstore-block-oops-panic-logger`

  - *Persistent storage for a kernel's "dying breath"*, Jake Edge, LWN, Mar 2011: `https://lwn.net/Articles/434821/`

- *Use ramoops for logging under Linux*, embear blog: `https://embear.ch/blog/using-ramoops`

- *XDA Basics: How to take logs on Android*, July 2021, G. Shukla: `https://www.xda-developers.com/how-to-take-logs-android/`

- Official kernel docs: *Linux Magic System Request Key Hacks*: `https://www.kernel.org/doc/html/latest/admin-guide/sysrq.html`

- Notifier chains:

    - *Notification Chains in Linux Kernel*: https://0xax.gitbooks.io/linux-insides/content/Concepts/linux-cpu-4.html

    - *The Crux of Linux Notifier Chains*, R. Raghupathy, January 2009: https://www.opensourceforu.com/2009/01/the-crux-of-linux-notifier-chains/

- Watchdogs and lockups:

    - *Linux Kernel Watchdog Explained*, 2018, Zak H: https://linuxhint.com/linux-kernel-watchdog-explained/

    - *IT log book: Linux – what are "CPU lockups"?*, January 2018: https://blog.seibert-media.com/2018/01/04/log-book-linux-cpu-lockups/

    - Official kernel documentation: *Using RCU's CPU Stall Detector*: https://www.kernel.org/doc/html/latest/RCU/stallwarn.html#using-rcu-s-cpu-stall-detector

    - *RUNNING FOREVER WITH THE RASPBERRY PI HARDWARE WATCHDOG*, D. Letz, July 2020: https://diode.io/raspberry%20pi/running-forever-with-the-raspberry-pi-hardware-watchdog-20202/

11
Using Kernel GDB (KGDB)

What if we could set breakpoints (even hardware break/watchpoints) on the kernel's or a module's code, single-stepping through it, viewing variables, and examining memory, as we easily do for application-space processes with the really well-known debugger **GNU Debugger (GDB)**? Well, that's exactly what **Kernel GDB (KGDB)** allows – it's a source-level debug tool for the Linux kernel (and modules)!

In this chapter, we're going to cover the following main topics:

- Conceptually understanding how KGDB works
- Setting up an ARM target system and kernel for KGDB
- Debugging the kernel with KGDB
- Debugging kernel modules with KGDB
- [K]GDB – a few tips and tricks

Technical requirements

With a few additions (which follow), the technical requirements and workspace remain identical to what's described in *Chapter 1, A General Introduction to Debugging Software*. The code examples can be found within the book's GitHub repository here: `https://github.com/PacktPublishing/Linux-Kernel-Debugging`.

In addition to the usual, you'll also need to install a few packages as well as a compressed root filesystem image we'll make use of later in the chapter:

1. The QEMU ARM and x86 emulator apps and a few miscellaneous packages (all in, they take up close to 400 MB of disk space):

    ```
    sudo apt install qemu-system-arm qemu-system-x86 lzop
    libncursesw5 libncursesw5-dev p7zip-full
    ```

*   Navigate to the book's GitHub repo's `ch11/` directory and download a compressed root filesystem image that we'll make use of later (in the *Debugging kernel modules with KGDB section*):

    ```
    cd <book_src>/ch11
    wget https://github.com/PacktPublishing/Linux-Kernel-
    Debugging/raw/main/ch11/rootfs_deb.img.7z
    ```

It gets downloaded (note that this particular file's rather large, weighing in at around 178 MB). However, as a meta-version of the `rootfs_deb.img.7z` file already exists, the actual downloaded file will be automatically named `rootfs_deb.img.7z.1`. So, after downloading, we need to now delete the original (dummy) file and rename the actual one to the proper name:

```
rm rootfs_deb.img.7z
mv rootfs_deb.img.7z.1 rootfs_deb.img.7z
```

(FYI, *Figure 11.8* reveals what the `ch11/` directory should ultimately look like; it will match when we later extract this image.)

For this chapter, we'll assume that you're familiar with basic GDB commands and running GDB in userspace (quick tip: Google "GDB cheat sheet"). Right, let's get started!

Conceptually understanding how KGDB works

KGDB is a source-level debugger, allowing you to debug kernel (and module) code at the level of the C source file(s)!

Hang on a moment though. In order for an application process such as GDB to debug the kernel, it will need to halt the kernel's execution upon it hitting a breakpoint and while single-stepping code paths within the kernel. How is that possible? What will run the GDB process (and the rest of the system) then?

The reality is that *GDB, supporting a client-server architecture, is used with two machines*: one, a host system where the client GDB program runs (the one we're used to working with); the other, the target system, where the GDB server component is embedded into the kernel itself! (Unlike typical client/server apps, the GDB server component is the smaller of the two, and the GDB client is the relatively large one – the regular GDB program you're used to using.)

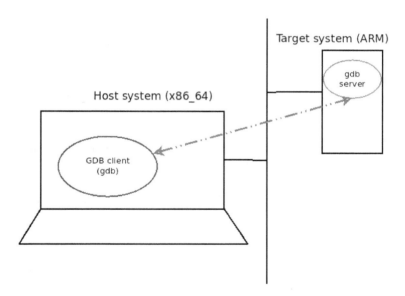

Figure 11.1 – Conceptual diagram showing how GDB works via a client-server architecture

The GDB client and server typically communicate over TCP/IP (using port 1234 by default), though communication across a serial port and other mechanisms are supported. The client sends the GDB command typed in by the user to the server. The server executes it on the target system, sending back the results to the client, which then displays them. The end result: we can remote debug a target system's kernel and modules, just as we debug user-mode apps, with GDB! We need to enable this support for the GDB server component within the kernel, of course – in effect, by enabling KGDB support.

> **FYI: JTAG Debuggers**
>
> JTAG debuggers (like the popular BDI2000/3000) also use a `gdb-server` component. Having a JTAG debugger with an embedded GDB server typically makes it easier and more stable to debug the kernel. Another advantage is you avoid multiplexing the serial port between the kernel debug session and the Linux console.

Great – let's move on to setting up KGDB on a target system.

Setting up an ARM target system and kernel for KGDB

When a Linux kernel is built, a couple of arch-specific kernel image files are generated: the uncompressed kernel image file, `vmlinux`, along with the compressed kernel image, found within the `arch/<your-arch>/boot` directory and named `bzImage` or `zImage`, and so on. The latter is always the image with which the Linux OS is booted. Both these kernel image files are rendered in the usual **Executable and Linker Format** (**ELF**), thus they're amenable to being used with a variety of tooling on Linux, including GDB. So, if we use GDB to interpret the kernel's uncompressed `vmlinux` file, it should work. Practically speaking, though, without debug symbolic information embedded within this file, it's a lot less useful. What we really require for kernel debug purposes using KGDB is the target's uncompressed `vmlinux` kernel image along with debug symbolic information and the kernel symbols within it. This is achieved by enabling the `CONFIG_DEBUG_INFO` kernel config option.

Building a minimal custom ARM Linux target system with SEALS

To try KGDB out, we'll need two machines. They don't have to be physical machines though; we can quite easily use a Linux VM as the target (or as the host)! Let's make it interesting, using our usual x86_64 Ubuntu guest (VM) as the host, and a QEMU emulated ARM32 Linux system (configured for KGDB) as the target!

You do realize though, that this will require a custom build of the target system. Any working Linux system requires a minimum of three (or four, depending on the CPU) components:

- A bootloader (here, QEMU serves as the bootloader, so we don't require anything else; Das U-Boot is a popular bootloader on many typical (ARM/PPC) embedded Linux systems, with GRUB being the preferred bootloader on x86).

- In the case of an ARM32/AArch64/PPC system, a **Device Tree Blob** (**DTB**) binary image (to pass onto the kernel at boot; used to interpret the hardware platform and load up appropriate drivers).

- A kernel image (this refers to the compressed kernel image); we shall soon configure and build it.

- A root filesystem.

We'll need to configure and build a custom kernel of course, as we'll require it configured for KGDB, along with debug symbols. Building the **root filesystem** (or **rootfs**), though, is a non-trivial task. Thus, to make it easier to build a custom ARM Linux system, I propose leveraging my **Simple Embedded ARM Linux System** (**SEALS**) project – it generates a very simplistic embedded Linux system indeed. The GitHub repo's here: `https://github.com/kaiwan/seals`. In a nutshell, this project allows you to configure the platform and the kernel, build the kernel, DTB, and (skeleton, bare minimum) BusyBox-based root filesystem images for an ARM platform that QEMU supports. (We default SEALS to use the ARM Versatile Express (VExpress) platform, based on an ARMv7 Cortex A9 multicore, with 512 MB RAM.) QEMU will *run* the system as a guest VM. You can run this QEMU-emulated ARM target VM within your x86_64 Ubuntu guest running on VirtualBox on a Windows/Linux/macOS host - nested virtualization, in effect!

Of course, there are umpteen ways to build yourself an embedded Linux system – the Yocto and Buildroot projects tend to be the de facto, much more powerful, and complete approaches. Also, you could always use an existing hardware/software platform (the popular Raspberry Pi and BeagleBone Black boards come to mind). For our purposes here, and to allow you to try it out without the need for specific hardware, I chose to use the much simpler SEALS project, to keep it simple.

Using the SEALS project requires a number of prerequisites. Most importantly, you'll need to install the QEMU emulator for ARM, a full-fledged x86_64-to-ARM32 toolchain, and a few miscellaneous packages. We don't have the bandwidth here to go into the details of configuring the SEALS project, choosing to focus instead on the topic of interest – configuring and building a kernel for KGDB. To better understand how to configure and use the SEALS project, I refer you to its wiki pages:

- *Welcome to the SEALS wiki*: `https://github.com/kaiwan/seals/wiki`

- *HOWTO Install required packages on the Host for SEALS*: `https://github.com/kaiwan/seals/wiki/HOWTO-Install-required-packages-on-the-Host-for-SEALS`

- **Detailed step-by-step instructions to use SEALS**: *SEALs HOWTO*: `https://github.com/kaiwan/seals/wiki/SEALs-HOWTO`

To get a sneak peek at how it looks when running, check out *Figure 11.4* and *Figure 11.5* and return here!

Configuring the kernel for KGDB

When configuring the kernel, you'd typically configure it as a *debug kernel*. We've covered this and the typical kernel debug options to employ right from the first chapter! Refer back if you need to.

Mandatory configs for KGDB support

Minimally, via the usual `ARCH=arm CROSS_COMPILE=<...>` make menuconfig UI, you'll require these kernel configs enabled (we are omitting the `CONFIG_` prefix):

- `DEBUG_KERNEL=y`: Selected when you select the `Kernel Hacking|Kernel debugging` Boolean menu option (this option could be auto-selected by default).

- `DEBUG_INFO=y`: The `Kernel Hacking|Compile-time checks and compiler options|Compile the kernel with debug info` option. This enables embedding kernel symbols and debug symbolic information (by compiling the kernel and modules with the `-g` compiler switch) in the uncompressed kernel image, `vmlinux`. Well, technically, this config option's not mandatory, but practically speaking, without debug symbols, GDB's not going to be very effective in helping you debug things.

- MAGIC_SYSRQ=y: Kernel Hacking | Generic Kernel Debugging Instruments | Magic SysRq key. Not strictly mandatory in all circumstances but often used in the KGDB/kdb context to, within a running system, issue a break into debugger command by writing g into /proc/sysrq-trigger. We recommend that all magic SysRq functionality is enabled by writing 1 into /proc/sys/kernel/sysrq.

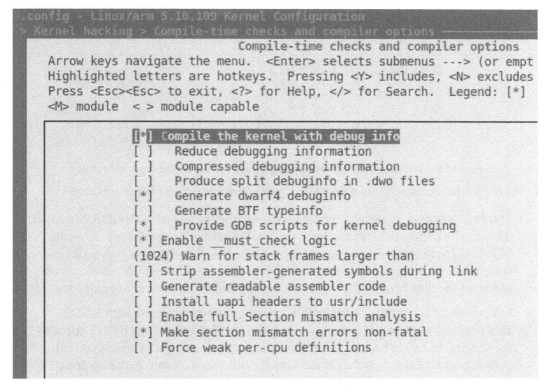

Figure 11.2 – Truncated screenshot showing the Kernel Hacking | Compile-time checks and compiler options kernel configuration menu

- In addition, turn on the relevant kernel config options pertaining to KGDB, whose menu falls here: `Kernel hacking | Generic Kernel Debugging Instruments | KGDB: kernel debugger`:

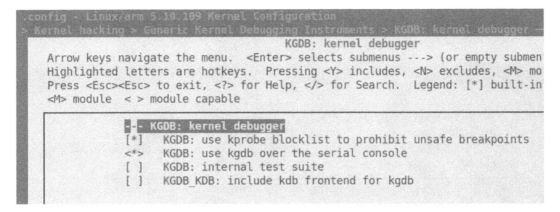

Figure 11.3 – Truncated screenshot showing the KGDB: kernel debugger submenu

Among the kernel configs just discussed, here are a few more to explicitly mention:

- `CONFIG_KGDB=y`: Enables support for debugging the kernel with GDB. Internally, the kernel will now contain the server-side GDB (aka gdbserver) code, allowing a GDB client to remotely connect to it (typically over Ethernet, though serial and other modes are supported) and send GDB commands to it, which it will execute and send the data back to the client, which displays it – remote debugging, in effect!

- `KGDB_SERIAL_CONSOLE=y`: Enables sharing the serial console with KGDB. This is actually one way by which you can connect from the remote GDB client to the server (the target kernel). To understand the various ways in which you can connect, do read through the official kernel doc here, focusing on the `kgdboc` kernel parameter: *Kernel parameter: kgdboc*: `https://www.kernel.org/doc/html/v5.10/dev-tools/kgdb.html#kernel-parameter-kgdboc`

- `KGDB_HONOUR_BLOCKLIST=y`: It is recommended this config is kept on. It prevents recursive traps by disallowing certain routines (the ones that can't be kprobed) from being set as breakpoints.

> **What's Kdb?**
>
> **Kdb** is a simple command-line-based debugger that you can employ on a serial console to break into the kernel and debug it in a limited way. You can inspect memory, CPU registers, variables, kernel log content, and so on... *But* you can't do source-level debugging with it. It has the advantage of not requiring two machines for debugging, but the downside of no source-level debugging either. The official kernel doc here covers it: *Using kdb*: `https://www.kernel.org/doc/html/v5.10/dev-tools/kgdb.html#using-kdb`.

What else? It can be useful to turn *off* the following kernel configs under the `General architecture-dependent options` menu:

- `CONFIG_STRICT_KERNEL_RWX`
- `CONFIG_STRICT_MODULE_RWX`

These being turned off enables GDB to be able to set software breakpoints on kernel and module functions (these configs only appear in the menu system when the kernel's configured in a certain way, so ignore this if they don't even appear in the menu – see their entries in `arch/Kconfig` for details). Further, using hardware breakpoints (we recommend using them) alleviate the need to set these configs.

Optional kernel configs for KGDB

In addition to the mandatory kernel config options just discussed, you can *optionally* set options such as the following:

- `FRAME_POINTER=y`: `Kernel hacking | Compile-time checks and compiler options | Compile the kernel with frame pointers`: Though marked as optional, it's really very useful – *do select it if your platform provides it* (interestingly, on the SEALS-generated ARM VExpress platform, it's not selected as a good alternate as `Kernel hacking | arm Debugging | ARM EABI stack unwinder` (`CONFIG_UNWINDER_ARM=y`) is selected as the default config for this platform).

- `DEBUG_INFO_SPLIT=y`: Helps significantly reduce the size by splitting and reusing `.dwo` debug info files.

- `DEBUG_INFO_BTF=y`: This generates de-duped **BPF Type Information (BTF)**. This could be useful for running eBPF in the future (requires `pahole` v1.16 or later installed).

- `GDB_SCRIPTS=y`: Sets up links to Python-based GDB helper scripts (the `lx-<foo>` ones), when loading `vmlinux` into GDB. This can greatly help when debugging the kernel or modules (see more in the official kernel doc: *Debugging kernel and modules via gdb*: `https://docs.kernel.org/dev-tools/gdb-kernel-debugging.html#debugging-kernel-and-modules-via-gdb`). We also cover using them briefly in the later *Setting up and using GDB scripts with CONFIG_GDB_SCRIPTS* section.

- `DEBUG_FS=y`: It's usually useful to have the debugfs pseudo filesystem present.

- The remaining of the other "usual" kernel debug infrastructure, much of which we've already covered, and has nothing to do directly with KGDB (or kdb).

- A quick (often useful) tip: When debugging the kernel (or modules) with KGDB, we really don't want hardware or software watchdogs interfering and tripping us up! Ensure they're disabled during your kernel debug sessions.

Now, enable the appropriate kernel configs for your KGDB-powered debug kernel and build it. If you're using SEALS, it employs a menu-driven system that will allow you to configure your chosen kernel – it will build and even run it (under QEMU of course). As a bonus, you can even turn on a KGDB mode option in your SEAL project board's `build.config` file. It's off by default (note that the source files regarding SEALS aren't in this book's GitHub repo; please clone SEALS from its GitHub repo mentioned previously):

```
$ grep KGDB build.config
KGDB_MODE=0 # make '1' to have qemu run with the '-s -S'
            # switch (waits for client GDB to 'connect')
```

> **Tip**
>
> I find that for our purposes, when building the kernel, if asked to enable *GCC plugins*, it's best to answer no.

Okay, let's actually test that it works (remember: be empirical – make no assumptions!).

Testing the target system

I shall be making use of the ARM32 virtual Linux platform generated via my SEALS project, which uses the QEMU emulator (qemu-system-arm). The SEALS project script generates all the required components to run and test an embedded Linux system (all components are in the SEALS VExpress board's staging region, a folder that the user sets in the board config file):

- Kernel images (in the staging folder, under the kernel source tree). Here, I've used the 5.10.109 kernel:

  - The compressed zImage file, meant for booting (location: <staging>/linux-5.10.109/arch/arm/boot/zImage).

  - The uncompressed vmlinux image with kernel and debug symbols present (location: <staging>/linux-5.10.109/vmlinux). This is meant for kernel debugging – we'll use it soon enough!

- The DTB image (required for boot). Here, it's the file <staging>/linux-5.10.109/arch/arm/boot/dts/vexpress-v2p-ca9.dtb. Modern ARM, ARM64, and PPC boards typically require a DTB image to correctly boot up.

- A very minimal (skeleton) root filesystem image (location: <staging>/images/rfs.img). In fact, the compressed kernel image and the DTB are also kept in the same images folder for convenience.

- As mentioned before, what about the bootloader? Don't we require one? Of course! Here, the QEMU emulator serves the role of the bootloader as well (in typical embedded Linux projects, *Das U-Boot* often tends to be the bootloader).

The powerful QEMU emulator ties all the pieces together. Here's a quick demo to show you how we can run the ARM32 VExpress board under QEMU. Once the SEALS build process is successfully done, this is the QEMU command I issue to have it run:

```
qemu-system-arm \
 -m 512 \
 -M vexpress-a9 -smp 4,sockets=2 \
 -kernel <...>/seals_staging_vexpress/images/zImage  \
 -drive file=/<...>/seals_staging_vexpress/images/rfs.
img,if=sd,format=raw \
 -append "console=ttyAMA0 rootfstype=ext4 root=/dev/mmcblk0
init=/sbin/init" \
 -nographic -no-reboot -audiodev id=none,driver=none \
 -dtb /<...>/seals_staging_vexpress/images/vexpress-v2p-ca9.dtb
```

Very briefly, here's how you interpret the arguments to qemu-system-arm:

- -m <MB>: Configure the amount of RAM available to the emulated machine.

- -M <machine-name>: Select the machine (or platform) to emulate. Run the command qemu-system-arm -M help to see all available machine/platform choices that QEMU can emulate.

- -kernel <path/to/kernel-img>: Use the specified image file as the (compressed, boot-time) kernel.

- -drive file=<path/to/rootfs-img>: Use the specified image file as the drive image. The if=sd option further specifies the interface as an SD card.

- -append "<kernel command-line parameters>": Pass the specified string as the boot parameters to the Linux kernel (after boot, doing cat /proc/cmdline reveals them).

- -dtb <path/to/DTB-file>: Use the specified file as the DTB (it's interpreted or "flattened" by the kernel at boot).

Of course, we're just scratching the surface here. QEMU is a powerful product with many more options available.

FYI, the SEALS code base has a bash script named `run-qemu.sh` that will run QEMU for you, allowing you to specify a regular or a `kgdb-mode` run. To get an idea of how it looks, check out the following (discontinuous) screenshots of my QEMU-emulated VExpress Cortex-A9 board booting up (as a guest VM) on the x86_64 host (which itself runs as a guest on my native host!). The first one shows the QEMU-emulated ARM Linux system booting up.

```
$ qemu-system-arm -m 512 -M vexpress-a9 -smp 4,sockets=2 -kernel /home/letsdebug/seals_staging/seals_stagi
ng_vexpress/images/zImage -drive file=/home/letsdebug/seals_staging/seals_staging_vexpress/images/rfs.img,
if=sd,format=raw -append "console=ttyAMA0 rootfstype=ext4 root=/dev/mmcblk0 init=/sbin/init " -nographic -
no-reboot -audiodev id=none,driver=none -dtb /home/letsdebug/seals_staging/seals_staging_vexpress/images/v
express-v2p-ca9.dtb
audio: Device lm4549: audiodev default parameter is deprecated, please specify audiodev=none
Booting Linux on physical CPU 0x0
Linux version 5.10.109 (letsdebug@dbg-LKD) (arm-none-linux-gnueabihf-gcc (GNU Toolchain for the A-profile
Architecture 10.3-2021.07 (arm-10.29)) 10.3.1 20210621, GNU ld (GNU Toolchain for the A-profile Architectu
re 10.3-2021.07 (arm-10.29)) 2.36.1.20210621) #1 SMP Wed Apr 6 18:58:58 IST 2022
CPU: ARMv7 Processor [410fc090] revision 0 (ARMv7), cr=10c5387d
CPU: PIPT / VIPT nonaliasing data cache, VIPT nonaliasing instruction cache
OF: fdt: Machine model: V2P-CA9
Memory policy: Data cache writealloc
Reserved memory: created DMA memory pool at 0x4c000000, size 8 MiB
OF: reserved mem: initialized node vram@4c000000, compatible id shared-dma-pool
cma: Reserved 16 MiB at 0x7f000000
Zone ranges:
  Normal   [mem 0x0000000060000000-0x000000007fffffff]
Movable zone start for each node
Early memory node ranges
  node   0: [mem 0x0000000060000000-0x000000007fffffff]
```

Figure 11.4 – Partial screenshot showing the QEMU command line highlighted and the printks from the emulated ARM32 kernel as it boots up

Next, a screenshot showing the target system boot being completed and we're on the shell:

```
VFS: Mounted root (ext4 filesystem) readonly on device 179:0.
Freeing unused kernel memory: 1024K
Checked W+X mappings: passed, no W+X pages found
Run /sbin/init as init process
random: crng init done
SEALS: /etc/init.d/rcS running now ...
mount: mounting none on /sys/kernel/debug failed: No such file or directory
EXT4-fs (mmcblk0): re-mounted. Opts: (null)
Generic PHY 4e000000.ethernet-ffffffff:01: attached PHY driver [Generic PHY] (mii_bus:phy_addr=4e000000.et
hernet-ffffffff:01, irq=POLL)
smsc911x 4e000000.ethernet eth0: SMSC911x/921x identified at 0xa08b0000, IRQ: 30
/bin/sh: can't access tty; job control turned off
ARM / $
ARM / $ cat /proc/version
Linux version 5.10.109 (letsdebug@dbg-LKD) (arm-none-linux-gnueabihf-gcc (GNU Toolchain for the A-profile
Architecture 10.3-2021.07 (arm-10.29)) 10.3.1 20210621, GNU ld (GNU Toolchain for the A-profile Architectu
re 10.3-2021.07 (arm-10.29)) 2.36.1.20210621) #1 SMP Wed Apr 6 18:58:58 IST 2022
ARM / $
ARM / $ nproc
4
ARM / $ cat /proc/cpuinfo
processor       : 0
model name      : ARMv7 Processor rev 0 (v7l)
BogoMIPS        : 454.65
Features        : half thumb fastmult vfp edsp neon vfpv3 tls vfpd32
CPU implementer : 0x41
CPU architecture: 7
CPU variant     : 0x0
CPU part        : 0xc09
CPU revision    : 0

processor       : 1
model name      : ARMv7 Processor rev 0 (v7l)
BogoMIPS        : 735.23
Features        : half thumb fastmult vfp edsp neon vfpv3 tls vfpd32
CPU implementer : 0x41
CPU architecture: 7
CPU variant     : 0x0
CPU part        : 0xc09
CPU revision    : 0
```

Figure 11.5 – Partial screenshot showing the continuation and completion of the boot and then the (BusyBox) shell prompt; the kernel version and some CPU information are shown

Cool – it works. Now that you've understood how to configure, build, and test drive a QEMU emulated ARM-based Linux system for KGDB, let's move on to the more interesting KGDB part of it – actually trying it out!

Debugging the kernel with KGDB

By now, I'll assume that you've configured and built a Linux target system suitable for KGDB (as described in detail in the previous section). It could be for any machine, including a guest system... Here, we'll continue to use the SEALS-generated ARM32 VExpress platform that we just set up as the target.

The intention here is to demo debugging the kernel early in the boot process with KGDB. To do so, the GDB server component within the target kernel will have to make it wait early in the boot process. This is so that the remote GDB client can connect to it. Linux supplies a boot parameter to do precisely this – it's named `kgdbwait`. To use it, you need to have a KGDB I/O driver built into the kernel image and specify which one via the `kgdboc` boot parameter (for example, `kgdboc=/dev/ttyS0`). You can also set it up later (on the console) by echoing the device name into the pseudofile `/sys/module/kgdboc/parameters/kgdboc`.

Here, we shan't delve into these details, mainly as they're very well documented in the official kernel documentation: *Kernel Debugger Boot Arguments*: `https://www.kernel.org/doc/html/v5.10/dev-tools/kgdb.html#kernel-debugger-boot-arguments`. Let's get onto running our target!

Running our target (emulated) ARM32 system

QEMU makes it easier. It provides a couple of parameters to deal with having the kernel wait for the GDB client to connect at early boot (effectively side-stepping the need to use the `kgdboc` boot parameter), from the `--help` option to `qemu-system-arm`:

```
-S     freeze CPU at startup (use 'c' to start execution)
-s     shorthand for -gdb tcp::1234
```

So, we run the QEMU target instance supplying these parameters. The guest target system then waits early in boot, waiting for a remote GDB client to connect to it. Let's begin. In one terminal window, run the target:

```
qemu-system-arm -m 512 \
 -M vexpress-a9 -smp 4,sockets=2 \
 -kernel <...>/seals_staging_vexpress/images/zImage \
 -drive file=<...>/seals_staging_vexpress/images/rfs.
img,if=sd,format=raw \
 -append "console=ttyAMA0 rootfstype=ext4 root=/dev/mmcblk0
init=/sbin/init nokaslr" \
 -nographic -no-reboot -audiodev id=none,driver=none \
 -dtb <...>/seals_staging_vexpress/images/vexpress-v2p-ca9.dtb
-S -s

< ... >
```

> **A Quick Tip**
>
> Within your `seals` directory, the `show_curr_config.sh` script will reveal detailed information about the current build, including the staging directory path, the toolchain, the kernel and BusyBox versions, and so on.

Due to the effect of the `-S` `-s` option switches, QEMU has the emulated target system patiently waiting... As you can see, it's pretty much the same QEMU command line as we employed when we tested our target system (in the *Testing the target system section*) except that we've now added the `-S` `-s` QEMU options as well as the `nokaslr` boot parameter to the kernel command line (**kernel address space layout randomization (KASLR)**). The random offsetting of the kernel base address in memory, an added kernel hardening measure, can further mess with kernel function identification, so it's best to disable it by passing `nokaslr` on the kernel command line. Another thing to mention: it's likely that QEMU won't work correctly if another QEMU instance is already running and also using the same hypervisor (typically KVM). Running a QEMU-emulated ARM (on x86_64) should pose no issue as it's pure software emulation being performed.

Running and working with the remote GDB client on the host system

Recall that the host system here is actually our x86_64 Ubuntu guest! No issues. In a new terminal window, run the following to have GDB - the "remote" client - execute:

```
$ arm-none-linux-gnueabihf-gdb -q <...>/seals_staging_vexpress/
linux-5.10.109/vmlinux

Reading symbols from <...>/seals_staging_vexpress/linux-
5.10.109/vmlinux...
```

Notice the following:

- We use the cross-compile toolchain version of GDB, not the native one!

- We pass the uncompressed kernel `vmlinux` image (the one with debug and symbolic info) as the parameter. GDB will read in all the symbols!

Next, connect to the target system (hey, we're remote debugging in effect!):

```
(gdb) target remote :1234
Remote debugging using :1234
0x60000000 in ?? ()
(gdb)
```

This is a key point in the process: once this command is successful, *we're all set to debug the kernel in the normal fashion (pretty much as you'd debug an app)*. Let's test drive kernel debugging a bit, by first setting up a couple of breakpoints:

```
(gdb) b panic
Breakpoint 1 at 0x80859840: file kernel/panic.c, line 178.
(gdb) b register_netdev
Breakpoint 2 at 0x80754bc8: file net/core/dev.c, line 10238.
```

> **Tip**
> Setting a breakpoint on a function that's executed very early during boot (for example, `start_kernel()`) can be problematic with the usual `break` command. Instead, try using GDB's `hbreak` command to set a hardware-assisted breakpoint. We'll be mostly doing just this from now on with GDB's `hbreak` command – that's short for *hardware-breakpoint*, not *heartbreak*.

Now issue GDB's `continue` command twice – by typing `continue` (or simply `c`) at the prompt once or twice. Very soon (assuming we're tracking the emulated VExpress ARM kernel), we should hit the `register_netdev()` function of the SMSC911x network driver. The situation after the second `c` command has taken effect is seen in this screenshot:

```
(gdb) c
Continuing.

Breakpoint 2, register_netdev (dev=dev@entry=0x81014800) at net/core/dev.c:10238
10238           if (rtnl_lock_killable())
(gdb) bt
#0  register_netdev (dev=dev@entry=0x81014800) at net/core/dev.c:10238
#1  0x8065dc70 in smsc911x_drv_probe (pdev=0x81186410) at drivers/net/ethernet/smsc/smsc911x.c:2504
#2  0x805c4d88 in platform_drv_probe ( dev=0x81186410) at drivers/base/platform.c:761
#3  0x805c29c8 in really_probe (dev=dev@entry=0x81186410, drv=drv@entry=0x80c66498 <smsc911x_driver+20>) at drivers/base/dd.c:564
#4  0x805c306c in driver_probe_device (drv=drv@entry=0x80c66498 <smsc911x_driver+20>, dev=dev@entry=0x81186410)
    at drivers/base/dd.c:752
#5  0x805c3350 in device_driver_attach (drv=drv@entry=0x80c66498 <smsc911x_driver+20>, dev=dev@entry=0x81186410)
    at drivers/base/dd.c:1027
#6  0x805c33d8 in __driver_attach (data=0x80c66498 <smsc911x_driver+20>, dev=0x81186410) at drivers/base/dd.c:1104
#7  __driver_attach (dev=0x81186410, data=0x80c66498 <smsc911x_driver+20>) at drivers/base/dd.c:1058
#8  0x805c0a88 in bus_for_each_dev (bus=<optimized out>, start=start@entry=0x0, data=data@entry=0x80c66498 <smsc911x_driver+20>,
    fn=fn@entry=0x805c3358 <__driver_attach>) at drivers/base/bus.c:305
#9  0x805c2330 in driver_attach (drv=drv@entry=0x80c66498 <smsc911x_driver+20>) at drivers/base/dd.c:1120
#10 0x805c1de8 in bus_add_driver (drv=drv@entry=0x80c66498 <smsc911x_driver+20>) at drivers/base/bus.c:622
#11 0x805c3f04 in driver_register (drv=0x80c66498 <smsc911x_driver+20>) at drivers/base/driver.c:171
#12 0x80102064 in do_one_initcall (fn=0x80b23760 <smsc911x_init_module>) at init/main.c:1214
#13 0x80b012b8 in do_initcall_level (command_line=0x81118200 "console", level=6) at init/main.c:1287
#14 do_initcalls () at init/main.c:1303
#15 do_basic_setup () at init/main.c:1323
#16 kernel_init_freeable () at init/main.c:1525
#17 0x80862328 in kernel_init (unused=<optimized out>) at init/main.c:1412
#18 0x80100148 in ret_from_fork () at arch/arm/kernel/entry-common.S:155
Backtrace stopped: previous frame identical to this frame (corrupt stack?)
(gdb) l
10233    */
10234   int register_netdev(struct net_device *dev)
10235   {
10236           int err;
10237
10238           if (rtnl_lock_killable())
10239                   return -EINTR;
10240           err = register_netdevice(dev);
10241           rtnl_unlock();
10242           return err;
```

Figure 11.6 – Screenshot – the stack backtrace showing we've hit the register_netdev() function of the SMSC911x network driver

Of course, you'll realize that this demo – entering a particular network driver – is very specific to this target board. Check out the fairly lengthy number of call frames on the kernel stack (seen via GDB's really useful `backtrace` (or `bt`) command)! The `list` (or `l` for short) command reveals a few source code lines of the `register_netdev()` function. Guess what: you can now simply issue the `next` (or `step`) command to single-step through kernel code! Have a go at it.

As short examples, let's now examine the parameter to `register_netdev()`:

```
(gdb) p dev
$1 = (struct net_device *) 0x818a0800
```

Let's dump this (large) structure, the content of struct net_device:

```
(gdb) p *dev
$2 = {name = "eth%d\000\000\000\000\000\000\000\000\000\000",
name_node = 0x0,    ifalias = 0x0, mem_end = 0, mem_start = 0,
base_addr = 0, irq = 30, state = 4,
   dev_list = {next = 0x0, prev = 0x0}, napi_list = {next
= 0x818a0e50, prev = 0x818a0e50}, unreg_list = {next =
0x818a083c, prev = 0x818a083c},
[...]
```

It works but can be difficult to read due to its sheer size. Make it better with the following:

```
(gdb) set print pretty
(gdb) p *dev
$3 = {
  name = "eth%d\000\000\000\000\000\000\000\000\000\000",
  name_node = 0x0,
  ifalias = 0x0,
  mem_end = 0,
  mem_start = 0,
  base_addr = 0,
  irq = 30,
  [...]
```

Now, issuing the continue command yet again has the ARM system run and complete booting (provided it doesn't hit any breakpoints or watchpoints, of course). To break back into the remote system, simply issue the SIGINT signal within GDB by typing ^C (that's *Ctrl* + *C*). You'll get the (gdb) prompt and can go ahead. The remote system will, if idle, typically be in the CPU idle state. Use the backtrace (bt) command to see the kernel-mode stack. A quick demo of this follows:

```
(gdb) c
Continuing.
^C
Program received signal SIGINT, Interrupt.
cpu_v7_do_idle () at arch/arm/mm/proc-v7.S:78
78      ret     lr
(gdb)
```

Awesome – you can see the power of source-level remote debugging with KGDB.

> **Tip**
> When done with the ARM QEMU VM, don't merely kill the QEMU process
> (or even just do `Ctrl` + `A` then `X` to have it exit); it's always better to
> correctly shut down a Linux system. Use the `[sudo] poweroff` (or
> equivalent) command to do so.

Good stuff! Let's now move on to the next major portion of this chapter – covering how
you can debug your (buggy) kernel modules.

Debugging kernel modules with KGDB

To debug a kernel module under KGDB, pretty much everything remains the same as
with debugging in-tree kernel code with GDB. The main difference is this: GDB can't
automatically *see* where the target kernel module's ELF code and data sections are in
(virtual) memory, as modules can be loaded and unloaded on demand – we need to tell it.
Let's get to how exactly we do so.

Informing the GDB client about the target module's locations in memory

The kernel makes the ELF section information of every kernel module available under
sysfs here: `/sys/module/<module-name>/sections/.*`. Do an `ls -a` on this
directory to see the so-called hidden files as well. For example, and assuming that the
`usbhid` kernel module is loaded up (you can run `lsmod` to check, of course), we can see
its sections (output truncated) with the following:

```
ls -a /sys/module/usbhid/sections/
./          [...] .rodata      .symtab [...] .bss        .init.
text [...] .text         [...]     .data [...] .text.exit [...]
.exit.text    [...]
```

Looking at the content of the files (as root, of course) beginning with a period (`.`), you'll
see the (kernel virtual) address where that section of the module is loaded into (kernel
virtual) memory. For example, a few of the sections of the `usbhid` module follow (this is
on my x86_64 Ubuntu 20.04 guest – I've reformatted the output a bit for readability):

```
cd /sys/module/usbhid/sections
cat .text .rodata .data .bss
```

```
0xffffffffc033b000    0xffffffffc0348060 0xffffffffc034e000
0xffffffffc0354f00
```

Now, we can feed this information to GDB via its `add-symbol-file` command! Specify the module's text section address first (the content of the `.text` pseudofile), followed by each individual section in the format `-s <section-name> <address>`. For example, with respect to the `usbhid` module example, we do this:

```
(gdb) add-symbol-file </path/to/>usbhid.ko 0xffffffffc033b000
\
        -s .rodata 0xffffffffc0348060 \
        -s .data 0xffffffffc034e000 \ [...]
```

To more or less automate this (it's a bit tedious to type it all in manually, right?), I make use of a cool script (slightly modified) from the venerable LDD3 book! Our copy's here: `ch11/gdbline.sh`. It works essentially by looping over most of the `.` files in `/sys/module/<module>/section`, printing out a GDB command string that we can simply copy-paste into GDB!

```
add-symbol-file <module-name> <text-addr> \
  -s <section> <section-addr> \
  -s <section> <section-addr> \ [...]
```

Do check it out (we'll cover using it with an example soon enough – read on!).

Step by step – debugging a buggy module with KGDB

As a demo, let's debug via KGDB a slightly modified – and very simple – version of our earlier `ch7/oops_tryv2` module. We call it `ch11/kgdb_try`. It uses a delayed workqueue (a workqueue whose worker thread begins execution only after a specified delay has elapsed). In the work function, we (very deliberately – very contrived) cause a kernel panic by performing an out-of-bounds write overflow to a stack memory buffer. Here are the relevant code paths. First, the init function, where the delayed workqueue is initialized and scheduled to run:

```
// ch11/kgdb_try/kgdb_try.c
static int __init kgdb_try_init(void)
{
    pr_info("Generating Oops via kernel bug in a delayed
workqueue function\n");
    INIT_DELAYED_WORK(&my_work, do_the_work);
```

```
    schedule_delayed_work(&my_work, msecs_to_jiffies(2500));
    return 0;         /* success */
}
```

Why do we use a delayed workqueue, with, as you can see, the delay set to 2.5 seconds? This is done just so you have sufficient time to add the module's symbols to GDB before the kernel Oops'es (you'll soon see us doing this)! The actual – and very contrived – bug is here, within the worker routine:

```
static void do_the_work(struct work_struct *work)
{
    u8 buf[10];
    int i;
    pr_info("In our workq function\n");
    for (i=0; i <=10; i++)
        buf[i] = (u8)i;
    print_hex_dump_bytes("", DUMP_PREFIX_OFFSET, buf, 10);
    [...]
```

The bug – the local buffer overflow that will occur when i reaches the value 10 (as, of course, the array has 10 elements only, 0 through 9, and we're attempting to access the non-existent eleventh element at buf[10]!) – though seemingly trivial, *caused my entire target system to simply freeze when run without KGDB!* This is because, internally, the kernel panicked! Try it out and you'll see... Of course, recollect that kernel memory checkers – remember KASAN! – will certainly catch bugs like this.

This time, to try something a little different from last time (debugging the kernel at early boot), we'll use *an x86_64 QEMU guest system as the target kernel* (instead of the ARM one we used previously). To do so, we'll set up a vanilla 5.10.109 kernel for KGDB, of course (as covered in the *Configuring the kernel for KGDB section*), and reuse (open source) code from here to set up the root filesystem (it's Debian Stretch): *[Linux Kernel Exploitation 0x0] Debugging the Kernel with QEMU*, K Makan, Nov 2020 (http://blog.k3170makan.com/2020/11/linux-kernel-exploitation-0x0-debugging.html). This blog article itself generates the rootfs using the Google syzkaller project! Do read through the article for details.

Here are the detailed steps to be carried out – read along and try it out for yourself.

Step 1 – preparing the target system's kernel, root filesystem, and test module on the host

This step involves a bit of work:

1. Configuring and building a (debug, KGDB-enabled) kernel for the target system (QEMU emulated x86_64)

2. Having a working root filesystem image for the target (so that we can store our module(s), log in, and so on)

3. Building the test module against the target kernel

Let's proceed!

Step 1.1 – configuring and building the target kernel

We'll keep it brief:

1. Download and extract the kernel source tree for an appropriate kernel. Let's use the 5.10.109 kernel (as it's within the 5.10 LTS series and matches the one we used for the ARM target). Keep the source tree in any convenient location on your system and note it (for the purposes of this demo, let's say you've installed the kernel source tree here: `~/linux-5.10.109`).

2. Configure the kernel in the usual manner (via the `make menuconfig` UI), taking into account the fact that you must enable support for KGDB and related items – we've covered this in detail in the *Configuring the kernel for KGDB section*. For your reference, I've kept my kernel config file here: `ch11/kconfig_x86-64_target`.

> **Tip**
>
> With recent 5.10 (or newer) kernels, the build could fail with an error such as this:
>
> ```
> make[1]: *** No rule to make target 'debian/
> canonical-revoked-certs.pem' , needed by certs/
> x509_revocation_list'
> ```
>
> A quick fix is to do this:
>
> ```
> scripts/config --disable SYSTEM_REVOCATION_KEYS
> scripts/config --disable SYSTEM_TRUSTED_KEYS
> ```
>
> Then, retry the kernel build.

- Build the kernel via `make -j[n] all`. The compressed kernel image (`arch/x86/boot/bzImage`) as well as the uncompressed kernel image with symbols (`vmlinux`) is generated. As this is all we require for this demo, we skip the (typical) remaining steps of modules and kernel/bootloader installation.

Here's my custom KGDB-enabled kernel images:

```
$ ls -lh arch/x86/boot/bzImage vmlinux
-rw-rw-r-- 1 osboxes osboxes 7.9M May  3 13:29 arch/x86/boot/
bzImage
-rwxrwxr-x 1 osboxes osboxes 240M May  3 13:29 vmlinux*
```

Let's move along...

Step 1.2 – obtaining a working root filesystem image for the target

We'll of course require a target root filesystem (or rootfs). Further, it will require having our test kernel module (compiled with the same target kernel) plus the `gdbline.sh` and `doit` wrapper scripts on it (we explain the purpose of the last one shortly). Now, building a rootfs from scratch isn't a trivial task, thus, to ease the effort, *we provide a fully functional root filesystem image* based on the Debian Stretch distro.

We covered downloading the compressed rootfs image file in the *Technical requirements* section (if you haven't yet done so, please ensure you download it now). Now extract it:

```
7z x rootfs_deb.img.7z
```

It will get extracted into a directory named `images/`. You now have the uncompressed and ready-to-use target rootfs binary image (of size 512 MB) here: `ch11/images/rootfs_deb.img`.

FYI, you can always edit the rootfs image by, on the host, loop mounting it (when it's not in use!), editing its content, then unmounting it (see *Figure 11.7*). Here, you don't need to do this yourself; it's been done and the target rootfs has been supplied to you.

We've kept all required files for the module debug demo on the target rootfs under the /myprj directory. As a quick sanity check, let's loop mount the target root filesystem image file and peek into it (ensure you create the mount point directory first, /mnt/tmp, here):

```
$ pwd
/home/osboxes/Linux-Kernel-Debugging/ch11
$ ls
gdbline.sh*  images/  kgdb_try/  rootfs_deb.img.7z  run_target.sh*
$ ls -l images/
total 524292
-rw-r--r-- 1 osboxes osboxes 536870912 May  3 17:20 rootfs_deb.img
$
$ sudo mount -o loop images/rootfs_deb.img /mnt/tmp
[sudo] password for osboxes:
$ ls /mnt/tmp/
bin/   dev/   home/  lib64/       media/  myprj/  proc/  run/   srv/   tmp/  var/
boot/  etc/   lib/   lost+found/  mnt/    opt/    root/  sbin/  sys/   usr/
$
$ ls /mnt/tmp/myprj/
doit*  gdbline.sh*  kgdb_try.ko
$
$ sudo umount /mnt/tmp
$
```

Figure 11.7 – Loop mounting and viewing content under our target root filesystem

Don't forget: only loop mount and edit the target rootfs when it's not in use via QEMU (or another hypervisor). Unmount it when done!

On our host system, here's what the directory tree structure under ch11/ should now look like:

Figure 11.8 – Screenshot showing the directory tree on the host Linux under ch11/

Right, let's continue.

Step 1.3 – building the module for the target kernel

One more step's required here: the test module (under `ch11/kgdb_try`) needs to be built and deployed on both the target and host systems. (Actually, it's already deployed on the target rootfs; we need to build it on our host.) So, `cd` to the `ch11/kgdb_try` directory and issue the `make` command to build it.

Importantly, the `Makefile` must take into account the fact that this module's built against the target 5.10.109 kernel (and not the native one)! So, we've changed the `KDIR` variable within the `Makefile` to reflect this location:

```
// ch11/kgdb_try/Makefile
#@@@@@@@@@@@@@ NOTE! SPECIAL CASE @@@@@@@@@@@@@@@@@@@
  # We specify the build dir as the linux-5.10.109 kernel src
tree; this is as
  # we're using this as the target x86_64 kernel and debugging
this module over KGDB
  KDIR ?= ~/linux-5.10.109
```

If the kernel's in a different location on your system, update the `Makefile`'s `KDIR` variable first and then build the module.

> **Note**
>
> If you make any changes in the `kgdb_try.c` source and rebuild, you'll need to update the module within the target rootfs as well, by loop mounting the rootfs image file, copying the new `kgdb_try.ko` module into its `/myprj` directory, and doing the unmount.

Good job! Let's move on to the next step...

Step 2 – target startup and wait at early boot

Start the x86_64 target (via QEMU). We expect you've installed `qemu-system-x86_64` by now (as advised in the *Technical requirements section*):

```
cd <book_src>/ch11
qemu-system-x86_64 \
  -kernel ~/linux-5.10.109/arch/x86/boot/bzImage \
  -append "console=ttyS0 root=/dev/sda earlyprintk=serial
rootfstype=ext4 rootwait nokaslr" \
```

```
-hda images/rootfs_deb.img \
-nographic -m 1G -smp 2 \
-S -s
```

For your convenience, the same command's available within a wrapper script here: `ch11/run_target.sh`. Simply run it, passing the kernel and rootfs image files as parameters.

> **Tip**
>
> Running QEMU with the `-enable-kvm` option switch can make guest execution (much!) faster. This requires hardware-level virtualization support of course (implying that CPU virtualization is enabled at the firmware/BIOS level). On the x86, you can check with `egrep "^flags.*(vmx|svm)" /proc/cpuinfo`. If there's no output, it isn't enabled and won't work. Also, this could fail if any other hypervisor is running and making use of KVM (your Ubuntu guest on VirtualBox perhaps); in effect, if nested virtualization isn't supported by KVM.

Right, the guest kernel starts and pretty much immediately waits, due to the effect of QEMU's `-S` option switch (see *Figure 11.9*).

Step 3 – host system remote GDB startup

On the host (which in our case is the Ubuntu x86_64 guest), let's set up the GDB client to debug the target system. So, `cd` into the target kernel source tree (here, we're taking it as being in `~/linux-5.10.109`). Run GDB, passing along the uncompressed 5.10.109 kernel image (`vmlinux`) as a parameter (see *Figure 11.10*), enabling GDB to read in all symbols. In addition, we employ the GDB initialization/startup file `~/.gdbinit` to define a simple macro (we cover GDB macros in the *GDB custom macros in its startup file* section). Here's the `connect_qemu` macro definition:

```
cat ~/.gdbinit
[...]
set auto-load safe-path /
define connect_qemu
  target remote :1234
  hbreak start_kernel
  hbreak panic
  #hbreak do_init_module
end
```

On startup, GDB will parse in its content, thus allowing us to run our custom macro `connect_qemu`, allowing us to connect to the target and set up a couple of hardware breakpoints (via GDB's `hbreak` command). Here are a few points regarding the GDB startup file content:

- The `set auto-load safe-path /` directive is to allow GDB to parse in and use various Python-based GDB helper scripts. We cover the details in the *Setting up and using GDB scripts with CONFIG_GDB_SCRIPTS* section.

- A tip, useful at times: adding the kernel function `do_fsync()` as a breakpoint is a convenience, allowing you to break into GDB by typing `sync` on the target command line.

- We add the `start_kernel()` hardware breakpoint here simply as a demo, for no other reason... It's pretty much the first C function hit as the kernel boots up!

- We have a commented-out hardware breakpoint on the function `do_init_module()`. This can be very helpful, allowing you to debug any module's init code path straight away (details follow in the *Debugging a module's init function* section).

> **Tip**
>
> Ensure you use **hardware breakpoints** (via GDB's `hbreak` command) for your key breakpoints, and not software watchpoints! The `info breakpoints` command (abbreviated as simply `i b`) will reveal all currently defined breakpoints and watchpoints.

A couple of screenshots will help clarify things. First, the state of the target kernel just after boot:

```
$ pwd
/home/osboxes/Linux-Kernel-Debugging/ch11
$ ls
gdbline.sh*  kconfig_x86-64_target  README.txt          run_target.sh*
images/      kgdb_try/              rootfs_deb.img.7z
$ ls -lh images/
total 513M
-rw-r--r-- 1 osboxes osboxes 512M May  4 07:36 rootfs_deb.img
$
$
$ ./run_target.sh ~/linux-5.10.109/arch/x86/boot/bzImage images/rootfs_deb.img
Note:
1. First shut down any other hypervisor instance that may be running
2. Once run, this guest qemu system will *wait* for GDB to connect from the host:
On the host, do:

$ gdb -q <linux-src-tree>/vmlinux
(gdb) target remote :1234

qemu-system-x86_64  -kernel /home/osboxes/linux-5.10.109/arch/x86/boot/bzImage
 -append console=ttyS0 root=/dev/sda earlyprintk=serial rootfstype=ext4 rootwait nokaslr
-hda images/rootfs_deb.img  -nographic -m 1G -smp 2  -S -s
WARNING: Image format was not specified for 'images/rootfs_deb.img' and probing guessed ra
w.
         Automatically detecting the format is dangerous for raw images, write operations
on block 0 will be restricted.
         Specify the 'raw' format explicitly to remove the restrictions.
```

Figure 11.9 – Target kernel waiting for the remote GDB client to connect to it

Here's a screenshot of running the GDB client on the host (from the kernel source tree location) and issuing our `connect_qemu` macro:

```
$ cd ~/linux-5.10.109/
$ ls
arch/           fs/        LICENSES/              modules.order    System.map
block/          include/   lsmod.now              Module.symvers   tools/
certs/          init/      MAINTAINERS            net/             usr/
COPYING         ipc/       Makefile               README           virt/
CREDITS         Kbuild     mm/                    samples/         vmlinux*
crypto/         Kconfig    modules.builtin        scripts/         vmlinux-gdb.py@
Documentation/  kernel/    modules.builtin.modinfo security/       vmlinux.o
drivers/        lib/       modules-only.symvers   sound/           vmlinux.symvers
$
$ gdb -q ./vmlinux
Reading symbols from ./vmlinux...
(gdb) connect_qemu
0x000000000000fff0 in exception_stacks ()
Hardware assisted breakpoint 1 at 0xffffffff8299df54: file init/main.c, line 850.
Hardware assisted breakpoint 2 at 0xffffffff81ad87f7: file kernel/panic.c, line 178.
(gdb) i b
Num     Type           Disp Enb Address            What
1       hw breakpoint  keep y   0xffffffff8299df54 in start_kernel at init/main.c:850
2       hw breakpoint  keep y   0xffffffff81ad87f7 in panic at kernel/panic.c:178
(gdb) █
```

Figure 11.10 – Host: within the kernel source tree, the remote GDB client connects to the target and sets up breakpoints

Fantastic – let's continue...

Step 4 – target system: install the module and add symbols to GDB

When debugging with KGDB, you'll need to `insmod` the (possibly buggy) module and add its symbols (as explained in the *Informing the GDB client about the target module's locations in memory* section). But – in this demo at least! – you need to do all this quickly, before it actually crashes! So, on the target rootfs, we have a simple wrapper script (`/myprj/doit`) to do the following:

1. Set the (target) kernel to panic on Oops.
2. `insmod` the module on the target system (the one running with the GDB server component, that is, with KGDB enabled, of course).
3. Execute our `gdbline.sh` script. It generates the key `add-symbol-file` GDB command! Quickly now...

4. We – quickly, before the kernel Oops'es and panics! – switch to the host system GDB and press ^C, interrupting (and stopping) the target kernel. (Whew, now we're safe.) We then copy-paste the GDB add-symbol-file command that was generated on the target, informing GDB about the module's symbols.

5. Add a hardware breakpoint for the routine of interest. Here, we run hbreak on our workqueue function do_the_work().

Here's the code of the target rootfs /myprj/doit script (which is itself already embedded within the target rootfs image):

```
echo 1 > /proc/sys/kernel/panic_on_oops
sudo insmod ./kgdb_try.ko
sudo ./gdbline.sh kgdb_try ./kgdb_try.ko
```

So, let's get going. First, have the target continue (type c) to boot up, log in to it (as required), and run this helper script to set things up. Of course, the target first hits the start_kernel() hardware breakpoint. Great – you can look around, then type c to have GDB continue the target. It boots up fully... (it can take a moment – be patient). The target kernel now asks you to log in. Here, simply pressing the *Enter* key is sufficient as we simply enter the Debian maintenance mode and work there – it's fine to do so:

Figure 11.11 – On the left is the target; in the right window is the GDB client process running on the host; we log in to the target kernel by pressing Enter

Now, a key part of this exercise: on the target root filesystem, cd to the /myprj directory and run our wrapper doit script. It runs, generating the output – the add-symbol-file command we must issue within GDB! You'll realize, of course, that the (buggy) kgdb_try.ko module is right now executing its code paths. As we're using a delayed workqueue, we've bought some time (2.5 s here) before the buggy do_the_work() code runs.

Quickly now! Switch to the host window where our client GDB process is running and press ^C (*Ctrl + C*). This has GDB break in - the target's execution is stopped, it's now frozen (whew!). This is important, as otherwise, the bug can trigger before we set up the breakpoint on our buggy module. In *Figure 11.12*, you can see our typing of ^C in the right-side host window. The following screenshot reveals the action:

```
root@syzkaller:~# cd /myprj/                                    $ ls
root@syzkaller:/myprj# ;s                                       arch/
bash: syntax error near unexpected token `;'                    block/
root@syzkaller:/myprj# ls                                       certs/
doit  gdbline.sh  kgdb_try.ko                                   COPYING
root@syzkaller:/myprj# cat doit                                 CREDITS
#!/bin/sh                                                        crypto/
# setup to panic on Oops                                        Documentati
echo 1 > /proc/sys/kernel/panic_on_oops                         $
sudo insmod ./kgdb_try.ko                                       $ gdb -q ./
sudo ./gdbline.sh kgdb_try ./kgdb_try.ko                        Reading sym
root@syzkaller:/myprj#                                          (gdb) conne
root@syzkaller:/myprj#                                          0x000000000
root@syzkaller:/myprj# ./doit          1                        Hardware as
sudo: unable to resolve host syzkaller: Connection refused      Hardware as
[  127.819717] kgdb_try: loading out-of-tree module taints ke  (gdb) i b
[  127.823922] kgdb_try: module verification failed: signatur  Num     Typ
[  127.838223] kgdb_try:kgdb_try_init():66: Generating Oops v  1       hw
sudo: unable to resolve host syzkaller: Connection refused      2       hw
Copy-paste the following lines into GDB                         (gdb) c
---snip---                                                      Continuing.
add-symbol-file ./kgdb_try.ko 0xffffffffc004a000 \
        -s .bss 0xffffffffc004d4c0 \                            Thread 1 hi
        -s .data 0xffffffffc004d000 \                           850    {
        -s .exit.text 0xffffffffc004a127 \                      (gdb) c
        -s .gnu.linkonce.this_module 0xffffffffc004d0c0 \       Continuing.
        -s .init.text 0xffffffffc0050000 \                      ^C       2
        -s .note.Linux 0xffffffffc004b024 \                     Thread 1 re
        -s .note.gnu.build-id 0xffffffffc004b000 \              0xfffffffff8
        -s .rodata 0xffffffffc004b148 \                         60
        -s .rodata.str1.1 0xffffffffc004b03c \                  (gdb) ▮
        -s .rodata.str1.8 0xffffffffc004b078
---snip---

root@syzkaller:/myprj# ▯
```

Figure 11.12 – (Truncated) screenshot: 1. run the doit script on the target (left window); 2. quickly switch to the right host window and interrupt (stop) the target with ^C

Great job! Now do the following:

1. From the target window (the left-side one in *Figure 11.12*), copy the output of our `gdbline.sh` script – the GDB `add-symbol-file` command and whatever follows (in effect, the content between the `---snip---` delimiters) – into the clipboard.

2. Switch back to the host window running the client GDB (the right-side one in *Figure 11.12*).

3. *Important!* `cd` to the directory where the kernel module's code is (GDB needs to be able to see it).

4. Paste the clipboard content – the complete `add-symbol-file <...>` command – into GDB. It prompts whether to accept this. Answer yes (`y`). GDB reads in the module symbols! See this in the (truncated) screenshot:

```
Press Enter for maintenance          60          asm volatile("sti; hlt": : :"memory");
(or press Control-D to continue):    (gdb) pwd
root@syzkaller:~#                    Working directory /home/osboxes/lkd_kernels/linux-5.10.109.
root@syzkaller:~#                    (gdb) cd ~/Linux-Kernel-Debugging/ch11/kgdb_try/
root@syzkaller:~# cd /myprj/         Working directory /home/osboxes/Linux-Kernel-Debugging/ch11/kgdb_try.
root@syzkaller:/myprj# ls            (gdb) add-symbol-file ./kgdb_try.ko 0xffffffffc004a000 \
doit gdbline.sh kgdb_try.ko              -s .bss 0xffffffffc004d4c0 \
root@syzkaller:/myprj#                   -s .data 0xffffffffc004d000 \
root@syzkaller:/myprj#                   -s .exit.text 0xffffffffc004a127 \
root@syzkaller:/myprj# ./doit            -s .gnu.linkonce.this_module 0xffffffffc004d0c0 \
sudo: unable to resolve host syzkaller: Connection refuse  -s .init.text 0xffffffffc0050000 \
[  15.441432] kgdb_try: loading out-of-tree module taint    -s .note.Linux 0xffffffffc004b024 \
[  15.442594] kgdb_try: module verification failed: sig     -s .note.gnu.build-id 0xffffffffc004b000 \
[  15.449505] kgdb_try:kgdb_try_init():66: Generating O      -s .rodata 0xffffffffc004b148 \
sudo: unable to resolve host syzkaller: Connection refuse  -s .rodata.str1.1 0xffffffffc004b03c \
Copy-paste the following lines into GDB                     -s .rodata.str1.8 0xffffffffc004b078 \
---snip---                           add symbol table from file "./kgdb_try.ko" at
add-symbol-file ./kgdb_try.ko 0xffffffffc004a000 \           .text_addr = 0xffffffffc004a000
    -s .bss 0xffffffffc004d4c0 \                             .bss_addr = 0xffffffffc004d4c0
    -s .data 0xffffffffc004d000 \                            .data_addr = 0xffffffffc004d000
    -s .exit.text 0xffffffffc004a127 \                       .exit.text_addr = 0xffffffffc004a127
    -s .gnu.linkonce.this_module 0xffffffffc004d0c0 \        .gnu.linkonce.this_module_addr = 0xffffffffc004d0c0
    -s .init.text 0xffffffffc0050000 \                       .init.text_addr = 0xffffffffc0050000
    -s .note.Linux 0xffffffffc004b024 \                      .note.Linux_addr = 0xffffffffc004b024
    -s .note.gnu.build-id 0xffffffffc004b000 \               .note.gnu.build-id_addr = 0xffffffffc004b000
    -s .rodata 0xffffffffc004b148 \                          .rodata_addr = 0xffffffffc004b148
    -s .rodata.str1.1 0xffffffffc004b03c \                   .rodata.str1.1_addr = 0xffffffffc004b03c
    -s .rodata.str1.8 0xffffffffc004b078                     .rodata.str1.8_addr = 0xffffffffc004b078
---snip---                           (y or n) y
                                     Reading symbols from ./kgdb_try.ko...
root@syzkaller:/myprj# ▯             (gdb) ▮
```

Figure 11.13 – (Truncated) screenshot showing how we cd and copy-paste the add-symbol-file command into the GDB process

Super! Now that GDB understands the module memory layout and has its symbols, simply add (hardware) breakpoints as required! Here, we just add the relevant one, the workqueue function:

```
(gdb) hbreak do_the_work
Hardware assisted breakpoint 3 at 0xffffffffc004a000: file /
home/osboxes/Linux-Kernel-Debugging/ch11/kgdb_try/kgdb_try.c,
line 43.
(gdb)
```

By the way, you'll recall we earlier enabled the kernel config GDB_SCRIPTS. This has several useful Python-based helper scripts become available during a GDB session kernel debug session (we cover this topic in more detail in the *Setting up and using GDB scripts with CONFIG_GDB_SCRIPTS* section). As an example, we issue the lx-lsmod helper to show all modules currently loaded (on the target kernel's memory):

```
(gdb) lx-lsmod
Address                 Module              Size   Used by
0xffffffffc004a000 kgdb_try                20480  0
(gdb)
```

Cool – its output is as expected. Notice how the kernel virtual address of where the module is loaded in memory (0xffffffffc004a000 here) perfectly matches the first parameter to the add-symbol-file command – it's the address of the module's .text (code) section!

Step 5 – debugging the module with [K]GDB

So, finally: we're all set up. We can now go ahead and debug the target module in the usual manner, setting breakpoints, examining data, and stepping through its code!

Within the host (client) GDB process, type c to continue. The target system resumes execution... Soon enough, the delay that we specified (2.5 s) before the workqueue function – do_the_work() – must run will elapse. The function will begin to execute, and immediately get trapped into via GDB (don't forget, we set up a hardware breakpoint on it in the previous step!):

```
(gdb) c
Continuing.

Thread 1 hit Breakpoint 3, do_the_work (work=0xffffffffc004d000 <my_work>)
    at /home/osboxes/Linux-Kernel-Debugging/ch11/kgdb_try/kgdb_try.c:43
43      {
(gdb) bt
#0  do_the_work (work=0xffffffffc004d000 <my_work>)
    at /home/osboxes/Linux-Kernel-Debugging/ch11/kgdb_try/kgdb_try.c:43
#1  0xffffffff811138bf in process_one_work (worker=worker@entry=0xffff8880035cc6c0,
    work=0xffffffffc004d000 <my_work>) at kernel/workqueue.c:2279
#2  0xffffffff81113aad in worker_thread (__worker=__worker@entry=0xffff8880035cc6c0)
    at kernel/workqueue.c:2425
#3  0xffffffff81119d34 in kthread (_create=0xffff8880035cbb00) at kernel/kthread.c:313
#4  0xffffffff81004562 in ret_from_fork () at arch/x86/entry/entry_64.S:296
#5  0x0000000000000000 in ?? ()
(gdb) l
38
39      /*
40       * Our delayed workqueue callback function
41       */
42      static void do_the_work(struct work_struct *work)
43      {
44              u8 buf[10];
45              int i;
46
47              pr_info("In our workq function\n");
(gdb)
48              for (i=0; i <=10; i++)
49                      buf[i] = (u8)i;
50              print_hex_dump_bytes("", DUMP_PREFIX_OFFSET, buf, 10);
```

Figure 11.14 – We continue: the hardware breakpoint's hit; we're in our do_the_work() function, single-stepping through its source; the buggy line 49 is highlighted

Looking at *Figure 11.14*, we examine the (kernel) stack with the bt (backtrace) GDB command – it's as expected. Next, let's do something interesting: we know the bug's in the loop when the local variable i reaches the value 10 (needless to say, in the C array, indices begin at 0, not 1). Now instead of single-stepping through the loop 10 times, we can set up a *conditional breakpoint*, telling GDB to stop execution when the value of i is, say, 8. This is easily achieved with the GDB command:

(gdb) b 49 if i==8

FYI, we cover more on this in the *Conditional breakpoints* section. So, let's proceed:

```
(gdb) b 49 if i==8
Breakpoint 5 at 0xffffffffc004a04c: file /home/osboxes/Linux-Kernel-Debugging/ch11/kgdb_try/kgd
b_try.c, line 49.
(gdb) c
Continuing.

Thread 2 hit Breakpoint 5, do the work (work=<optimized out>)
    at /home/osboxes/Linux-Kernel-Debugging/ch11/kgdb_try/kgdb_try.c:49
49                          buf[i] = (u8)i;
(gdb) p i
$8 = 8
(gdb) p/x buf
$9 = {0x0, 0x1, 0x2, 0x3, 0x4, 0x5, 0x6, 0x7, 0xff, 0xff}
(gdb) n
48                  for (i=0; i <=10; i++)
(gdb) display i
1: i = 8
(gdb) n
49                          buf[i] = (u8)i;
1: i = 9
(gdb)
48                  for (i=0; i <=10; i++)
1: i = 9
(gdb) p/x buf
$10 = {0x0, 0x1, 0x2, 0x3, 0x4, 0x5, 0x6, 0x7, 0x8, 0x9}
(gdb) n
49                          buf[i] = (u8)i;
1: i = 10
```

Figure 11.15 – Screenshot showing how we set up a conditional breakpoint on line 49 and single-step through the module's code

We have GDB continue. The conditional breakpoint is hit... It works: the value of i is 8 (to begin with). Notice how I used the display i GDB command to have GDB always display the value of the variable i (after every step (s) or next (n) GDB command). Look at *Figure 11.15* carefully: we find that, though the bug's hit (when i reaches the value 10), execution seems to continue. Yes, for just a short while. The kernel's built-in stack overflow detection code paths do kick in soon enough – and guess what: *the kernel panics quite spectacularly!* The parameter to panic() is a string – the reason for the panic. Clearly, it's due to kernel stack corruption! The following figure shows all this clearly:

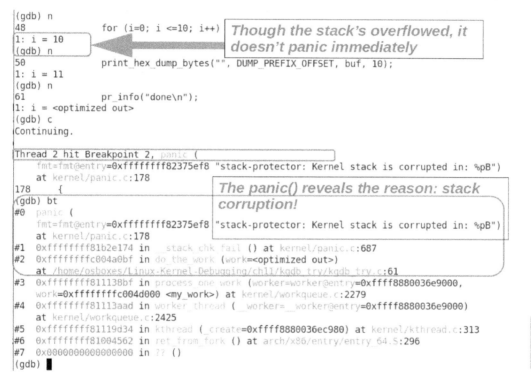

Figure 11.16 – Screenshot revealing the actual bug and the subsequent kernel panic; the panic message reveals it's kernel stack corruption

When we have GDB continue the target's execution (by typing c); the panic message details are now seen in the target system console window:

```
root@syzkaller:/myprj# [   21.428961] kgdb_try:do_the_work():47: In our workq function
[   21.484495] kgdb_try:do_the_work():61: done
[   21.495971] Kernel panic - not syncing: stack-protector: Kernel stack is corrupted in: do_the_work+0xbf/0xc5 [kgdb_try]
[   21.496133] CPU: 0 PID: 253 Comm: kworker/0:4 Tainted: G          OE     5.10.109-kgdb2 #1
[   21.496133] Hardware name: QEMU Standard PC (i440FX + PIIX, 1996), BIOS 1.13.0-1ubuntu1.1 04/01/2014
[   21.496133] Workqueue: events do_the_work [kgdb_try]
[   21.496133] Call Trace:
[   21.496133]  dump_stack+0x74/0x92
[   21.496133]  panic+0x101/0x2e3
[   21.496133]  ? do_the_work+0xbf/0xc5 [kgdb_try]
[   21.496133]  __stack_chk_fail+0x14/0x20
[   21.496133]  do_the_work+0xbf/0xc5 [kgdb_try]
[   21.496133]  process_one_work+0x1ef/0x390
[   21.496133]  worker_thread+0x4d/0x3f0
[   21.496133]  kthread+0x114/0x150
[   21.496133]  ? process_one_work+0x390/0x390
[   21.496133]  ? kthread_park+0x90/0x90
[   21.496133]  ret_from_fork+0x22/0x30
[   21.496133] Kernel Offset: disabled
[   21.496133] ---[ end Kernel panic - not syncing: stack-protector: Kernel stack is corrupted in: do_the_work+0xbf/0xc5 [kgdb_try] ]---
```

Figure 11.17 – Target system: the kernel panic message on the console

Great going!

There's a nagging issue though: how do you debug a module's early init code in KGDB? That's what we cover next!

Debugging a module's init function

Here, for the purpose of a simple-yet-interesting demo, we used a delayed workqueue. Once the delay elapsed (2.5 s here), the buggy workqueue function executed and resulted in an Oops (and a subsequent panic). We could debug it with KGDB! But, think about this: in a project, what if the module's init function *doesn't* use a delayed workqueue, just a regular workqueue? Then, the workqueue function will run almost immediately, before you have time to set up a breakpoint on it! How do we debug such situations?

The key, is to be able to debug the early module initialization code itself, allowing you to then single-step through it. This can be achieved by setting up a breakpoint on the module's init function itself. Hang on – this may not work out. Think: the setting of the breakpoint has to happen after the insmod command is issued but by the time you type hbreak kgdb_try_init (or whatever), the bug could trigger!

So, here's a workable solution: *set a (hardware) breakpoint on the kernel infrastructure code that performs the actual work when invoking a module's init function* – the do_init_ module(struct module *mod) function. This can be done at any time, even as part of the connect_qemu (or equivalent) macro that we set up! Then, once the breakpoint's hit, proceed debugging the code from when the breakpoint's hit. You can even check which module's being loaded by looking up the pointer to the module structure (it's the single parameter passed to the do_init_module() function), then running the set print pretty gdb command, followed by p *mod. Then, look for the structure member named name. What a surprise – it reveals the name of the module! I also found the module name this way:

```
(gdb) x/s mod.mkobj.kobj.name
0xffff8880036eb770:    "kgdb_try"
```

Neat.

> **Exercise**
> Debug a kernel module's init function using the method just described.

On a project, this whole process – debugging a module via KGDB – can be very powerful: you saw how we can single-step through the module's code, employ the backtrace command to see the (kernel) stack in detail, and examine memory (with GDB's x command) and variable values (and even change variables with GDB's set variable command!). In a later section (*Using GDB's TUI mode*), we even show how you can single-step through assembly code! All this can result in getting valuable insight into the code's behavior, ultimately (fingers crossed!) helping you find the root cause of that annoying bug.

Awesome going – let's wind up this chapter with a few useful tips and tricks.

[K]GDB – a few tips and tricks

GDB is a large and powerful program with many features. We'll describe a few top-of-mind tips and tricks here that can prove useful when using GDB (and KGDB).

Setting up and using GDB scripts with CONFIG_GDB_SCRIPTS

Dating back to the 4.0 kernel, Linux provides several Python-based helper scripts – seen as additional GDB commands – to help debug the kernel (and kernel modules). Their code's within the kernel source in scripts/gdb.

Enable them by setting CONFIG_GDB_SCRIPTS=y. Once enabled, it's advisable to put the following line in your GDB startup file ~/.gdbinit:

```
add-auto-load-safe-path <...>/scripts/gdb/vmlinux-gdb.py
```

Or, simpler but more permissive, add-auto-load-safe-path /. This has GDB parse this Python script within the kernel source tree (scripts/gdb/vmlinux-gdb.py) and thus recognize the Python-based GDB helper scripts – a very useful thing! All the helper scripts are prefixed with lx- whereas the helper functions are prefixed with lx_. Once set up, we can see them and a one-line explanation of what they're for by employing the apropos lx command (the word *apropos* means with reference to or concerning – it's a GDB keyword):

```
(gdb) apropos lx
function lx_clk_core_lookup -- Find struct clk_core by name
function lx_current -- Return current task.
function lx_device_find_by_bus_name -- Find struct device by bus and name (both strings)
function lx_device_find_by_class_name -- Find struct device by class and name (both strings)
function lx_module -- Find module by name and return the module variable.
function lx_per_cpu -- Return per-cpu variable.
function lx_rb_first -- Lookup and return a node from an RBTree
function lx_rb_last -- Lookup and return a node from an RBTree.
function lx_rb_next -- Lookup and return a node from an RBTree.
function lx_rb_prev -- Lookup and return a node from an RBTree.
function lx_task_by_pid -- Find Linux task by PID and return the task_struct variable.
function lx_thread_info -- Calculate Linux thread_info from task variable.
function lx_thread_info_by_pid -- Calculate Linux thread_info from task variable found by pid
lx-clk-summary -- Print clk tree summary
lx-cmdline --  Report the Linux Commandline used in the current kernel.
lx-configdump -- Output kernel config to the filename specified as the command
lx-cpus -- List CPU status arrays
lx-device-list-bus -- Print devices on a bus (or all buses if not specified)
lx-device-list-class -- Print devices in a class (or all classes if not specified)
lx-device-list-tree -- Print a device and its children recursively
lx-dmesg -- Print Linux kernel log buffer.
lx-fdtdump -- Output Flattened Device Tree header and dump FDT blob to the filename
lx-genpd-summary -- Print genpd summary
lx-iomem -- Identify the IO memory resource locations defined by the kernel
lx-ioports -- Identify the IO port resource locations defined by the kernel
lx-list-check -- Verify a list consistency
lx-lsmod -- List currently loaded modules.
lx-mounts -- Report the VFS mounts of the current process namespace.
lx-ps -- Dump Linux tasks.
lx-symbols -- (Re-)load symbols of Linux kernel and currently loaded modules.
lx-timerlist -- Print /proc/timer_list
lx-version --  Report the Linux Version of the current kernel.
```

Figure 11.18 – Screenshot showing the output from the GDB apropos lx command – kernel Python-based helper scripts made available, when CONFIG_GDB_SCRIPTS=y

These helper scripts certainly make mundane debugging tasks – such as looking up the kernel log (lx-dmesg), looking up the kernel command line (lx-cmdline) or the modules currently loaded (lx-lsmod), defined I/O memory locations (lx-iomem), and so on – a lot easier. (We mention just a few here explicitly; *Figure 11.18* shows the complete list on the installation.) To get brief help on a helper script, type help lx-<scriptname> in GDB.

Note that this facility requires GDB 7.2 or later. The official kernel documentation provides some help and examples of using them here: https://docs.kernel.org/dev-tools/gdb-kernel-debugging.html#examples-of-using-the-linux-provided-gdb-helpers. Do check it out and try them out yourself.

KGDB target remote :1234 command doesn't work on physical systems

This can and does happen – the target remote :1234 GDB command fails to have the GDB client connect to the target system. Setting up KGDB between physical hardware can be frustrating at times! *Check and recheck the connection between the host and the target!* First of all, without worrying about the low-level details of setting up KGDB, ensure that you can send and receive packets (a simple message will suffice) from the host to the target system and vice versa. For example, you can do the following (as root, and assuming the serial connection is on /dev/ttyUSB0):

```
[Host] echo "hello, target" > /dev/ttyUSB0
[Target] cat /dev/ttyUSB0
```

You should see the hello, target message on the target. Test the other way around as well; *both must work.* If the connection isn't okay, the target remote :<port#> command will fail or return quirky error/warning messages (stuff such as warning: unrecognized item "timeout" in "qSupported" response).

Also (and as mentioned here: https://stackoverflow.com/a/36861909/779269), the USB-to-serial type of connection is fine to have on the host system but not necessarily on the target. The target system must have a direct serial (COM) port interface (or Ethernet/wireless interfaces). See the *Further reading* section for more on this topic.

Setting the system root with sysroot

When debugging foreign binaries and kernels, which can happen either when performing remote debugging or using KGDB, keep in mind the following points:

- Use the toolchain-specific GDB and not the native one; for example, running `arm-none-linux-gnueabihf-gdb -q <...>/vmlinux` and not `gdb -q vmlinux`.

- Setting GDB's `sysroot` variable (and/or the `solib-search-path` variable) to correctly point to the root of the filesystem upon which the binaries and libraries for the target reside. This is important when you're on a host system performing remote debugging and you need to correctly reference the target root filesystem which will be on the host (as well).

Aside from the kernel support provided via KGDB, GDB natively provides several useful features. We discuss a few of them in the following sections.

Using GDB's TUI mode

GDB has a **Text User Interface** (**TUI**) mode, where, instead of just the usual command-line interface, the terminal window is split into two or three horizontally tiled panes (it internally uses curses-based libraries and APIs to achieve its somewhat graphic-like capabilities – peek at *Figure 11.19*).

The point here is that using GDB in TUI mode can be quite empowering for the developer/debug session! To see why let's do a quick GDB TUI session. We'll debug the kernel in much the same way as in the *Debugging the kernel with GDB* section, except that here, we run GDB with the `-tui` option switch (also, FYI, we're running it on our x86_64 Ubuntu guest with a custom 5.10.53 debug kernel):

```
gdb -tui -q linux-5.10.53/vmlinux
[...]
```

Initially, only two horizontally tiled window panes show up. To enable the third, as well as to cycle between the various displays, press Ctrl-x-2 (that is, press *Ctrl* + *X* followed by pressing *2*)! The supported window displays are the CPU registers view (with changing registers being highlighted) pane, the source/assembly code pane, and the GDB command prompt window pane. Do try this out yourself!

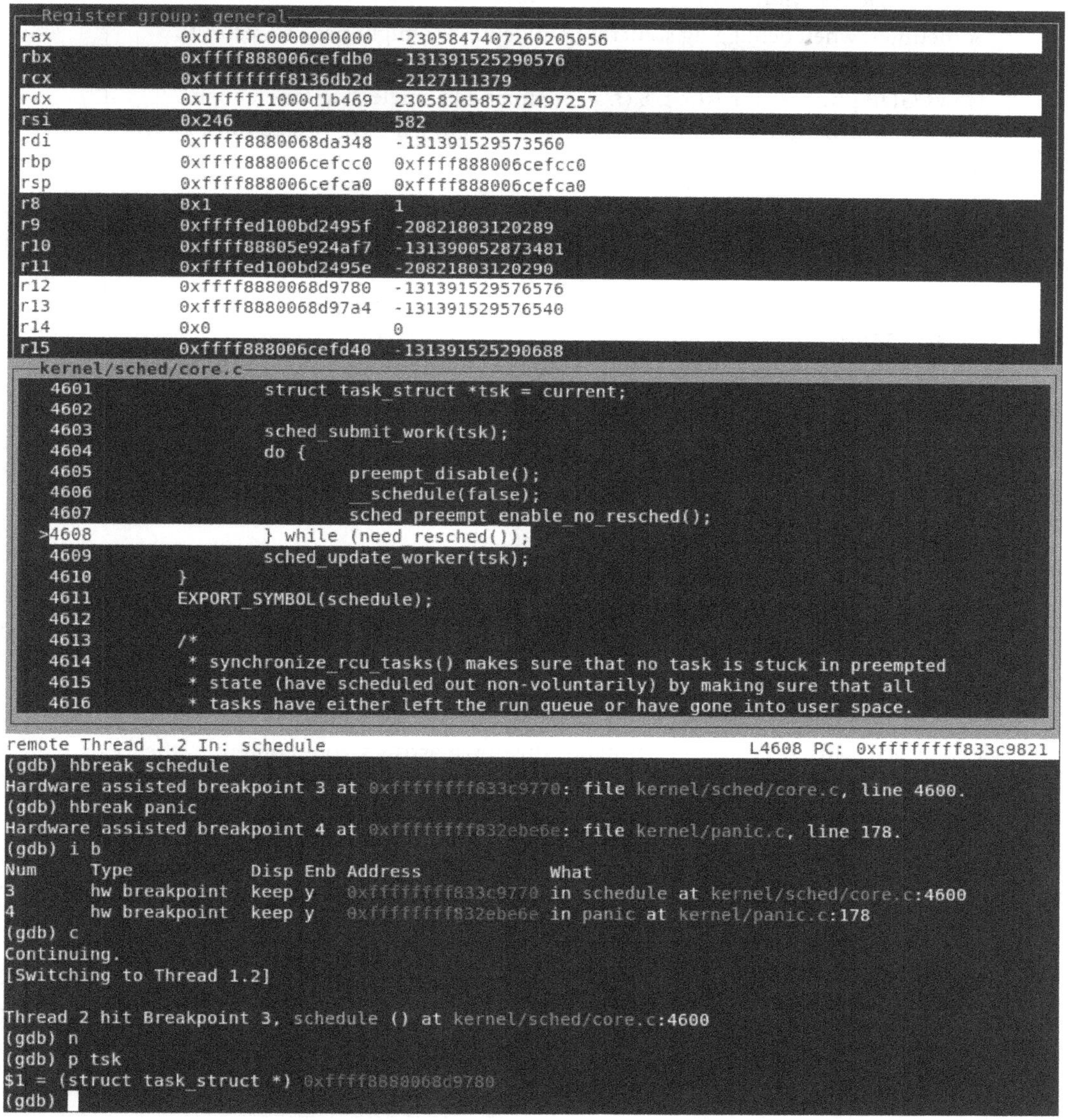

Figure 11.19 – Screenshot showing GDB running in TUI mode; note the horizontally tiled panes – the CPU registers, source code, and GDB command window pane (in this instance)

Viewing the source code, the corresponding assembly, as well as CPU registers simultaneously – and in sync – can be very powerful when in a debug session! GDB's TUI mode makes it behave almost like an IDE (typically a GUI-driven one familiar to most developers)...

> **Tip – Single-Stepping at the Level of Assembly, Disassembling Code**
>
> Sometimes, when debugging something in depth, we might need to resort to working at the level of the machine, single-stepping through assembly code. How? GDB provides the `stepi` (`si`) command to single-step exactly one machine instruction! (You can also specify `si N` to tell GDB to step through N assembly-level instructions.)
>
> Further, GDB is powerful – it also has a separate disassemble (`disas`) command, even allowing for mixed assembly-source output (via the `/m` or `/s` modifier; note that using the `-s` option is recommended). Reference: `https://sourceware.org/gdb/onlinedocs/gdb/Machine-Code.html`.

Here's a GDB TUI mode quick-reference table for your convenience:

Keyboard shortcut or command	TUI mode action
`Ctrl-x 2`	Cycle the content of the horizontally tiled windows – between the CPU registers view, source/assembly code view, and GDB command prompt.
`Ctrl-p`	Recall the previous command (the up arrow scrolls content in the focus window).
`Ctrl-n`	Recall the next command (the down arrow scrolls content in the focus window).
`fs next`	Switch the keyboard focus to the next window pane (useful to scroll content via the arrow keys in a pane).
`tui <cmd>`	`<cmd>` is one of `enable`, `disable`, or `reg`. `enable` and `disable` are self-explanatory; `reg` is explained next.
`tui reg <tab><tab>`	Display all possible CPU register display modes; the output is arch-dependent. On the x86_64, we get `all    float    general    mmx    next    prev    restore    save    sse    system    vector`.
`winheight`	Adjust the specified window's height (format: `winheight WINDOW-NAME [+ | -] NUM-LINES`).

Table 11.1 – Summary of various GDB TUI mode commands and shortcuts

Do ensure you try GDB in the powerful TUI mode – you won't regret it!

What to do when the <value optimized out> GDB response occurs

What do you do when, upon attempting to print a variable's value (for example, (gdb) p i), GDB responds with a cryptic <value optimized out> message? This often happens with local variables and function parameters. It's usually a symptom of compiler optimization with the compiler avoiding the use of stack memory to store locals and/ or parameters, using appropriate CPU **General Purpose Registers** (**GPRs**) instead. So, if you know enough about the CPU ABI, you can hopefully figure out where it stored the variable's value! How do we come to know this? By studying the CPU ABI for your processor! We've already covered this in some detail back in *Chapter 4, Debug via Instrumentation – Kprobes*, in the *Understanding the basics of the Application Binary Interface (ABI)* section. Do take another look.

> **Tip**
>
> Additionally, enabling the CONFIG_DEBUG_INFO_DWARF4 kernel config can have a positive effect. The kernel documents the fact that using this option *significantly improves the success of resolving variables in gdb on optimized code*.

Also, running GDB in TUI mode (as just described) can help you see things more clearly and thus help resolve variable/parameter values.

GDB convenience routines

GDB provides some built-in convenience routines. Some of them require GDB to have Python support:

- $_memeq(): Check whether two buffers are the same for a given length.
- $_regex(): A (Python-based) regular expression matching function.
- $_strlen(str): Returns the length of the string str.

There are several more – look them up here: https://sourceware.org/gdb/current/onlinedocs/gdb/Convenience-Funs.html#Convenience-Funs.

GDB custom macros in its startup file

When you're deep in a debug session, life outside pretty much ceases to matter (smiley!). Often, we repeatedly type in the same commands over and over, (easily) several dozen times a day. Isn't there a shortcut? Yes, indeed: *GDB macros* – effectively your custom-built GDB commands or shortcuts – really help here! They're very easy to set up – just put whatever GDB commands you'd like executed between `define <macro-name>` `[...]` `end`. Simply define them within your GDB startup file, `~/.gdbinit`. This ensures they're auto-parsed in at GDB startup and available to you. A couple of examples follow:

```
// ~/.gdbinit
# connect and set up breakpoints
define connect_qemu
    target remote :1234
    hbreak start_kernel
    hb panic
    hb do_init_module
    b do_fsync
end
# xs - examine stack
define xs
    printf "x/8x $sp\n"
    x/8x $sp
    printf "x/8x $sp-12\n"
    x/8x $sp-12
end
```

So, for example, in GDB, simply typing `connect_qemu` (as we've already seen) at the prompt will have GDB execute all the commands under this macro (somewhat analogous to running GDB in batch mode – another useful thing for you to look up!). Also, more sophisticated GDB (sample) macros can be found here: `https://www.kernel.org/doc/Documentation/admin-guide/kdump/gdbmacros.txt`

Fancy breakpoints and hardware watchpoints

Besides the "usual" hardware/software breakpoints, GDB supports **conditional breakpoints, temporary breakpoints**, and **hardware watchpoints**. All of these can be really useful!

Conditional breakpoints

As an example, say you're debugging the code in a loop. The loop executes, shall we say, 100,000 times. You suspect the bug lies in the last or second-last loop iteration (your typical *off-by-one* bugs!). What are you to do? Single-stepping through the loop thousands of times won't be very productive, right? Set up a conditional breakpoint! GDB's condition command is the way to go:

```
(gdb) help condition
Specify breakpoint number N to break only if COND is true.
Usage is 'condition N COND', where N is an integer and COND is
an expression to be evaluated whenever breakpoint N is reached.
```

Alternatively, you can use this syntax:

```
[h]break <loc> if COND
```

We showed an example of using a conditional breakpoint with our kgdb_try module (in the previous section).

Temporary breakpoints

Sometimes, you just want a breakpoint set up temporarily. The tbreak command sets up a breakpoint that will work *only once*. After that, it's automatically cleaned up. You don't have to remember to use the disable N or delete N GDB commands – useful when in a hurry.

Hardware watchpoints

Hardware watchpoints are a means to have GDB stop the execution of the target when something happens to a variable or expression you specify (it internally makes use of special CPU debug registers, making it a lot faster than traditional software (conditional) breakpoints). What can or should occur to trigger a hardware watchpoint? The changing (write to) of a given variable, or even the reading of a given variable - in effect, the read/write of any valid memory - can trigger them! Hardware watchpoints do require processor support (typically, all modern processors do support them). Find out with the following:

```
(gdb) show can-use-hw-watchpoints
Debugger's willingness to use watchpoint hardware is 1.
```

Ah, it's supported. There are essentially three hardware watchpoint GDB commands:

- `watch <var/expression>`: Sets up a hardware watchpoint that triggers when the variable/expression changes (is written to).

- `rwatch <var/expression>`: Sets up a hardware watchpoint that triggers when the variable/expression is read.

- `awatch <var/expression>`: Sets up a hardware watchpoint that triggers when there is either a read or a write on the variable/expression.

On being triggered, GDB will immediately stop execution and show details: which thread hit the watchpoint, the hardware watchpoint number, name, and, if the value has changed, the old and new value of the variable/expression being watched (or, if it's a read watchpoint, the current value is displayed).

For example, let's set up a hardware watchpoint on the kernel variable `jiffies_64` (an unsigned 64-bit quantity that's incremented on every timer interrupt – in code, we usually employ the `jiffies` macro to look it up safely). Let's try it out:

```
(gdb) help watch
Set a watchpoint for an expression.
Usage: watch [-l|-location] EXPRESSION
A watchpoint stops execution of your program whenever the value of
an expression changes.
If -l or -location is given, this evaluates EXPRESSION and watches
the memory to which it refers.
(gdb) watch jiffies_64
Hardware watchpoint 3: jiffies_64
(gdb) i b
Num     Type           Disp Enb Address            What
1       hw breakpoint  keep y   0xffffffff8299df54 in start_kernel at init/main.c:850
        breakpoint already hit 1 time
2       hw breakpoint  keep y   0xffffffff81ad87f7 in panic at kernel/panic.c:178
3       hw watchpoint  keep y                      jiffies_64
(gdb) c
Continuing.

Thread 1 hit Hardware watchpoint 3: jiffies_64

Old value = 4294892296
New value = 4294892297
do_timer (ticks=ticks@entry=1) at kernel/time/timekeeping.c:2269
2269            calc_global_load();
(gdb) bt
#0  do_timer (ticks=ticks@entry=1) at kernel/time/timekeeping.c:2269
#1  0xffffffff8119dd32 in tick_periodic (cpu=cpu@entry=0) at kernel/time/tick-common.c:93
#2  0xffffffff8119dd75 in tick_handle_periodic (dev=0xffff888003451400)
    at kernel/time/tick-common.c:111
#3  0xffffffff8108e808 in timer_interrupt (irq=<optimized out>, dev_id=<optimized out>)
    at arch/x86/kernel/time.c:57
#4  0xffffffff8116b465 in handle_irq_event_percpu (desc=desc@entry=0xffff88800352c800,
    flags=flags@entry=0xffffc90000003f84) at kernel/irq/handle.c:156
#5  0xffffffff8116b5c3 in handle_irq_event_percpu (desc=desc@entry=0xffff88800352c800)
    at kernel/irq/handle.c:196
#6  0xffffffff8116b64b in handle_irq_event (desc=desc@entry=0xffff88800352c800)
    at kernel/irq/handle.c:213
```

Figure 11.20 – Demo of setting a hardware watchpoint on the kernel jiffies_64 variable

Glance at the preceding figure: we show the output of `help watch`, add the hardware watchpoint on the `jiffies_64` variable, then show the current breakpoints (which includes this hardware watchpoint). Then, we continue the execution of the kernel. The watchpoint triggers soon enough, displaying the old and new value of `jiffies_64` (as a timer interrupt fired, it increments by 1)! We're at the location in the code where the variable underwent a change. We can easily examine how we got there by using the `backtrace` (`bt`) command, seeing the kernel stack (it's truncated in the screenshot).

Think about these [K]GDB debug features – they can be precious! Often, we have a hunch that a particular variable being changed is what's leading to the bug. Hardware watchpoints can help you verify your hunch.

Miscellaneous GDB tips

Here are a few more quick tips:

- At the GDB prompt, you can type the first few letters of a keyword – or even when specifying a variable or function name – and press the *Tab* key twice. GDB will attempt to autocomplete it (if there are multiple matches, it will show all and you can repeat typing in a few more characters). Note, though, that this can slow GDB down.

- Recall what we learned back in *Chapter 7, Oops! Interpreting the Kernel Bug Diagnostic*, in the *Using GDB to help debug the Oops* section: given a function name and an offset into it (as is reported when an Oops triggers), GDB's `list` command can be leveraged to show a particular line of source code. The syntax is: `list *<function>+<offset>`.

- You can run shell commands from within GDB; for example, to run `ls -l` within it, issue the command `shell ls -l` at the GDB prompt.

- There's plenty more for you to read up on and try out: GDB's `record / reverse*` command set, pretty printers, setting breakpoints on machine code (`b *<addr>`), and so on...

And we're done!

Summary

Good going on completing this very hands-on chapter (I really hope you went through it in a hands-on manner)!

We began with a brief conceptual understanding of how KGDB works; it's down to GDB employing a client/server architecture (the target kernel imbibes the GDB server component, with the remote GDB client running on the host system). Next, we set up and tested a full-fledged ARM32 target (via the SEALS project and QEMU). We used it as the target system when demonstrating how to use KGDB to debug in-tree kernel code, right from early boot.

We then moved on to understanding how you can use KGDB to debug kernel modules, often what driver developers (and similar) have to do on projects and products. This time we demonstrated it using a QEMU-emulated x86_64 as the target system.

We rounded off this chapter with a few useful [K]GDB tips and tricks!

Admittedly, getting KGDB set up for remote debugging on actual hardware can be a bit tedious! Ensuring that the serial connection (often over USB via an RS232 TTL UART adapter on the host) is working in both directions is key... see the *Further reading* section for links on helping fix these (pretty annoying) issues and more on this topic.

Take the time to work on KGDB in a hands-on fashion... When done, guess what? We're down to the wire now – the last chapter awaits! There, we'll wrap up this book by peeking into a few more kernel debugging approaches and techniques. I'll see you there!

Further reading

- (Humor!) *The GDB Song*: `https://www.gnu.org/music/gdb-song.en.html`

- Official kernel doc: *Using kgdb, kdb and the kernel debugger internals*: `https://www.kernel.org/doc/html/v5.10/dev-tools/kgdb.html`

- *How to use KGDB*, Timesys: `https://linuxlink.timesys.com/docs/how_to_use_kgdb`

- Excellent material (covers muxing the serial port for GDB and the console): *Using Serial kdb / kgdb to Debug the Linux Kernel - Douglas Anderson*, Google, Oct 2019, YouTube: `https://www.youtube.com/watch?v=HBOwoSyRmys`

- *Linux Kernel Exploitation 0x0] Debugging the Kernel with QEMU*, K Makan, Nov 2020: `http://blog.k3170makan.com/2020/11/linux-kernel-exploitation-0x0-debugging.html`

- *KGDB/KDB over serial with Raspberry Pi*, B Kannan, May 2018; Yocto-biased: https://eastrivervillage.com/KGDB-KDB-over-serial-with-RaspberryPi/

- Blog article; also shows setting up KGDB for libvirt and Vagrant: *USING 'GDB' TO DEBUG THE LINUX KERNEL*, D Robertson, Nov 2019: https://www.starlab.io/blog/using-gdb-to-debug-the-linux-kernel

- Blog article: *A KDB / KGDB SESSION ON THE POPULAR RASPBERRY PI EMBEDDED LINUX BOARD*, kaiwanTECH, July 2013: https://kaiwantech.wordpress.com/2013/07/04/a-kdb-kgdb-session-on-the-popular-raspberry-pi-embedded-linux-board/

- *5 Easy Ways to Reduce Your Debugging Hours*, Dr G Law, Dec 2021: https://undo.io/resources/gdb-watchpoint/5-ways-reduce-debugging-hours/

- Man page on *kdb(8) Built-in Kernel Debugger for Linux*: https://manpages.org/kdb/8

- *Merging kdb and kgdb*, Jake Edge, LWN, Feb 2010: https://lwn.net/Articles/374633/

- Debugging KGDB-serial connection and other issues:

  - *KGDB remote debugging connection issue via USB and Serial connection*: https://stackoverflow.com/a/36861909/779269

  - *Breakpoints not being hit in remote Linux kernel debugging using GDB*: https://stackoverflow.com/questions/28165812/breakpoints-not-being-hit-in-remote-linux-kernel-debugging-using-gdb

  - *Breakpoints not working for GDB while debugging remote arm target on qemu*: https://stackoverflow.com/questions/70874764/breakpoints-not-working-for-gdb-while-debugging-remote-arm-target-on-qemu

  - *Debugging ARM kernels using fast interrupts*, Daniel Thompson, LWN, May 2014: https://lwn.net/Articles/600359/

- Red Hat Developer series on GDB:

  - *The GDB developer's GNU Debugger tutorial, Part 1: Getting started with the debugger*, Seitz, RedHat Developer, Apr 2021: https://developers.redhat.com/blog/2021/04/30/the-gdb-developers-gnu-debugger-tutorial-part-1-getting-started-with-the-debugger

- *The GDB developer's GNU Debugger tutorial, Part 2: All about debuginfo*, Seitz, RedHat Developer, Jan 2022: `https://developers.redhat.com/articles/2022/01/10/gdb-developers-gnu-debugger-tutorial-part-2-all-about-debuginfo`

- *Printf-style debugging using GDB, Part 3*, Beutnner, RedHat Developer, Dec 2021: `https://developers.redhat.com/articles/2021/12/09/printf-style-debugging-using-gdb-part-3`

- Using GDB in TUI mode:

  - From the GDB manual: *GDB Text User Interface*: `https://sourceware.org/gdb/onlinedocs/gdb/TUI.html`

  - *Debug faster with gdb layouts (TUI)*, YouTube video: `https://www.youtube.com/watch?v=mm0b_H0KIRw`

- FYI, interesting: *A kernel debugger in Python: drgn*, Jake Edge, LWN, May 2019: `https://lwn.net/Articles/789641/`.

12
A Few More Kernel Debugging Approaches

At the outset, back in *Chapter 2*, *Approaches to Kernel and Driver Debugging*, we covered various approaches to kernel debugging. (In that chapter, *Table 2.5* provides a quick summary of kernel debug tools and techniques versus the various types of kernel defects; take another look.) The previous chapters of this book have covered many (if not most) of the tools and techniques mentioned, but certainly not all of them.

Here, we intend to merely introduce ideas and frameworks not covered so far (or just briefly mentioned) that you might find useful when debugging Linux at the level of the OS/drivers. We have neither the intent nor the bandwidth (space/pages) to go into these topics in depth, but feel free to use the links in the *Further reading* section to go deeper! Nevertheless, there are important topics covered in this chapter.

In this chapter, we're going to cover the following main topics:

- An introduction to the kdump/crash framework
- A mention on performing static analysis on kernel code
- An introduction to kernel code coverage tools and testing frameworks
- Miscellaneous – using `journalctl`, assertions, and warnings

An introduction to the kdump/crash framework

When a userspace application (a process) crashes, it's often feasible to enable the kernel **core dump** feature; this allows the kernel to capture relevant segments (mappings) of the process **virtual address space** (**VAS**), and write them to a file that is traditionally named `core`. On Linux, the name – and indeed various features – are now settable (look up the man page on `core(5)` for details). How does this help? You can later examine and analyze the core dump using the **GNU debugger** (**GDB**) (the syntax is `gdb -c core-dump-file original-binary-executable`); it can help to find the root cause of the issue! This is called *post-mortem analysis*, as it's done upon the **dead body** of the process, which is the core dump image file.

That's great, but wouldn't it be useful to be able to do the same with the kernel? This is precisely what the **kernel dump** (**kdump**) infrastructure provides – the ability to collect and capture the entire kernel memory segment (the kernel VAS) when the kernel crashes! Furthermore, a powerful userspace open source app (tool), **crash**, allows you to perform post-mortem analysis upon the kdump image, helping to find the root cause of the issue!

Why use kdump/crash?

Why use kdump/crash when we know how to analyze an **Oops** and **kernel panic**, use KGDB, KASAN, KCSAN, and so on? There are several reasons:

- Tooling such as debug instrumentation (printk), KASAN, UBSAN, KCSAN, and KGDB are typically effective and enabled on a debug kernel. When your software is running in production and fails with a kernel-level issue, they are usually disabled and so don't help much.

- Even having the Oops/panic diagnostic (the complete kernel log when the Oops occurred) might not be sufficient to find the root cause of a deep kernel bug. For example, you might require all frames of the kernel-mode stack(s) in question, not just the one where the crash occurred, as well as the content of kernel memory – in effect, the state of all kernel data.

- Only kdump enables capturing all of this, in production. And `crash` lets you analyze it.

There's a downside to using kdump: it implies reserving some fairly significant amounts of system RAM and even possibly flash/disk memory space; this can be impractical, especially on some types of embedded systems.

Understanding the kdump/crash basic framework

Still interested? There are essentially two parts to using kdump/`crash`:

1. Setting up the kernel to capture the kernel memory image if it does crash (Oops or panic); this involves configuring the primary kernel to enable kdump, and the setup to launch a so-called **dump-capture kernel** via a special `kexec` mechanism if a crash/Oops/panic does occur.

2. Installing the `crash` utility on the dev/debug/host system; it takes the kdump image as one of its parameters. Learn how to use it to help debug the kernel/module issue(s).

Setting up and using kdump to capture the kernel image on crash

We won't attempt to go into details here as we lack the space to do so and as it's well-documented in the official kernel documentation, *Documentation for Kdump - The kexec-based Crash Dumping Solution*, here: `https://www.kernel.org/doc/html/latest/admin-guide/kdump/kdump.html#documentation-for-kdump-the-kexec-based-crash-dumping-solution`. If you are setting up kdump, I'd urge you to check this document out in detail. Do note, though, that many Linux distributions – especially the enterprise-class ones such as Red Hat, CentOS, SUSE, and Ubuntu – have their own wrappers around setting up kdump (special config files, packages, and modes, for example); look up the documentation for your distribution as required.

Distilled down, the kdump setup process goes like this:

- Install the `kexec-tools` (via source or distribution package).
- Configure one or two kernels:
  - The primary kernel (configured for kdump; runs in the usual manner)
  - A dump-capture kernel

In architectures that support relocatable kernels (`i386`, `x86_64`, `arm`, `arm64`, `ppc64`, and `ia64`, as of this writing), the primary kernel can also work as the dump-capture kernel (yay!). Look up the kernel configuration details here: `https://www.kernel.org/doc/html/latest/admin-guide/kdump/kdump.html#system-kernel-config-options`.

Right, continuing with the kdump activation process, follow these steps:

- When booting, pass the `crashkernel=size@offset` kernel command-line parameter appropriately to the primary kernel; this reserves a portion of RAM (details can be found in the kernel documentation here: `https://www.kernel.org/doc/html/latest/admin-guide/kdump/kdump.html#crashkernel-syntax`).
- The primary kernel, as part of the startup, employs an architecture-specific way to load the dump-capture kernel into the reserved memory region, via the `kexec` utility; details can be found at `https://www.kernel.org/doc/html/latest/admin-guide/kdump/kdump.html#load-the-dump-capture-kernel`.

That's it! The primary kernel will quite literally **warm-boot** (preserving RAM content) into the dump-capture kernel when a *trigger point* is hit; as of writing, these include the following:

- `panic()`: Setting the `kernel.panic_on_oops` sysctl to 1 ensures that the dump-capture kernel is booted into when the kernel Oops'es (recommended in production).
- `die()` and `die_nmi()`
- Magic SysRq's c command (when enabled, of course): This allows you to test the kdump feature by forcing a NULL pointer dereference and therefore, a kernel Oops by doing this: `echo c > /proc/sysrq-trigger`, as root.

Do note, though, that kdump is essentially useless if the reboot into the dump-capture kernel can't happen, perhaps due to hardware issues.

```
kmembugs_test.c:113:9: error: Returning pointer to local
variable 'name' that will be invalid when returning.
[returnDanglingLifetime]
return (void *)name;
        ^

kmembugs_test.c:113:17: note: Array decayed to pointer here.
return (void *)name;
        ^

kmembugs_test.c:105:16: note: Variable created here.
volatile char name[NUM_ALLOC];
[...]
```

Bang on target (do reread the code to see for yourself)!

As another example of how static analyzers can help, the kernel's `checkpatch.pl` Perl script is, in many ways, very specific to the Linux kernel and attempts to enforce the Linux kernel code style guidelines, which is very important to follow when submitting a patch (the guidelines are here: `https://www.kernel.org/doc/html/latest/process/coding-style.html`). A couple of quick examples to show you the value of running `checkpatch.pl` on your module's source code; here, I run it on our `ch5/kmembugs_test/kmembugs_test.c` source file, by leveraging our `Makefile`, invoking the appropriate target via `make`:

```
make checkpatch
[...]
WARNING: Using vsprintf specifier '%px' potentially exposes
the kernel memory layout, if you don't really need the address
please consider using '%p'.
#134: FILE: kmembugs_test.c:134:
+#ifndef CONFIG_MODULES
+    pr_info("kmem_cache_alloc(task_struct) = 0x%px\n",
+        kmem_cache_alloc(task_struct, GFP_KERNEL));
[...]
WARNING: unnecessary cast may hide bugs, see http://c-faq.com/
malloc/mallocnocast.html
#312: FILE: kmembugs_test.c:312:
+    kptr = (char *)kmalloc(sz, GFP_KERNEL);
```

These warnings are valuable (the first one, on a security aspect, while the second is of the usual type); do pay attention to them!

A significant issue with many static analysis tools is the problem of *false positives* – issues raised by the tool that turn out to be non-issues for the developer; it is a thorn in the side. Nevertheless, using static analysis as part of the development workflow is critical and must be incorporated by the team.

An introduction to kernel code coverage tools and testing frameworks

Code coverage is tooling that can identify which lines of code get executed during a run and which lines of code don't. Tools such as **GNU coverage (gcov)**, and **kcov** and frontend tools such as **lcov** can be very valuable in gleaning this key information.

Why is code coverage important?

Here are a few typical reasons why you should (I'd go so far as to say *must*) perform code coverage:

- **Debugging**: To help identify code paths that are never executed (error paths are pretty typical), thereby making it clear that you need test cases for them (to then catch bugs that lurk in such regions).

- **Testing/QA**: Identify test cases that work and, more to the point, ones that need to be written in order to cover lines of code that never get executed, as, after all, *100% code coverage is the objective*!

- They can help with (minimal) kernel configuration. Seeing that certain code paths are never taken perhaps implies you don't require the configuration that uses them (this can be off the mark; take care to ensure it's really not required before disabling it).

Let's dig deeper into the area of interest here – the first point, debugging. To illustrate the point, we take a simple pseudocode example of an error code path within regular code:

```
p = kzalloc(n, GFP_KERNEL);
if (unlikely(!p)) {  [...] } // let's assume this alloc is fine
foo();  // assume it all goes well here
q = kzalloc(m, GFP_KERNEL);
if (unlikely(!q)) {  // if this allocation fails ...
    ret = do_cleanup_one();
    if (!ret) /* ... and if this is true, then we end up with a
memory leak!!! */
```

```
        return -ENOSPC;
    kfree(p);
    return -ENOMEM;
}
```

If you don't have a (negative) test case where the value `ret` is NULL, then that code path – the one where we return an error value *but fail to free the previously allocated memory buffer first – never gets run*; therefore, it never gets tested. Then, even powerful dynamic analysis tools, such as KASAN, SLUB debug, kmemleak, and so on, *cannot catch the leakage bug, as they never run the code path!* This illustrates why 100% code coverage is key to a successful product or project.

Tip – Fault Injection

So, how exactly do we create a (negative) test case to test error paths (such as in the previous simple example)? Also, kernel-level allocations via the slab cache (`kmalloc()`, `kzalloc()`, and similar), *pretty much never fail,* yet we're taught to always check and write code for the failure case (there are corner cases where they can fail; please, always check for the failure case!); but how do we test that code? The kernel has a **fault-injection framework** to help with precisely this! It's important as only when you run the code can you catch potential bugs (except for static analyzers). The official kernel documentation covers the kernel fault-injection framework in detail (*Fault injection capabilities infrastructure*: `https://www.kernel.org/doc/html/latest/fault-injection/fault-injection.html#fault-injection-capabilities-infrastructure`); do check it out, and look in the *Further reading* section for more on this topic.

Although `gcov` is a userspace tool, it's also used for Linux kernel (and module) coverage analysis. When used in the context of the Linux kernel, the `gcov` coverage data is read off `debugfs` pseudofiles (under `/sys/kernel/debug/gcov`). The mechanics of using `gcov` to perform kernel-level code coverage are definitively covered in the official kernel documentation here: `https://www.kernel.org/doc/html/latest/dev-tools/gcov.html#using-gcov-with-the-linux-kernel`. Tools such as `lcov` are frontends to `gcov`; they provide useful features such as generating HTML-based code coverage reports (they work in the usual manner, whether used for user or kernel-space reporting).

As experienced folk in the industry know, many customers' **service level agreements (SLAs)** or contracts will mandate 100% (or close to it) code coverage being documented and signed off.

A brief note on kernel testing

Testing/QA is a key part of the software process. Although the aphorism *testing can reveal the presence of bugs but not their absence* is, unfortunately, true, giving testing its due, by using state-of-the-art Linux kernel testing tools and frameworks, you can indeed root out (and thus let you subsequently fix) many, if not most, OS- and driver-level bugs. It's a key thing to do; ignore testing at your peril!

As explained in the official kernel documentation (*Kernel Testing Guide*: https://www.kernel.org/doc/html/latest/dev-tools/testing-overview.html#kernel-testing-guide), there are two major types of test infrastructure within the Linux kernel, which differ in how they're used. Besides them, a technique called **fuzzing** turns out to be a key and powerful means to catch those difficult-to-tease-out bugs; read on!

Linux kernel selftests (kselftest)

This is a collection of user-mode scripts and programs (with a few modules thrown in as well); you'll find them within the kernel source tree under tools/testing/selftests. The approach here is more of a black box one; **kselftest** is appropriate when testing or verifying large-ish features of the kernel, using well-defined user-to-kernel interfaces (system calls, device nodes, pseudofiles, and such). To see how to use kselftest and run it, refer to the official kernel documentation here: https://www.kernel.org/doc/html/latest/dev-tools/kselftest.html#linux-kernel-selftests.

Linux kernel unit testing (KUnit)

These tend to be smaller self-contained test cases that are part of the kernel code and so understand internal kernel data structures and functions (therefore, more in line with *unit testing*). We've already covered using **KUnit** test cases to test the powerful KASAN memory checker; refer back to *Chapter 5, Debugging Kernel Memory Issues – Part 1,* and the *Using the kernel's KUnit test infrastructure to run KASAN test cases* section. KUnit is covered in depth (including how to write your own test cases) in the official kernel documentation here: https://www.kernel.org/doc/html/latest/dev-tools/kunit/index.html#kunit-linux-kernel-unit-testing.

Test results are often generated in a well-known form – the **Test Anything Protocol** (**TAP**) format is used by apps and the kernel. There are going to be cases, though, where the original protocol doesn't align well with kernel requirements; so, the kernel community has evolved a **kernel TAP** (**KTAP**) format for reporting. The official kernel documentation has the details: https://docs.kernel.org/dev-tools/ktap.html#the-kernel-test-anything-protocol-ktap-version-1.

What is fuzzing?

There are other means by which both apps and kernel code can be (very effectively!) tested – a powerful one is called **fuzzing**. Essentially, fuzzing is a test technique, a framework, where the **program under test** (**PUT**) is fed (semi) randomized input (the monkey-on-the-keyboard technique!); this often leads to it failing and/or triggering bugs in subtle ways, not commonly caught by more traditional testing techniques. Fuzzing can be especially helpful in catching security vulnerabilities, which tend to be the typical memory bugs. (We covered these in some detail in *Chapter 5, Debugging Kernel Memory Issues – Part 1*, and *Chapter 6, Debugging Kernel Memory Issues – Part 2*).

There are many well-known **fuzzers**; among them are **American Fuzzy Lop** (**AFL**), **Trinity**, and **syzkaller**. For the Linux kernel, syzkaller (also known as **syzbot** or **syzkaller robot**) is perhaps the best-known de facto continuously running (unsupervised) fuzzer on the kernel codebase; it has already found and reported hundreds of bugs (`https://github.com/google/syzkaller#documentation`). The syzkaller web dashboard, showing reported bugs of the upstream kernel and other interesting statistics, is available here: `https://syzkaller.appspot.com/upstream`. Do check it out.

> **Where Does kcov Fit In?**
>
> Fuzzers internally mutate interesting test cases into many more test cases. To do their job well, they require good code coverage tooling so that they can prioritize which mutated test cases will likely yield the most interesting results. For the Linux kernel, this is where `kcov` comes in – it's a code coverage tool that *"exposes kernel code coverage information in a form suitable for coverage-guided fuzzing (randomized testing)."*

Would you like to learn more and even try some hands-on kernel fuzzing? Check out the following (quite non-trivial) exercise.

> **Exercise**
>
> Try out fuzzing a portion of the Linux kernel with AFL! To do so, read the excellent *A gentle introduction to Linux Kernel fuzzing* tutorial and follow along: `https://blog.cloudflare.com/a-gentle-introduction-to-linux-kernel-fuzzing/`. (Also see `https://github.com/cloudflare/cloudflare-blog/blob/master/2019-07-kernel-fuzzing/README.md`.)

Well, almost there! Let's round off this chapter with a few miscellaneous areas.

Miscellaneous – using journalctl, assertions, and warnings

The modern framework for system initialization on Linux is considered to be **systemd** (although, by now, it's been in use on Linux for over a decade). It's a very powerful framework, although it does have its share of detractors as well. One thing you'll notice regarding systemd is that it's a pretty invasive system! On many (if not most) Linux distributions, besides providing a robust initialization framework (via service units, targets, and such), systemd takes over many activities, replacing their original counterparts, such as system logging, the udevd userspace daemon service, network services startup/shutdown, core dump management, watchdog, and so on. Also, with systemd, apps can be carefully tuned to operate within specified system resource limits by leveraging the powerful kernel **control groups** (**cgroups**) framework.

Looking up system logs with journalctl

As our central topic is debugging, we'll briefly look at only the *logging* aspect of systemd. Logging is often a straightforward way to get insight into what exactly was happening on the system before a bug occurred (and, if lucky, even during and after).

A feature of systemd logging is that it maintains the logs of both userspace app (and system daemon) processes as well as the kernel log. By using its frontend to view and filter log messages – journalctl – we can pretty intuitively gain insight into what's going on at any moment. This is largely because journalctl automatically, and by default, *displays all logs in chronological order* – those of user-mode processes as well as those of all kernel components (the core kernel itself and drivers/modules), in short, all printk-type messages.

```
-1 093<...>cb3 Fri [...] IST—Fri 2022-05-06 ... IST
 0 d72<...>a8b Fri [...] IST—Mon 2022-05-09 ... IST
```

This output informs us that this particular system has been booted 83 times. The integer value in the left-most column is *how many boots ago*, so the last boot (the current session) is the left column value 0 (the negative numbers imply earlier boots, in chronological order; so, -1 implies the boot before this one). Nice.

journalctl – a few useful aliases

There are just too many option switches to journalctl to discuss here. To keep it short but still useful, here are a few *aliases* to journalctl that you might find useful. I typically put these into a startup script and source them at login:

```
# jlog: current boot only, everything
alias jlog='journalctl -b --all --catalog --no-pager'
# jlogr: current boot only, everything, *reverse* chronological
order
alias jlogr='journalctl -b --all --catalog --no-pager
--reverse'
# jlogall: *everything*, all time; --merge => _all_ logs merged
alias jlogall='journalctl --all --catalog --merge --no-pager'
# jlogf: *watch* log, 'tail -f' mode
alias jlogf='journalctl -f'
# jlogk: only kernel messages, this boot
alias jlogk='journalctl -b -k --no-pager'
```

Using the journalctl -f variant can be particularly useful to literally *watch* logs as they appear in real time. Also, simply use the -k option switch to show kernel printks.

You can do more with journalctl; filtering logs based on flexibly stated since and/or until-type keywords. For example, let's say you want to see all logs since 11 a.m. today but only until an hour ago (let's say it's 1 p.m. now, lunch right?). You could do so like this:

```
journalctl -b --since 11:00 --until "1 hour ago"
```

There are several variations too; powerful stuff, indeed!

Assertions, warnings, and BUG() macros

Assertions are a way to test assumptions. In userspace, the `assert()` macro serves the purpose. The parameter to `assert()` is the Boolean expression to test – if `true`, execution continues as usual (within the calling process or thread); if `false`, the assertion fails. This makes it invoke the `abort()` function, causing the process to die accompanied by a noisy `printf` message conveying the fact that the assertion failed (it will display the filename and line number as well as the failing assertion's expression).

Assertions are in effect a code-level debug tool, helping us achieve something very important (that I have tried to emphasize throughout the book): *do not make assumptions; be empirical.* Assertions allow us to test those assumptions. As a silly example, let's say a signal handler within a process sets an integer x to the value 3; in another function, `foo()`, we're assuming it's set to 3. Hey, that can be dangerous! Instead, we test our assumption with an assertion and then proceed on our merry way:

```
static int foo(void) {
    assert(x == 3);
    bar(); [...]
}
```

Now, you can see that an assertion is a way to say what you expect; if what's expected doesn't actually happen at runtime, you'll be notified! That's very useful.

So, why don't we use the same idea within the kernel? Wouldn't that be useful? It would, but there's a problem: we can't realistically have the kernel abort if the assertion fails, can we? Well, actually, we can: it's what macros such as `BUG_ON()` (and friends) do. So, some kernel/driver authors write their own version, in effect, a custom `assert` macro; here's an example (from a block driver named `sx8`):

```
// drivers/block/sx8.c
#define assert(expr) \
        if(unlikely(!(expr))) { \
        printk(KERN_ERR "Assertion failed! %s,%s,%s,line=%d\n", \
    #expr, __FILE__, __func__, __LINE__); \
        }
```

Nice and simple, and an effective way to check assumptions! This driver invokes its custom `assert` macro a few times; here's one example:

```
assert(host->state == HST_PORT_SCAN);
```

> **Exercise**
>
> Look up the kernel code for the definition of `BUG_ON()`. You'll see it's a macro that invokes the `BUG()` macro when the condition comes `true`. Guess what? The (arch-specific) `BUG()` macro typically invokes a `printk` specifying the location of the code and then calls `panic("BUG!")`.

Don't lightly invoke any of the `BUG*()` macros; you only call them when you have an unrecoverable situation, when there's no way out, when you must panic. A better alternative, perhaps, is using one of the many `WARN*()` type macros found within the kernel; they cause a warning-level printk to be emitted to the kernel log when the condition (passed as a parameter) is `true`! Thus, the `WARN*()` macros are perhaps the closest built-in kernel equivalent to the user-mode `assert()` macro. Do realize, though, that even the `WARN*()` macros spell out that a significant situation exists within the kernel – again, don't invoke them unnecessarily!

Summary

How awesome! Congratulations on completing this, the final chapter, and this book!

Here, you got an introduction to some remaining kernel debugging approaches – things we perhaps mentioned but hadn't covered elsewhere. We began by mentioning the powerful kdump/`crash` framework. Kdump allows capturing the complete kernel image (the trigger typically being a kernel crash/Oops /panic), and the `crash` userspace utility helps you (post-mortem) analyze it.

Static analyzers can play a really useful role in discovering potential bugs and security vulnerabilities. Don't ignore them; learn to leverage them!

The importance of code coverage was then delved into for a bit (along with a brief mention of how the kernel's fault-injection framework helps in setting up negative test cases, having control actually going to those pesky and possibly buggy error code paths). We briefly examined the kernel testing framework landscape; you saw that the kernel selftests and KUnit frameworks are the typical ones used to cover a lot of ground. Don't forget the powerful fuzzing technique though – Google's syzbot (syzkaller robot) uses it to its advantage to automatically and continually fuzz the Linux kernel, teasing out many bugs!

We finished the chapter with a quick mention of how system (and app) logs can be examined and filtered using the powerful `journalctl` frontend. Testing your assumptions by employing a custom kernel-space `assert` macro, and a mention of using the `WARN*()` and `BUG*()` macros, completed the discussion here.

A key point I'd like to (re)emphasize upon completion of this book is one of Fred Brook's well-known aphorisms: *there is no silver bullet*. In effect, what's meant is that *one tool or debugging technique or analysis type cannot and will not catch all possible bugs; use several.* These include compiler warnings (`-Wall` and `-Wextra`), static and dynamic analyzers (KASAN and others), dynamic debug printks, kprobes, lockdep, KCSAN, ftrace and trace-cmd, KGDB/kdb, and custom panic handlers. Our so-called *better* `Makefile` (for example, `https://github.com/PacktPublishing/Linux-Kernel-Debugging/blob/main/ch3/printk_loglevels/Makefile`) tries to enforce exactly this discipline by having several targets. Take the trouble to use them!

So, you're at the end? No, it's really more like the beginning, but armed with precious, useful, and practical tools, techniques, tips, and knowledge (this is our sincere hope!). Go forth, my friend!

Further reading

- Kdump:

    - *Documentation for Kdump - The kexec-based Crash Dumping Solution*: `https://www.kernel.org/doc/html/latest/admin-guide/kdump/kdump.html`

    - *Marian Marinov - Analyzing Linux kernel crash dumps*, YouTube presentation, December 2016: `https://www.youtube.com/watch?v=wcId2Y9bM-M`

    - (Biased to the Fedora distribution) *Using Kdump for examining Linux Kernel crashes*, Pratyush Anand, June 2017: `https://opensource.com/article/17/6/kdump-usage-and-internals`

    - *Linux Kernel Debugging, Kdump, Crash Tool Basics Part-1*, Linux Kernel Foundation, YouTube video tutorial: `https://www.youtube.com/watch?v=6l0ulgv1OJ4`

    - *How to use kdump to debug kernel crashes*, January 2022: `https://fedoraproject.org/wiki/How_to_use_kdump_to_debug_kernel_crashes`

- Using the `crash` app to interpret and debug the kdump image:

  - Probably the best, a white paper on `crash` by its lead developer and maintainer, David Anderson: `https://crash-utility.github.io/crash_whitepaper.html`; this even includes a pretty deep case study: `https://crash-utility.github.io/crash_whitepaper.html#EXAMPLES`

  - *Introduction to Linux Kernel Crash Analysis* – Alex Juncu, YouTube video, February 2016: `https://www.youtube.com/watch?v=w8XnnG68rqE`

  - *Analysing Linux kernel crash dumps with crash - The one tutorial that has it all*, Dedoimedo, June 2010: `https://www.dedoimedo.com/computers/crash-analyze.html`

- Static analysis tools:

  - *Checking the Linux Kernel with Static Analysis Tools*, Steven J. Vaughan-Nichols, The New Stack, June 2021: `https://thenewstack.io/checking-linuxs-code-with-static-analysis-tools/`

  - *Static analysis in GCC 10*, Red Hat Developer, March 2020: `https://developers.redhat.com/blog/2020/03/26/static-analysis-in-gcc-10`

  - List of tools for static code analysis: `https://en.wikipedia.org/wiki/List_of_tools_for_static_code_analysis`

  - *Smatch Static Analysis Tool Overview*, Dan Carpenter, Oracle blog, December 2015: `https://blogs.oracle.com/linux/post/smatch-static-analysis-tool-overview-by-dan-carpenter`

- Fuzzing:

  - *A gentle introduction to Linux Kernel fuzzing*, Marek Majkowski, Cloudflare blog, October 2019: `https://blog.cloudflare.com/a-gentle-introduction-to-linux-kernel-fuzzing/`

  - Also see: `https://github.com/cloudflare/cloudflare-blog/blob/master/2019-07-kernel-fuzzing/README.md`

  - *Fuzzing Linux Kernel*, Andrey Konovalov, Senior Software Engineer, Google; video presentation, March 2021: `https://www.linuxfoundation.org/webinars/fuzzing-linux-kernel/`

- *Fuzzing Applications with American Fuzzy Lop (AFL)*, A Priya, medium, June 2020: `https://medium.com/@ayushpriya10/fuzzing-applications-with-american-fuzzy-lop-afl-54facc65d102`

- Fault injection:

  - *Fault injection capabilities infrastructure*: `https://www.kernel.org/doc/html/latest/fault-injection/fault-injection.html#fault-injection-capabilities-infrastructure`

  - Old but a useful introduction: *Injecting faults into the kernel*, Jon Corbet, LWN, November 2006: `https://lwn.net/Articles/209257/`

  - A modern approach with BPF: *BPF-based error injection for the kernel*, Jon Corbet, November 2017: `https://lwn.net/Articles/740146/`

  - *FIFA: A Kernel-Level Fault Injection Framework for ARM-Based Embedded Linux System*, Eunjin Jeong, et al, IEEE, March 2017: `https://ieeexplore.ieee.org/abstract/document/7927960`

- Logs with systemd's `journalctl`:

  - *journalctl(1) — Linux manual page*: `https://man7.org/linux/man-pages/man1/journalctl.1.html`

  - *How to Check Logs Using journalctl*, F Civaner, March 2021: `https://www.baeldung.com/linux/journalctl-check-logs`

- Finally, the *LWN Kernel Index* (precious! Be sure to bookmark it): `https://lwn.net/Kernel/Index/`

Index

Symbols

A

B

`Packt.com`

Subscribe to our online digital library for full access to over 7,000 books and videos, as well as industry leading tools to help you plan your personal development and advance your career. For more information, please visit our website.

Why subscribe?

- Spend less time learning and more time coding with practical eBooks and Videos from over 4,000 industry professionals

- Improve your learning with Skill Plans built especially for you

- Get a free eBook or video every month

- Fully searchable for easy access to vital information

- Copy and paste, print, and bookmark content

Did you know that Packt offers eBook versions of every book published, with PDF and ePub files available? You can upgrade to the eBook version at `packt.com` and as a print book customer, you are entitled to a discount on the eBook copy. Get in touch with us at `customercare@packtpub.com` for more details.

At `www.packt.com`, you can also read a collection of free technical articles, sign up for a range of free newsletters, and receive exclusive discounts and offers on Packt books and eBooks.

Other Books You May Enjoy

If you enjoyed this book, you may be interested in these other books by Packt:

Linux Kernel Programming Part 2 - Char Device Drivers and Kernel Synchronization

Kaiwan N Billimoria

ISBN: 9781801079518

- Get to grips with the basics of the modern Linux Device Model (LDM)
- Write a simple yet complete misc class character device driver
- Perform user-kernel interfacing using popular methods
- Understand and handle hardware interrupts confidently
- Perform I/O on peripheral hardware chip memory
- Explore kernel APIs to work with delays, timers, kthreads, and workqueues
- Understand kernel concurrency issues
- Work with key kernel synchronization primitives and discover how to detect and avoid deadlock

Mastering Linux Device Driver Development

John Madieu

ISBN: 9781789342048

- Explore and adopt Linux kernel helpers for locking, work deferral, and interrupt management
- Understand the Regmap subsystem to manage memory accesses and work with the IRQ subsystem
- Get to grips with the PCI subsystem and write reliable drivers for PCI devices
- Write full multimedia device drivers using ALSA SoC and the V4L2 framework
- Build power-aware device drivers using the kernel power management framework
- Find out how to get the most out of miscellaneous kernel subsystems such as NVMEM and Watchdog

Packt is searching for authors like you

If you're interested in becoming an author for Packt, please visit authors. packtpub.com and apply today. We have worked with thousands of developers and tech professionals, just like you, to help them share their insight with the global tech community. You can make a general application, apply for a specific hot topic that we are recruiting an author for, or submit your own idea.

Share Your Thoughts

Now you've finished *Linux Kernel Debugging*, we'd love to hear your thoughts! Scan the QR code below to go straight to the Amazon review page for this book and share your feedback or leave a review on the site that you purchased it from.

https://packt.link/r/1801075034

Your review is important to us and the tech community and will help us make sure we're delivering excellent quality content.